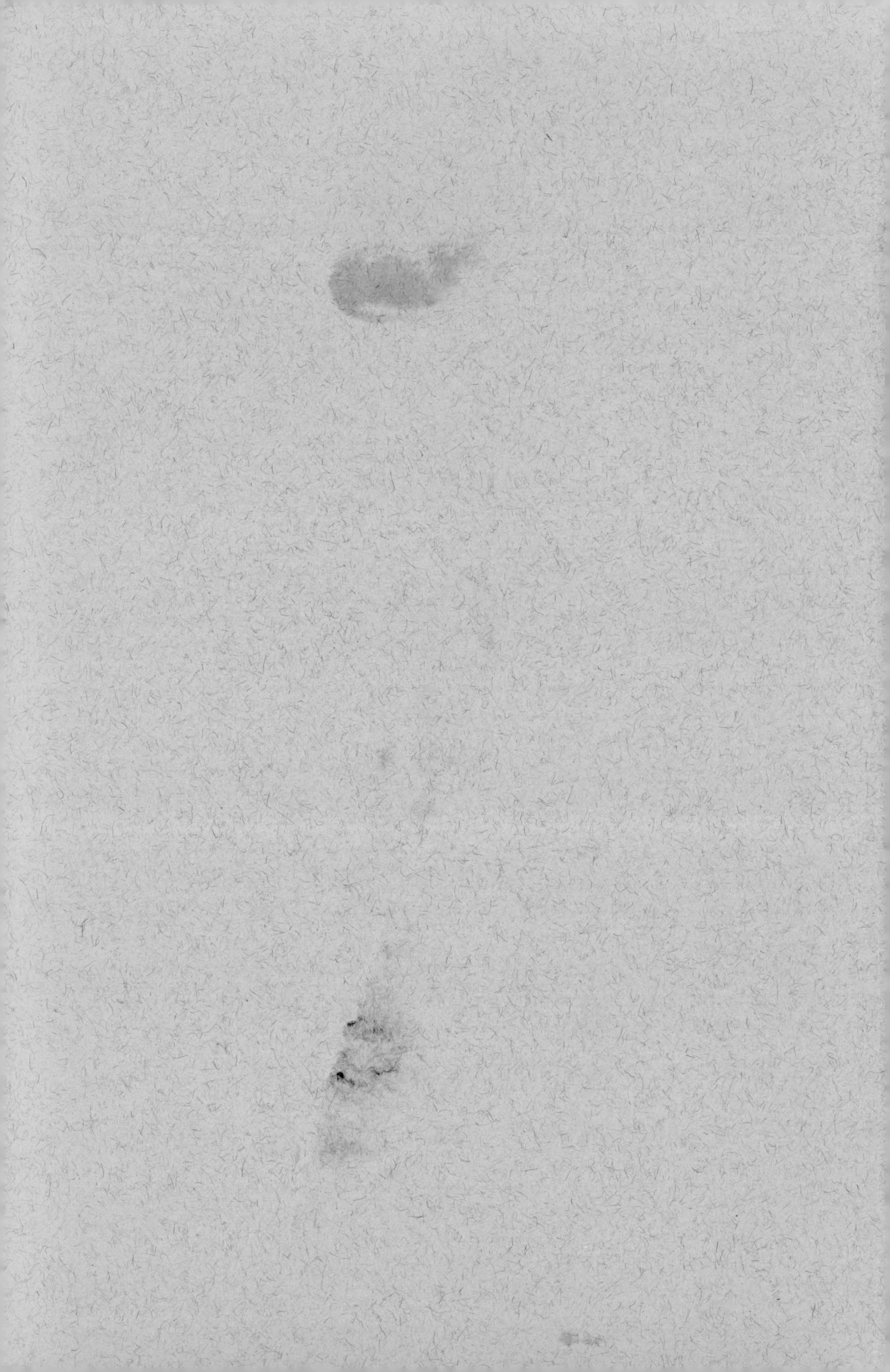

MTP International Review of Science

Alkaloids

MTP International Review of Science

Publisher's Note

The MTP International Review of Science is an important new venture in scientific publishing, which we present in association with MTP Medical and Technical Publishing Co. Ltd. and University Park Press, Baltimore. The basic concept of the Review is to provide regular authoritative reviews of entire disciplines. We are starting with chemistry because the problems of literature survey are probably more acute in this subject than in any other. As a matter of policy, the authorship of the MTP Review of Chemistry is international and distinguished; the subject coverage is extensive, systematic and critical; and most important of all, new issues of the Review will be published every two years.

In the MTP Review of Chemistry (Series One), Inorganic, Physical and Organic Chemistry are comprehensively reviewed in 33 text volumes and 3 index volumes, details of which are shown opposite. In general, the reviews cover the period 1967 to 1971. In 1974, it is planned to issue the MTP Review of Chemistry (Series Two), consisting of a similar set of volumes covering the period 1971 to 1973. Series Three is planned for 1976, and so on.

The MTP Review of Chemistry has been conceived within a carefully organised editorial framework. The over-all plan was drawn up, and the volume editors were appointed, by three consultant editors. In turn, each volume editor planned the coverage of his field and appointed authors to write on subjects which were within the area of their own research experience. No geographical restriction was imposed. Hence, the 300 or so contributions to the MTP Review of Chemistry come from many countries of the world and provide an authoritative account of progress in chemistry.

To facilitate rapid production, individual volumes do not have an index. Instead, each chapter has been prefaced with a detailed list of contents, and an index to the 10 volumes of the MTP Review of Organic Chemistry (Series One) will appear, as a separate volume, after publication of the final volume. Similar arrangements will apply to the MTP Review of subsequent series.

Butterworth & Co. (Publishers) Ltd.

**Organic Chemistry
Series One**
Consultant Editor
D. H. Hey, F.R.S.
*Department of Chemistry
King's College, University of London*

Volume titles and Editors

1 STRUCTURE DETERMINATION IN ORGANIC CHEMISTRY
Professor W. D. Ollis, F.R.S.,
University of Sheffield

2 ALIPHATIC COMPOUNDS
Professor N. B. Chapman,
Hull University

3 AROMATIC COMPOUNDS
Professor H. Zollinger, *Swiss Federal Institute of Technology, Zurich*

4 HETEROCYCLIC COMPOUNDS
Dr. K. Schofield, *University of Exeter*

5 ALICYCLIC COMPOUNDS
Professor W. Parker, *University of Stirling*

6 AMINO ACIDS, PEPTIDES AND RELATED COMPOUNDS
Professor D. H. Hey, F.R.S. and
Dr. D. I. John,
King's College, University of London

7 CARBOHYDRATES
Professor G. O. Aspinall, *Trent University, Ontario*

8 STEROIDS
Dr. W. F. Johns, *G. D. Searle & Co., Chicago*

9 ALKALOIDS
Professor K. Wiesner, F.R.S.,
University of New Brunswick

10 FREE RADICAL REACTIONS
Professor W. A. Waters, F.R.S.,
University of Oxford

INDEX VOLUME

Organic Chemistry
Series One

Consultant Editor
D. H. Hey, F.R.S.

MTP International Review of Science

Volume 9
Alkaloids

Edited by **K. Wiesner, F.R.S.**
University of New Brunswick

Butterworths · London
University Park Press · Baltimore

THE BUTTERWORTH GROUP

ENGLAND
Butterworth & Co (Publishers) Ltd
London: 88 Kingsway, WC2B 6AB

AUSTRALIA
Butterworths Pty Ltd
Sydney: 586 Pacific Highway 2067
Melbourne: 343 Little Collins Street, 3000
Brisbane: 240 Queen Street, 4000

NEW ZEALAND
Butterworths of New Zealand Ltd
Wellington: 26–28 Waring Taylor Street, 1

SOUTH AFRICA
Butterworth & Co (South Africa) (Pty) Ltd
Durban: 152–154 Gale Street

ISBN 0 408 70283 4

UNIVERSITY PARK PRESS

U.S.A. and CANADA
University Park Press
Chamber of Commerce Building
Baltimore, Maryland, 21202

Library of Congress Cataloging in Publication Data

Wiesner, K., 1919–
   Alkaloids.

  (Organic chemistry, series one, v. 9) (MTP
international review of science)
   1. Alkaloids.
QD251.2.074 vol. 9 [QD421] 547′.008s [547′.72]
73–4238
ISBN 0–8391–1037–5

First Published 1973 and © 1973
MTP MEDICAL AND TECHNICAL PUBLISHING CO. LTD.
St. Leonard's House
St. Leonardgate
Lancaster, Lancs.
and
BUTTERWORTH & CO. (PUBLISHERS) LTD.

Filmset by Photoprint Plates Ltd., Rayleigh, Essex
Printed in England by Redwood Press Ltd., Trowbridge, Wilts
and bound by R. J. Acford Ltd., Chichester, Sussex

Consultant Editor's Note

The subject of Organic Chemistry is in a rapidly changing state. At the one extreme it is becoming more and more closely involved with biology and living processes and at the other it is deriving a new impetus from the extending implications of modern theoretical developments. At the same time the study of the subject at the practical level is being subjected to the introduction of new techniques and advancements in instrumentation at an unprecedented level. One consequence of these changes is an enormous increase in the rate of accumulation of new knowledge. The need for authoritative documentation at regular intervals on a world-wide basis is therefore self-evident.

The ten volumes in Organic Chemistry in this First Series of biennial reviews in the MTP International Review of Science attempt to place on record the published achievements of the years 1970 and 1971 together with some earlier material found desirable to assist the initiation of the new venture. In order to do this on an international basis Volume Editors and Authors have been drawn from many parts of the world.

There are many alternative ways in which the subject of Organic Chemistry can be subdivided into areas for more or less self-contained reviews. No single system can avoid some overlapping and many such systems can leave gaps unfilled. In the present series the subject matter in eight volumes is defined mainly on a structural basis on conventional lines. In addition, one volume has been specially devoted to methods of structure determination, which include developments in new techniques and instrumental methods. A further separate volume has been devoted to Free Radical Reactions, which is justified by the rapidly expanding interest in this field. If there prove to be any major omissions it is hoped that these can be remedied in the Second Series.

It is my pleasure to thank the Volume Editors who have made the publication of these volumes possible.

London   D. H. Hey

Preface

This is the first in a series of biennial volumes on alkaloids. To review alkaloid chemistry in articles which are interesting to read and not mere lists of references is difficult because of the immense breadth and variety of the subject. Some of the most complex terpenes and steroids may be found among the alkaloids besides compounds based on a great variety of entirely different building principles. For this reason a single volume cannot be expected to review alkaloid chemistry exhaustively nor can the individual articles be homogeneous in scope and general concept. The easiest and most satisfying for review are topics where a breakthrough has occurred in a narrow field and the investigators responsible for this breakthrough tell a self-contained story in somewhat greater detail. The excellent articles on Ochotensine and Tuberostemonine are examples of this type. Secondly, we have a number of more general authoritative reviews written in all cases by experts who have made significant contributions to the field (Lycopodium Alkaloids, Indole Alkaloids, Diterpene Alkaloids, Steroid Alkaloids and Amaryllidaceae Alkaloids). The article on Benzylisoquinoline and Homo-benzylisoquinoline Alkaloids written by one of the leading investigators in the field is somewhat different. Because of the immense breadth and variety of the subject, detail had to be sacrificed to a much greater extent. On the other hand, a bird's eye view is necessary to bring into focus the fascinating inter-relationships in this mushrooming group of substances. The discussion of biosynthesis is included in the individual articles.

The only exception is the Biogenesis of Indole Alkaloids. Because of the explosive progress and special interest of this topic a separate chapter has been devoted to it and we have been fortunate to have secured one of the pioneers in the field as author.

New Brunswick   K. Wiesner

Contents

1
The Lycopodium Alkaloids Including Synthesis and Biosynthesis

WILLIAM A. AYER
University of Alberta, Edmonton

1.1 INTRODUCTION

Although the presence of alkaloids in the genus *Lycopodium* (family *Lyco-podiaceae*) was first noted in 1881, extensive investigation was not commenced until the 1940s, and it was not until the late 1950s and early 1960s that the structures of these alkaloids began to be clarified. Comprehensive reviews of these developments are available[1-3]. It is the intention in this review to survey progress in the area during the past 6 years. Previous to the mid-1960s, studies were mainly concerned with isolation and structure determination. The more recent past has seen not only further progress in this area but also the first major accomplishments in synthesis and the first informative studies on the biosynthesis of the alkaloids.

As a basis for discussion, the alkaloids will be divided into nine groups according to carbon–nitrogen skeletons. The numbering system employed is adapted from that first suggested by Wiesner[1].

1.2 THE LYCOPODINE GROUP

Members of this group have the carbon–nitrogen skeleton of lycopodine (1).

1.2.1 Isolation and structural studies

The early structural studies on lycopodine have been well summarised[2]. During these studies it was found that lycopodine methiodide does not undergo a normal Hofmann elimination but instead affords a tetracyclic base originally formulated[4] as (2). Recently it has been shown[5] that this reaction involves a rather unusual reorganisation of the carbon–nitrogen skeleton leading to structure (3) for the Hofmann base. The transformation may be rationalised in terms of a mechanism involving abstraction of the proton from C-4 and formation of the anti-Bredt intermediate (4), followed by transannular hydride shift from C-9 to C-13 and subsequent C-4 to C-9 bond formation[5]. A similar rearrangement occurs with clavolonine (5) methiodide[5, 6]. Whether or not dihydrolycopodine (6) methiodide, lacking the C-5 carbonyl, undergoes normal elimination to give a methine such as (7) seems not to have been investigated.

Serratidine, an alkaloid isolated from a variety of *L. serratum*, has been shown[8] to be hydroxylated at the bridge-head position 7 and to possess structure (8). The related hydroxyl-free compound, anhydrolycodoline (9), recently isolated[9] from *L. alopecuroides L.*, and lucidioline (10)[10] belong to this 11,12-unsaturated series. Catalytic hydrogenation of the $\Delta^{11,12}$ double bond in (8), (9), and (10) leads to a mixture of C-12 epimers[8-10] in which the isomer having the 'unnatural' 12α-configuration predominates (α and β are used in the steroid sense with the C-8, C-15, C-14 bridge defined as β). In the case of (10) it has been shown that the proportion of 12β-epimer is increased when the hydrogenation is carried out in the presence of acid[10].

Alkaloid L23, first isolated[11] in 1946, has been shown[12] to be 12α-hydroxy-lycopodine (11), and is the only reported naturally occurring member of the lycopodine group with a 12α-configuration. Dehydration of (11) gives (9), first obtained by dehydration of lycodoline, the C-12 epimer of alkaloid L23. Compounds differing in configuration at C-12 may be distinguished either by the appearance of Bohlmann bands[13] in the infrared spectra of the 12α compounds[14], or in the case of compounds with a carbonyl group at C-5 this distinction may be made by u.v. and c.d. measurements[12]. Compounds with the 12β configuration show a non-classical σ coupled p interaction between the nitrogen lone pair and the carbonyl group which gives

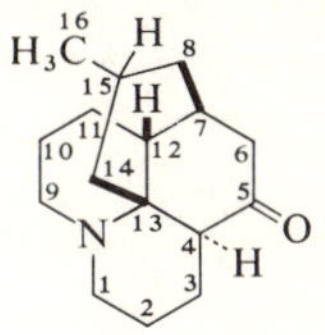

(1)

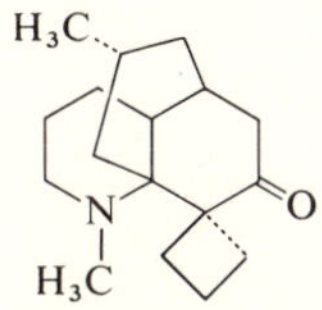

(2)

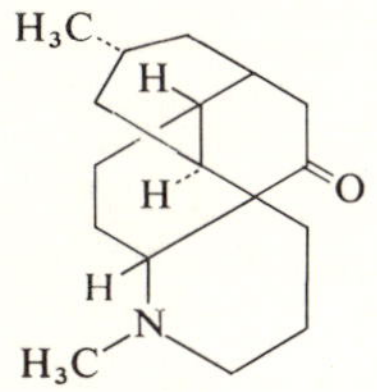

(3)

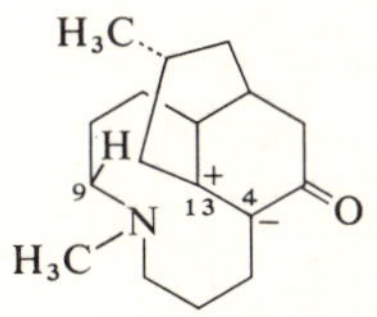

(4)

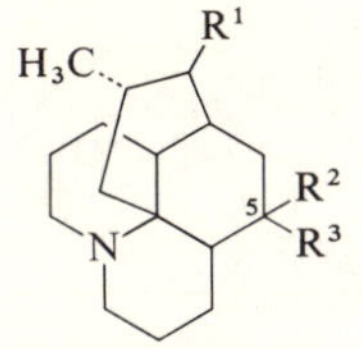

(5) R^1 = OH, R^2 = R^3 = O
(6) R^1 = R^3 = H, R^2 = OH

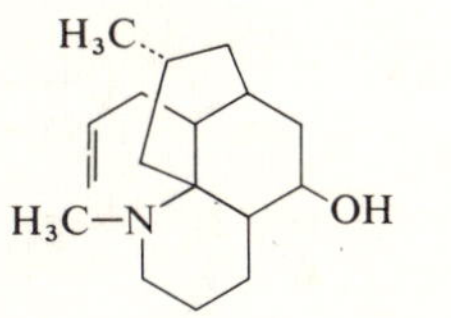

(7)

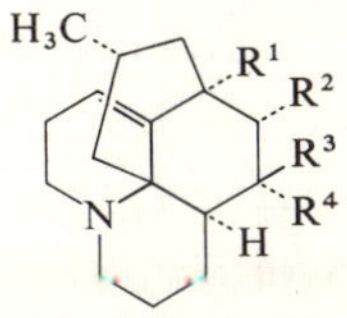

(8) R^1 = OH, R^2 = H, R^3 = R^4 = O
(9) R^1 = R^2 = H, R^3 = R^4 = O
(10) R^1 = R^4 = H, R^2 = R^3 = OH

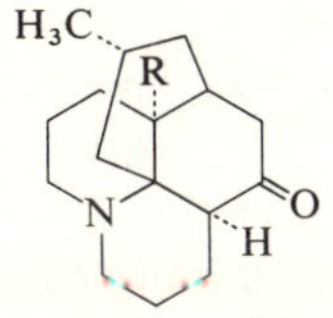

(11) R = OH
(12) R = H

rise to an absorption band detectable in the 220–230 nm region of the u.v. or c.d. spectrum. This absorption band is not present when the nitrogen is protonated. The n→π* transition in C-5 ketones of the 12β series shows octant-like behaviour when the nitrogen is unprotonated, but antioctant behaviour when it is protonated; in the 12α series the behaviour is antioctant in both cases[12, 15]. A set of useful rules has been developed to correlate the spectral (u.v. and c.d.) properties of such aminocyclohexanones[15].

Structure (11) has also been assigned to the alkaloid pseudoselagine[16], first isolated from *L. selago*[17]. Comparison of the published data[12, 16] indicates that pseudoselagine and alkaloid L23 are identical, although attempts to dehydrate pseudoselagine to the anhydro compound (9) were reported to be unsuccessful[16]. The name isolycodoline has been suggested[16] for (11).

Lycopodine and other known[2] members of the lycopodine group have recently been isolated from *L. alopecuroides*[9], *L. selago*[16], *L. inundatum*[18], and *L. alpinum*[19].

1.2.2 Synthesis

During the period under review the first two successful syntheses of lycopodine (1) have been reported[20, 21]. Previous to this 12α-lycopodine (12), which is not a naturally occurring alkaloid but which is available by hydrogenation of anhydrolycodoline (9), was successfully synthesised[22]. The

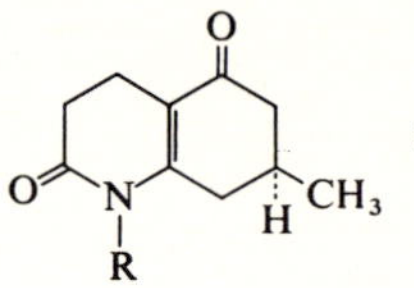

(13)

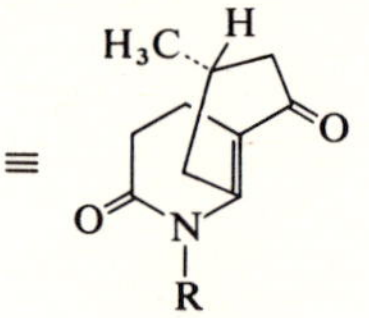

(13a)

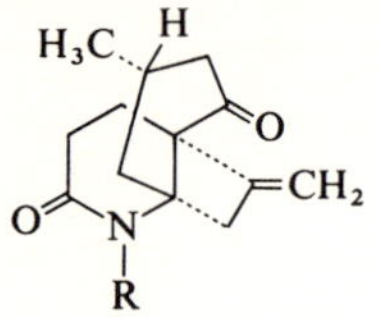

(14)

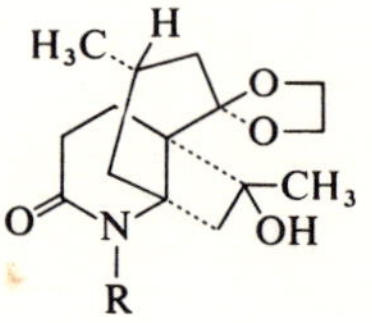

(15)

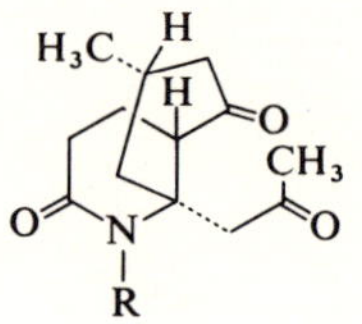

(16)

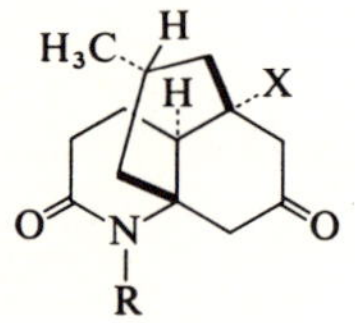

(17) X = OH
(18) X = Cl

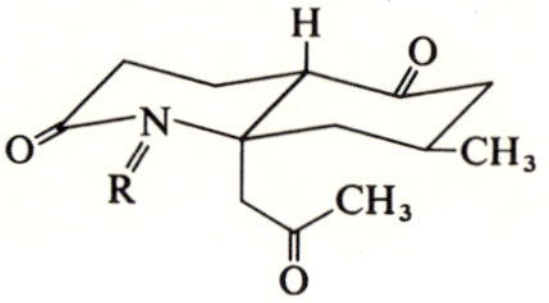

(16a)

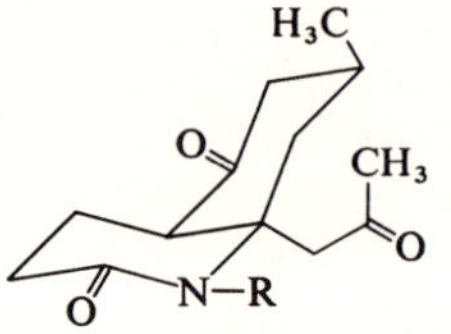

(16b)

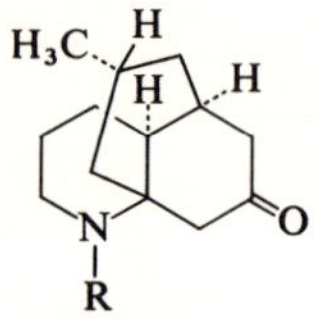

(19) R = H
(20) R = COCH = CH_2

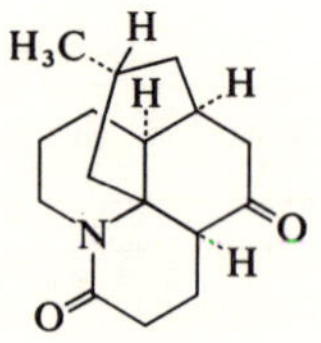

(21)

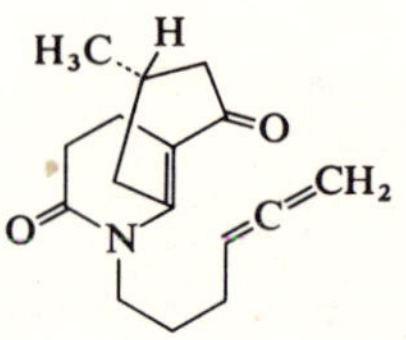

(22)

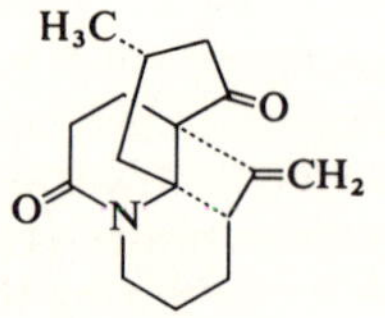

(23)

starting material for this synthesis is the bicyclic ketolactam (13) (= 13a, R = H) which is readily available from dihydro-orcinol and acrylonitrile. Benzylation at nitrogen followed by photochemical addition of allene gave in 30% yield the tricyclic compound (14) (R = CH_2Ph), where addition of allene had occurred on the side opposite to the potential C-15 methyl group. Ketalisation of (14) (R = CH_2Ph) followed by epoxidation and selective reduction of the epoxide utilising lithium borohydride furnished the alcohol (15) (R = CH_2Ph). Treatment of (15) with dilute aqueous acid effected deketalisation and the resulting β-hydroxyketone underwent retroaldol cleavage to the diketone (16) (R = CH_2Ph), which, when treated with dilute base, gave the tricyclic compound (17) (R = CH_2Ph). In this last step the relative configuration at C-12 is established. Although the stereo-

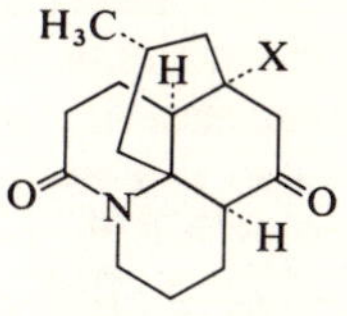

(24) X = Cl
(25) X = H
(27) X = OH

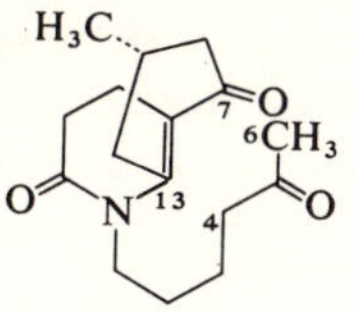

(26)

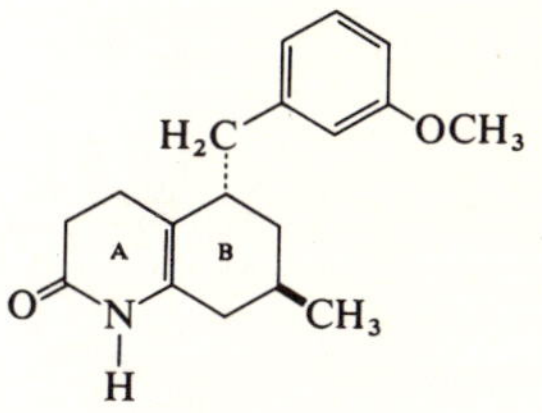

(28)

$\equiv$

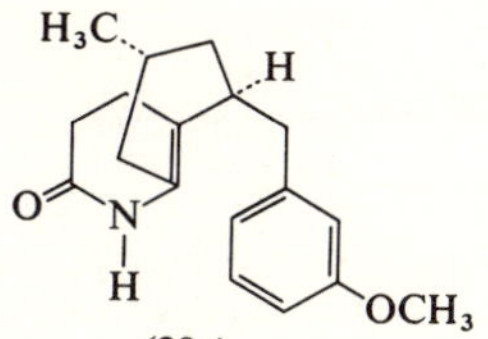

(28a)

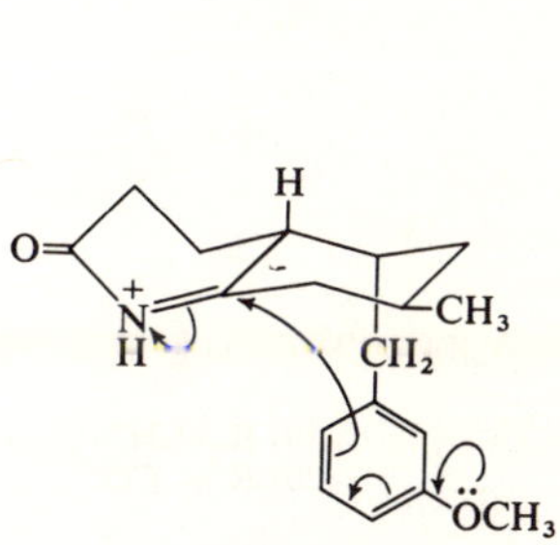

(29)

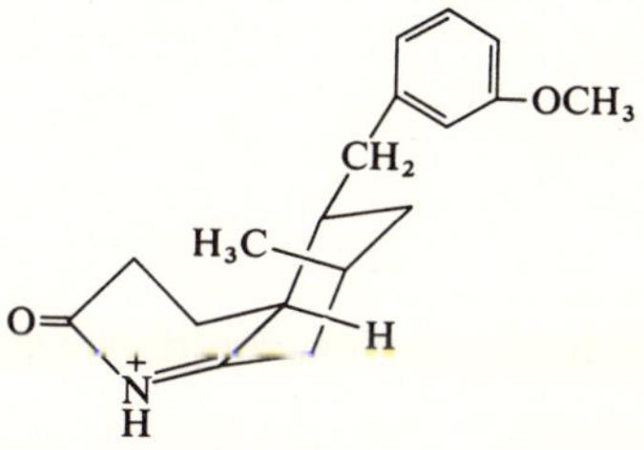

(30)

chemistry at C-12 in (16) has not been established, it does not seem unreasonable that it may indeed have the desired *trans* stereochemistry (16a). However, the transition state leading to cyclisation appears to be less crowded in the *cis*-epimer (16b) and since both (16a) and (16b) may be present in equilibrium during the base-catalysed cyclisation the more facile cyclisation of (16b) may account for the observed result.

Treatment of (17) (R = CH_2Ph) with phosphorus pentachloride provided the bridge-head chloro compound (18) (R = CH_2Ph) which after ketalisation was reduced first with sodium–ammonia (removal of chlorine and benzyl) and then with lithium aluminium hydride (lactam → amine). Deketalisation provided the ketone (19). Treatment of (19) with acrylyl chloride gave (20) which underwent cyclisation to (21) when heated in the presence of acid.

Reduction of (21) with LAH followed by oxidation with Jones' reagent gave racemic 12α-lycopodine (12). The low yield (30%) in the photocyclisation (13) → (14) is due, in part at least, to the fact that allene also adds in the reverse sense. A dramatic improvement in the overall yield has been achieved[23] by a process in which this difficulty is surmounted and, simultaneously, ring C is constructed. Alkylation of (13) (R = H) with 6-bromohexa-1,2-diene gave (22) which on irradiation afforded the tetracyclic compound (23) in 70% yield. When (23) was carried through a sequence of reactions similar to that described for the transformation of (14) into (18), the ketolactam (24) was obtained in 28% overall yield. Reductive removal of the bridge-head chlorine using zinc dust gave (25), which was transformed to 12α-lycopodine (12) in the same way as described for the ketolactam (21). This same paper[23] describes an even shorter synthesis of the 12α-lycopodine system. Alkylation of (13) (R = H) with the ethylene ketal of 6-bromohexan-2-one followed by deketalisation gave the bicyclic compound (26). Treating (26) with warm dilute methanolic sodium methoxide brought about internal Michael

(31) R^1 = OCH_3, R^2 = H
(32) R^1 = H, R^2 = OCH_3

(33)

(34)

(35)

addition (C-4 to C-13) followed by internal aldol condensation (C-6 to C-7) to give in 62% yield[24] the tetracyclic compound (27), previously converted to 12α-lycopodine.

The difficulties encountered in establishing the natural configuration at C-12 have been overcome in the two syntheses of lycopodine itself. In one approach[20, 25] this was achieved by acid-catalysed cyclisation of the racemic hexahydroquinolone (28) (= 28a). Protonation of (28) may occur either on the same side as the methyl group to give the acylimmonium ion (29) or on the opposite side to give the ion (30). Intramolecular electrophilic attack by the acylimmonium ion on the anisole ring is possible in (29) but in (30) ring B would have to attain an energetically unfavourable twist conformation before cyclisation could occur. Treatment of (28) with phosphoric acid in formic acid gave the desired 12β compound (31) in 55% yield along with 29% of the *ortho* substitution product (32). Two routes to compound (28) have been developed[20, 25], only one of which[25] will be described here. Copper-catalysed 1,4-addition of *m*-methoxybenzylmagnesium bromide to

5-methyl-2-cyclohexenene proceeds with axial addition of the Grignard reagent, establishing the *trans*-relationship of the methyl and *m*-methoxybenzyl groups. The enolate (33) was alkylated with allyl bromide in the presence of hexamethylphosphoramide. The resulting compound (34) was transformed (hydroboration, oxidation) to the acid (35), the methyl ester of which when heated with ammonia–methanol gave (28).

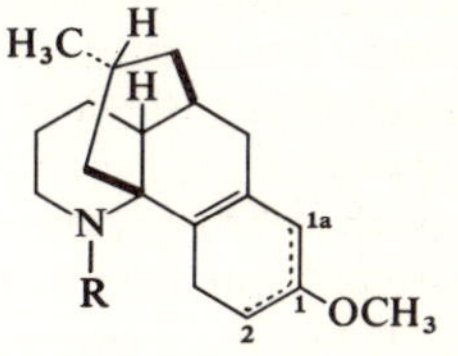

(36) $\Delta^{1,2}$, R = H
(37) $\Delta^{1,1a}$, R = H
(38) $\Delta^{1,1a}$, R = CO$_2$CH$_2$CCl$_3$

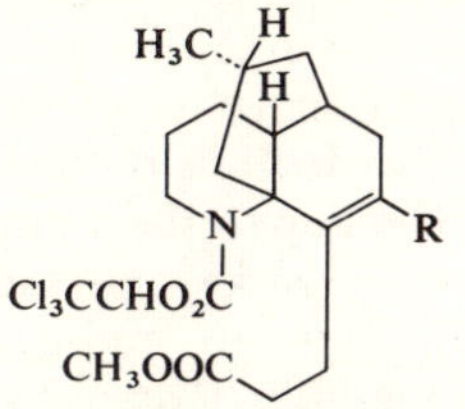

(39) R = CHO
(40) R = OCHO

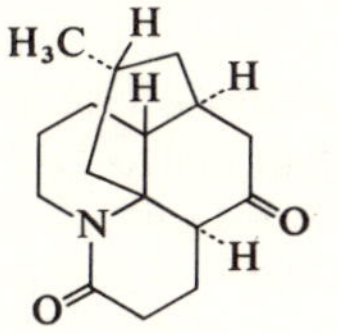

(41)

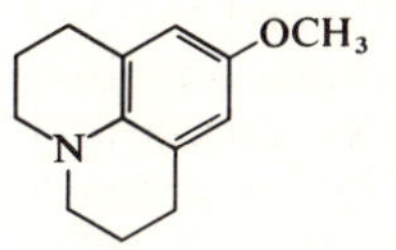

(42)

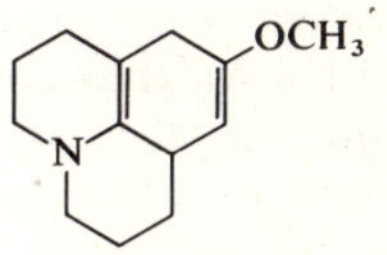

(43)

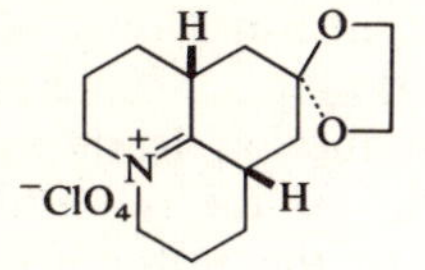

(44)

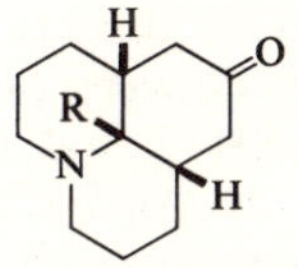

(45)

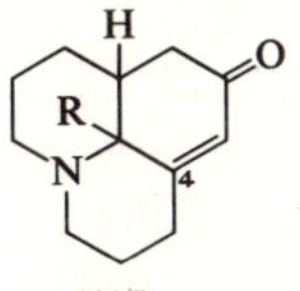

(46)

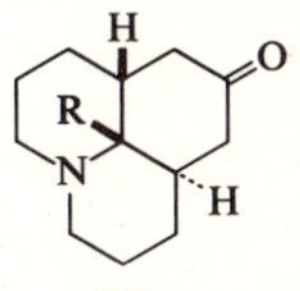

(47)

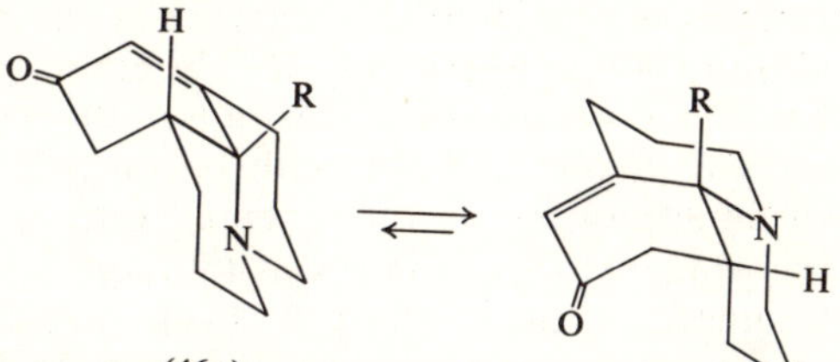

(46a)

(46b)

The transformation of compound (31) into *dl*-dihydrolycopodine (6) involved an interesting sequence of reactions. Lithium aluminium hydride reduction of the amide followed by Birch reduction gave the dihydroanisole (36), which on hydrolysis gave the β,γ-unsaturated ketone. The plan was to isomerise the β,γ-unsaturated ketone to the α,β-unsaturated compound and remove carbon 1a by means of ozonolysis. However, all attempts to isomerise the β,γ-unsaturated ketone were unsuccessful, presumably because the non-bonded interactions developed in the α,β-unsaturated system, which are not present when the carbon–carbon double bond is endocyclic in the bicyclo[3.3.1] system, more than offset the energy gain due to conjugation. Treatment of (36) with strong base, however, gave the isomeric enol ether (37), which was then acylated with trichloroethyl chlorocarbonate to give (38). Ozone attacked only the less hindered 1,1a double bond in (38) and yielded after esterification aldehyde (39) which on modified Baeyer–Villager oxidation gave the enol ester (40). Methanolysis of (40) followed by removal of the N-protecting group and lactam formation led to racemic lycopodine lactam (41). Hydride reduction gave *dl*-dihydrolycopodine (6). Dihydrolycopodine, which is itself a naturally occurring alkaloid, may be oxidised to lycopodine (1)[2].

The second synthesis of lycopodine[21] illustrates a different approach to the solution of the stereochemical problem at C-12. The starting material in this case was the readily accessible 9-methoxyjulolidine (42). This was transformed by Birch reduction to the dihydro compound (43) which when treated with ethylene glycol–perchloric acid gave the immonium perchlorate (44). Addition of Grignard reagents to (44) occurs[26] on the less hindered 'convex face' to give, after hydrolysis of the ketal grouping, all-*cis* hexahydro-julolidines of type (45). The key to the solution of the C-12 problem was the discovery[26] that α,β-unsaturated ketones (46), derived from (45) by bromination and dehydrobromination are reduced by lithium–ammonia to give mainly the *cis–trans* ketones (47). Because of the conformational mobility at nitrogen compounds such as (46), trigonal at C-4, are reminiscent of *cis*-decalin systems and are free to assume either of the two possible *all*-chair conformations (46a) or (46b). The Bohlmann criteria[13] mentioned earlier may be used to distinguish not only between the isomers (45) and (47) but also between (46a) and (46b). It has been shown that conformation (46b) is preferred and thus reduction leads mainly to the *cis–trans* isomers (47). In the synthesis of lycopodine the addition of 2-methyl-3-methoxypropyl-magnesium chloride to the immonium salt (44) gave racemic (45) (R = $CH_2CH(CH_3)CH_2OCH_3$) which was then transformed via the unsaturated ketones (46) (since R contains a centre of asymmetry, a pair of diastereo-isomers is obtained) to a mixture of the racemic diastereoisomeric ketones (48) and (49), epimeric at C-15. The ethers (48) and (49) could not be separated, but the corresponding alcohols (50) and (51), prepared by treating the mixture of (48) and (49) with boron tribromide, were readily separable by chromatography. In the racemic series it was not possible to determine at this stage which alcohol was (50) and which was (51). However, using the Grignard reagent derived from optically active 1-chloro-2-methyl-3-methoxypropane of known R-configuration it has been possible to prepare (45) (R = $CH_2CH(CH_3)CH_2OCH_3$) in optically active form[27]. When the optically active (45) is subjected to the *cis–cis* to *cis–trans* isomerism via (46) this, after ether cleavage, leads to a mixture of (50) which has the 'natural' configuration, and (51) which is in the enantiomeric series (i.e. the mirror image of (51) as drawn). Since 6-ketones in the natural series show negative Cotton effects[28] in the o.r.d. the alcohol showing a negative Cotton effect may be assigned structure (50) and that showing a positive Cotton effect is the enantiomer of (51)[27]. Racemic compound (50) was used to complete the synthesis[21]. In order to prevent intramolecular ring-closure on nitrogen during the cyclisation step it was necessary to transform the amine by permanganate oxidation to the corresponding lactam (52). Transformation of (52) to the mesylate followed by treatment with potassium t-butoxide in t-butyl alcohol effected ring closure (36% yield) to give the racemic ketolactam (53) which had already been prepared in optically active form from lycopodine (1). The optically active material was used as a natural relay to complete the synthesis. Hydride reduction of (53), followed by Jones' oxidation gave ketone (54) which was oxidised by selenium dioxide to the diosphenol (55), also available by similar oxidation of lycopodine. Modified Wolff–Kishner reduction of (55) gave lycopodine (1), as well as anhydrodihydrolycopodine (56), also a naturally occurring alkaloid.

Two interesting and similar approaches to the synthesis of the lycopodine skeleton which involve the construction of the bicyclo[3.3.1]nonane system

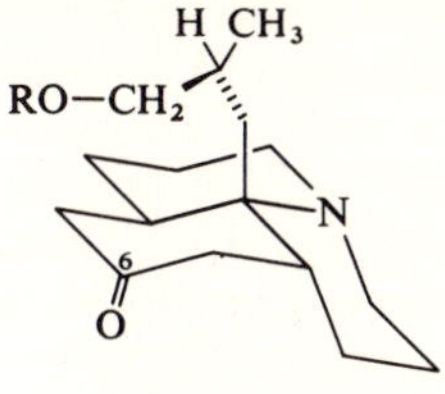

(48) R = CH$_3$
(50) R = H

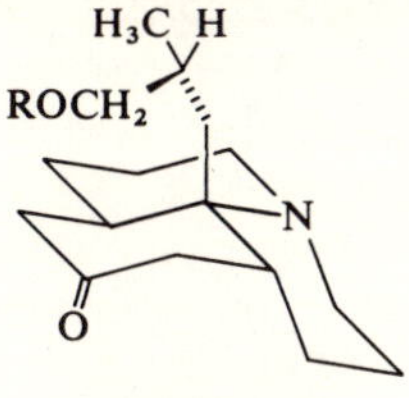

(49) R = CH$_3$
(51) R = H

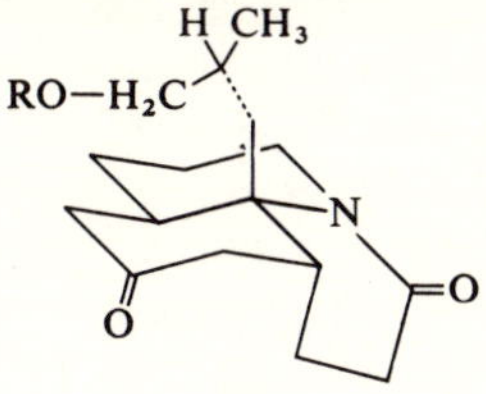

(52)

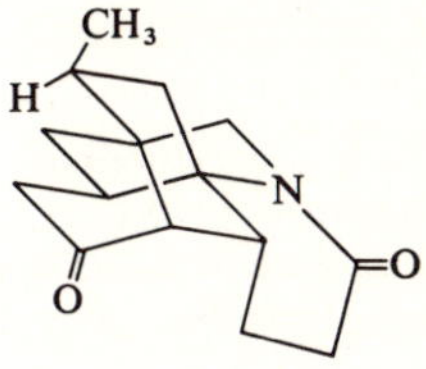

(53)

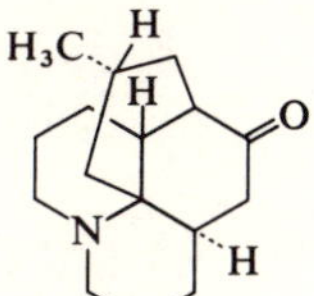

(54)

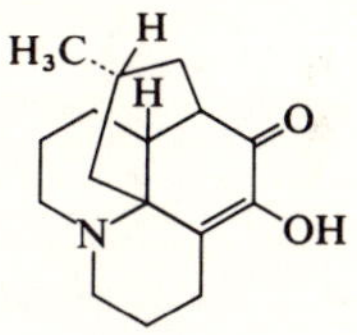

(55)

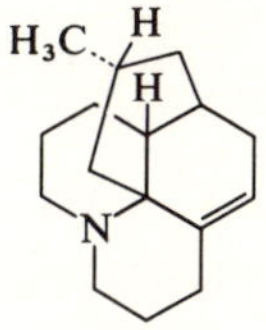

(56)

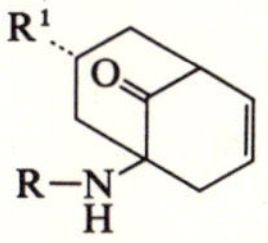

(57) R = R^1 = H
(58) R = COCOCH$_3$, R^1 = H
(60) R = H, R^1 = CH$_3$

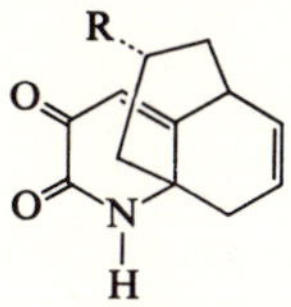

(59)

of rings B and D followed by addition of ring A have been reported[29, 30]. In one case[29] the bicyclic aminoketone (57) was constructed and acylated with pyruvyl chloride to give (58) which was cyclised using sodium hydride to give the tricyclic compound (59) (R = H). In the other case[30] the methyl substituted bicyclic compound (60) was subjected to a similar cyclisation process to yield (59) (R = CH$_3$). Reductive transformations of both (59) and (60) are reported[29, 30] but as yet neither has been transformed to a tetracyclic alkaloid.

The biosynthesis of lycopodine is discussed in Section 1.11.

1.3 THE ANNOTININE GROUP

Annotinine (61) is the only representative of this particular skeleton.

1.3.1 Isolation and structural studies

The structural studies on annotinine have been well reviewed[1-3]. Recently it has been found[31] that treatment of annotinine (61) with a concentrated

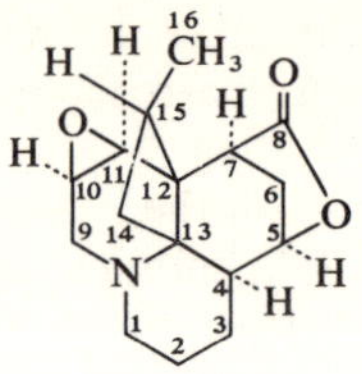

(61)

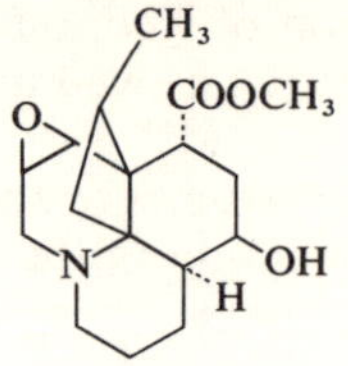

(62)

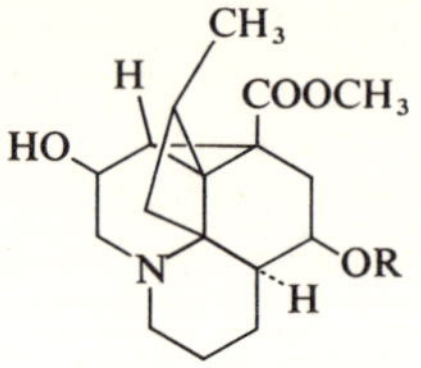

(63) R = H
(64) R = COCH$_3$

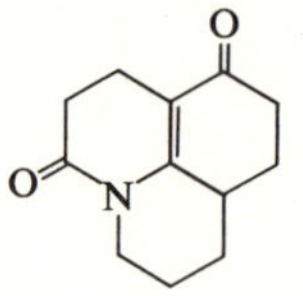

(65)

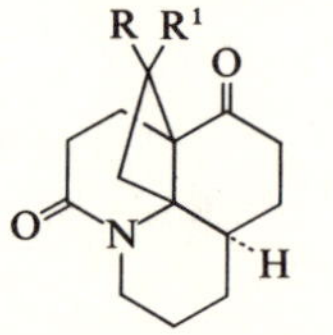

(66) R = R^1 = CH$_2$
(67) R = H, R^1 = CH$_3$

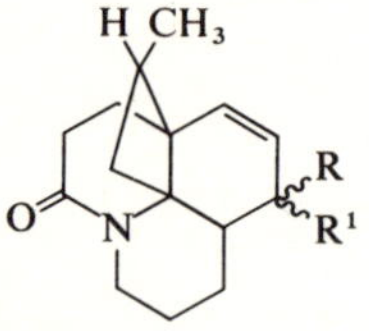

(68) R = R^1 = H
(69) R = H, R^1 = OAc
(70) R = H, R^1 = OH
(71) R = R^1 = O

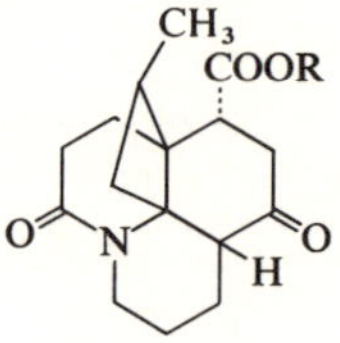

(72) R = CH$_3$
(73) R = H

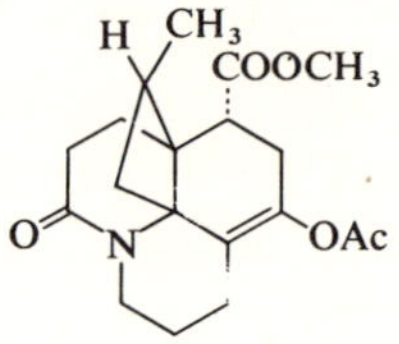

(74)

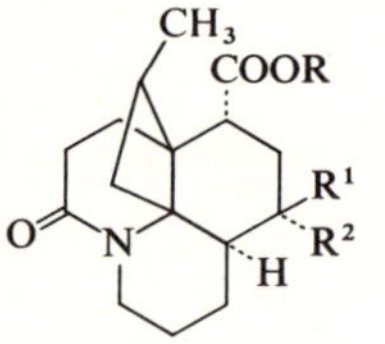

(75) R^1 = OH, R^2 = H
(76) R^1 = H, R^2 = OH

solution of potassium methoxide in methanol at 60–70 °C leads, via methyl epiannotinate (62), to the pentacyclic compound (63) containing a spiro[3.2]-hexane system. The structure of (63) was established[31] by x-ray diffraction studies on the hydrobromide of the monoacetyl derivative (64).

1.3.2   Synthesis

Annotinine (61) was the first Lycopodium alkaloid to yield to structural analysis. It was also the first alkaloid of the group to be prepared by total synthesis[32, 33]. A key step in the synthesis was the photochemically-induced addition of allene to the tricyclic compound (65), the preparation of which was achieved by heating $\Delta^{8,\,9}$-octahydroquinol-7-one with acrylic acid. The reaction of (65) with allene. gave, in better than 50% yield, the adduct (66) possessing the requisite cis–trans ring fusions and with the proper orientation of the methylene group for further modification to the methyl group of annotinine. This latter transformation was accomplished by catalytic hydrogenation of the ethylene ketal of (66), the ketal serving to hinder addition from that side and leading to the stereospecific production of the desired C-15 epimer (67).

Borohydride reduction of (67) followed by mesylation of the resulting 7β-ol and elimination of methanesulphonic acid utilising potassium t-butoxide in dimethyl sulphoxide gave the olefin (68). This was transformed into the unsaturated ketone (71) via the sequence (69) (SeO_2–HOAc oxidation), (70) (hydrolysis), (71) (Sarett oxidation). A carboxylate grouping was introduced at C-7 by hydrocyanation followed by methanolysis to the racemic ester (72), identical except for rotation with optically-active ester (72) of established stereochemistry[1] prepared from annotinine. Racemic acid (73) was resolved via the brucine salt to give, after esterification, the totally synthetic natural enantiomer (72).

The lactone ring of annotinine was constructed by converting (72) to the enol acetate (74) and reducing this with sodium borohydride to a mixture of

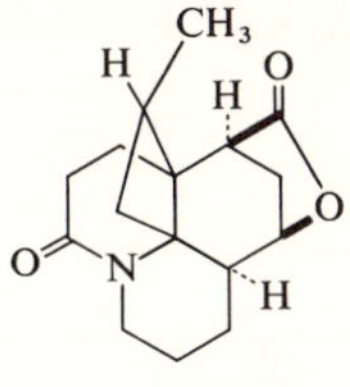

(77) 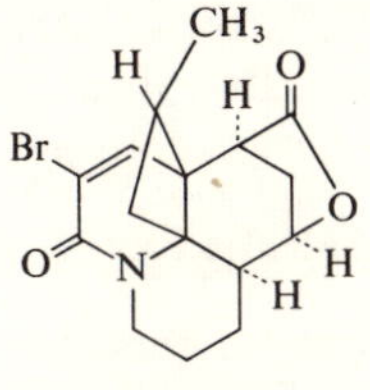   (78) 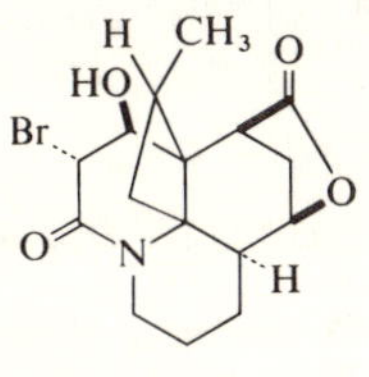   (79)

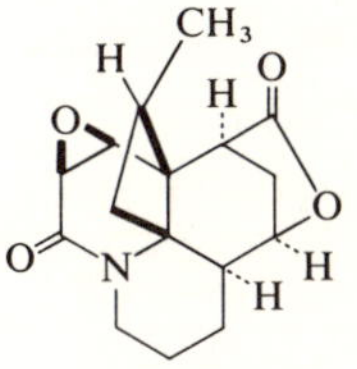

(80) 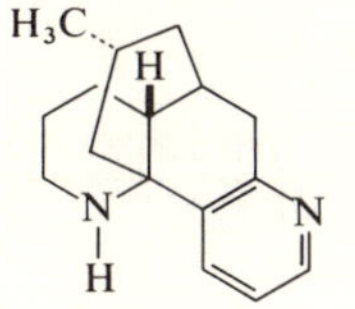   (81) 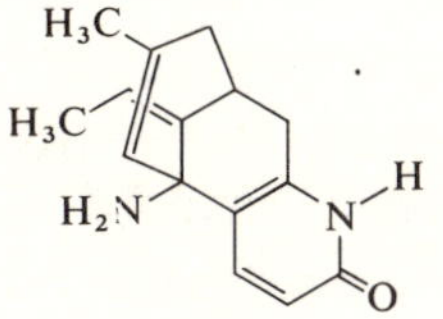   (82)

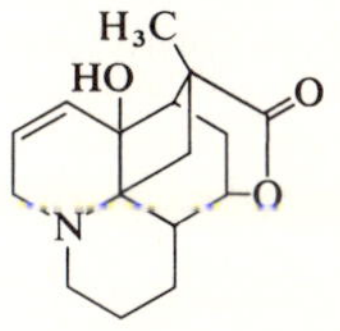

(83) 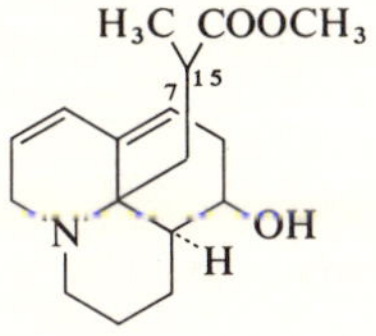   (84) 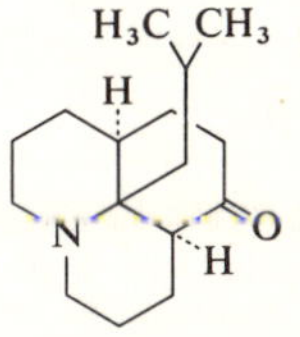   (85)

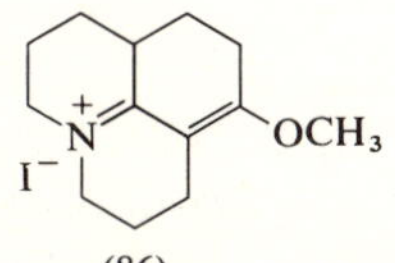

(86)

the alcohols (75) (R = CH_3) and (76) (R = CH_3). Preparation of the enol acetate was necessary since direct reduction of (72) gives the alcohols epimeric with (75) and (76) at C-4. The esters (75) and (76) were not separated but were hydrolysed to the corresponding acids and the mixture of acids (75) (R = H) and (76) (R = H) was heated in benzene containing toluene-p-sulphonic acid. Under these conditions it is possible to equilibrate the carboxyl group at C-7 and, in the case of the C-7 epimer of (75), effect lactonisation to (77).

Treatment of (77) with *N*-bromosuccinimide in CCl$_4$ led, presumably by dibromination and dehydrobromination, to the unsaturated bromo compound (78) which was hydrated utilising aqueous hydrobromic acid to the bromohydrin (79). Treatment of the bromohydrin with aqueous bicarbonate yielded oxoannotinine (80) which had previously been shown [34] to undergo unexpectedly easy acid-catalysed hydrogenolysis to annotinine (61) in the presence of hydrogen and platinum.

1.4   THE LYCODINE GROUP

Members of this group have the carbon–nitrogen skeleton of lycodine (81). Selagine (82) is regarded as a member of this group.

1.4.1   Isolation and structural studies

No new members of this group have been reported during the period under review. The occurrence of α- and β-obscurine in *L. selago*[16] and of de-*N*-methyl-α-obscurine in *L. alpinum*[19] has been noted.

1.4.2   Synthesis

Since lycopodine (1) has been transformed[35] into lycodine (81) the synthesis of lycopodine represents, in a formal sense, the synthesis of lycodine.

1.5   THE ANNOTINE GROUP

Members of this group possess the carbon–nitrogen skeleton of annotine (83). Lyconnotine (84) lacking the C-7, C-15 bond of annotine, is placed in this group.

1.5.1   Isolation and structural studies

No new alkaloids of this type have been reported. The mass spectra of annotine (83) and some of its derivatives have been studied in detail[69].

1.5.2   Synthesis

The synthesis of compound (85), a transformation product of lyconnotine (84) has been reported[36]. The key step in this synthesis was the addition of isobutyl-lithium to the immonium double bond in (86). Hydrolysis of the resulting enol ether gave (85).

1.6   THE CERNUINE GROUP

Members of this group have the carbon–nitrogen skeleton of cernuine (87) and are numbered as shown in (87).

1.6.1   Isolation and structural studies

A paper describing the determination of the constitution of cernuine (87) and lycocernuine (88) has appeared[37]. This aspect of cernuine chemistry has been reviewed[2, 3]. The stereochemistry shown in (87) and (88) has been proposed for these alkaloids on the basis of the following arguments[38]. N.M.R. measurements on *O*-acetyl-lycocernuine (89) indicate that the

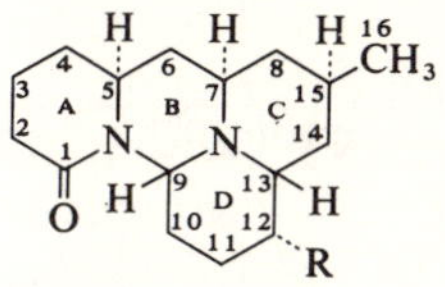

(87) R = H
(88) R = OH
(89) R = OAc
(94) R = H, $\Delta^{12, 13}$

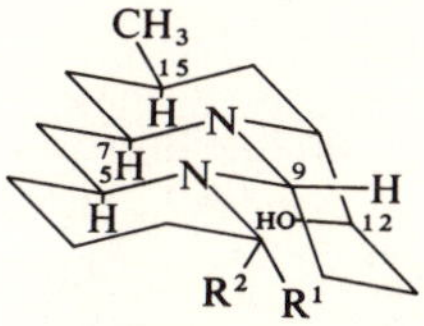

(90) R¹ = R² = O
(92) R¹ = R² = H

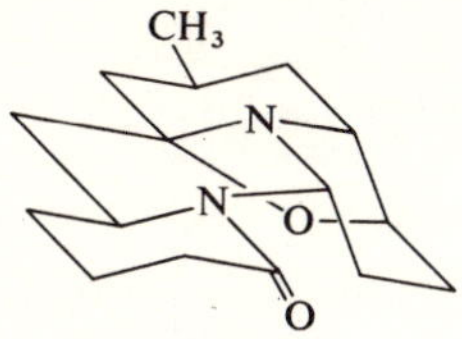

(91)

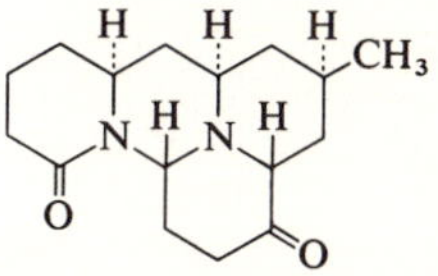

(93)

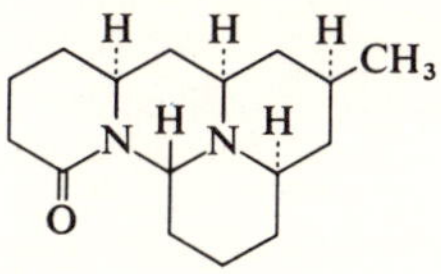

(95)

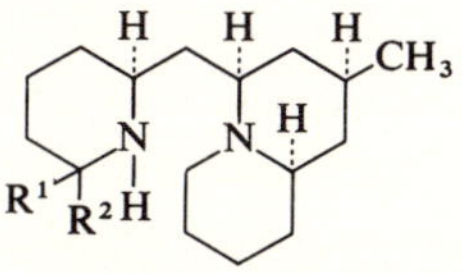

(96) R¹ = R² = O
(99) R¹ = R² = H

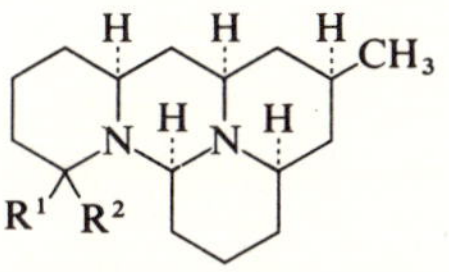

(97) R¹ = R² = O
(100) R¹ = R² = H

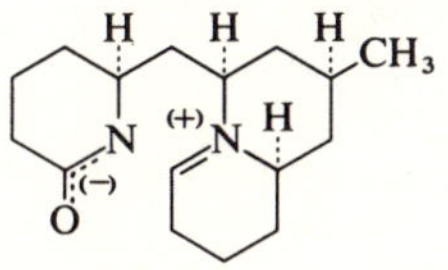

(98)

oxygen function is axial to ring D and i.r. measurements show that the hydroxyl group in lycocernuine is not hydrogen bonded to the amino nitrogen. This can only be the case if rings C and D form a *cis*-quinolizidine system as in stereostructure (90) for lycocernuine. Consistent with this, lycocernuine does not show Bohlmann bands in the i.r. The proton at C-9, deshielded by both nitrogens, is well resolved in the n.m.r. and since it shows an axial–axial coupling with one of the C-10 protons it is assigned the relative stereochemistry depicted in (90). Mercuric acetate oxidation, which requires an *anti* co-planar arrangement of the nitrogen lone pair and an α-hydrogen, occurs towards C-7, establishing the axial nature of the hydrogen at C-7. In the case of lycocernuine the product was the oxazolidine (91). Since dihydro-deoxycernuine (92), the LiAlH₄ reduction product of cernuine, shows Bohlmann bands in the i.r. the AB ring system must be a *trans*-quinolizidine system and thus the configuration at C-5 must be as depicted in (90) and (92). The assignment of configuration at C-15 is based mainly on the synthetic studies discussed in Section 1.6.2. The absolute configurational assignment is based on application of the Horeau method[39] to lycocernuine which indicated that C-12 has the R-configuration. Application of the Octant rule to dehydrolycocernuine (93) led to the same assignment of absolute configuration[38].

Dihydrodeoxycernuine (92) has been isolated from *L. cernuum*[37].

1.6.2 Synthesis

During the structural studies[37, 38] on cernuine it was discovered that catalytic hydrogenation of anhydrolycocernuine (94) in non-protic solvent afforded allocernuine (95), the C-13 epimer of cernuine, whereas in protonic solvents the hydrogenolysis product (96) was obtained. Sodium borohydride reduction of (95) also afforded (96). Allocernuine (95), when heated in methanol, is isomerised to epiallocernuine (97). The formation of both (96) and (97)

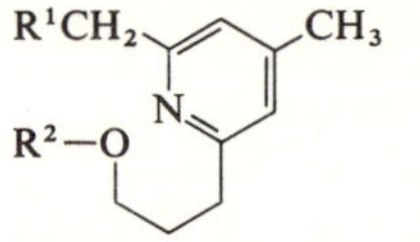

(101) R^1 = R^2 = H
(102) R^1 = Li, R^2 = THP

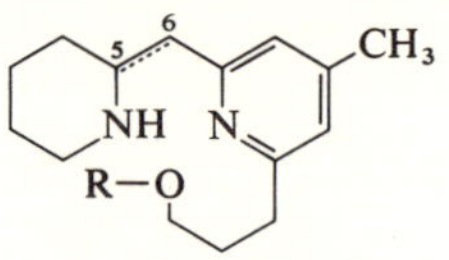

(103) R = THP, $\Delta^{5,6}$
(104) R = H

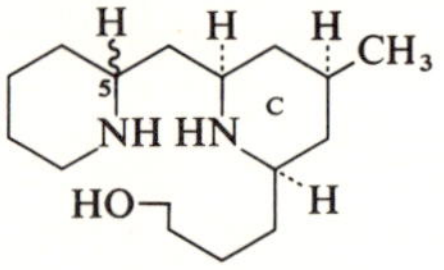

(105)

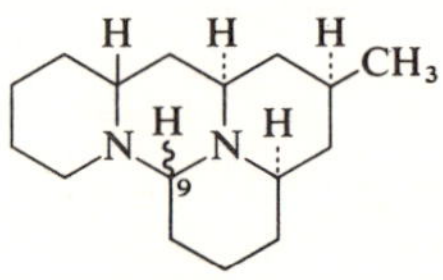

(106)

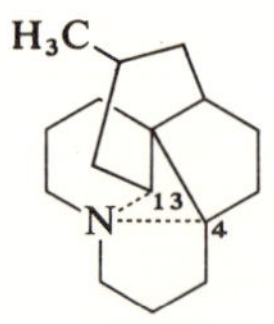

(107)

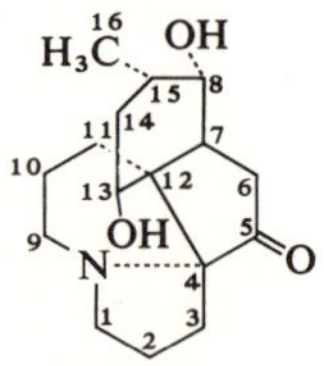

(108)

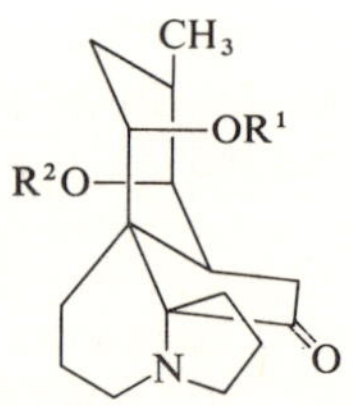

(108a) R^1 = R^2 = H
(110) R^1 = R^2 = Ac
(112) R^1 = Ac, R^2 = H

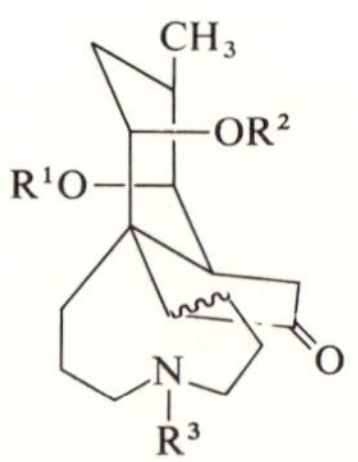

(109) R^1 = R^2 = R^3 = Ac

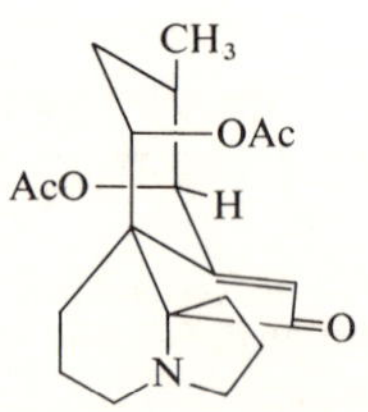

(111)

may be interpreted as occurring via the zwitterionic intermediate (98). LiAlH$_4$ reduction of (96) and (97) gave the dihydrodeoxy compounds (99) and (100), respectively. The synthesis[40] of the racemic forms of compounds (99) and (100) (dihydrodeoxyepiallocernuine) provides additional evidence for the constitutional and stereochemical assignments outlined above.

Hydroboration of 2-(3-butenyl)-4,6-dimethylpyridine, prepared by allylation of the lithium derivative of 2,4,6-collidine, gave the alcohol (101) which after formation of the tetrahydropyranyl ether and treatment with phenyl-lithium gave the lithium compound (102). Condensation of (102) with 2-ethoxy-3,4,5,6-tetrahydropyridine afforded the enamine (103) which after hydrogenation at atmospheric pressure and subsequent hydrolysis yielded (104). Reduction of the pyridine ring in (104) was accomplished in two ways; by high-pressure hydrogenation over rhodium–charcoal and by reduction with sodium in boiling isoamyl alcohol. In each case a mixture of the racemic C-5 epimers (105) was obtained. The stereochemistry of ring C was assigned on the grounds that the chemical reduction should lead to the more stable all-*cis* arrangement and that catalytic hydrogenation should also give the all-*cis* compound. Heating the mixture (105) with hydrobromic acid brought about cyclisation to (99) and its C-5 epimer which were separated by gas chromatography. The C-5 epimers (105) were separated as the di-*N*-formyl derivatives. When the individual alcohols (105) were oxidised with Sarett's reagent, the resulting aldehydes cyclised spontaneously. One of the epimers gave *dl*-dihydrodeoxyepiallocernuine (100) in very high yield. The other furnished a mixture of the C-9 epimers (106)[40].

The biosynthesis of cernuine is discussed in Section 1.11.

1.7   THE SERRATININE GROUP

Compounds possessing the skeleton (107) are assigned to this group. Most members possess an additional bond (either N to C-4 or N to C-13).

1.7.1   Isolation and structural studies

L. serratum Thumb var. *serratum* F. *serratum* has proved to be a rich source of alkaloids[41, 42]. Lycopodine (1), clavolonine (5), lycodoline (C-12 epimer of (11)), lycodine (81) and serratidine (8) have been isolated along with a series of alkaloids possessing skeletons different from the groups discussed thus far. As the result of a very elegant chemical study[43–45], summarised below, serratinine, one of the major alkaloids of this species has been shown to possess structure (108) (= 108a). An x-ray crystallographic study[46] has verified the constitution of serratinine and established the configuration at C-4, a feature which the chemical study failed to define.

The presence of the α-aminoketone functionality in serratinine, indicated by basicity measurements, was demonstrated by reductive cleavage with zinc–acetic anhydride to give *O,O,N*-triacetyl*chano*dihydroserratinine (109). Serratinine readily forms a benzylidene derivative indicating the carbonyl group is flanked on one side by a methylene group. Reduction of (109) with sodium borohydride gave an alcohol which on dehydration afforded a trisubstituted olefin, indicating that in the *chano* compound the carbonyl group is flanked by methine and thus in serratinine itself by a quaternary carbon. Selenium dioxide oxidation of diacetylserratinine (110) gave the α,β-unsaturated ketone (111) possessing (as indicated by n.m.r.) an allylic secondary acetoxyl function. Dehydration of monoacetylserratinine II (112), prepared by selective hydrolysis of (110), gave a trisubstituted olefin (113) containing a vinylic methyl group. On selenium dioxide oxidation (113) yielded the conjugated dienone (114). These results served to locate one of the hydroxyl groups and the *C*-methyl group relative to one another and relative to the carbonyl group. Hydrolysis of (113) followed by oxidation gave a β,γ-unsaturated ketone which was isomerised by basic alumina to the α,β-unsaturated ketone (115), still possessing a vinylic methyl group and thus establishing that the second hydroxyl group in serratinine is in a 1,3-

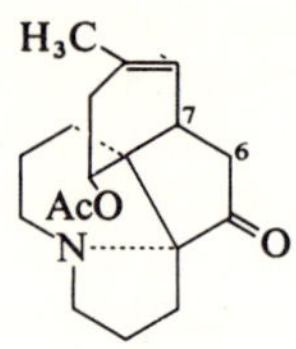

(113)
(114) $\Delta^{6,7}$

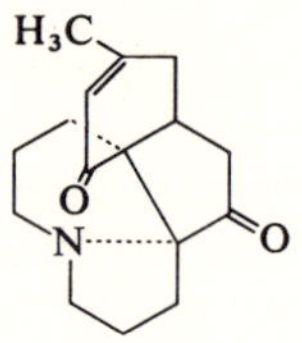

(115)

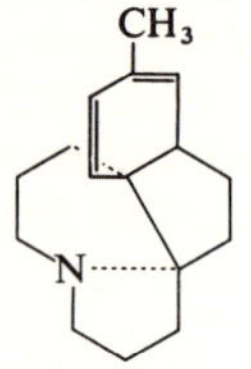

(116)

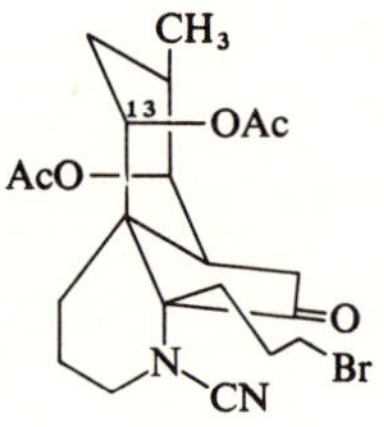

(117)

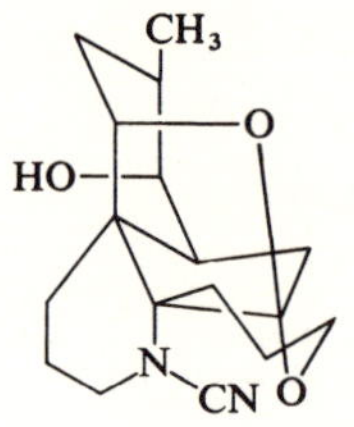

(118)

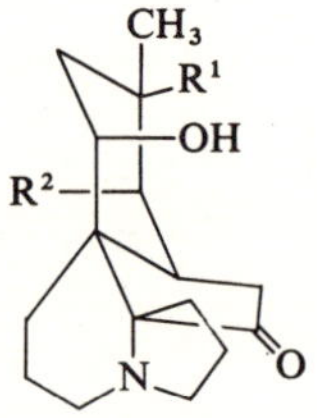

(119) R^1 = OH, R^2 = H
(120) R^1 = R^2 = OH

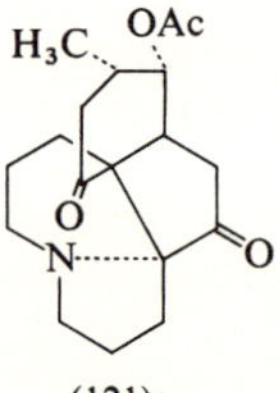

(121)

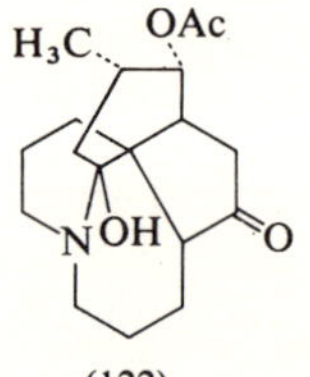

(122)

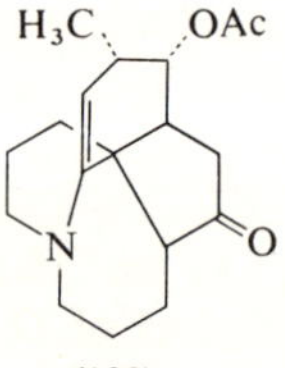

(123)

relationship with the *C*-methyl group. Evidence for the size of the hydroxylated ring was obtained by dehydration of deoxoserratinine ((108), carbonyl replaced by CH_2) which furnished the diene (116), λ_{max} 266 nm. Diels–Alder addition of diethyl acetylenedicarboxylate to (116) gave an adduct which on heating underwent a reverse Diels–Alder reaction to give diethyl 4-methylphthalate, possible only if the original ring was six-membered and substituted as shown in (116).

Application of the Hofmann and von Braun degradation sequences[42] delineated the environment of the nitrogen in serratinine and led uniquely to the constitution (108). The von Braun degradation of diacetylserratinine (110) led to a novel reaction sequence which provided valuable evidence concerning the stereochemistry of the alkaloid[45]. The von Braun cleavage product (117) when treated with aqueous base was transformed to the ketal (118), a reaction possible only if the oxygenated rings are *cis*-fused and the C-13 hydroxyl is axially oriented. Final definition of the complete relative and absolute stereochemistry was achieved by an x-ray crystallographic study of 13-acetyl-8-*p*-bromobenzoylserratinine[46]. A detailed study of the mass spectra of serratinine and its derivatives indicates that mass spectrometry may be used to detect this particular skeleton[47]. This technique was utilised[48] in the determination of the structures of serratine (119) and serratanidine (120). Serratine has been correlated with serratinine by conversion to both (113) and (115) and serratanidine has been transformed into 8-dehydroserratinine[48]. Oxidation of the olefin (113) with hydrogen peroxide–formic acid yielded serratanidine (120).

An interesting reaction in the serratinine series which provided the key to the solution of the structures of several other alkaloids involves the

(124)   (125)   (126)

zinc–acetic acid reduction of the diketone (121) derived from serratinine[49, 50]. The products were the carbinolamine (122) (acetylaposerratinine) and the enamine (123) (acetylanhydroaposerratinine) formed by reductive cleavage of the α-aminocyclopentanone system and subsequent transannular carbinolamine formation (to give (122)) and dehydration (to (123)). It was observed that the grouping $>$N—C=CH—CH(CH$_3$)CH(OAc)$-$which is

present in (123) is readily detectable by n.m.r. spin-decoupling studies. This grouping is also present in the alkaloid serratinidine, shown to possess structure (124) and synthesised from acetylanhydroaposerratinine (123)[50]. Fawcettidine, an alkaloid of *L. fawcettii*, has the related structure (125)[50, 51]. Alolycopine, an alkaloid present in *L. alopecuroides* is (126)[52]. Catalytic hydrogenation of *O*-acetylalolycopine (126, H = Ac) gives acetylanhydroaposerratinine (123). Fawcettidine (125) also occurs in *L. alopecuroides*[53].

The *L. fawcettii* alkaloid fawcettimine is the carbinolamine (127)[54]. Dehydration of fawcettimine with phosphorus oxychloride gives fawcettidine (125). Fawcettimine has been shown to exist in solution predominantly as the carbinolamine, although acetylation occurs on nitrogen to give *N*-acetyl-fawcettimine (128).

1.7.2   Biogenetic considerations

The possible biogenetic relationship between members of the serratinine group and those of the lycopodine group has been pointed out[45]. Rearrangement of the 12-hydroxylycopodine skeleton as portrayed in (129) would lead to a carbinolamine such as fawcettimine (127). Opening of the carbinolamine followed by C-4 to N bond formation leads to the serratinine skeleton. Although attempts to execute this rearrangement on various 12-hydroxylycopodine derivatives have been unsuccessful, the synthetic compound (130) when treated with hydrochloric acid–acetic acid at 100 °C

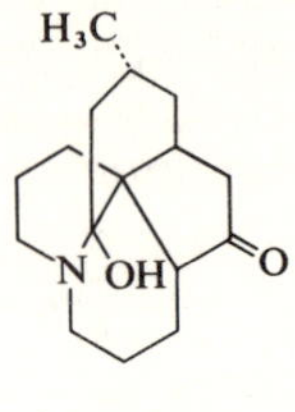

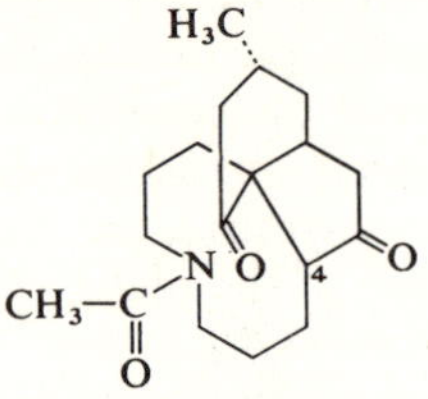

undergoes hydrolysis, rearrangement, and dehydration to give the cyclopentanone derivative (131)[55].

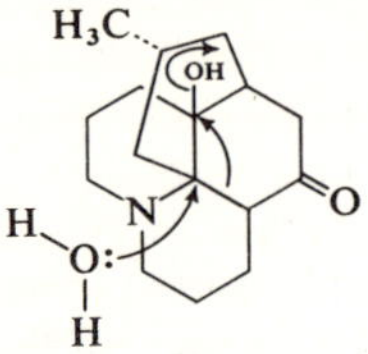

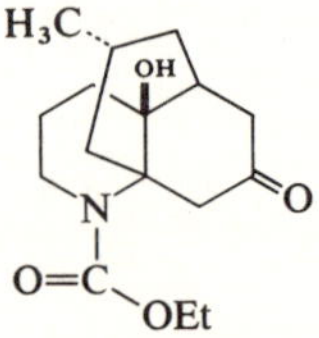

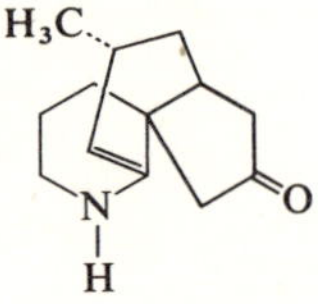

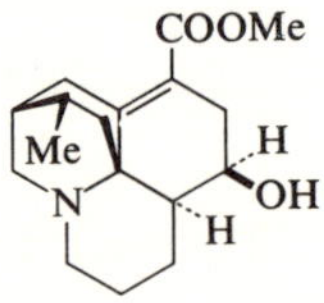

1.8   THE ANNOPODINE GROUP

Annopodine (132) is the only alkaloid discovered to date which has this particular skeleton.

1.8.1   Isolation and structural studies

Annopodine is one of the minor alkaloids occurring in *L. annotinum* L. and has been shown to possess structure (132) by an x-ray crystallographic study of the hydrobromide[56]. The absolute configuration is that depicted in (132).

1.9   THE ALOPECURINE GROUP

Members of this group possess the pentacyclic skeleton exemplified by debenzoylalopecurine (133). This skeleton is the lycopodine skeleton with an additional bond between C-4 and C-10 (133) = (133a).

1.9.1   Isolation and structural studies

Alopecurine (134), the only benzoylated Lycopodium alkaloid reported to date, was the first member of this group of pentacyclic alkaloids to be isolated[9]. The debenzoylated compound (133) also occurs in *L. alopecuroides*[9] as does the acetyl derivative (135)[53]. X-ray crystallographic studies on the hydrobromide of debenzoylalopecurine (133) and on the methobromide of alopecurine (134) established the structure and absolute stereochemistry of these substances[53]. An interesting feature arising from the x-ray study was the finding that ring C in these alkaloids exists in the twist conformation in

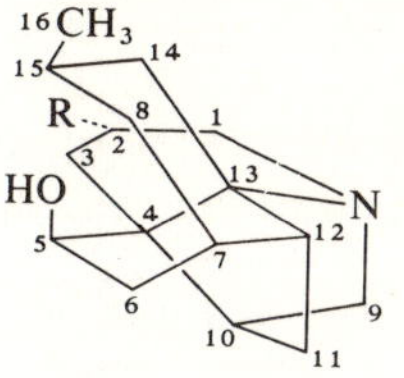

(133) R = OH
(134) R = OCOPh
(135) R = OCOCH$_3$
(136) R = H

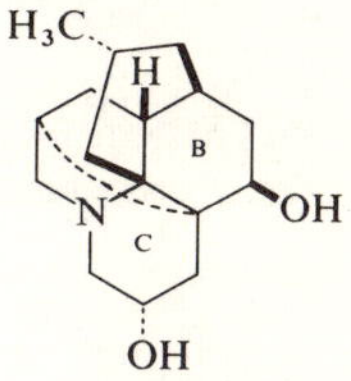

(133a)

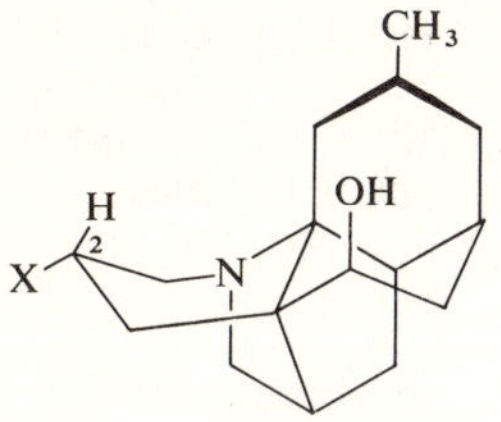

(137)

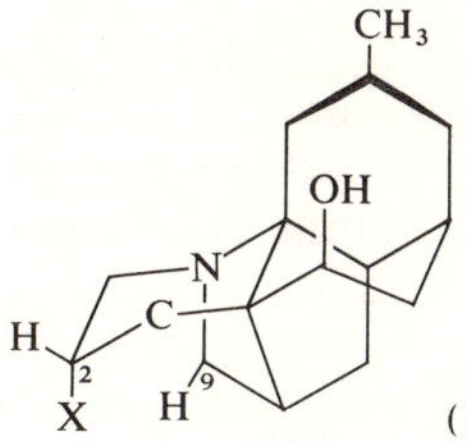

(138)

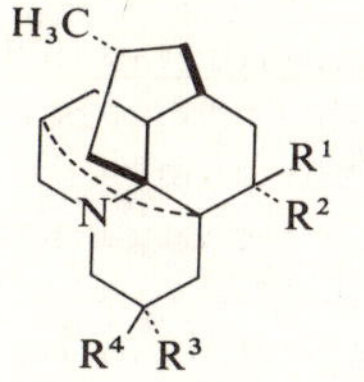

(139) R^1 = R^2 = O, R^3 = OH, R^4 = H
(140) R^1 = OH, R^2 = H, R^3 = R^4 = O
(141) R^1 = R^2 = O, R^3 = R^4 = H

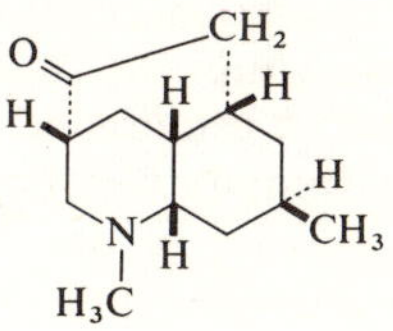

(142)

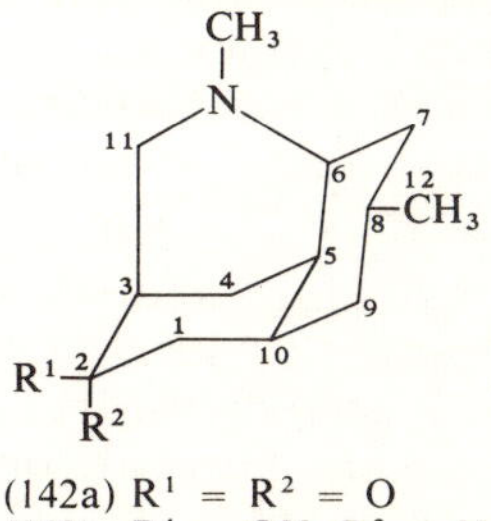

(142a) R^1 = R^2 = O
(143)   R^1 = OH, R^2 = H

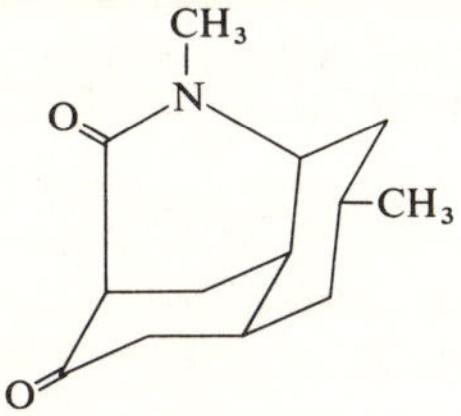

(144)

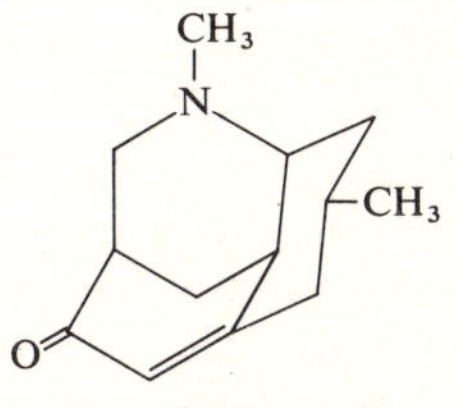

(145)

the crystal, and n.m.r. studies[53] indicate that this is also the preferred conformation in solution. Apparently the non-bonded interaction between the oxygen substituent at C-2 and the exo hydrogen at C-9 is sufficient to tip the energy balance in favour of the twist conformation. When the C-2 oxygen function is replaced by hydrogen as in the *L. alopecuroides* alkaloid lycopecurine (136), x-ray studies show that the chair conformation is favoured[57]. The perspective drawings (137) and (138) illustrate this situation more clearly. When X is hydrogen, (138) is favoured but when X is hydroxyl or benzoyloxy, (137) is preferred.

Pentacyclic alkaloids of the alopecurine type have been isolated from *L. inundatum*[18]. Inundatine (139), isoinundatine (140), and dehydrolyco-pecurine (141) occur in this species. Oxidation of (139) and (140) leads to the same diketone as obtained from debenzoylalopecurine (133). Reduction of (141) with sodium borohydride gives lycopecurine (136)[18].

The pentacyclic alkaloids of this group all show a high-frequency methylene vibration at $1500 \pm 10 \text{ cm}^{-1}$ in the infrared which may be used to detect the presence of this tightly-bridged skeleton[18, 57].

1.10   THE LUCIDULINE GROUP

This group is defined as possessing the carbon–nitrogen skeleton of luciduline (142) = (142A). The numbering system is based on the 2-decalone system present in luciduline.

1.10.1   Isolation and structural studies

Luciduline is a weakly basic alkaloid, originally designated alkaloid L21, occurring in *L. lucidulum*[11, 58]. Sodium borohydride reduction gives the equatorial alcohol dihydroluciduline (143), which is also naturally occurring[59]. The environment of the carbonyl group in luciduline was determined in the following way. Luciduline readily exchanges the hydrogens on C-1 for deuterium under acid conditions. Reduction of the dideuterated compound with sodium borohydride furnishes 1,1-dideuteriodihydroluciduline. The n.m.r. spectrum of the dideuterated alcohol revealed that the proton geminal to the hydroxyl was still coupled to a vicinal hydrogen, suggesting that the carbonyl was flanked by a methine carbon the hydrogen of which was not readily exchangeable. This methine proton was shown to be coupled to a methylene proton α to nitrogen. Oxidation ($KMnO_4$–acetone) of this latter methylene to carbonyl gave a lactam (144) absorbing at 1640 cm^{-1}, indicating that the nitrogen-containing ring is at least six-membered. Selenium dioxide oxidation of luciduline yielded an α,β-unsaturated ketone (145) which showed no β-protons in the n.m.r. These and related experiments served to define the part structure —CH—CH_2—C(=O)—$\overset{*}{CH}$—CH_2—N(CH_3) —CH—, where the asterisk denotes a bridge-head carbon atom, and when combined with the finding that selenium dehydrogenation of luciduline gave 2,6-dimethylnaphthalene led to the formulation (142) (stereochemistry of the *C*-methyl undefined) for luciduline. An x-ray crystallographic study on *O-p*-bromobenzoyldihydroluciduline revealed the orientation of the *C*-methyl group and defined the absolute configuration[58].

The rather interesting biogenetic implications of this unique skeleton are discussed in the next section.

1.11   BIOSYNTHESIS

An early hypothesis of biogenesis of the Lycopodium alkaloids suggested that the 16-carbon framework found in most of the alkaloids might arise from the combination of two 8 carbon polyketide chains and the appropriate nitrogen source[60]. The genesis of the lycopodine (1) skeleton by this pathway is sketched in (146). Condensation of two polyketide chains as shown in (147) would give rise to the lycodine (81) skeleton and the cernuine (87) skeleton may be derived as shown in (148). Later it was recognised that the latter two skeletons could also be derived by dimerisation of two 2-propyl-piperidine units as shown in (149) for the generation of the lycodine (81) skeleton[61, 62]. 2-Propylpiperidine is a naturally occurring alkaloid (coniine) and has been shown[63] to be derived via a polyacetate pathway. In the case

(146)   (147)   (148)

(149)   (150)   (151)

of the structurally similar alkaloid pelletierine (150), however, only the side chain originates from acetate, the piperidine ring being derived from lysine[64]. Recent studies have been directed toward distinguishing which, if either, of these alternate pathways is actually followed in the biosynthesis of the Lycopodium alkaloids.

Both pathways mentioned would lead to the incorporation of acetate into lycopodine. The polyketide route (146) predicts that feeding 1-^{14}C-acetate would lead to lycopodine labelled at C-1, C-3, C-5, C-7, C-9, C-11, C-13 and C-15 (these carbons are indicated by an asterisk in (151)), while administration of 2-^{14}C-acetate would give lycopodine labelled at the even-numbered carbons. Since the carboxyl group of acetate gives rise to only C-2′ of pelletierine, 1-^{14}C-acetate by the pelletierine pathway would lead to lycopodine labelled only at C-7 and C-15, while 2-^{14}C-acetate (C-1′ and C-3′ of pelletierine) would give lycopodine labelled at C-6, C-8, C-14 and C-16 (carbons indicated by ■ in (151)), assuming that lycopodine arises via a lycodine skeleton (see (149)) with subsequent C-5, N-cleavage, nitrogen extrusion, and cyclisation between C-1 and the ring A nitrogen. When 1-^{14}C-acetate was administered to *L. tristachyum*, radioactive lycopodine was obtained. Degradation showed that C-5 was unlabelled and that the acetic acid obtained by Kohn–Roth oxidation (C-15, C-16) contained approximately one-half of the total activity[62]. C-2 labelled acetate provided lycopodine in which approximately one-quarter of the activity was present at C-15 and C-16. The polyketide hypothesis requires that only one-eighth of the activity be located at C-15, C-16 regardless of whether C-1 or C-2 labelled acetate is used in the feeding experiment.

Since the results with acetate tended to preclude a polyketide origin for the alkaloids, the next experiments were designed to test the validity of the pelletierine scheme[62, 65]. The piperidine ring of pelletierine originates from lysine[64]. Both 2-^{14}C-lysine and 6-^{14}C-lysine were incorporated into lycopodine and *in each case* one-quarter of the activity was located at C-5 and one-quarter was at C-9. This finding is consistent with a biosynthesis proceeding through two pelletierine-like units (149) provided the lysine passes through a symmetrical intermediate which equilibrates C-2 and C-6. This would lead to labelling at C-1, C-5, C-9 and C-13 and thus the observed distribution at C-5 and C-9. The same distribution of activity was observed when 1-^{14}C-cadaverine (1,5-diaminopentane), a symmetrical compound potentially available *in vivo* from lysine, was incorporated into lycopodine[62].

Δ^1-Piperideine (152), an intermediate between lysine and pelletierine and

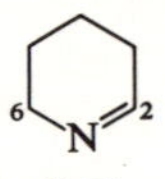

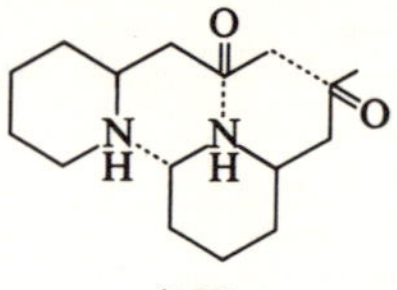

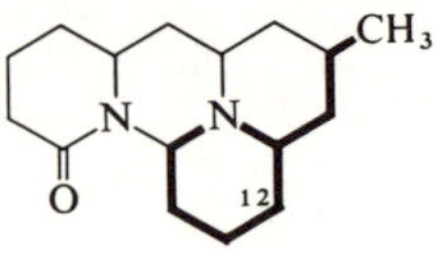

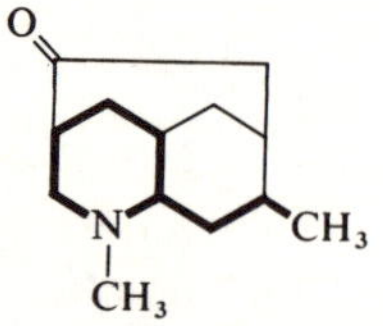

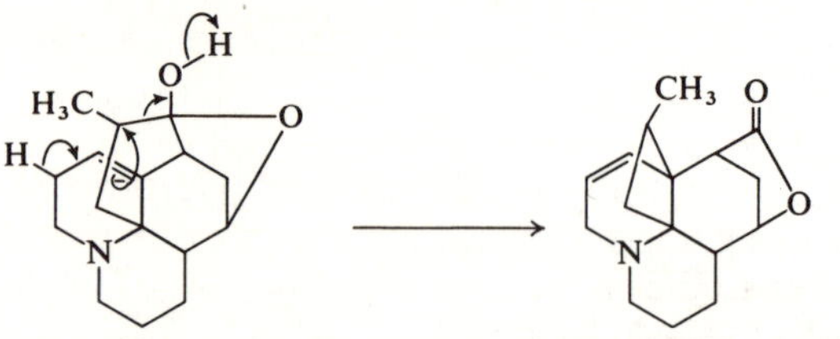

one which may be derived from cadaverine, was also incorporated into lycopodine[65]. This precursor, however, is incorporated in a non-symmetrical fashion. Thus 2-[14]C-Δ^1-piperideine furnishes lycopodine labelled only at C-5 and (presumably) C-13, whereas 6-[14]C-Δ^1-piperideine gives rise to labelling at C-9 and (presumably) C-1. These results are consistent with the sequence lysine $\rightarrow$ cadaverine $\rightarrow$ Δ^1-piperideine. Since both 'halves' of the alkaloid are equally labelled in these experiments, the possibility that lycopodine does arise by a dimerisation process of two C_8N units at some stage appeared attractive and a plausible hypothesis involving pelletierine (150) as the monomeric unit was put forward[61]. When this hypothesis was tested by administering variously labelled pelletierine to *L. tristachyum*, a very intriguing result was obtained[65]. It was found that the C-9 to C-16 portion (heavy lines in (151)) of lycopodine was indeed derived from pelletierine, but the C-1 to C-8 portion was not. Similarly, it has been shown that in the case of cernuine (87), the skeleton of which contains two unrearranged 'pelletierine units' (153), only the C-9 to C-16 portion (heavy lines in (154)) originates from pelletierine, although both 'halves' are generated from lysine and acetate as in the biosynthesis of lycopodine[66, 67]. The lycocernuine (12-hydroxycernuine, (88)) obtained from *L. cernuum* plants which had been fed the radioactive precursors and which produced labelled cernuine was devoid of activity, suggesting that lycocernuine is biosynthesised much more slowly than cernuine.

Very recent work[68] has established that pelletierine is an obligatory intermediate in the biosynthesis of lycopodine and it has been shown by dilution with inactive carrier that pelletierine is present in very low concentration in *L. tristachyum*. A scheme[65] involving 2-allylpiperidine rather than pelletierine as the 'monomeric' intermediate has been tested[68]. It was found that 2-allylpiperidine was incorporated but, as with pelletierine, entered only the C-9 to C-16 portion of lycopodine.

The study of the biosynthesis of the C_{16} Lycopodium alkaloids has thus arrived at a rather intriguing point. It seems clear that eight of the sixteen carbons are derived from a 'pelletierine unit', and that the other eight are derived from pelletierine precursors (lysine and acetate) but the mode of incorporation of these precursors into the C-1 to C-8 portion of the alkaloids remains to be determined. It is interesting to note that the C_{12} skeleton of luciduline (142) contains a single pelletierine unit (heavy lines in (155)) together with a 4 carbon straight chain the origin of which will possibly shed some light on the mode of formation of the C-1 to C-8 portion of the C_{16} compounds.

It has been suggested[1] that the annotinine (61) skeleton may arise via the lycopodine skeleton. The sequence (156) to (157) illustrates a possible biogenetic relationship between acrifoline (156), a member of the lycopodine group, and a compound (157) which on epoxidation of the double bond would lead to annotinine (61).

References

1. Wiesner, K. (1962). *Fortschritte der Chemie Organischer Naturstoffe*, Vol. 20, 271. (Vienna: Springer-Verlag).

2. MacLean, D. B. (1968). *Alkaloids*, **10**, 305

3. MacLean, D. B. (1970). *Chemistry of the Alkaloids*, 469. (S. W. Pelletier, editor). (New York: Van Nostrand Reinhold)

4. Harrison, W. A., Curcumelli-Rodostamo, M., Carson, D. F., Barclay, L. R. C. and MacLean, D. B. (1961). *Can. J. Chem.*, **39**, 2086

5. Chin-You, N., MacLean, D. B., Prakash, A. and Calvo, C. (1971). *Can. J. Chem.*, **49**, 3240

6. Burnell, R. H. and Taylor, D. R. (1961). *Tetrahedron*, **15**, 173

7. Inubushi, Y., Tsuda, Y., Ishii, H., Sano, T., Hosokawa, M and Harayama, T. (1964). *Yakugaku Zasshi*, **84**, 1108

8. Inubushi, Y., Harayama, T., Akatsu, M. and Ishii, H. (1968). *Chem. Commun.*, 1138

9. Ayer, W. A., Altenkirk, B., Valverde-Lopez, S., Douglas, B., Raffauf, R. F. and Weisbach, J. A. (1968). *Can. J. Chem.*, **46**, 15

10. Ayer, W. A. and Altenkirk, B. (1969). *Can. J. Chem.*, **47**, 499

11. Manske, R. H. F. and Marion, L. (1946). *Can. J. Res.*, **24B**, 57

12. Ayer, W. A., Altenkirk, B., Burnell, R. H. and Moinas, M. (1969). *Can. J. Chem.*, **47**, 449

13. Bohlmann, F. (1958). *Chem. Ber.*, **91**, 2157

14. Ayer, W. A. and Iverach, G. G. (1964). *Can. J. Chem.*, **42**, 2514

15. Hudec, J. (1970). *Chem. Commun.*, 829

16. Rodewald, W. J. and Grynkiewicz, G. (1968). *Roczniki Chem.*, **42**, 465

17. Achmatowicz, O. and Rodewald, W. (1956). *Roczniki Chem.*, **30**, 233

18. Braekman, J. C., Hootele, C. and Ayer, W. A. (1971). *Bull. Soc. Chim. Belges*, **80**, 83

19. Miller, N., Mees, F. and Braekman, J. C. (1971). *Phytochem.*, **10**, 1931
20. Stork, G., Kretchmer, R. A. and Schlessinger, R. H. (1968). *J. Amer. Chem. Soc.*, **90**, 1647
21. Ayer, W. A., Bowman, W. R., Joseph, T. C. and Smith, P. (1968). *J. Amer. Chem. Soc.*, **90**, 1648
22. Dugas, H., Hazenberg, M. E., Valenta, Z. and Wiesner, K. (1967). *Tetrahedron Lett.*, 4931
23. Wiesner, K., Musil, V. and Wiesner, K. J. (1968). *Tetrahedron Lett.*, 5643
24. Wiesner, K. Personal communication
25. Stork, G. (1968). *Pure Appl. Chem.*, **17**, 383
26. Ayer, W. A., Bowman, W. R., Cooke, G. A. and Soper, A. C. (1966). *Tetrahedron Lett.*, 2021
27. Ayer, W. A. and Smith, P. Unpublished results
28. Ayer, W. A. and Law, D. A. (1962). *Can. J. Chem.*, **40**, 2088
29. Colvin, E., Martin, J., Parker, W. and Raphael, R. A. (1966). *Chem. Commun.*, 596
30. Horii, Z., Kim, S., Imanishi, T. and Ninomiya, I. (1968). *Chem. Pharm. Bull.*, **16**, 2107
31. Ayer, W. A., Masaki, N. and Flanagan, R. Unpublished results
32. Wiesner, K. and Poon, L. (1967). *Tetrahedron Lett.*, 4937
33. Wiesner, K., Poon, L., Jirkovský, I. and Fishman, M. (1969). *Can. J. Chem.*, **47**, 433
34. Betts, E. E. and MacLean, D. B. (1957). *Can. J. Chem.*, **35**, 211
35. Anet, F. A. L. and Rao, M. V. (1960). *Tetrahedron Lett.*, No. 20, 9
36. Valenta, Z., Deslongchamps, P., Ellison, R. A. and Wiesner, K. (1964). *J. Amer. Chem. Soc.*, **86**, 2533
37. Ayer, W. A., Jenkins, J. K., Valverde-Lopez, S. and Burnell, R. H. (1967). *Can. J. Chem.*, **45**, 433
38. Ayer, W. A., Jenkins, J. K., Piers, K. and Valverde-Lopez, S. (1967). *Can. J. Chem.*, **45**, 445
39. Horeau, A. and Kagan, H. B. (1964). *Tetrahedron*, **20**, 2431
40. Ayer, W. A. and Piers, K. (1967). *Can. J. Chem.*, **45**, 451
41. Inubushi, Y., Tsuda, Y., Ishii, H., Sano, T., Hosokawa, M. and Harayama, T. (1964). *Yakugaku Zasshi*, **84**, 1108
42. Inubushi, Y., Ishii, H., Yasui, B., Harayama, T., Hosokawa, M., Nishino, R. and Nakahara, Y. (1967). *Yakugaku Zasshi*, **87**, 1394
43. Inubushi, Y., Ishii, H., Yasui, B., Hashimoto, M. and Harayama, T. (1968). *Chem. Pharm. Bull.*, **16**, 82
44. Inubushi, Y., Ishii, H., Yasui, B., Hashimoto, M. and Harayama, T. (1968). *Chem. Pharm. Bull.*, **16**, 92
45. Inubushi, Y., Ishii, H., Yasui, B. and Harayama, T. (1968). *Chem. Pharm. Bull.*, **16**, 101
46. Nishio, K., Fujiwara, T., Tomita, K., Ishii, H., Inubushi, Y. and Harayama, T. (1969). *Tetrahedron Lett.*, 861
47. Inubushi, Y., Ibuka, T., Harayama, T., and Ishü, H. (1968). *Tetrahedron*, **24**, 3541
48. Inubushi, Y., Harayama, T., Akatsu, M., Ishii, H. and Nakahara, Y. (1968). *Chem. Pharm. Bull.*, **16**, 2463
49. Yasui, B., Ishii, H., Harayama, T., Nishino, R. and Inubushi, Y. (1966). *Tetrahedron Lett.*, 3967
50. Ishii, H., Yasui, B., Nishino, R., Harayama, T. and Inubushi, Y. (1970). *Chem. Pharm. Bull.*, **18**, 1880
51. Ishii, H., Yasui, B., Harayama, T. and Inubushi, Y. (1966). *Tetrahedron Lett.*, 6215
52. Ayer, W. A. and Altenkirk, B. (1969). *Can. J. Chem.*, **47**, 2457
53. Ayer, W. A., Altenkirk, B., Masaki, N. and Valverde-Lopez, S. (1969). *Can. J. Chem.*, **47**, 2449
54. Inubushi, Y., Ishii, H., Harayama, T., Burnell, R. H., Ayer, W. A. and Altenkirk, B. (1967). *Tetrahedron Lett.*, 1069
55. Horii, Z., Kim, S., Imanishi, T. and Momose, T. (1970). *Chem. Pharm. Bull.*, **18**, 2235
56. Ayer, W. A., Iverach, G. G., Jenkins, J. K. and Masaki, N. (1968). *Tetrahedron Lett.*, 4597
57. Ayer, W. A. and Masaki, N. (1971). *Can. J. Chem.*, **49**, 524
58. Ayer, W. A., Masaki, N. and Nkunika, D. S. (1968). *Can. J. Chem.*, **46**, 3631
59. Ball, L. F. (1971). *Ph. D. Thesis*, University of Alberta, 101
60. Conroy, H. (1960). *Tetrahedron Lett.*, No. 10, 34

61. Gupta, R. N., Castillo, M., MacLean, D. B., Spenser, I. D. and Wrobel, J. T. (1968). *J. Amer. Chem. Soc.*, **90**, 1360
62. Castillo, M., Gupta, R. N., MacLean, D. B. and Spenser, I. D. (1970). *Can. J. Chem.*, **48**, 1893
63. Leete, E. (1964). *J. Amer. Chem. Soc.*, **86**, 2509
64. Gupta, R. N. and Spenser, I. D. (1969). *Phytochem.*, **8**, 1937
65. Castillo, M., Gupta, R. N., Ho, Y. K., Maclean, D. B. and Spenser, I. D. (1970). *Can. J. Chem.*, **48**, 2911
66. Gupta, R. N., Ho, Y. K., Maclean, D. B. and Spenser, I. D. (1970). *Chem. Commun.*, **409**
67. Ho, Y. K., Gupta, R. N., MacLean, D. B. and Spenser, I. D. (1971). *Can. J. Chem.*, **49**, 3352
68. Brackman, J. C., Gupta, R. N., MacLean, D. B. and Spenser, I. D. (1972). *4th Natural Products Symposium, University of the West Indies, Abstracts of Papers*, 14. (Kingston)
69. MacLean, D. B. and Curcumelli-Rodostamo, M. (1966). *Can. J. Chem.*, **44**, 611

2
The Total Synthesis of the Indole Alkaloids

J. P. KUTNEY

University of British Columbia, Vancouver

2.1   INTRODUCTION

The indole alkaloids comprise a very large family of natural products and a great deal of research has gone into their structural elucidation, synthesis and biosynthesis. The area of isolation and structure elucidation has attracted the interests of chemists since the early 1900s and extensive discussions of such investigations can be found in the series entitled 'The Alkaloids' edited by Manske and Holmes. With the advent of sophisticated instrumentation which has greatly simplified the structural problem and at least to a certain extent reduced somewhat the challenge inherent in the earlier investigations, this area has received less attention in recent years. It is pertinent to note that researchers are now able to solve problems of great complexity and in this regard the area of the bisindole alkaloids has recently seen dramatic progress. A recent and detailed review of these investigations has appeared[1] and the interested reader will find the essential information here.

In more recent times, particularly during this last decade, the areas of synthesis and biosynthesis have seen unparalleled advances and it is appropriate to emphasise these areas in discussing recent investigations in this field. Since another chapter in this series will emphasise the biosynthetic investigations, I will focus the present discussion entirely on progress which has been made over the last 5 years or so in the synthesis of these natural products.

For purposes of clarity and coherence in presenting the various synthetic endeavours in the different families of this large class, I have chosen to organise the material in terms of currently accepted biosynthetic inter-relationships. Detailed reviews on biosynthesis, which discuss the pertinent experiments, are now available[2–6]. On this basis, for example, the syntheses of the corynanthe series are presented first since these alkaloids are now considered as the biosynthetic precursors of the Strychnos, Aspidosperma and Iboga families.

2.2   THE CORYNANTHEINE FAMILY

The first total synthesis of a representative of the corynantheine group was reported by van Tamelen[7, 8].

The synthesis is based on a reductive alkylation of tryptamine with a suitable acyclic compound (6) to yield the basic tetracyclic ring system. The acyclic condensing unit was obtained through a method described by Preobrazhenskii[9]. A Michael condensation of diethylglutaconate (1) with ethyl cyanoacetate (2) yielded the diester (3) which on alkylation with ethyl iodide allows formation of the cyanotriester (4). Selective saponification followed by pyrolytic decomposition of the resulting acid (5) gave the desired cyano diester (6). Precautions are required in this step since (4) undergoes sodium ethoxide induced elimination of ethyl ethylcyanoacetate, regenerating (1) which in turn undergoes self condensation to (7).

The cyano diester (6) was condensed with tryptamine (8) under reductive conditions. Unfortunately the desired lactam (9) was obtained in low yield (13%) since the major product resulted from direct reduction of (6) and subsequent cyclisation to the lactam (11). Also contained in the reaction mixture was compound (10) formed by reductive alkylation of tryptamine

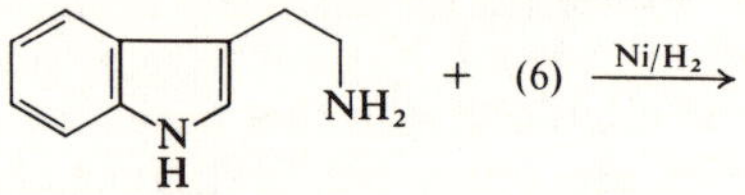

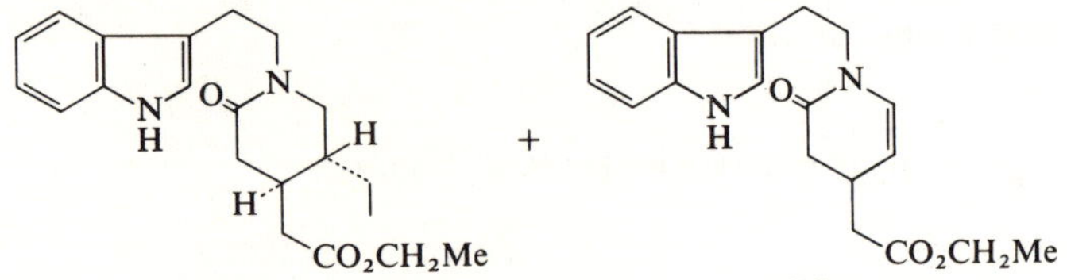

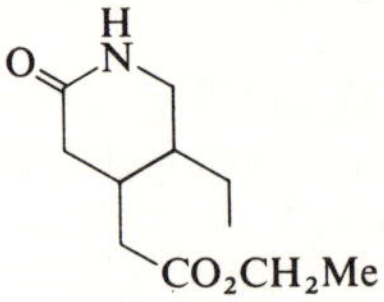

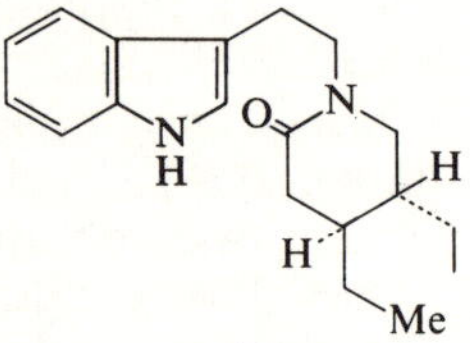

with (3) which had been carried through the sequence as an impurity. Saponification of the resultant mixture ((9) and (10)) and separation of the corresponding acids yielded after esterification of the latter, the pure esters (9) and (10).

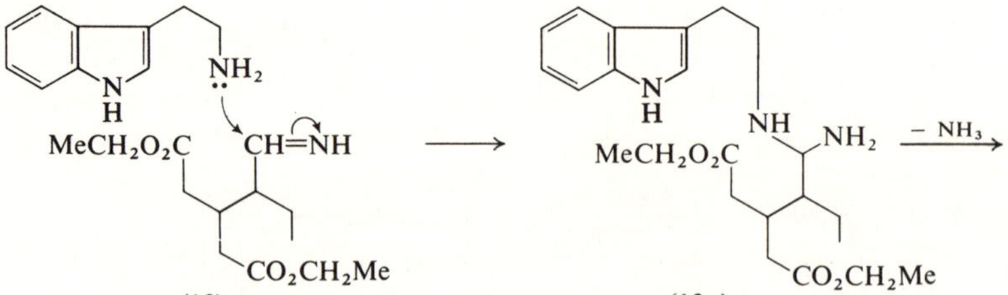

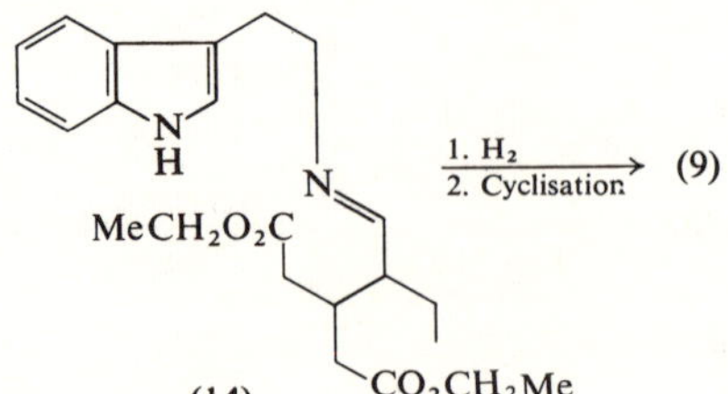

The structure and stereochemistry of the required lactam (9) was established by conversion to a known compound (12) employing a reliable and proven degradative sequence[10].

Although several plausible alternatives may be proposed for the reductive alkylation step, it appears that the above pathway is most attractive.

Formation of the tetracyclic quaternary iminium chloride (15) is achieved via Bischler–Napieralski conditions and the necessary stereochemistry at C-3 results from catalytic reduction of this product. *Trans* esterification of the ethyl ester (16) to the methyl ester, formylation to the hydroxymethylene derivative (17), and reaction of the latter with diazomethane afforded (±)-dihydrocorynantheine (17a).

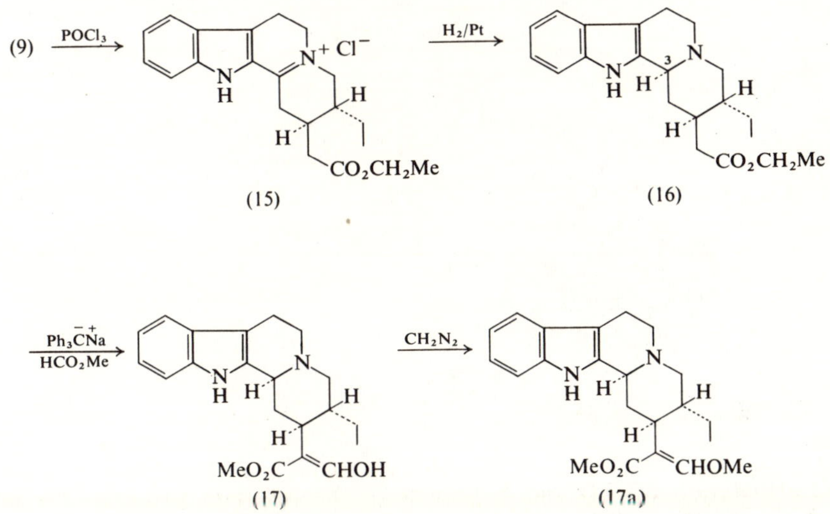

In a complementary series of investigations van Tamelen describes the total synthesis of (±)-corynantheine (25a)[11, 12]. In this sequence a key tetracyclic intermediate previously employed in the ajmalicine synthesis (see later) was utilised.

Thus a Mannich type condensation of tryptamine (8), formaldehyde and the keto triester (18) led to the lactam (19). Cyclisation of the latter under Bischler–Napieralski conditions followed by catalytic reduction yielded the tetracyclic compound (20) which upon acid treatment afforded (21) as a mixture of *cis* and *trans* isomers.

The final stages of the sequence reveal the generation of the required vinyl side chain via thermal decomposition of sulphonylhydrazone salts, a method known to produce olefinic compounds via a carbenoid mechanism[13]. The tosylhydrazone of the *trans* isomer (22) was treated with sodium methoxide in diglyme and a mixture of the olefins (23) and (24) was obtained.

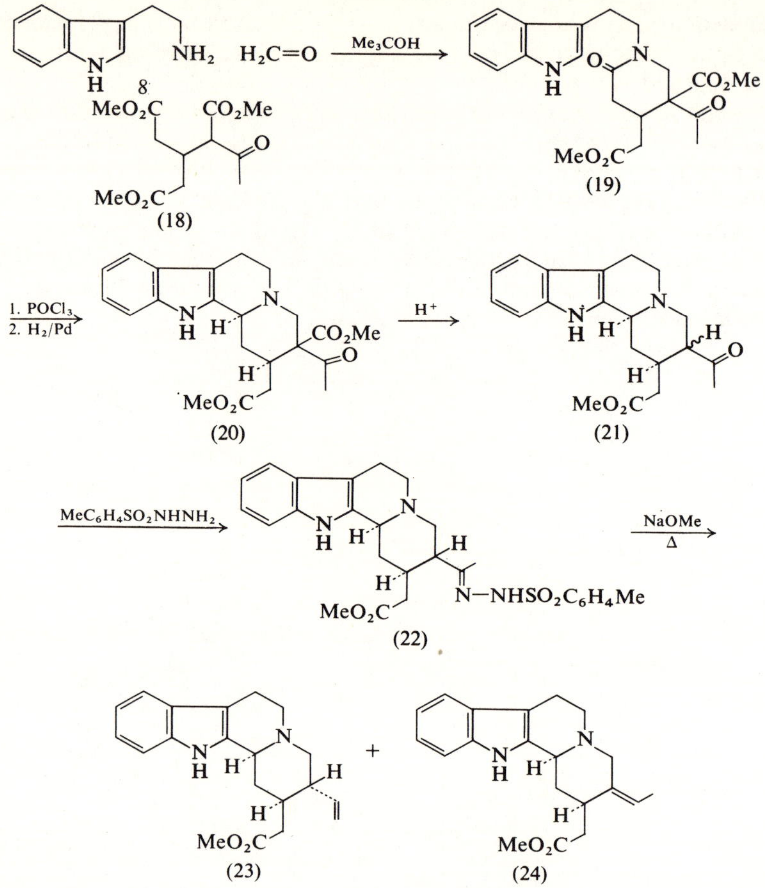

(18)

Me_3COH

(19)

1. $POCl_3$
2. H_2/Pd

(20)

H^+

(21)

$MeC_6H_4SO_2NHNH_2$

(22)

NaOMe
Δ

(23) + (24)

Formylation of (23) and treatment of the resultant product (25) gave (±)-corynantheine (25a).

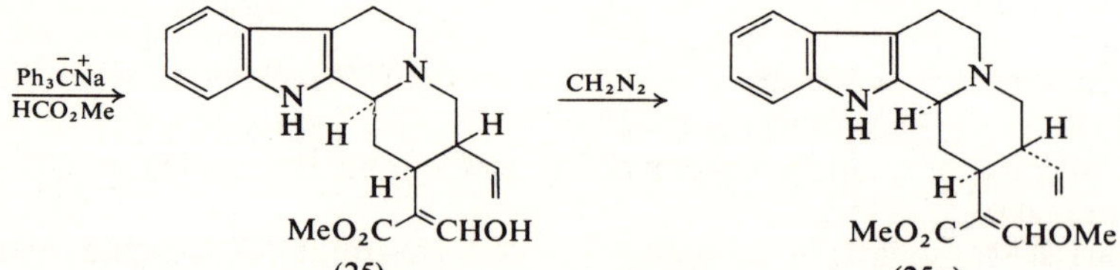

(23)

Ph_3CNa / HCO_2Me

(25)

CH_2N_2

(25a)

A rather different approach to the synthesis of corynantheine has been reported by Autrey[14]. In this sequence, yohimbone (26), a commercially available substance and one which has also been totally synthesised and resolved[15] was selected as the starting material. This compound was converted to 18-formylyohimban-17-one (27) by previously published procedures and this latter substance was then employed in an interesting series of reactions which result in an overall fragmentation of the yohimbine skeleton to the required corynantheine ring system.

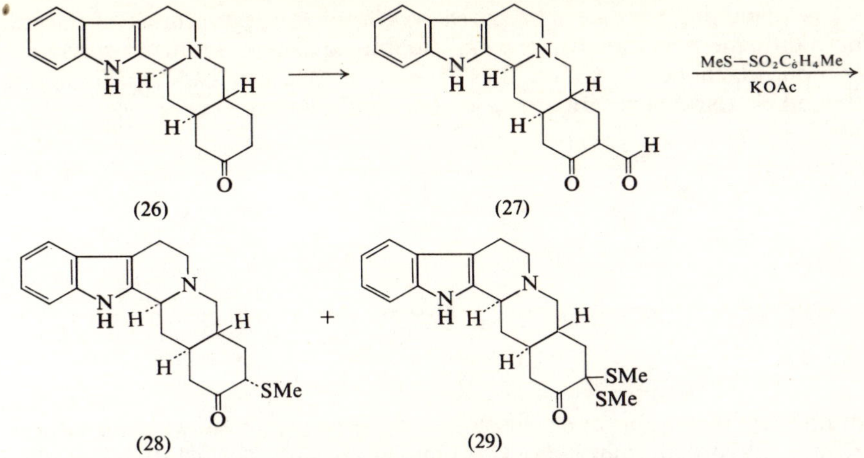

(26) (27)

(28) + (29)

Introduction of the thiomethyl group was accomplished by reaction with methyl thiotosylate and subsequent deformylation of the resultant product by the base. The undesired disubstituted product (29), presumably formed by further reaction of the intermediate anion (30) with methyl thiotosylate, could be suppressed so that the overall yield of the monomethioxy compound

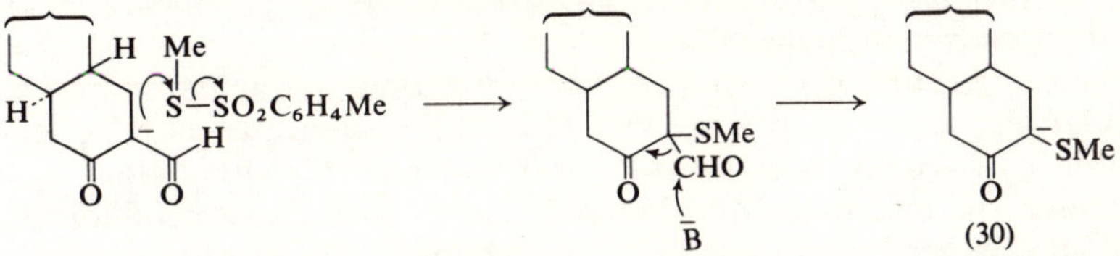

(30)

was in the range of 57–63 %. Oximation of (28) and Beckmann fragmentation of the oxime (31) provided the enol thioether which now possesses the required tetracyclic skeleton and the essential stereochemistry. Desulphurisation of the thioether proceeded smoothly with Raney nickel to corynanthenitrile and the latter was readily converted to methyl corynantheate (33) under Fischer esterification conditions.

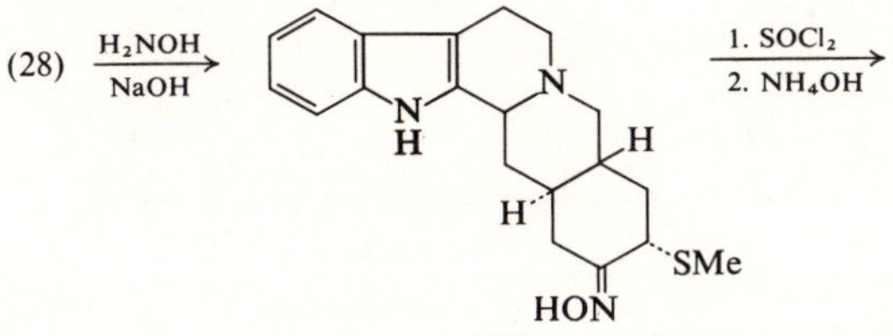

(31)

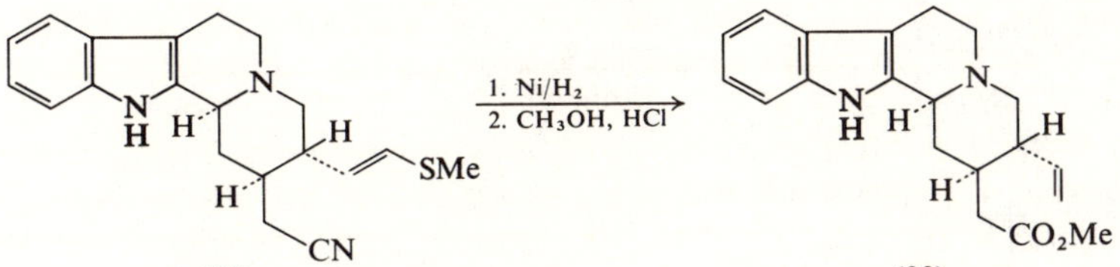

(32) (33)

The remaining two steps required to complete the synthesis of natural corynantheine are already known. An improvement in the formylation reaction, which yields desmethylcorynantheine (34), was made by employing ethyl ether–tetrahydrofuran as a solvent combination.

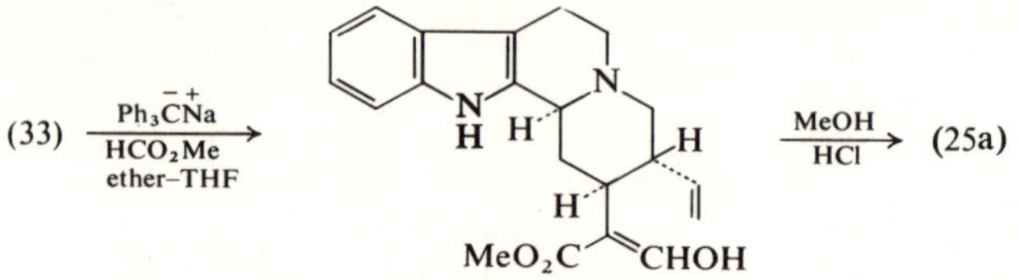

$$(33) \xrightarrow[\substack{\text{HCO}_2\text{Me} \\ \text{ether–THF}}]{\text{Ph}_3\overset{-}{\text{C}}\overset{+}{\text{Na}}} (34) \xrightarrow[\text{HCl}]{\text{MeOH}} (25a)$$

Another synthetic approach developed by Weisbach[16] allows the synthesis of both dihydrocorynantheine (17a) and its isomer, corynantheidine (44a) from a common intermediate. The starting material (35), available from tryptamine, was converted to the esters (36) and (37) by reaction with trimethyl and triethylphosphonoacetate respectively. Catalytic reduction of these substances yielded the tetracyclic esters (38) and (39) possessing the desired stereochemistry at C-15. Formylation of (38) and methylation of the latter according to procedures already discussed previously afforded (±)-dihydrocorynantheine (17a).

The required stereochemistry of corynantheidine was achieved by a series of oxidation–reduction procedures. Thus the unsaturated ethyl ester (40) available by alkoxide treatment of (37) was reacted with iodine–sodium acetate and the resultant salt (41) obtained as a mixture of perchlorate and iodide was reduced directly to (42). Conversion of the latter to methyl ester (43), followed by formylation (44) and methylation completed the synthesis.

The alternative and more direct sequence (37 → (41), employing mercuric acetate as oxidant, is not apparently advantageous in terms of yield.

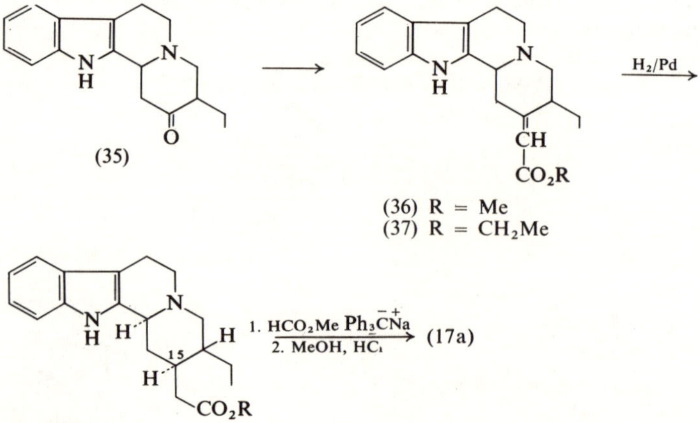

(35)

(36) R = Me
(37) R = CH$_2$Me

(38) R = Me
(39) R = CH$_2$Me

Then:

$(37) \xrightarrow{\ -OR\ }$

Hg(OAc)$_2$

CO$_2$CH$_2$Me

(40)

I$_2$, NaOAc

(41) $^+X^-$ CO$_2$CH$_2$Me

$\xrightarrow[\text{NaOCH}_2\text{Me}]{\text{H}_2/\text{Pd}}$

(42)   CO$_2$CH$_2$Me

$\xrightarrow[\text{MeOH}]{\text{NaOMe}}$

(43)   CO$_2$Me

$\longrightarrow$ (44) $\longrightarrow$

(44a)   MeO$_2$C   CHOMe

The first total synthesis of the optically active, natural isomer of corynantheidine was reported by Szantay[17, 18]. The starting material was similar to that utilised in the previous synthesis. Condensation of (45) with methyl cyanoacetate provided (46) in 66% yield. Borohydride reduction of the latter and methanolysis of the resultant nitrile (47) allowed isolation of the diester (48). An interesting controlled reduction converts the diester to the α-hydroxymethylene derivative (49) which, in the form of its sodium salt, is methylated in essentially quantitative yield to racemic corynantheidine. Resolution of this racemate with dibenzoyl-(+)-tartaric acid completed the synthesis of (−)-corynantheidine.

In a series of investigations, Wenkert has developed a new procedure for the construction of the indoloquinolizidine skeleton so common to a large family of indole alkaloids. It is based on a partial reduction of 1-[β[(3-indolyl)ethyl]-3-acyl pyridinium salts and the subsequent acid-catalysed cyclisation of the resultant 2-piperideines. This approach has led to a number of syntheses among which is included the synthesis of (±)-corynantheidine[19].

Alkylation of methyl 4-carbomethoxymethyl-5-ethylnicotinate (51) with tryptophyl bromide (50) yielded the pyridinium salt which was isolated and characterised as the perchlorate (52). Hydrogenation of the latter produced the tetrahydropyridine (53). Two procedures for the cyclisation and decarbo-alkoxylation of such tetrahydropyridines had been previously developed and employed in the present case. Alkaline hydrolysis of (53) followed by re-

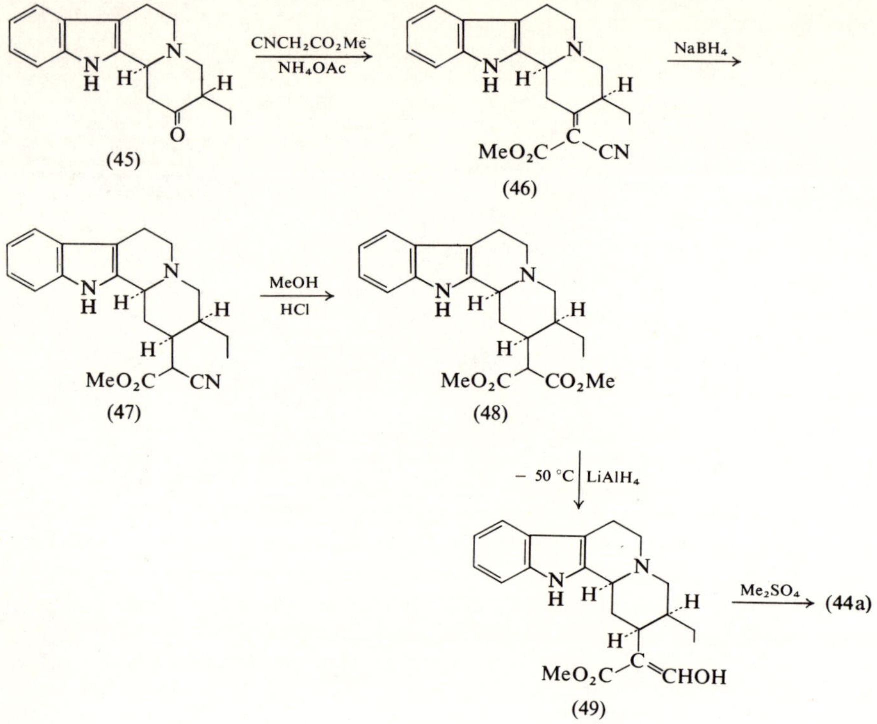

$CNCH_2CO_2Me$
NH_4OAc

(45)

MeO_2C C CN

(46)

$NaBH_4$

(47)

MeO_2C CN

$MeOH$
HCl

(48)

MeO_2C CO_2Me

$-50\ ^\circ C$ $LiAlH_4$

(49)

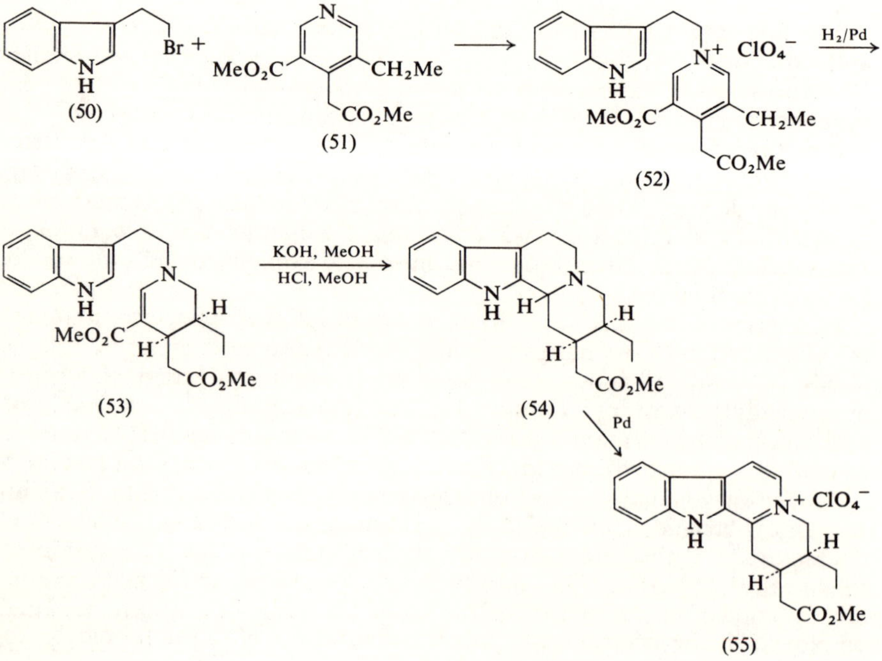

MeO_2C C $CHOH$

Me_2SO_4 (44a)

(50)

Br +

(51)

MeO_2C CH_2Me
CO_2Me

(52)

N^+ ClO_4^-

MeO_2C CH_2Me
CO_2Me

H_2/Pd

(53)

MeO_2C CO_2Me

$KOH, MeOH$
$HCl, MeOH$

(54)

CO_2Me

Pd

(55)

N^+ ClO_4^-

CO_2Me

esterification with methanolic acid led to the tetracyclic ester (54). Palladium dehydrogenation converted the latter to the tetradehydro compound (55) which could be characterised in the form of its perchlorate salt. Sodium borohydride reduction of (55) afforded (56).

The alternate route to (55) involved acid-induced cyclisation of (53), acid hydrolysis of the resultant tetracyclic ester (56), dehydrogenation and re-esterification.

The two remaining reactions necessary to complete the synthesis of ($\pm$)-corynantheidine, namely formylation and methylation, were already in hand from the above-mentioned syntheses.

Ziegler[20] has reported a synthesis of dihydrocorynantheol (67) and 3-epi-dihydrocorynantheol (68) employing a Claisen rearrangement to elaborate the necessary functionality at C-15 and C-20 of the tetracyclic skeleton. In this sequence 6-chloronicotinic acid (57) is converted to the required 2-chloro-5-(α-hydroxyethyl) pyridine (61) by standard methods and this substance is then condensed with tryptophyl bromide to yield the pyridinium salt (62) in 27% yield. Reduction of (62) with sodium borohydride provided a diastereo-isomeric mixture of allylic alcohols (63).

Further functionalisation of (63) involved the application of a Claisen rearrangement previously studied by Eschenmoser[21].

Reaction of the mixture of allylic alcohols (63) with dimethylacetamide dimethylacetal gave rise to a diastereoisomeric mixture of amides ((64) and (65)). Saponification of this mixture followed by esterification afforded (66) as a mixture of double-bond isomers. Finally hydride and catalytic reduction

provided a mixture of dihydrocorynantheol (67) and 3-epi-dihydrocoryn-
antheol (68).

(63) $\longrightarrow$

$\xrightarrow[\text{2. MeOH, HCl}]{\text{1. KOH}}$

(64) C—3αH
(65) C—3βH

$\xrightarrow[\text{2. Pd/H}_2]{\text{1. LiAlH}_4}$

(66)

(67) C—3αH; C—20βH
(68) C—3βH; C—20βH

Recent studies on the structure elucidation of quaternary bases from
Hunteria eburnea Pichon have revealed the first natural examples of corynantheinoid variants in which both possible isomers at N_b are evident. During
these investigations in which the structures, including absolute configuration,
of hunterburnine α- and β-methochlorides ((69) and (70)) were being evaluated,
interesting syntheses of several corynantheine derivatives were also achieved.
For example, 10-methoxydihydrocorynantheol (72), ochrosandwine (73)
and dihydrocorynantheol (67) were synthesised from quinine (71)[22].

2.3 THE AJMALICINE FAMILY

The rather large class of ring-E heterocyclic indole alkaloids is obviously
closely related in a biogenetic sense to the corynantheinoid members
discussed above. For example, vincoside (74) upon hydrolysis of the glycosidic
linkage yields (75) which after unexceptional cyclisations would provide
ajmalicine (77) one of the well-known members in this series. It is therefore
appropriate to discuss the synthesis of this alkaloid at this time.

The total synthesis of ajmalicine (77) in racemic form, represents the first
synthesis of a member of the ring-E heterocyclic indole alkaloids[23, 24]. In this
study the acyclic component (82) corresponding to the nontryptophan
portion of the alkaloid and already employed in a previously discussed
synthesis of (±)-corynantheine, was prepared according to the alternate
pathways discussed below ((78) → (82) and (79 → (84) → (82)).

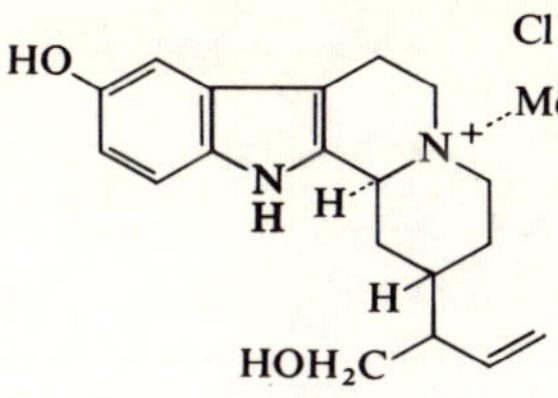

(69)

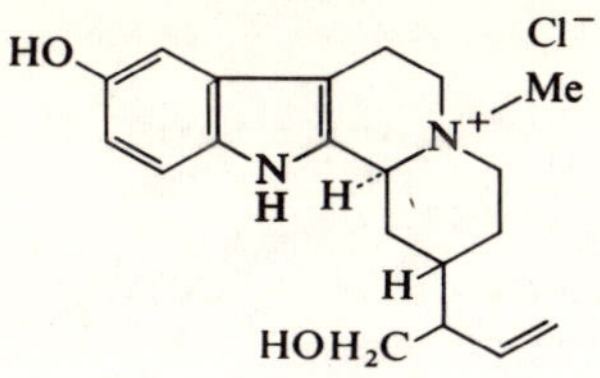

(70)

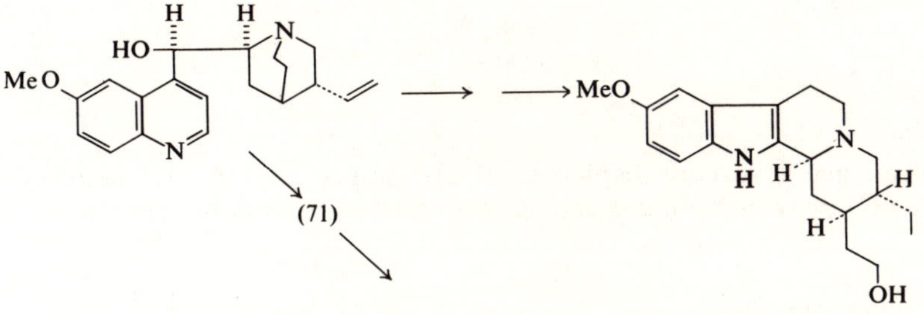

(71)

(72)

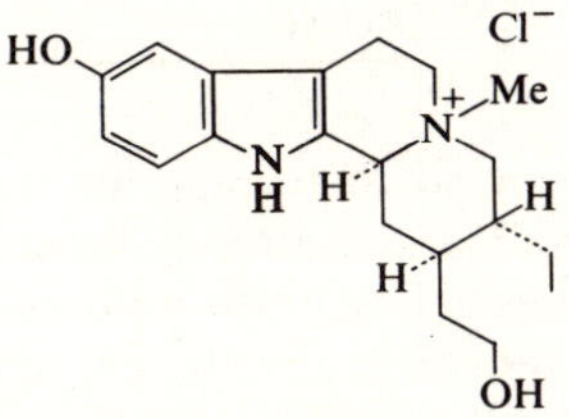

(73)

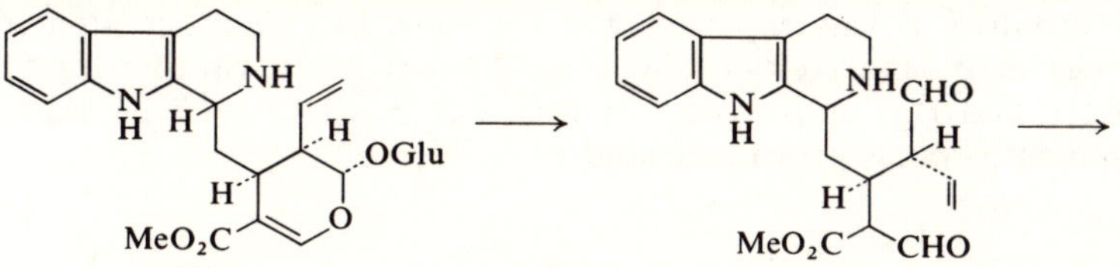

(74)

(75)

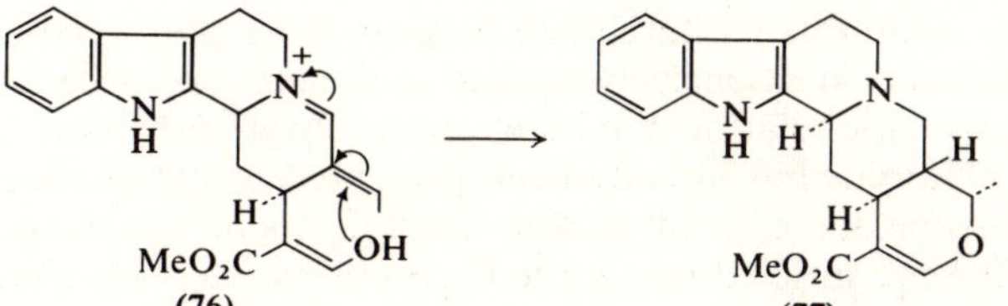

(76)

(77)

$$CO_2Me\text{—}(78) \xrightarrow[\text{Ni}]{H_2} CO_2Me\text{—}(79) \xrightarrow{Ac_2O} CO_2Me\text{—}(80) \xrightarrow{\Delta} CO_2Me\text{—}(81)$$

(78)   (79)   (80)   (81)

$$Me\text{—}CO_2Me \longrightarrow (82) + (83)$$

(82)   (83)

The sequence involving displacement of halogen in dimethyl β-chloro-glutarate (84) provided a more convenient and better yielding route to (82).

$$(79) \xrightarrow{PCl_5} (84) \xrightarrow{Me\text{—}CO_2Me} (82)$$

(84)

A simple Mannich reaction of tryptamine, formaldehyde and the keto triester (82) yielded, in the manner already described previously, the lactam (85) which was converted as before to the keto ester (86). This compound was then elaborated via the δ-lactone (87) to the ajmalicine system. Low temperature reduction of (86) provided a nearly quantitative yield of the lactone (87).

Although the isomeric composition of the lactone product mixtures was not determined, it was expected that the *trans* isomer predominated. The α-hydroxymethylene lactone (88) was obtained by a routine formylation procedure and this substance was then subjected to an acyl lactone rearrangement process previously studied by Korte[25].

2.4 THE SARPAGINE-AJMALINE FAMILY

The ajmaline–sarpagine family of alkaloids has been extensively investigated from a structural point of view[26, 27] and much of the chemistry associated with this group was undoubtedly developed because of its close relationship with the Rauwolfia bases so extensively studied in the last two decades.

The apparent biogenetic relationship between the corynantheinoid family and this group of alkaloids has already been presented in the introductory section and indeed some of the synthetic steps in the ajmaline syntheses to be discussed below take the implied biosynthetic reactions into consideration.

The sarpagine group has the characteristic bridged skeleton portrayed in

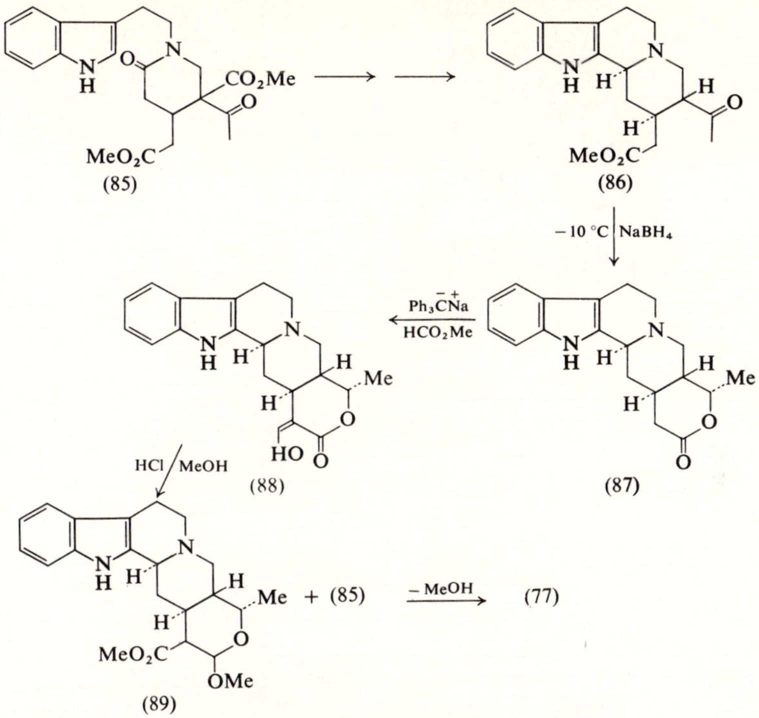

(90) while the additional bridging to the β-position of the indole system in (90) allows the elaboration of the dihydroindole skeleton typical of the ajmaline family (91).

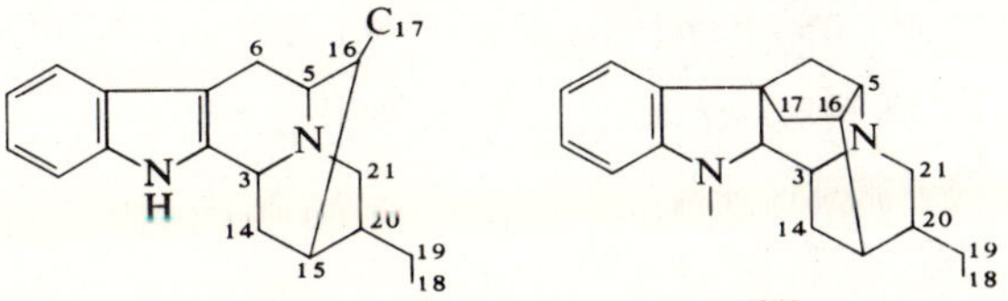

Specific members of the sarpagine group may be exemplified by vellosimine (92), normacusin B (93), sarpagine (94) and polyneuridine (95) while tetraphyllicine (96), vincamedine (97) and ajmaline (98) are typical members of the ajmaline family.

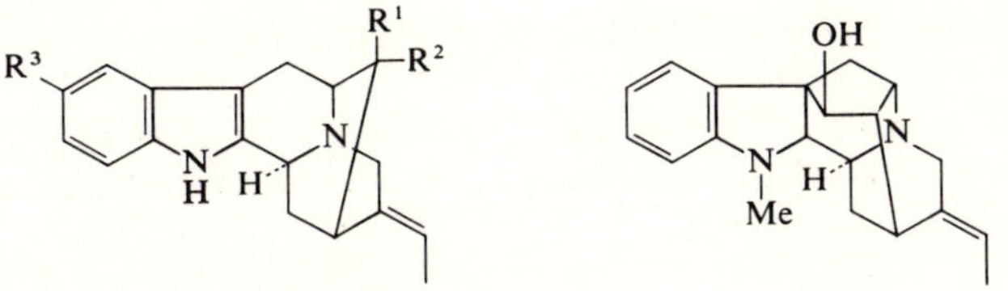

(92) $R^1 = R^3 = H$; $R^2 = CHO$
(93) $R^1 = R^3 = H$; $R^2 = CH_2OH$
(94) $R^1 = H$; $R^2 = CH_2OH$; $R^3 = OH$
(95) $R^1 = CH_2OH$; $R^2 = CO_2Me$; $R^3 = H$

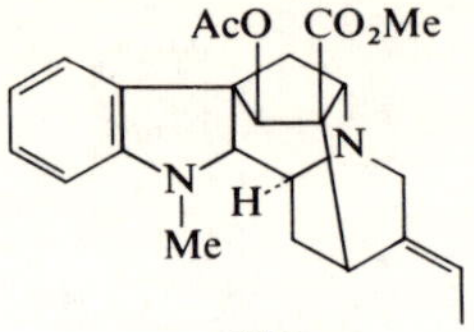

AcO CO₂Me
(97)
OH 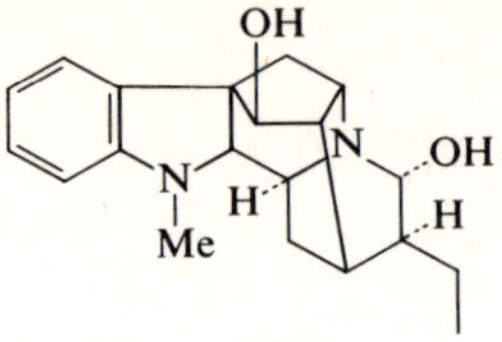
OH
Me H
(98)

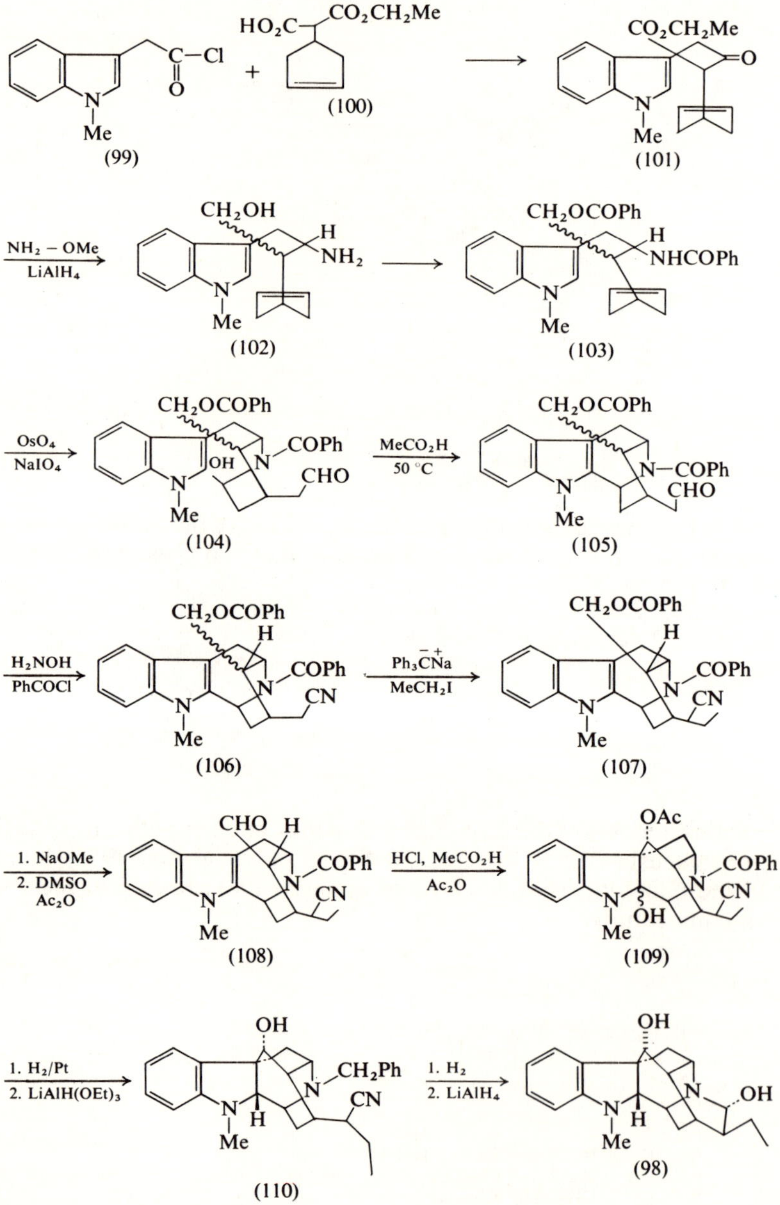

HO₂C CO₂CH₂Me
(100)
+
Me
(99)
CO₂CH₂Me
O
Me
(101)

NH₂ – OMe
LiAlH₄
CH₂OH
H
NH₂
Me
(102)
CH₂OCOPh
H
NHCOPh
Me
(103)

OsO₄
NaIO₄
CH₂OCOPh
OH
N—COPh
CHO
Me
(104)
MeCO₂H
50 °C
CH₂OCOPh
N—COPh
CHO
Me
(105)

H₂NOH
PhCOCl
CH₂OCOPh
H
N—COPh
CN
Me
(106)
Ph₃CNa
MeCH₂I
CH₂OCOPh
H
N—COPh
CN
Me
(107)

1. NaOMe
2. DMSO
   Ac₂O
CHO H
N—COPh
CN
Me
(108)
HCl, MeCO₂H
Ac₂O
OAc
N—COPh
CN
OH
Me
(109)

1. H₂/Pt
2. LiAlH(OEt)₃
OH
N—CH₂Ph
CN
Me
(110)
1. H₂
2. LiAlH₄
OH
N
OH
H
Me
(98)

The first total synthesis of ajmaline was reported by Masamune[28]. The sequence employed in this investigation is as shown.

Condensation of N-methyl-3-indoleacetyl chloride (99) with the magnesium chelate of ethyl hydrogen Δ^3-cyclopentenylmalonate (100) afforded the keto ester (101) in 80% yield. Reaction of the latter with methoxyamine followed by reduction provided a 2:1 mixture of the epimeric amino alcohols (102) which could be readily separated and characterised as their diacetyl and dibenzoyl derivatives. Since both epimeric series are useful and they are interconvertible at a later stage of the synthesis no differentiation of these stereoisomers is made in this discussion. The dibenzoyl derivative (103) on reaction with osmium tetroxide and sodium metaperiodate afforded the aldehyde mixture (104) in essentially quantitative yield. Acid-induced cyclisation provided (105) which is converted to the nitrile (106) by dehydration of the oxime derivative. Base-catalysed alkylation allowed introduction of the ethyl group at the carbon atom bearing the nitrile function. At this point the structures and stereochemistry of several synthetic intermediates (107) and the alcohols obtained by hydrolysis of (107) were compared with the corresponding degradation products of ajmaline. The epimeric aldehydes (108) obtained from Moffatt oxidation were interconvertible by means of alumina and therefore both epimeric series could be utilised.

A number of previous investigations had already revealed that the desired β,β-disubstituted skeleton characteristic of the ajmaline molecule, could be achieved through the reaction of the electron-rich indole ring and an appropriate electron-deficient centre (111) → (112)).

$$(111) \longrightarrow (112)$$

Application of this principle allows conversion of (108) to (109). Catalytic hydrogenolysis and hydride reduction of the amide function in (109) provided a good overall yield of (110). Removal of the N-benzyl group leads to the secondary amine and since the latter has been converted to ajmaline by use of lithium aluminium hydride[29], the synthesis of this alkaloid is now complete.

A second synthesis of ajmaline has been reported by van Tamelen[30]. This approach attempts to utilise chemical operations which are likely to prevail in some of the biosynthetic steps leading to this natural product. The indolic material, N-methyl tryptophan (118) is exposed to reaction with an appropriate C_9 unit (117) which after the required carbonyl-amine condensations leads to the tetracyclic indole skeleton (121). This latter substance then serves as the fundamental template for the remaining elaboration to ajmaline.

The synthesis of the essential C_9 unit starts with $(\pm)$-α $(\Delta^3$-cyclopentenyl)-butyric acid (113) and proceeds according to the sequence (113) → (117).

HO_2C, Me

(113)

1. $LiAlH_4$
2. $Ph \cdot CH_2Cl$ KOH, 100 °C

$PhCH_2O$, Me

(114)

OsO_4
C_5H_5N

$PhCH_2O$, Me

HO   OH

(115)

$(MeO)_2CO$
NaOMe

$PhCH_2O$, Me

O   O
‖
O

(116)

1. H_2/Pd
2. CrO_3

OHC, Me

O   O
‖
O

(117)

Reductive alkylation of *N*-methyltryptophan with the aldehyde (117) provided the expected amino acid which without further purification was saponified to the required diol (119). This substance was now suitable for generating the intermediate dialdehyde (120) which undergoes spontaneous Pictet–Spengler cyclisation to the tetracyclic indole derivative (121). This approach involving dialdehyde formation by cycloalkane-1,2-diol cleavage followed by cyclisation with an appropriate tryptamine system was first applied by van Tamelen in a yohimbine synthesis[31]. It has since seen application in other syntheses including the Masamune synthesis discussed above.

CO_2H

NH_2

Me

(118)

+ (117)

1. H_2
2. Pd/C
KOH, MeOH

CO_2H

NH

Me   Me

HO   OH

(119)

$NaIO_4$

CO_2H

NH

Me   Me

OHC   CHO

(120)

CO_2H

N
Me   H   H

H

Me

CHO

(121)

It has been postulated previously that a plausible biosynthetic route linking the tetracyclic corynantheine skeleton as portrayed in (122), for example, with the bridged system characteristic of the ajmaline series could involve a cyclisation reaction of an activated site in the side chain with an electrophilic centre as shown in the conversion (122) → (123).

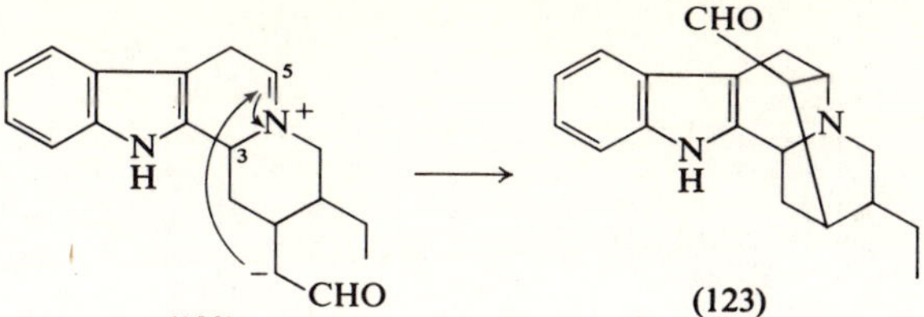

(122)

(123)

It is clear that generation of the iminium system as revealed in (122) required specific reaction at C-5 and therefore direct dehydrogenation of a tetrahydrocarboline derivative is not plausible since the conjugated and thermodynamically more stable Δ^3-dehydro-β-carboline would result (see (124) → (125).

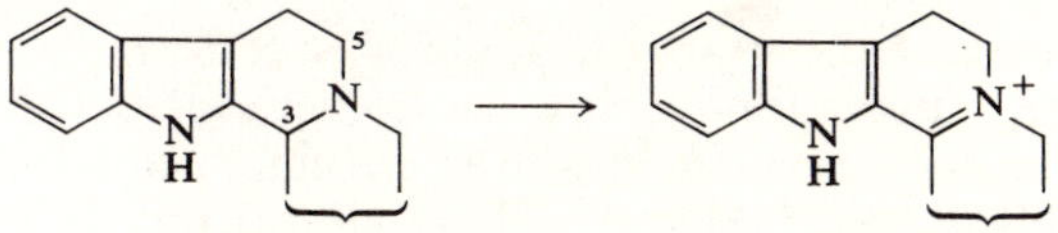

(124)

(125)

For this purpose these workers studied, initially in a model system, a decarbonylation reaction which places the double bond at the required site. The conversion (126) → (127) was evaluated by reducing the resultant product (127) with sodium borodeuteride to the 3-monodeuterio tetrahydrocarboline derivative (128).

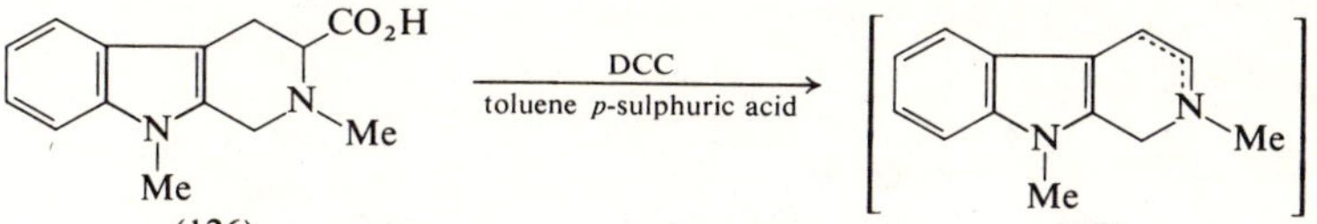

(126)

DCC

toluene *p*-sulphuric acid

(127)

NaBD₄

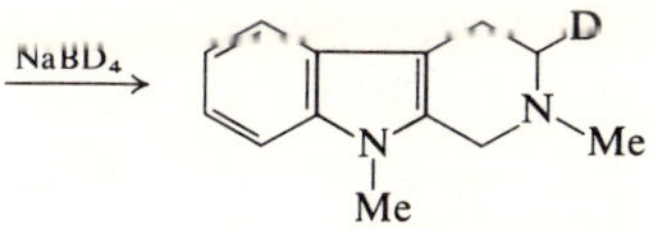

(128)

Under the above conditions, the amino acid (121) was converted to the iminium salt (129) which, without isolation, was cyclised spontaneously to (±)-deoxyajmalal-B (130). This substance was resolved by use of D-camphor--10-sulphonic acid to provide the optically active specimen.

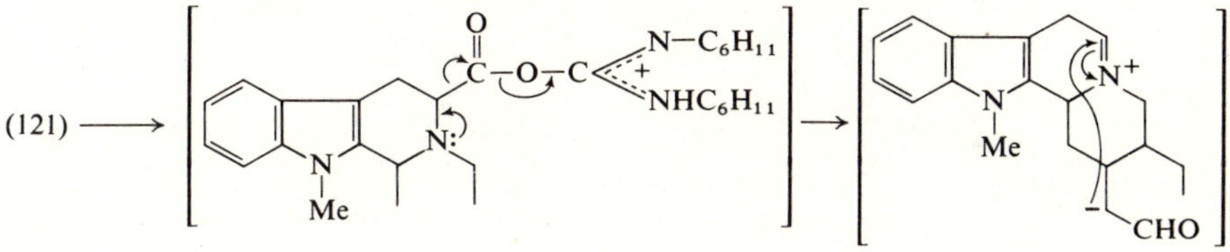

(121) →

(129)

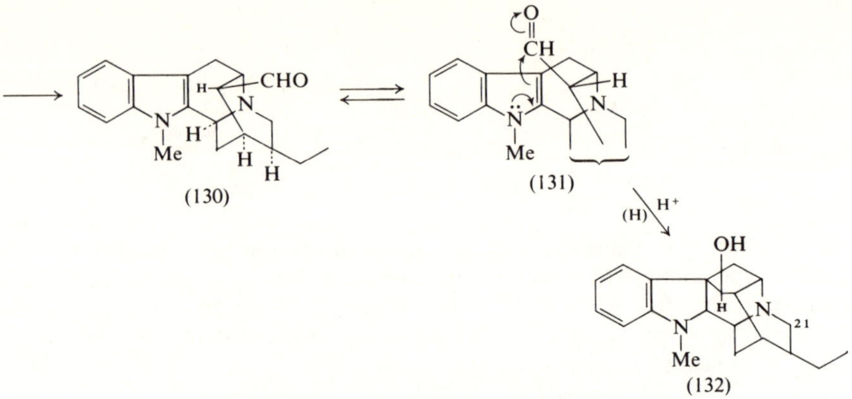

(130)   (131)

(132)

The completion of the synthetic route was performed by utilisation of appropriate relay substances. Either the isomeric deoxyajmalal-A (131) or (130) on reaction with acetic acid–sodium acetate provides an equilibrium mixture (about 15% of (131) and 85% of (130) from which the authentic deoxyajmalal-A could be isolated. Reductive cyclisation of the latter according to a published procedure[32] provided deoxyajmaline (132). Functionalisation of the latter at C-21 was accomplished by the phenyl chloroformate ring opening–oxidative ring closure method reported by Hobson and McCluskey[33] whereupon ajmaline (98) was finally obtained.

A Japanese group[34] recently completed the total synthesis of isoajmaline (145), an alkaloid which is isomeric with ajmaline at C-20 and C-21. The successful sequence is outlined here.

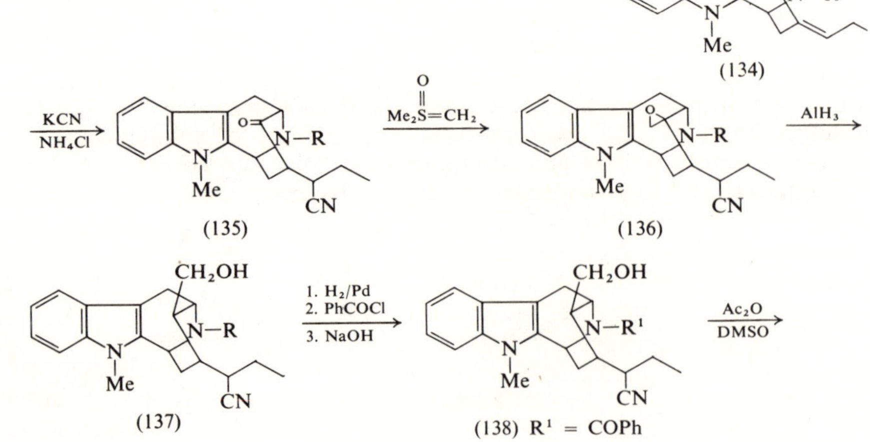

(132) R = CH₂Ph   (133)

(134)

(135)   (136)

(137)   (138) R¹ = COPh

(139) → (140) Al$_2$O$_3$

Ac$_2$O, MeCO$_2$H, HCl →

(141) → (142) Pt, HCl

1. Et$_3$O$^+$BF$_4^-$ 2. NaBH$_4$ →

(143) → (144) H$_2$/Pd → (145)

The starting β-keto ester (132), available from ($\pm$)-tryptophan via a previously published procedure[35] was converted to the unsaturated ketone (134) by standard procedures. This latter substance then served as the necessary template for further elaboration to the ajmaline skeleton. The hydrocyanation reaction associated with the conversion (134) → (135), proceeded with a high degree of stereoselectivity and a 70% yield of the keto nitrile (135) was obtained. Oxirane ring formation (136) according to the Corey method was achieved in 50% yield while the ring cleavage to the primary alcohol (137) was accomplished in 54% yield with aluminium hydride. The mixed hydride HAlCl$_2$, successful in a model series allowed conversion in only 20% yield. It is interesting to note that the oxidative step, (138) → (139), normally difficult due to the facile oxidation of the indole ring, could be accomplished in high yield (80%) by means of DMSO–acetic anhydride. The isomeric aldehydes obtained by alumina treatment of (139) could not be separated but the resultant alcohols ((138) and its isomer) prepared by NaBH$_4$ reduction of (139) and (140) could be isolated in pure form.

The structures of these alcohols were ascertained by comparison of the appropriate alcohols obtained by the known degradation of natural ajmaline. Spectroscopic comparison revealed that the synthetic materials belong to the isoajmaline series, that is, the stereochemistry of the nitrile group appeared to be isomeric to that of the naturally derived materials. The total synthesis is isoajmaline then proceeded from this point employing the aldehyde (140), obtained from natural sources, as a relay substance.

Acid catalysed ring closure of (140) in a manner similar to that employed in the previous syntheses afforded, after reduction, the acetate (142). Removal of the acetate and *N*-benzoyl groups was achieved by first treating (142) with the Meerwein reagent followed by borohydride reduction and finally catalytic debenzylation. During these reactions the nitrile group remains intact and the nitrile-alcohol (144) was obtained. Since conversion of (144) into isoajmaline had already been previously described, a formal synthesis of this alkaloid was complete.

An extension of the above route also allows a synthesis of ajmaline[36]. The ketone (133) via its pyrrolidine enamine was treated with chlorqaceto-nitrile and the resulting product (146) was converted to the oxirane (147) in the manner previously described. Elaboration of the latter substance by essentially the reaction sequence mentioned above leads to a compound (106) which proved identical with the intermediate previously obtained by Masamune.

2.5 THE 2-ACYL INDOLE FAMILY

The 2-acyl indole alkaloids represent now a large family although the recognition of this structural type and its abundance in a variety of apocynaceous plants is largely the result of a concentrated effort within the last decade[37]. Biogenetically it is rather clear that these alkaloids are related to the corynantheine and sarpagine families although direct experimental evidence is lacking. Thus if one considers the transannular cyclisation of a corynantheinoid system, as for example the alkaloid geissoschizine (148) in the manner presented under discussion of the sarpagine–ajmaline group and then ring fusion of the latter occurs, the 2-acyl indole skeleton is readily discernible (e.g. (148) → (149) → (150)).

Thus far the only synthetic work in this area has been reported by Shioiri and Yamada[38]. In this investigation, the first synthesis of 1-methyl-16-

demethoxycarbonyl-20-desethylidene vobasine (161), according to the following sequence, is described.

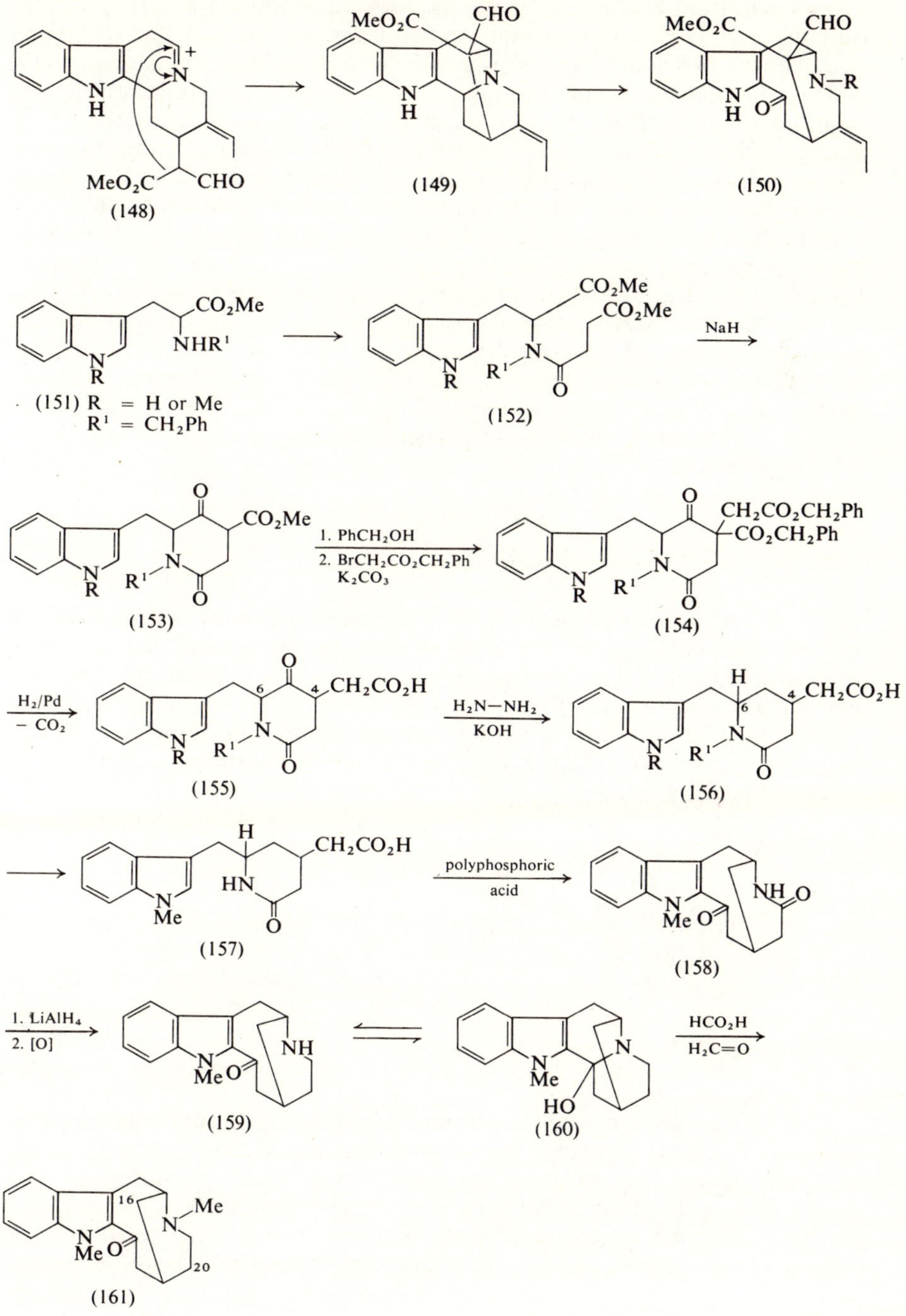

The starting tryptophan derivative (151), in the parent series or possessing the *N*-methyl group, was transformed to the keto ester (152) by reaction with 3-chloroformyl propionate under Schotten–Baumann conditions. Dieckmann cyclisation of the resulting esters was studied under a variety of conditions but NaH gave the optimum yields. *Trans*-esterification and alkylation with benzyl bromoacetate furnished a mixture of diastereoisomers (154) which upon debenzylation and decarbonylation allowed conversion to the required ketoacids (155). For the purpose of this discussion only the 4,6-*cis* isomer will be utilised for the remaining elaboration to the final product. Huang–Minlon reduction of (155) and standard debenzylation conditions provided (157) which upon cyclisation with polyphosphoric acid allowed construction of the requisite 2-acyl indole skeleton. Removal of the amide carbonyl and mild oxidation of the expected amino alcohols with chromium trioxide in pyridine furnished the corresponding ketone (159) which exhibited a behaviour characteristic of such systems. The facile transannular cyclisation of this amino ketone destroys the carbonyl chromophore, as evidenced by infrared and ultraviolet spectroscopy, and clearly the carbinolamine (160) is the predominant species in solution. However, reaction of (160) under Clarke–Eschweiler conditions forces ring opening and the final product (161) is obtained.

2.6   THE YOHIMBINE FAMILY

The yohimbine alkaloids were extensively investigated from structural and synthetic points of view during the 1950s when the pharmacological potential of reserpine (163) and related analogues was clearly evident. Such investigations brought forth a number of synthetic routes to these substances. The parent alkaloid yohimbine (162) was synthesised by van Tamelen in 1958 [31] two years after Woodward announced his stereospecific synthesis of reserpine (163) [39, 40].

(162)   (163)

These alkaloids which differ considerably in the different arrangement of asymmetric centres, portray the requirement for different strategy in the synthetic plan. The necessary synthetic development of the D/E *trans* fusion of the yohimbine skeleton cannot be easily modified to accommodate the D/E *cis* fused system present in reserpine. In fact this stereochemical variation was approached in rather different ways in the above mentioned routes.

The large majority of the alkaloids in this family possess the more stable D/E *trans* system and therefore careful stereochemical control in the early phases is not an essential requirement in these instances.

Ernest and Kakac[41] have successfully applied the principles of the Woodward synthesis in preparing appropriate D/E *cis* fused analogues and then modifying the route for the synthesis of D/E *trans* derivatives in the yohimbane series. Thus the diol (165) available from the known lactone (164) was readily converted to the aldehydo acid (166) and the latter on equilibration provided a mixture of (166) and (167). When the latter substance was subjected to the steps outlined below, (±)-apoyohimbane (172) was obtained.

(164) $\xrightarrow{\text{6 steps}}$ (165) $\xrightarrow[\text{Ba(ClO}_3)_2]{\text{OsO}_4}$ (166) $\xrightarrow{\text{H}^+}$

(167) $\xrightarrow[\substack{\text{2. tryptamine}\\\text{3. NaBH}_4}]{\text{1. esterify}}$ (168) $\xrightarrow[\text{OH}^-]{\text{MeOH}}$

(169) $\xrightarrow[\text{2. NaBH}_4]{\text{1. POCl}_3}$ (170) $\xrightarrow[\text{2. CH}_2\text{N}_2]{\text{1. Hydrolysis}}$

(171) $\xrightarrow[\text{2. MeOH, OH}^-]{\text{1. MeC}_6\text{H}_4\text{SO}_2\text{Cl}}$ (172)

A similar consideration, namely the preponderance of *trans* 1,2-disubstituted cyclohexane derivatives under appropriate equilibrating conditions, led to another synthesis of ($\pm$)-yohimbane[42]. Thus the crude acid aldehyde (175), obtained as an equilibrium mixture and assumed to exist mainly in the *trans* stereochemistry was condensed with tryptamine and the resultant lactam (176) was converted to ($\pm$)-yohimbane (177). ($+$)-Yohimbane was obtained from the racemic mixture by resolution with ($-$)-dibenzoyltartaric acid.

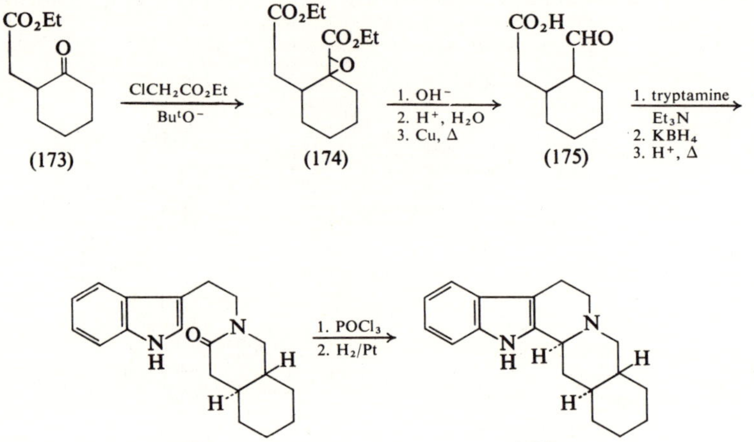

(173)   (174)   (175)

(176)   (177)

A substantial number of recent syntheses in this family utilise the cyclisation of appropriate iminium salts, a method first described by Robinson and Potts[43] and then subsequently studied extensively by the Wenkert group[44, 45]. In this way ($\pm$)-pseudoyohimbane (179) was prepared[46]. Unfortunately the commonly employed oxidant, mercuric acetate, attacks the decahydro-isoquinoline derivative (178) in a rather indiscriminate manner and a mixture of yohimbane isomers is obtained.

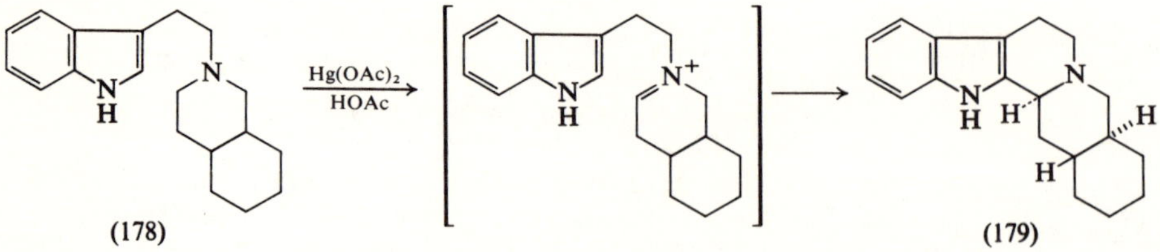

(178)   (179)

A convenient method for generating the requisite imine systems for subsequent elaboration to the yohimbane and related series is provided via hydride reduction of appropriate pyridinium or isoquinolinium derivatives. Several recent examples will portray the utility of this approach. Beisler[47] reports a high yielding short synthesis of several alkaloids in the gambirtannine series. In this study the isoquinolinium bromide (182) prepared by condensing 5-carbomethoxy isoquinoline (180) with tryptophyl bromide (181), is subjected to sodium borohydride reduction in a multiphase system (methanol–water–ether) containing a high concentration of cyanide ion. The initially formed enamine is trapped by cyanide and the postulated intermediate (183) receives minimum contact with the reducing medium since it passes into the ether layer as soon as it is formed. In this way undesirable side reactions are minimised and a good yield of the desired intermediate is obtained. This procedure was first employed by Fry[48] in the synthesis of dihydropyridines.

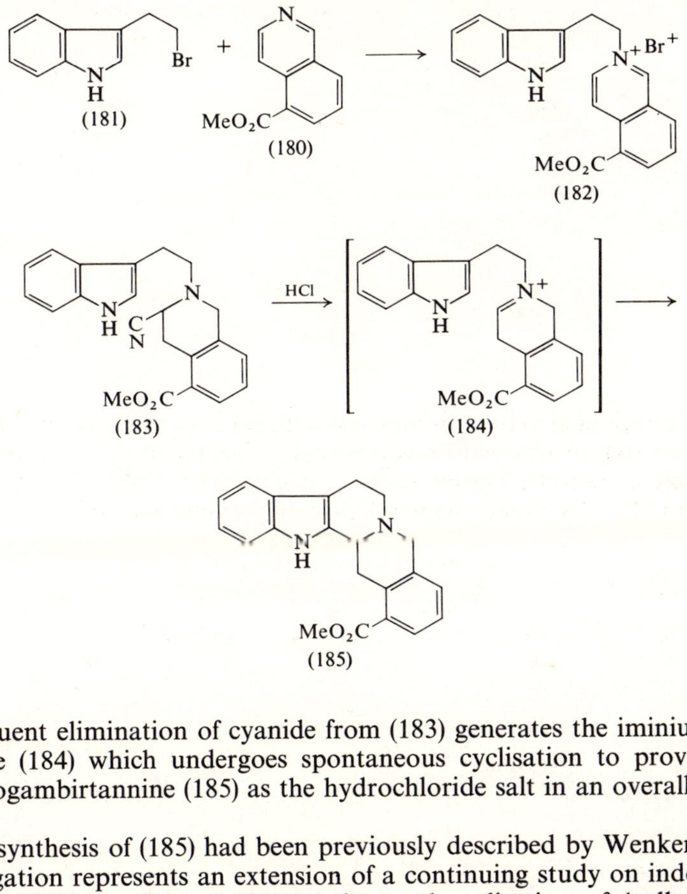

Subsequent elimination of cyanide from (183) generates the iminium intermediate (184) which undergoes spontaneous cyclisation to provide (±)-dihydrogambirtannine (185) as the hydrochloride salt in an overall yield of 83%.

The synthesis of (185) had been previously described by Wenkert[49]. This investigation represents an extension of a continuing study on indole alkaloid syntheses in which hydrogenation and cyclisation of 1-alkyl-3-acyl-pyridinium salts occupies a crucial role. In this instance the reduction of the appropriate isoquinolinium salt was shown to follow a pattern observed previously in the pyridinium series. The successful pathway to (185) follows the sequence (186) → (187) → (188) → (185).

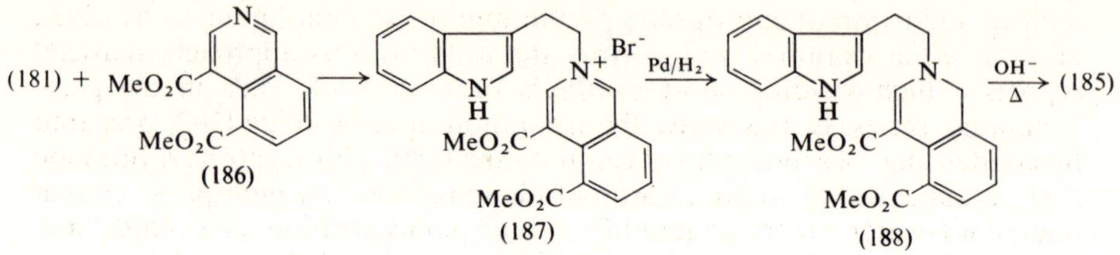

$(181) + $ (186)

(187) $\xrightarrow{Pd/H_2}$ (188) $\xrightarrow[\Delta]{OH^-}$ (185)

A synthesis of the alkaloid alstoniline (191) has been recently described by Beisler[50] in which his approach mentioned above has been applied.

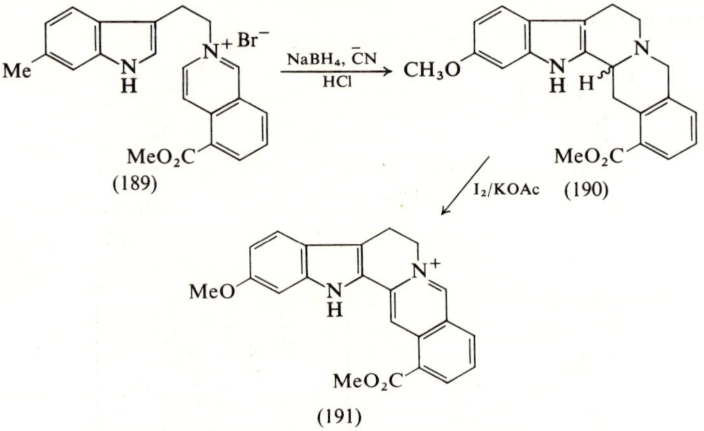

(189) $\xrightarrow[HCl]{NaBH_4,\ \overline{C}N}$ (190)

$I_2/KOAc$ (191)

Another approach to the yohimbine system has been reported by Szantay[51]. The crucial step in this synthesis involves a Dieckmann cyclisation of the appropriate tetracyclic diester (194) or nitrile ester (198). The tetracyclic keto ester (192) was already available from a previous study[52].

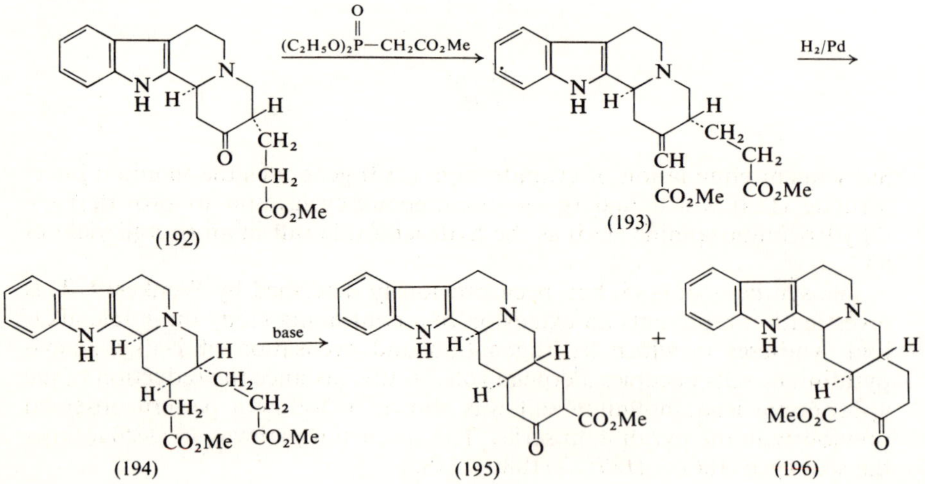

(192) $\xrightarrow{(C_2H_5O)_2\overset{O}{\overset{\|}{P}}-CH_2CO_2Me}$ (193) $\xrightarrow{H_2/Pd}$

(194) $\xrightarrow{base}$ (195) $+$ (196)

A detailed investigation of the Dieckmann cyclisation of (194) was made and a variety of reaction conditions was employed. Unfortunately the cyclisation proceeded predominantly in the undesired direction to provide (195) as the major product. The desired keto ester, yohimbinone (196), was obtained in only 10–15% yield. This latter substance on reduction provided a 1:3 mixture of yohimbine (162) and β-yohimbine (197).

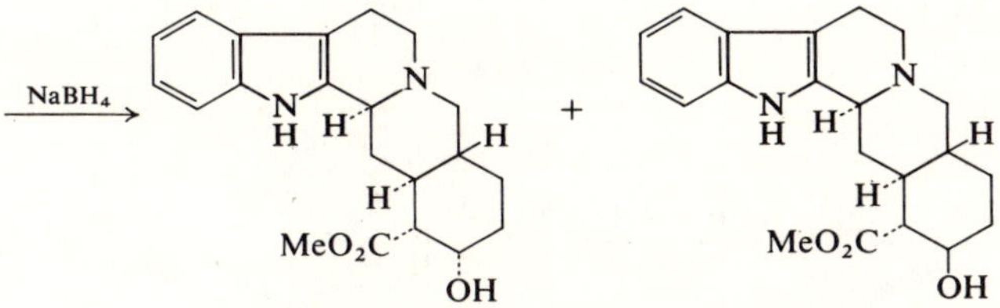

(162) (197)

When (192) was reacted with the appropriate phosphonate, the expected nitrile ester (198) was obtained. Saturation of the double bond and ring closure of the resultant product under basic conditions provided the keto nitrile (199) in 52% yield. Reduction of the carbonyl group in the latter allowed the isolation of three nitrile alcohols (200), (201), (202), two of which were then elaborated to the alkaloid systems, (200) → (162) and (201) → (197).

(192)

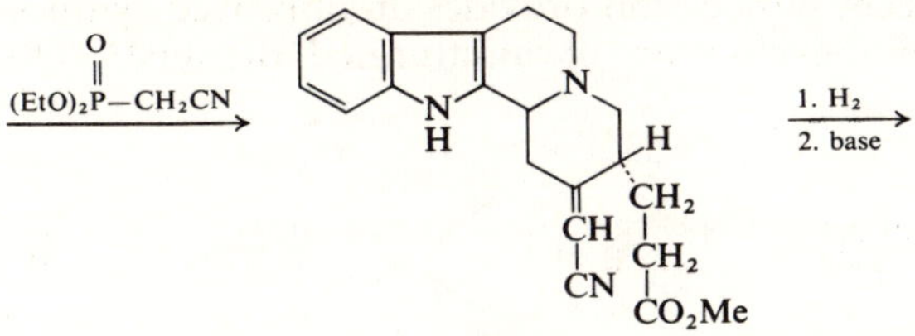

(198)

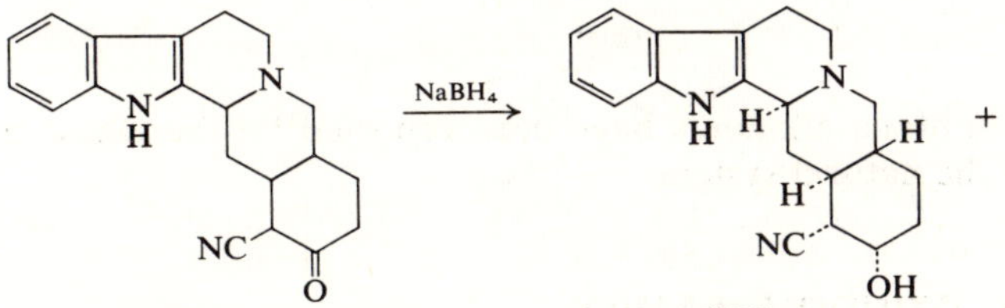

(199) (200)

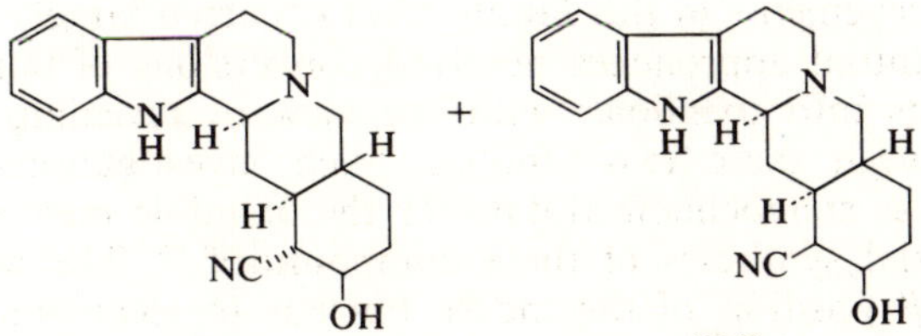

(201) (202)

Then

(200) $\xrightarrow[\text{H}_2\text{O}_2]{\text{NaOH, MeOH}}$ (203) $\xrightarrow[\Delta]{\text{HCl}}$ (162)

(203)

and

(201) $\xrightarrow[\text{H}_2\text{O}_2]{\text{NaOH, MeOH}}$ (204) $\xrightarrow[\Delta]{\text{HCl}}$ (197)

(204)

The synthetic racemic products (162) and (197) were resolved by employing *N*-acetyl-L-leucine for ($\pm$)-yohimbine and L-camphor sulphonic acid for ($\pm$)-β-yohimbine as the resolving agents. In this way ($+$)-yohimbine and ($-$)-β-yohimbine were obtained.

A more recent publication provides an improved synthesis of yohimbine by Dieckmann cyclisation of unsaturated diesters[53]. Other approaches

(193) $\xrightarrow{\text{NaH, THF}}$ (205) $\xrightarrow{\text{H}_2/\text{Pd}}$ (206)

(205)   (206)

to the yohimbine alkaloids have been reported[54–57] but these have not as yet led to the natural systems.

2.7 THE OXINDOLE FAMILY

The synthetic achievements in this family have occurred largely within the last decade. The initial approaches involved conversions of tetrahydro-β-carboline alkaloids into oxindoles with the view of revealing the inter-relationships between these two families. Such investigations provided structural as well as stereochemical data for the oxindole members and in turn allowed partial syntheses of these compounds[58, 59]. The well known reactivity of the β position of the indole nucleus to electrophilic agents allowed the utilisation of this property in the conversion of the appropriate indole alkaloids by means of t-butyl hypochlorite to the resulting chloro-indolenines and subsequent rearrangement of the latter intermediates to the oxindole systems. In this manner, ajmalicine (207) leads to mitraphylline (208), while corynantheine (209) and dihydrocorynantheine (210) give rise to corynoxeine (211) and rhyncophylline (212).

These types of rearrangements have received renewed interest in the last few years since it has been postulated that the *in vivo* rearrangement of the Corynanthe series to the Strychnos family in the plants occurs via similar processes.

Several other approaches[60, 61] have employed the appropriate hydroxy-tryptamine derivatives directly to prepare derivatives of the rhyncophylline series. Thus Ban[60] was able to condense 1-methyl-2-hydroxytryptamine hydrochloride (213) with the aldehyde (214) to provide the base (215).

A similar sequence has served to provide a synthesis of (±)-8-oxovincatine[62]. Thus the reaction of (213) with dimethyl 4-ethyl-4-formylpimelate (216) gave in 70% yield the dilactam 8-oxovincatine (217) as a mixture of two racemates apparently differing in configuration at C-5. Reduction of the separated lactams produced a diol (218) which was identical with that obtained from the alkaloid, vincatine.

The extensive investigations of van Tamelen and his group concerning biogenetic type syntheses have led to a total synthesis of rhyncophyllol (230) and (±)-isorhyncophyllol (235)[63, 64]. In the belief that many of the structural types inherent in the numerous indole families are derived by varying modes of cyclisation of natural intermediates in which certain reactive sites have been transformed enzymatically to specific oxidation levels, van Tamelen devised a sequence in which the crucial step in the synthesis of (230) involved a cyclisation of the appropriate oxindole dialdehyde (228).

(228) → (229) → (230)

A modification of the above sequence completed the synthesis of $(\pm)$-isorhyncophyllol (235).

(224) $\xrightarrow[\text{NaIO}_4]{\text{OsO}_4}$ (231) $\longrightarrow$

(232) $\xrightarrow{\text{LiAlH}_4}$ (233) $\xrightarrow{\text{Me}_3\text{COCl}}$

(234) $\xrightarrow{\text{H}_2/\text{Pd}}$ (235)

2.8   THE STRYCHNOS FAMILY

A major synthetic effort by the Woodward group culminated in the first total synthesis of strychnine in 1954[65, 66]. Since that time several other syntheses of members within this class have appeared.

In general the approach utilised is a direct application of a fragmentation reaction first applied by Dolby[67] in the tetrahydrocarboline system and then by Freter[68, 69]. Dolby demonstrated that certain dihydrocorynantheine derivatives undergo C–D ring cleavage by the action of hot acetic anhydride containing sodium acetate (236) → (237).

(236) $\xrightarrow[\text{NaOAc, }\Delta]{\text{Ac}_2\text{O}}$ (237) R = CHOMe

Freter[68] also observed that treatment of 1-phenyl-2-methyltetrahydro-β-carboline (238) with acetic anhydride gave the 1,2-seco compound (239). Subsequent investigations[69] in his laboratory demonstrated the generality of this approach.

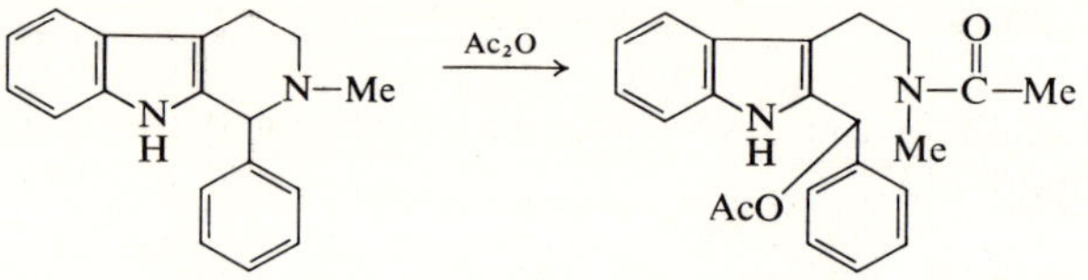

(238)   (239)

Harley-Mason has employed this reaction in providing syntheses of several members of the *Strychnos* family. In the synthesis[70] of racemic tubifoline (244), tubifolidine (246) and condyfoline (245) the hexahydroindoloindolizine (240) was reacted with α,α'-dichlorobutyric anhydride to provide the amide ester (241). The conversion of the latter to the natural systems was then performed as shown in the sequel:

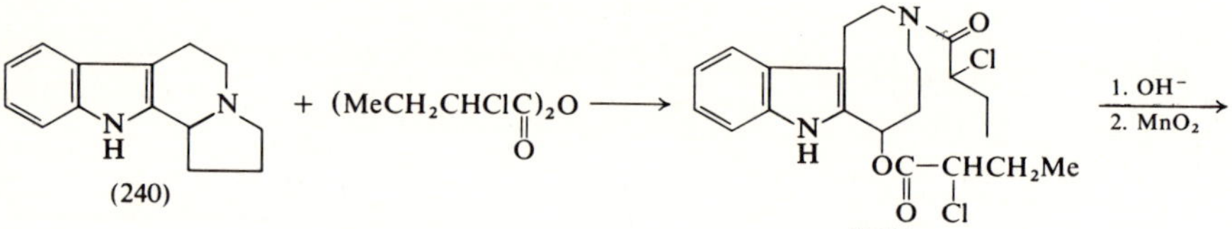

(240)   + (MeCH$_2$CHClC)$_2$O ⟶ (241)

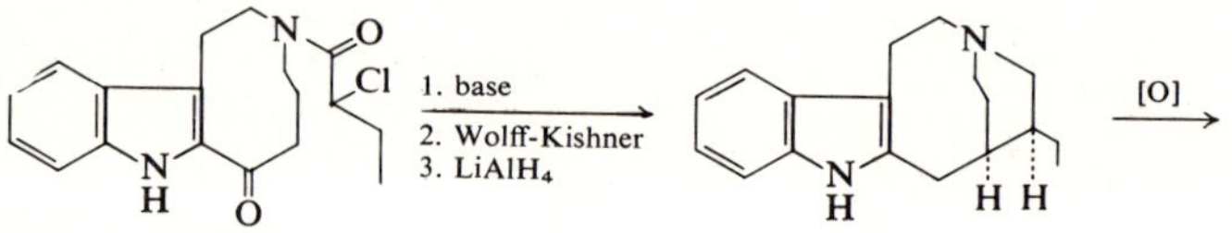

(242)   1. base   2. Wolff-Kishner   3. LiAlH$_4$ ⟶ (243)   [O] ⟶

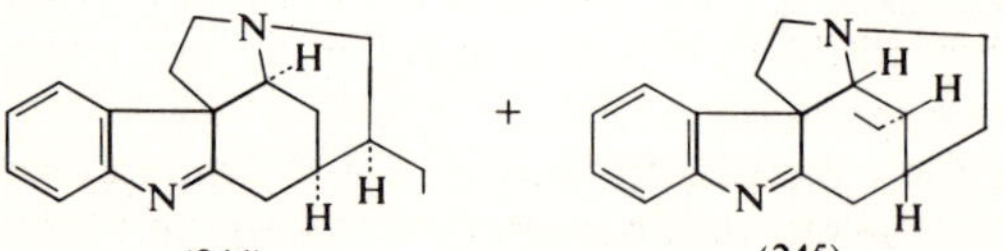

(244)   +   (245)

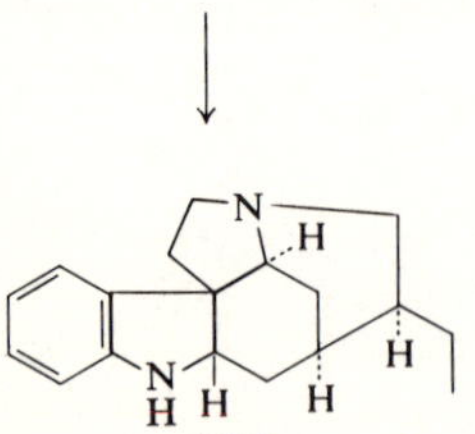

(246)

An extension of the above investigations, employing the keto amide (247) as the starting material, allowed completion of the synthesis of racemic geissoschizoline (251)[71]. Here again as in the above synthesis, the crucial cyclisation of the medium sized ring intermediate (249) was accomplished by a procedure previously developed by Schmid[72].

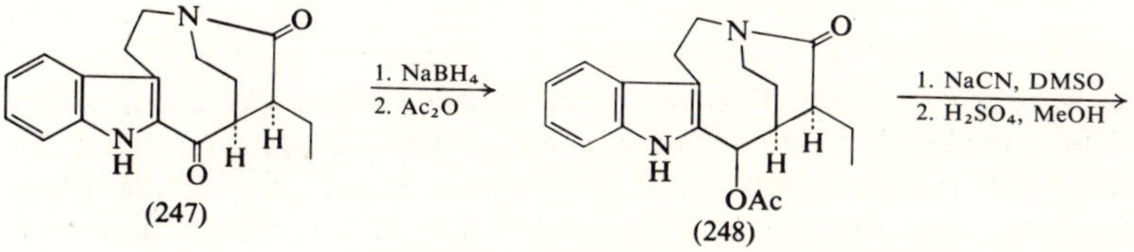

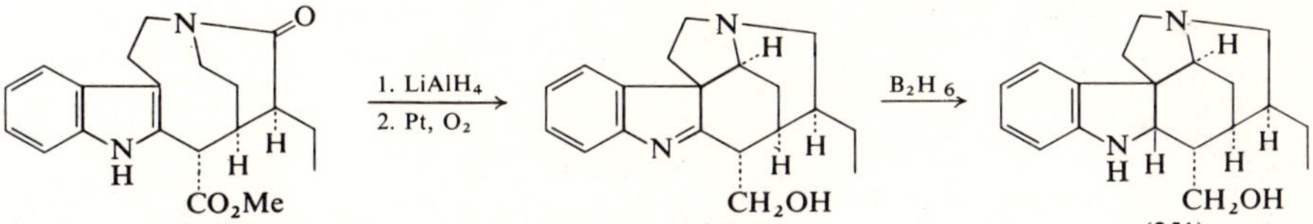

Transannular cyclisation of the tetracyclic ester amide (249) provided a low yield of racemic tubotaiwine (252)[73].

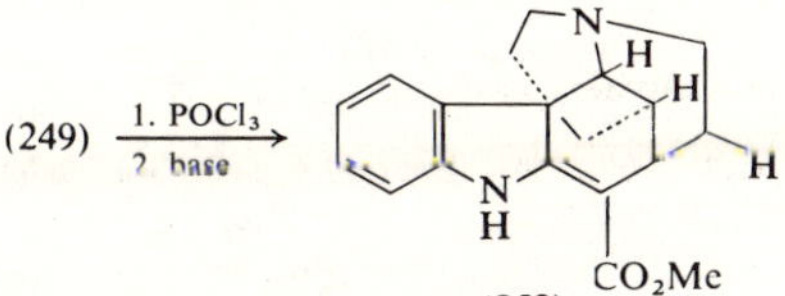

An improved synthesis of geissoschizoline as well as a synthesis of racemic dihydronorfluorocurarine (256) was subsequently reported by Harley-Mason[74]. The approach is identical to those discussed above. The starting keto amide (247) is converted via routine reactions to the *Strychnos* derivative (256).

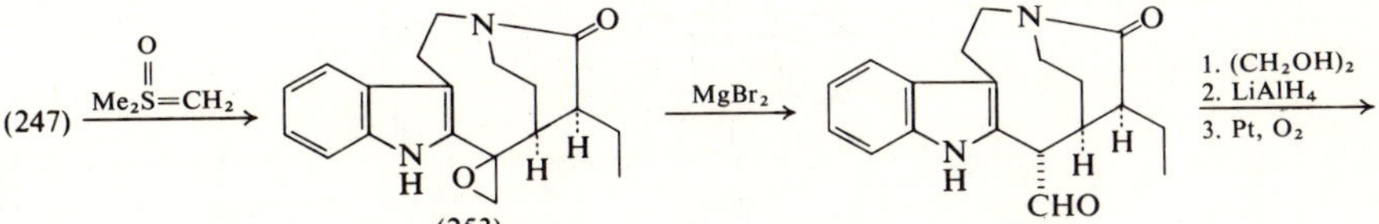

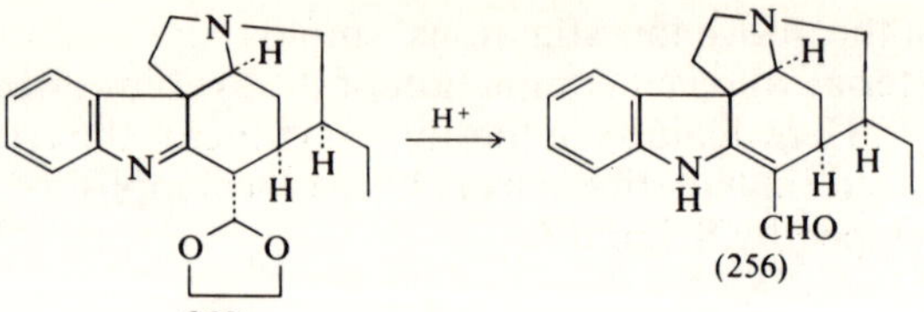

(255)

$H^+ \longrightarrow$

CHO

(256)

Finally a synthesis of racemic fluorocurarine iodide (260), a Calabash curare alkaloid was completed[75]. The authors simply apply the previous reactions with appropriate modifications. The tetracyclic indole derivative (240), employed earlier, was reacted with the anhydride of 2-bromo-3-methoxy-butyric acid and the resultant product (257) now bears the necessary functionality for subsequent ring closure and elimination to provide the essential ethylidene side chain. The remainder of the sequence is straightforward.

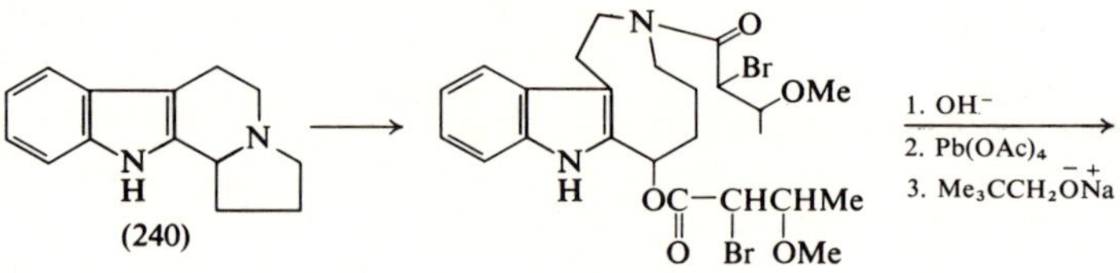

(240)

1. OH^-
2. $Pb(OAc)_4$
3. $Me_3CCH_2\overset{-}{O}\overset{+}{N}a$

(257)

(258)

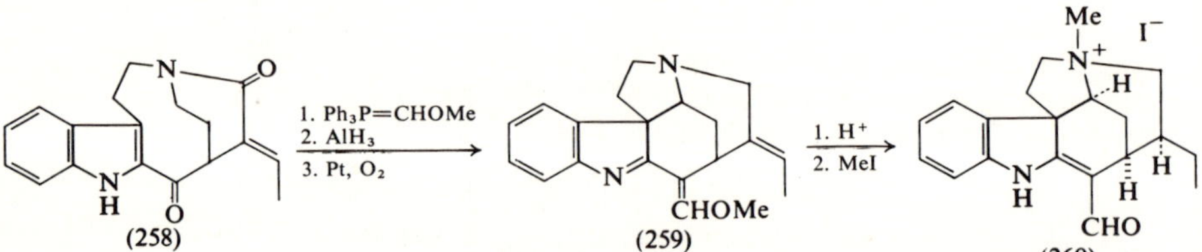

1. $Ph_3P{=}CHOMe$
2. AlH_3
3. Pt, O_2

(259)

1. H^+
2. MeI

(260)

Two groups have reported syntheses of the *Strychnos* alkaloid, retuline (261)[76, 77]. Both synthetic endeavours started with optically active materials derived from the akuammicine and strychnine series.

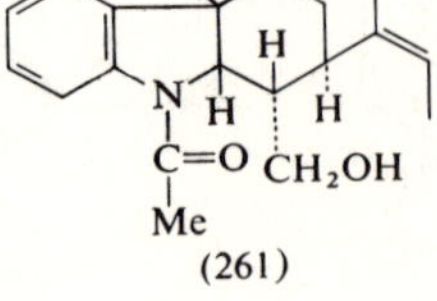

(261)

2.9 THE ASPIDOSPERMA FAMILY

During the past decade there has been an enormous effort by various groups to achieve syntheses of alkaloids within this family. It will not be possible, within the limitations in scope and length of this chapter, to discuss in detail all the synthetic schemes which have been published.

Along with the synthetic endeavours an extensive series of investigations directed at the isolation and structure elucidation of alkaloids have revealed the presence of the fundamental template normally associated with the Aspidosperma family in various plant genera other than those of Aspidosperma (for example Vinca, Kopsia, etc.). Consequently for the sake of clarity the present discussion will concern itself with the various synthetic schemes which lead to alkaloids possessing the rearranged non-tryptophan C_{10} unit (261-1) normally regarded as fundamental in the biosynthetic considerations of the Aspidosperma family. The two main skeleta contained in the large number of natural systems can be exemplified by quebrachamine (262) and aspidospermidine (263).

(261-1)   (262)   (263)

The first total synthesis of the natural Aspidosperma system aspidospermine (273) was reported by Stork[78]. This approach utilised the Fischer indole synthesis in achieving the construction of the desired pentacyclic skeleton. The requisite tricyclic intermediate (271) was obtained by effectively employing the pyrrolidine enamine reaction, also developed previously in Stork's laboratory, in the initial stages of the sequence.

$MeCH_2CH_2CHO$ (264) → [1. pyrrolidine; 2. $CH_2{=}CHCO_2Me$; 3. HOAc] → (265) → [1. pyrrolidine; 2. $MeCCH{=}CH_2$; 3. HOAc] → (266)

(267) → [1. $(CH_2OH)_2$; 2. NH_4OH] ... [1. $LiAlH_4$; 2. H^+, H_2O; 3. base] → (268) → [$ClCH_2C{-}Cl$]

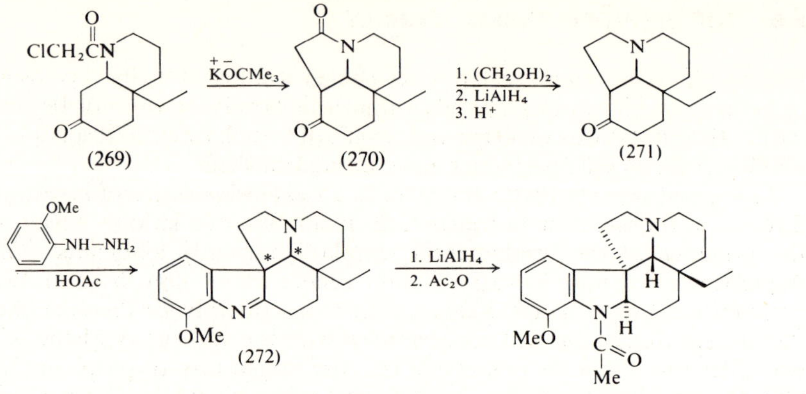

(269)   (270)   (271)

(272)   (273)

The stereochemistry of the various bicyclic and tricyclic intermediates utilised was left undefined since the authors felt that the indolenine (272) being formed under equilibrating conditions would lead to equilibration at the two asterisked centres via a reverse Mannich reaction. There was literature precedent for this situation[79]. Since the conformational expression for aspidospermine (273) denotes the most stable arrangement at the various asymmetric centres, such an equilibration process would be expected to lead eventually to the desired natural stereochemistry. The racemic substance (273) thus obtained was identical with the natural alkaloid.

Fischer indole cyclisation of (271) employing phenylhydrazine allowed extension of the synthesis to the quebrachamine system (262). The reductive cleavage with potassium borohydride had been previously applied by Biemann[80] in a structural elucidation study.

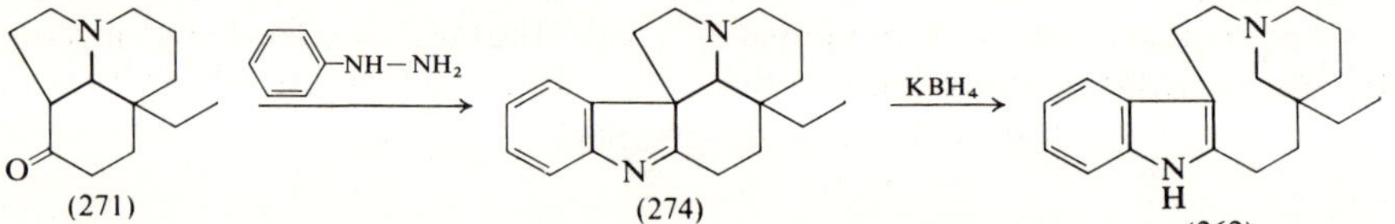

(271)   (274)   (262)

Another total synthesis of aspidospermine was reported by Ban and co-workers[81]. An interesting situation developed during this study since during their planned objective in similarly utilising the Fischer indole synthesis in the later steps of the synthesis, these workers developed a different pathway to a bicyclic intermediate (279) identical in gross structure to (268). However, the physical and chemical data for the compound available from Ban's experiments clearly revealed that this intermediate was different from that obtained by Stork.

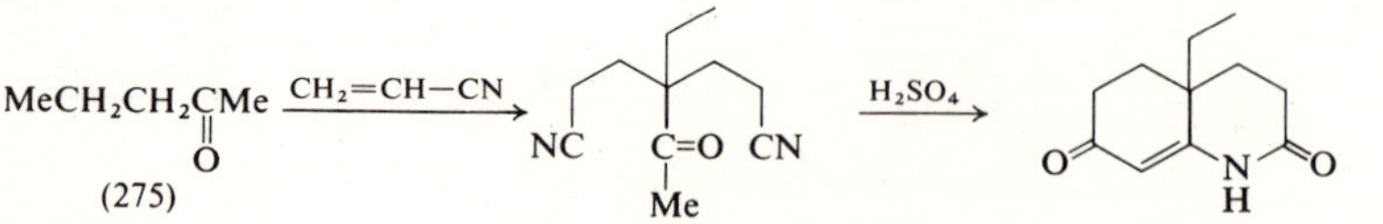

(275)   (276)   (277)

Then:

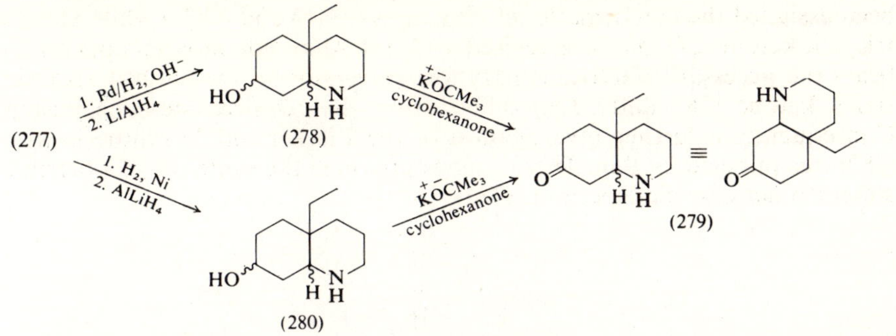

(277)

(278)

(279)

(280)

Conversion of (279) to (282) and the latter to (±)-aspidospermine (273) in exactly the manner outlined above in the Stork synthesis (268) → (273) provided conclusive evidence that the respective sequences yielded a series of intermediates which were diastereoisomeric.

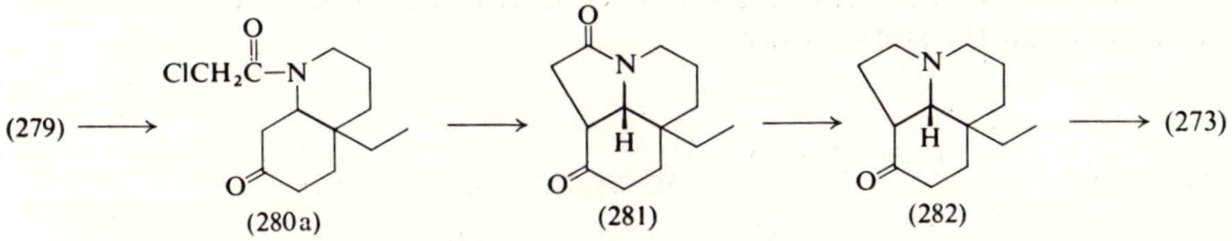

(279) ⟶

(280a)

(281)

(282)

⟶ (273)

A detailed spectroscopic study allowed the assignment of the conformational structures (270a) and (281a) to the tricyclic keto amides (270) and (281) in the respective studies.

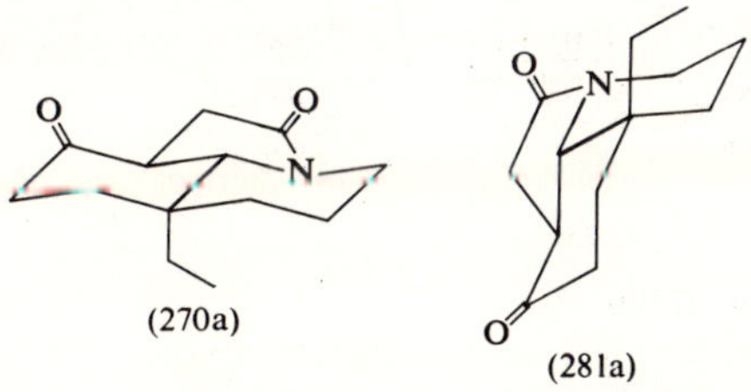

(270a)

(281a)

The conversion of both (271) and (282) to (±)-aspidospermine supports Stork's contention that equilibration occurs during the Fischer indole cyclisation and therefore the stereochemical centres in the various intermediates are of no consequence in determining the end product.

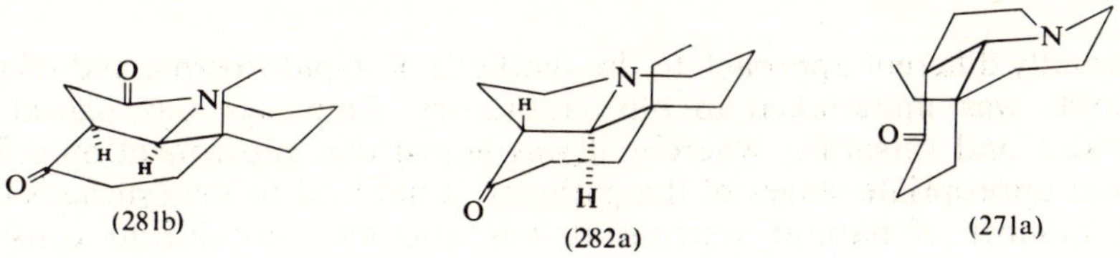

(281b)

(282a)

(271a)

In more recent studies by Ban[82] the tricyclic ketones (281) and (282) have been assigned the conformational structures (281b) and (282a) while Stork's tricyclic ketone (271) has been revised to (271a). Although the latter apparently bears the necessary stereochemistry for conversion to the natural alkaloid, Ban's ketone (282 and 282a) differs from the required stereochemistry. Consequently a further investigation of the Fischer indole synthesis with (282) was pursued by Ban. These studies provided the synthesis of a racemic stereoisomer of aspidospermine (283).

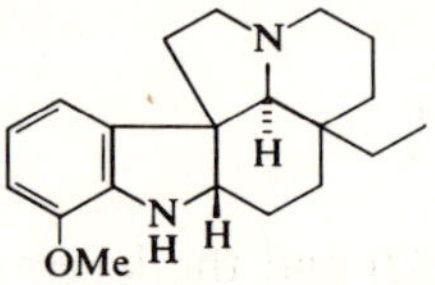

(283)

Another study by Kuehne[83] has also provided the tricyclic intermediate of gross structure (271) but possessing the stereochemistry portrayed in (271a) and (290). The sequence starts with proline ethyl ester (284) and proceeds through the steps shown.

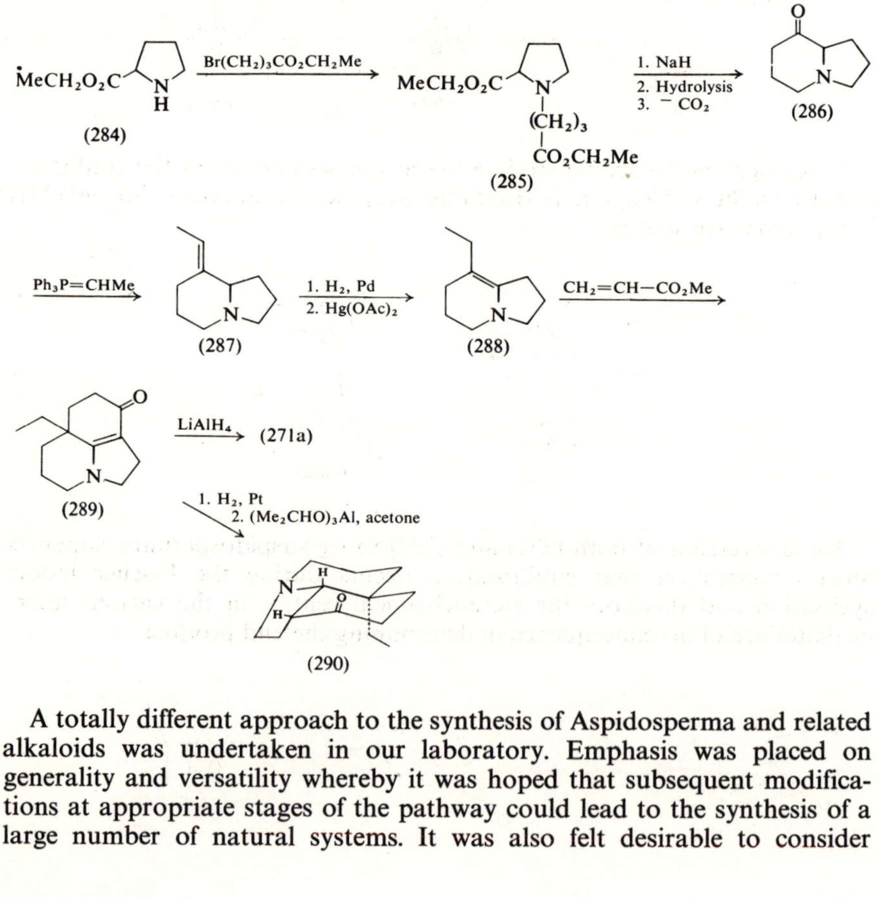

(284)

(285)

(286)

(287)

(288)

(289)

$\xrightarrow{\text{LiAlH}_4}$ (271a)

(290)

A totally different approach to the synthesis of Aspidosperma and related alkaloids was undertaken in our laboratory. Emphasis was placed on generality and versatility whereby it was hoped that subsequent modifications at appropriate stages of the pathway could lead to the synthesis of a large number of natural systems. It was also felt desirable to consider reactions which may have significance in biosynthesis, i.e. in the *in vivo* construction of these alkaloids by the living plant systems. For this purpose we selected the possible utilisation of a transannular cyclisation reaction of appropriate nine-membered ring intermediates since it seemed to satisfy most of our requirements. This study[84-88] revealed the general usefulness of such an approach for the synthesis of a variety of alkaloids. Thus nine-membered ring systems of the cleavamine (291) and quebrachamine (262) series were shown to cyclise in a completely stereospecific manner to generate the necessary asymmetry for the natural systems (294) and (296).

Utilisation of the ester derivatives (297) and (302) of the above systems then provided an alteration in the course of the cyclisation process and a direct entry into the Aspidosperma (299) and (303) and Iboga (301) alkaloids.

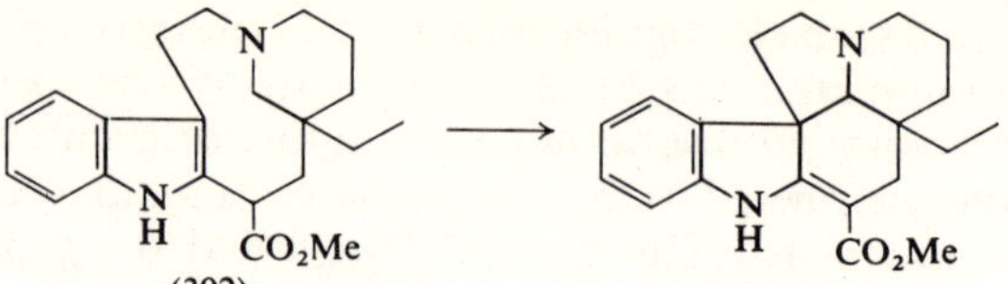

(302)   (303)

The crucial role occupied by the appropriate nine-membered ring intermediates for the total syntheses of a variety of these alkaloids demanded an
investigation into their synthesis. The approach selected involved the generation of the medium sized ring system by means of a reductive cleavage in the
last step of the synthesis. A preliminary study in our laboratory had revealed
the generality of this reaction (304) → (305).

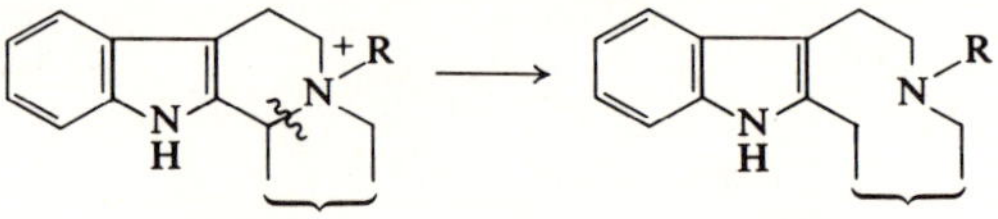

(304)   (305)

A new total synthesis of ($\pm$)-quebrachamine (261) and ($\pm$)-aspidospermi-
dine (296) was accomplished by means of the following scheme[89, 90].

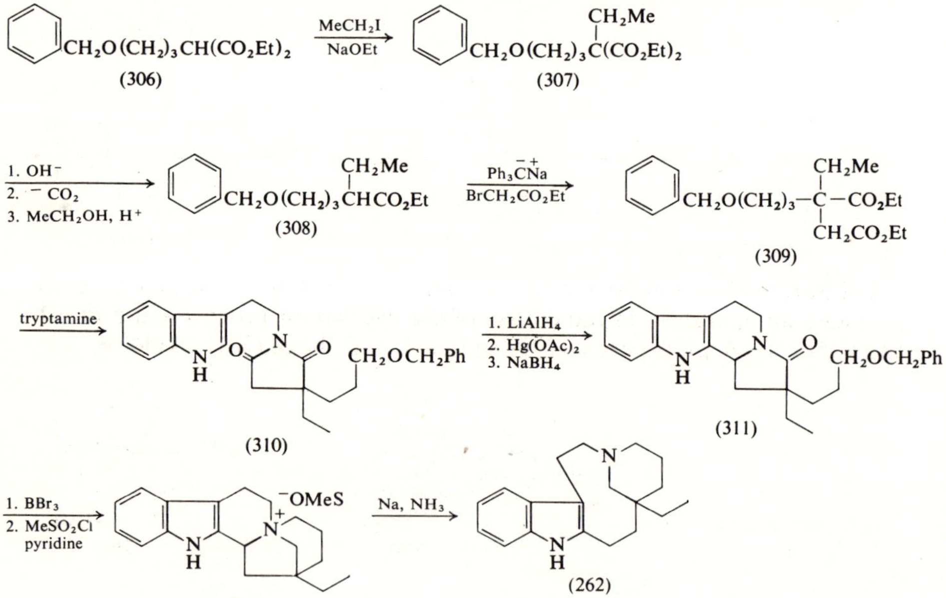

(306)   (307)

(308)   (309)

(310)   (311)

(312)   (262)

The total synthesis of ($\pm$)-quebrachamine (262) also completed the synthesis of ($\pm$)-aspidospermidine (296) in view of the previously established
transannular cyclisation of the former to the latter (see (262) → (295) → (296)).

Although the yields in the above conversions were generally satisfactory
one of the reactions involving mercuric acetate in the oxidative cyclisation

of the tricyclic amine to the tetracyclic base (311) proceeded in only 30–40%
yield. Subsequent modifications in the sequence[91] obviated this difficulty
and provided a vastly improved sequence to these alkaloids. The modified
scheme involves condensation of tryptamine with the aldehydo-ester (315)
whereupon the more reactive aldehyde function directs the cyclisation in
only one manner. In this way the desired tetracyclic lactam (316) is formed in
90% yield.

The versatility of the above approach to the synthesis of various alkaloids
bearing the Aspidosperma template but isolated from Vinca plants was now
demonstrated. The quaternary mesylate (312) was exposed to nucleophilic
displacement by cyanide ion with the resulting generation of the nine-
membered ring system possessing the nitrile function at C-3 (317) and (318).
Subsequent conversion of this function to the ester group, present in many
alkaloids, opened a synthetic entry to all of these systems. During reaction
at C-3 both possible stereoisomers at this site were obtained. To illustrate
the stereochemical consequences in these conversions the most plausible
conformational expressions for these compounds are depicted (317)–(320).

(317) R¹ = R³ = H; R² = CN
(318) R¹ = R² = H; R³ = CN

(319) R¹ = R³ = H; R² = CO₂Me
(320) R¹ = R² = H; R³ = CO₂Me

The synthetic compounds (319) and (320) represent (±)-vincadine and
(±)-epivincadine respectively. The latter compound was also prepared
from the natural alkaloid (319) by isomerisation of the ester function.

The synthesis of ($\pm$)-vincadine also completes the total synthesis of ($\pm$)-vincaminoreine (319, R = Me) and ($\pm$)-vincaminorine (320, R = Me) in view of the already known interconversions[92].

The previously established cyclisation now made possible the total synthesis of the pentacyclic series exemplified by vincadifformine (321) and minovine (322) from vincadine and vincaminoreine respectively.

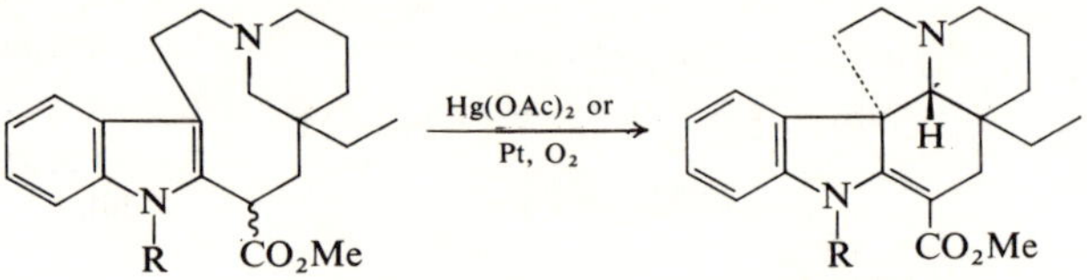

$$\xrightarrow{\text{Hg(OAc)}_2 \text{ or Pt, O}_2}$$

(321) R = H
(322) R = Me

There are a number of alkaloids which possess oxygen functions, particularly methoxyl groups, in the aromatic ring and it was clear that the above approach could be expanded to encompass these systems as well. Indeed reaction of 6-methoxytryptamine (323) with the aldehydo ester (315) provided a 70% yield of the tetracyclic lactam (324). Further elaboration of this intermediate in the manner discussed above allowed the total synthesis of 16-methoxy vincadine (325) and 16-methoxyepivincadine (326). These latter substances possess the skeletal features of vincaminoridine, an alkaloid isolated from *Vinca minor* L. but no stereochemical assignment is available from the published data nor could an authentic sample be obtained.

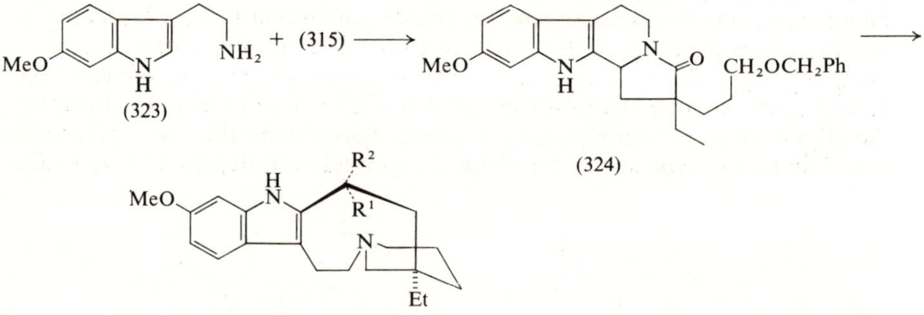

(323)

(324)

(325) R^1 = CO$_2$Me; R^2 = H
(326) R^1 = H; R^2 = CO$_2$Me

An interesting rearrangement reaction has been employed by Harley-Mason to complete a total synthesis of ($\pm$)-aspidospermidine (296). In the initial study[93] the tetracyclic lactam (327) was shown to rearrange to an *Aspidosperma* skeleton (328) which upon simple reduction provided 3-methylaspidospermidine (329).

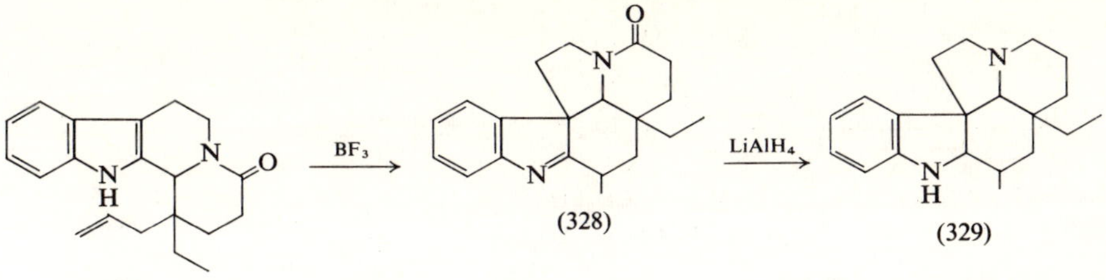

(327) (328) (329)

An extension of this work led to synthesis of the natural system (296)[99].

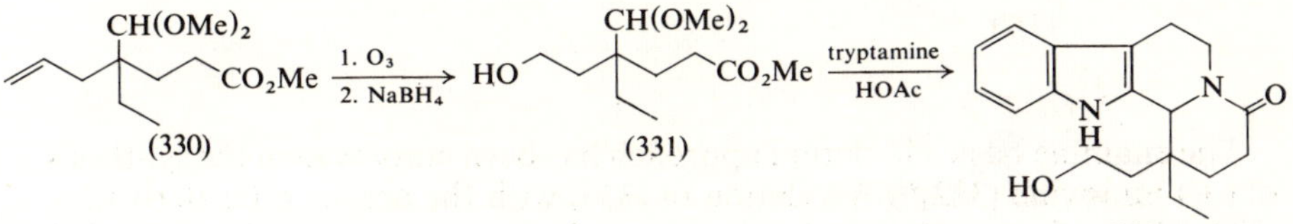

(330) (331) (332)

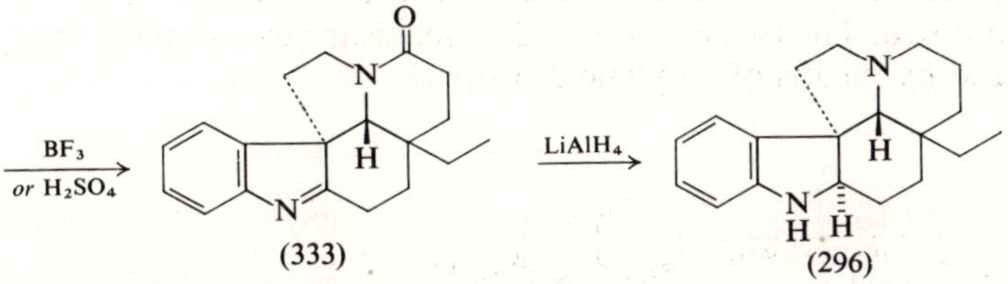

(333) (296)

Still another approach to quebrachamine and the pentacyclic Aspidosperma alkaloids was undertaken by Ziegler[95]. In one of the crucial steps
of the synthesis the alkylation of an appropriate enamine, 1-benzyl-3-ethyl-
1,4,5,6-tetrahydropyridine (334) is invoked. Thus in the quebrachamine
synthesis[95], this enamine, prepared according to the scheme shown, undergoes alkylation with methyl haloacetates to provide (334) which upon
debenzylation and reaction with β-indolylacetyl chloride provided the
lactam ester (335). The cyclisation of (335) proceeds in high yield (85%) to
(336) but unfortunately conversion of the latter to quebrachamine (262)
was a very low yielding process. The double bond in the accompanying
product, 3,4-dehydroquebrachamine (337) was very unreactive to reduction.

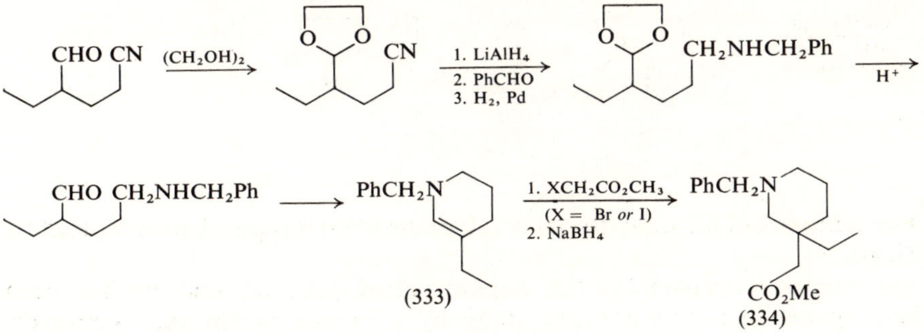

(333) (334)

(335)

(336)

+ (262)

(337)

The enamine basic alkylation approach has been employed in the synthesis of (±)-minovine (322)[96]. Alkylation of (333) with the acrylic ester derivative (388), allows the resultant iminium intermediate to undergo facile cyclisation at the β-position of the indole nucleus and the tetracyclic indole derivative (339) is obtained. The two carbon bridge necessary to complete the synthesis was induced by means of ethylene dibromide.

(338)

(339) $R^1 = CO_2Me$; $R^2 = H$
(340) $R^1 = H$; $R^2 = CO_2Me$

(322)

(343)

The synthesis of 6,7-dehydroquebrachamine (343) has also been reported by Ziegler[97].

A related approach to the Aspidosperma alkaloid nucleus has been investigated in a collaborative study by a French group and Wenkert[98].

Here again the resulting iminium intermediate generated during the catalytic reduction of the pyridine ring undergoes cyclisation at the β-position of the indole nucleus to provide a tetracyclic intermediate (346) similar in structure to that obtained by Ziegler. Introduction of the two carbon side chain (347) was also performed in a manner similar to Ziegler's. Unfortunately the conversion (346) → (347) proceeds in only 10% yield.

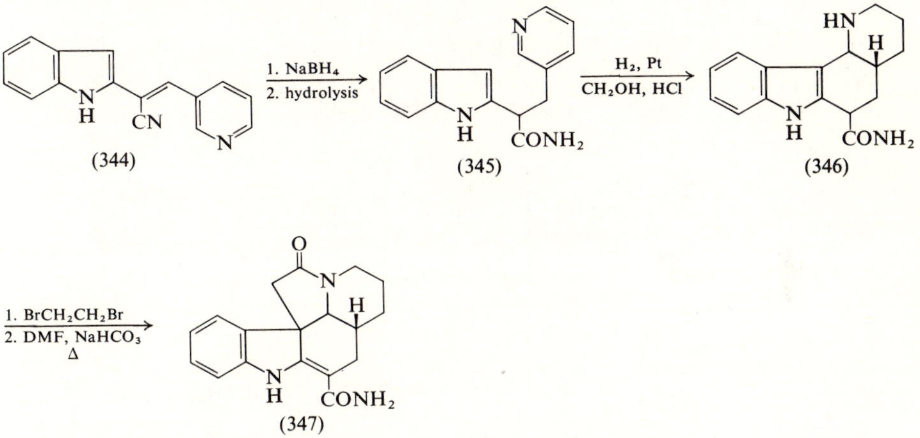

A synthesis of (±)-tabersonine (351) by Ziegler[99] utilises, in essence, his previously developed scheme for the construction of the required tetracyclic amino alcohol (348). The later stages of the synthesis are essentially a repetition of the chemistry previously involved in our synthesis of the monomeric Vinca alkaloids, for example, minovine (322) and vincadifformine (321)[90, 91, 100] (compare the conversion (349) → 350) → (351) with (312) → (317), (318) → (319), (320) → (321), (322).

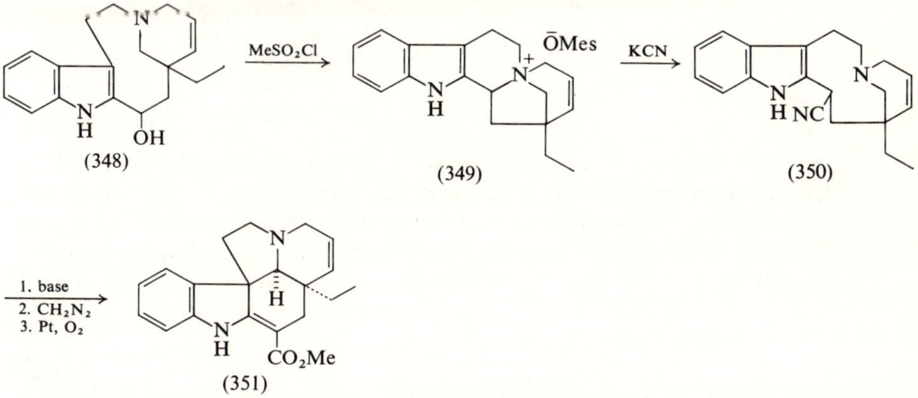

Stevens[101] has reported a fundamentally different approach which involves the acid-catalysed thermal rearrangement of a cyclopropyl imine (352) to an appropriately substituted 2-pyrroline (353) as a crucial stage. This intermediate was then converted according to the scheme shown to the hydro-lulolidine (357) which had already been established as a synthetic precursor in the approaches employed by Stork and Ban in their syntheses of (±)-aspidospermine.

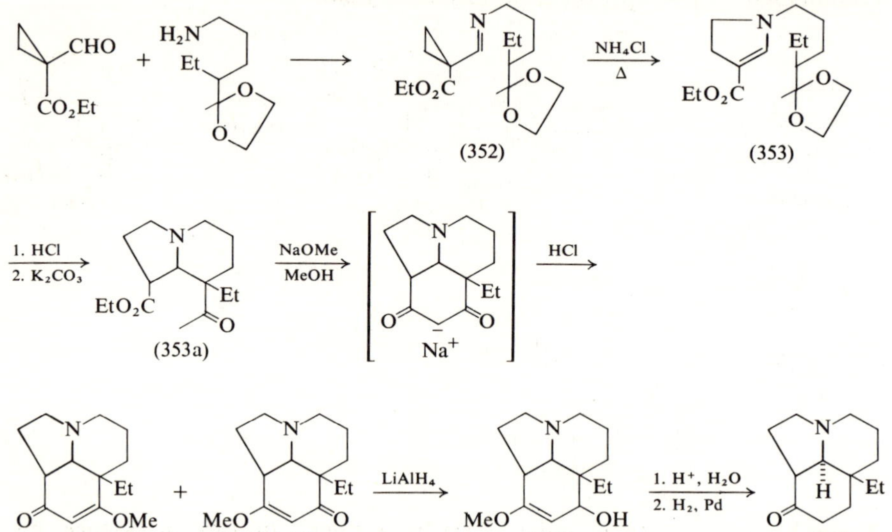

(352)   (353)

(353a)   Na⁺

(354) + (355)   (356)   (357)

Ban[102] has made an attempt to extend his previous method to the synthesis of C-21 oxygenated Aspidosperma alkaloids, as for example, limaspermine (358) and haplocine (359).

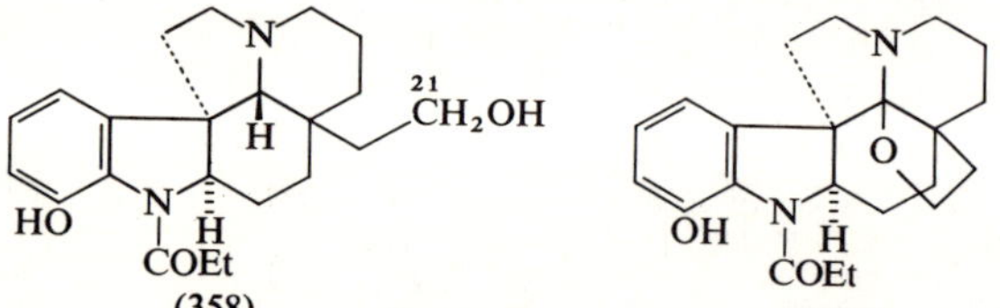

(358)   (359)

These investigations which employ 5-phenoxypentan-2-one (360) as starting material proceed via an eleven-step sequence to (±)-21-phenoxy-isopalosine (361). Unfortunately this latter intermediate when treated with hydrobromic acid, with the intention of cleaving the ether linkage, undergoes quaternisation to (362) thereby suggesting to the authors that the stereo-

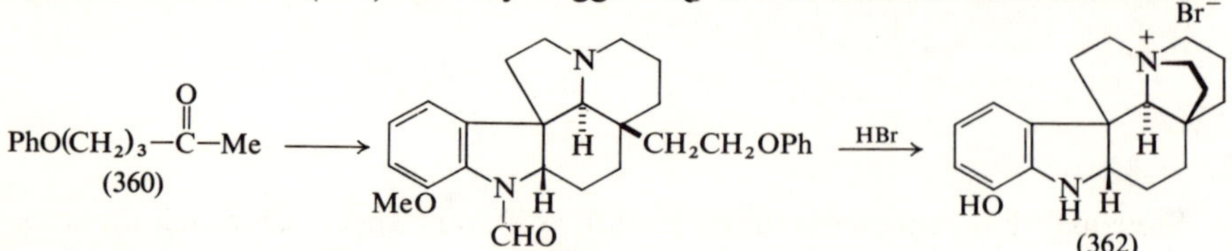

(360)   (361)   (362)

chemistry inherent in their synthetic compounds is not compatible with that of the natural alkaloid.

Buchi[103] has recently completed the synthesis of vindorosine (363), a penta-cyclic highly oxygenated Aspidosperma alkaloid which bears close resemblance to vindoline (364), the latter being a major alkaloid in *Vinca rosea* Linn and the monomeric template in the oncolytic alkaloid vincaleukoblastine.

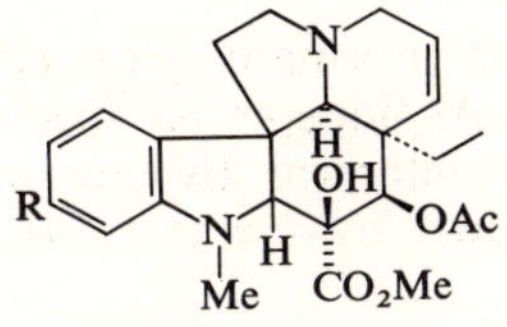

(363) R = H
(364) R = OMe

The sequence starts with (365) which undergoes an interesting boron trifluoride catalysed cyclisation to (366), the latter process involving a different mechanistic pathway than that envisaged in Harley-Mason's previous rearrangements with this reagent[104, 93]. Conversion of (366) to the pentacyclic systems (367) and (368) follows via appropriate condensations and base catalysed alkylation of the resultant conjugated ketone. An interesting hydroxylation of the keto ester (369) is performed by means of an oxygen–hydrogen peroxide treatment under basic conditions. The authors propose the rationale shown in (372) → (373) → (370).

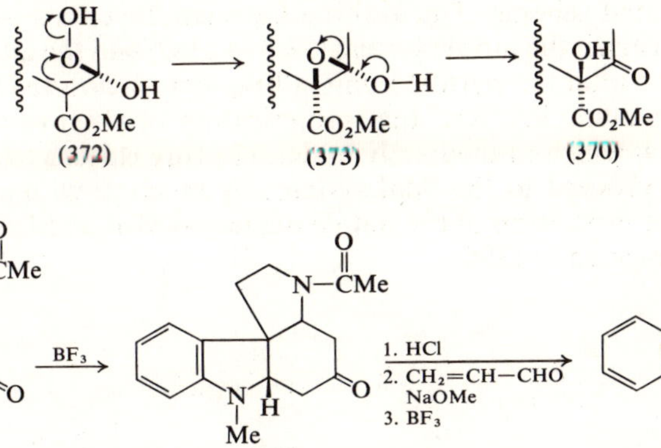

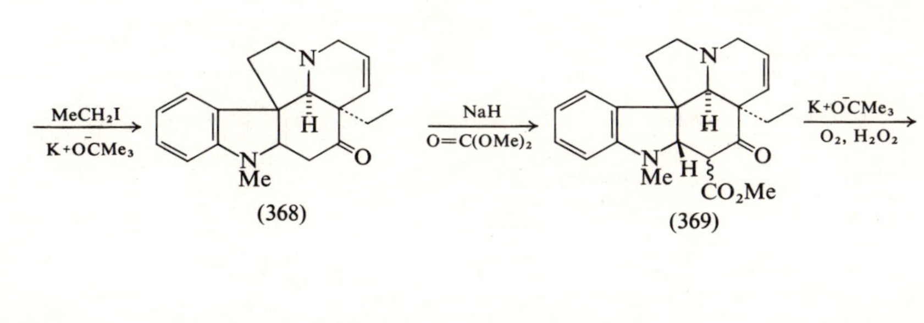

(372)   (373)   (370)

(365)   (366)   (367)

(368)   (369)

(370) $\xrightarrow[-\ 70\ °C]{\text{LiAlH}_4}$ (371) $\xrightarrow{\text{Ac}_2\text{O}}$ (363)

There is a series of alkaloids normally grouped under the Aspidosperma family, since they occur in Aspidosperma species, which do not bear the fundamental structure characterised by the members discussed above. Examples of these are shown by uleine (374), dasycarpidone (375) and apparicine (376).

(374) R = CH$_2$
(375) R = O

(376)

Several syntheses of these skeletal types are now on hand in the recent literature. Joule and co-workers[105, 106, 107] describe syntheses of the racemic forms of dasycarpidone (375), 3-epidasycarpidone (375, epimeric ethyl group), uleine (374) and 3-epiuleine (374, epimeric ethyl group) according to the indicated scheme. The authors have employed a Fischer indole synthesis to construct the indole system (381) → (382) and the latter, upon further elaboration within the pyridine unit, allows conversion to (383). The crucial step in the scheme involved the isomerisation of (383) to the intermediate enamine (384) which immediately undergoes ring closure to (385). The latter is shown to proceed to the final systems by reaction with acetic acid. Ring closure at the β-position of the indole nucleus is visualised as proceeding via intermediates such as (386).

(377) $\xrightarrow{\text{SeO}_2}$ (378) $\xrightarrow[\substack{\text{2. CH}_3\text{OH} \\ \text{3. MeCH}_2\text{CO}_2\text{Me} \\ \text{NaH}}]{\text{1. SOCl}_2}$ (379) $\xrightarrow{\text{H}_2\text{SO}_4,\ \text{HOAc}}$

(380) $\xrightarrow[\text{2. PhN}_2{}^+\text{Cl}^-]{\text{1. pyrrolidine}}$ (381) $\xrightarrow{\text{H}_3\text{PO}_4}$ (382)

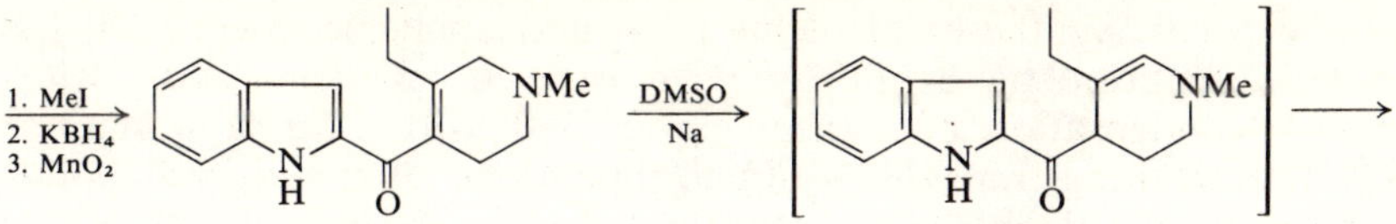

(383)

(384)

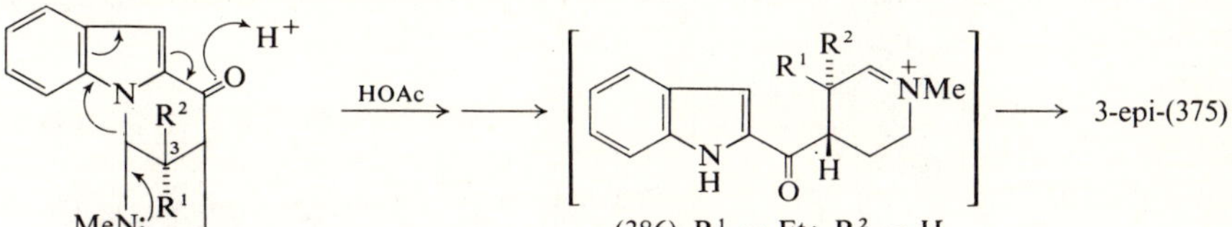

HOAc ⟶ ⟶   (386) R^1 = Et; R^2 = H   ⟶ 3-epi-(375)

(385) R^1 = Et; R^2 = H

epimerises at C-3

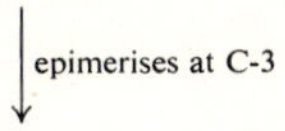

(375)

Extension of the above to the uleine series (374) simply involved converting the above two ketones (375 and its 3-epi-isomer) into the corresponding methylene derivatives. Thus,

$$(375) \xrightarrow{Ph_3P=CH_2} (374)$$

and   and 3-epi-(375) $\xrightarrow[CH_2I_2]{Mg(Hg)}$ 3-epi-(374)

Dolby[108, 109] also succeeded in synthesising these alkaloids at about the same time and has employed an entirely different approach. Dolby's synthetic scheme takes advantage of the high reactivity of indole systems to electrophilic attack at both the α and β positions. Thus an appropriately substituted piperidine derivative (387) is attached at these positions to the indole itself. Vilsmier condensation followed by reduction provides (388) and the latter after saponification and ring closure affords (375) and 3-epi (375).

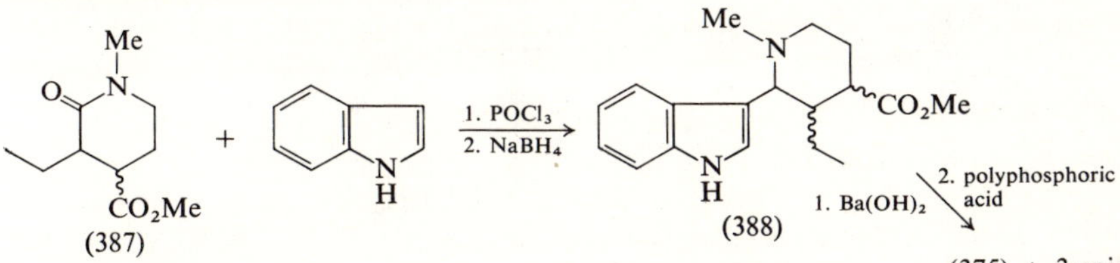

(387)   + $\xrightarrow[2.\ NaBH_4]{1.\ POCl_3}$ (388)   1. Ba(OH)$_2$ \ 2. polyphosphoric acid ⟶ (375) + 3-epi-(375)

The conversion to uleine (374) also followed a different pathway from that established by Joule.

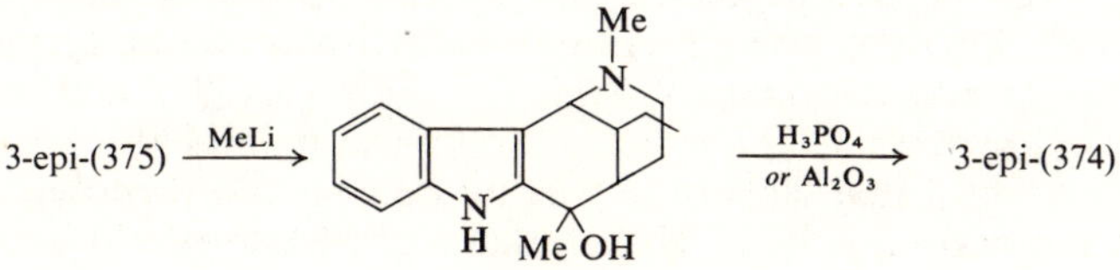

3-epi-(375) $\xrightarrow{MeLi}$ (389) $\xrightarrow[or\ Al_2O_3]{H_3PO_4}$ 3-epi-(374)

Stereospecific syntheses of uleine (374) and epiuleine (3-epi-(374) have also been reported by Buchi[110]. The scheme for epiuleine starts with a Mannich condensation of 3-formylindole (390) with 1-aminohexan-3-one (391). The resulting *trans* disubstituted piperidone (392) is then elaborated via the acetylenic alcohol (393) to the final product, the crucial cyclisation to the required exocyclic olefin (396) being performed by means of boron trifluoride.

In the uleine sequence, the acetate (394) was chosen as the starting material.

Kametani and his group have also succeeded in synthesising several members of the dasycarpidone–uleine series. The approach employed[111–113] involves, in the key step, condensation of indolyl-magnesium bromide (398) with the appropriate nicotinate N-oxide (for example, (399)) and subsequent conversion of the pyridine unit in the latter (for example, (400)) to the reduced piperidine system. Final closure of this intermediate with polyphosphoric acid yields the desired system. The synthesis of (±)-de-ethyldasycarpidone (402) is typical of the approach.

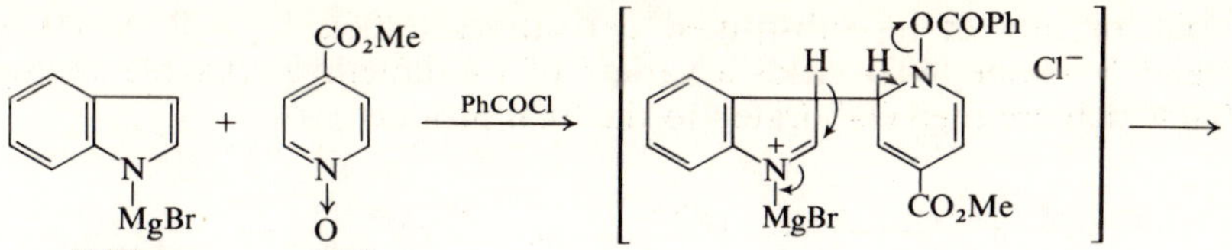

(398) + (399) → (PhCOCl) → [intermediate] →

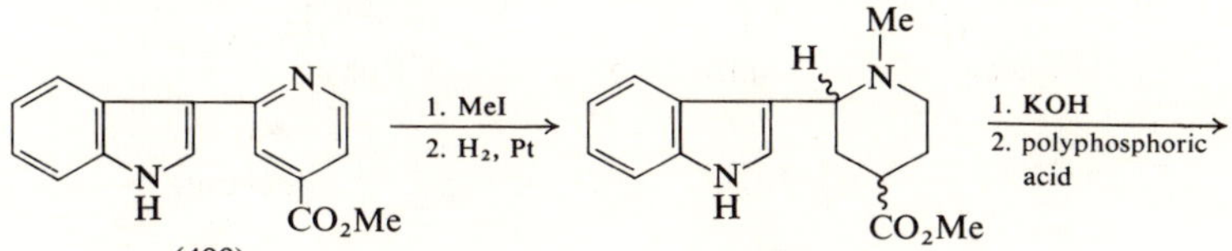

(400) — 1. MeI, 2. H_2, Pt → (401) — 1. KOH, 2. polyphosphoric acid →

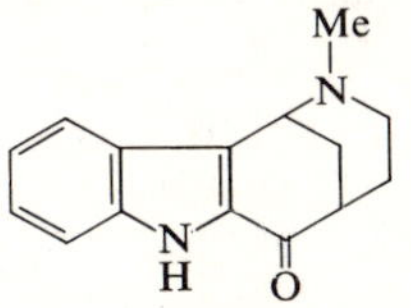

(402)

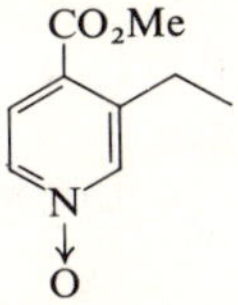

(403)

Extension of the above employing methyl 3-ethylisonicotinate 1-oxide (403), completed the syntheses of dasycarpidone (375), 3-epidasycarpidone, uleine (374), 3-epiuleine, as well as the various other isomers in this series. Finally there is yet another family of alkaloids which in their fundamental structure lack the normal tryptamine unit but which are best considered under the Aspidosperma series since they occur in Aspidosperma, as well as other species. These are best exemplified by the alkaloids olivacine (404) and ellipticine (405).

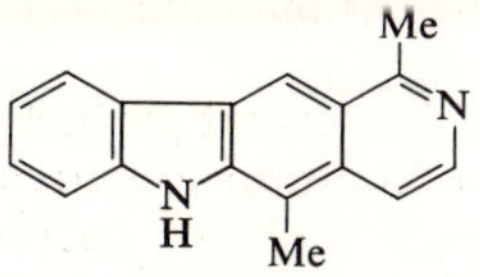

(404)

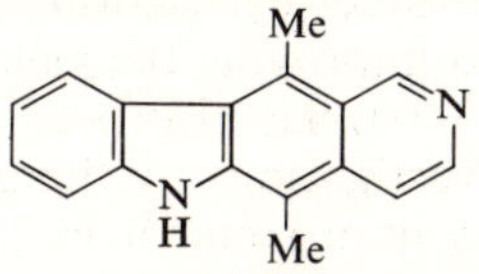

(405)

Both of these alkaloids and their respective derivatives attained interest beyond their chemistry, since in appropriate biological evaluation in various tumour systems they exhibited some anti-tumour activity.

The first synthesis of ellipticine by Woodward[114] was followed by several other successful syntheses[115–118]. The five-step scheme employed by Cranwell and Saxton[116] has been improved in the more recent investigations of the Australian group[118] and this approach has been adapted for large scale preparation of a wide variety of ellipticine analogues for purposes of anti-tumour evaluation. It is therefore appropriate to discuss this pathway. In essence, versatility is provided in the initial phases since condensation of indole, or any of its substituted derivatives ((406), $R^1 \neq R^2 \neq H$), with hexane-2,5-dione (407) yields a variety of 1,4-dimethylcarbazole analogues (408) which are then elaborated to the final product.

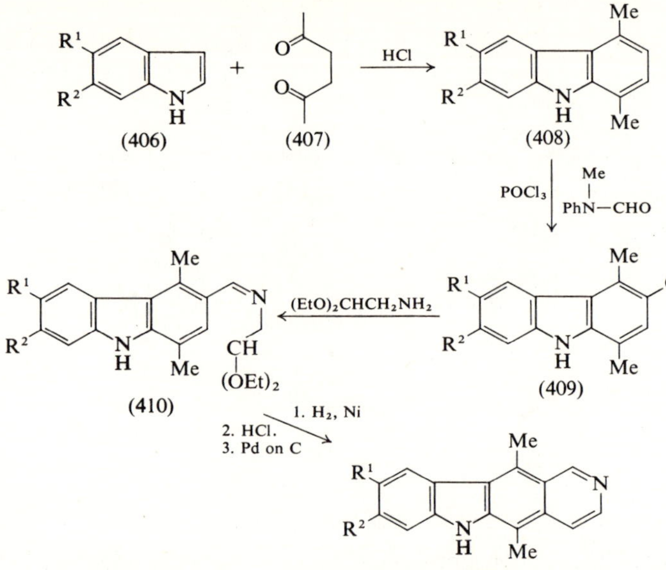

In the initial work Cranwell and Saxton attempted to cyclise the azomethine (410) under conditions normally employed in the Pomeranz–Fritsch isoquinoline synthesis (sulphuric acid, arsenic pentoxide, polyphosphoric acid) but were unable to succeed. It was therefore necessary to reduce (410), then cyclise and finally dehydrogenate the latter to the final product.

The Australian workers[118] by utilising orthophosphoric acid were able to effect the direct cyclisation of (410), thus eliminating two steps in the synthesis and improving the yield.

Olivacine (404) has also been synthesised by several groups[119–121]. Again the latest investigations by the group at the Stanford Research Institute[121] constitute an improvement in the Swiss synthesis[119] for purposes of large scale preparation for anti-tumour testing. A brief outline of the synthesis follows.

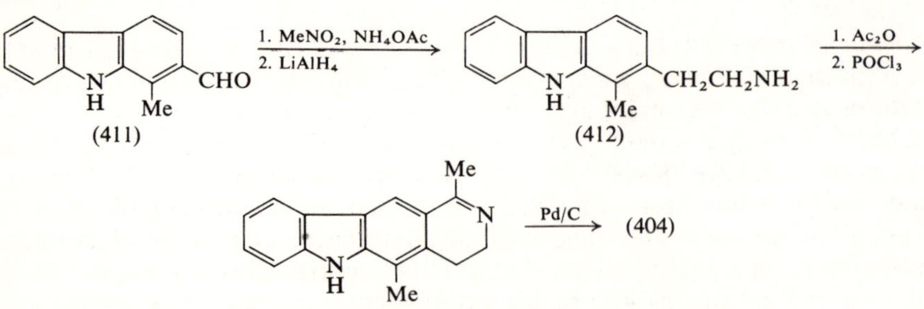

2.10 THE EBURNAMINE–VINCAMINE FAMILY

This family of alkaloids possesses the same fundamental C_{10} non-tryptophan
unit (414) already noted in the Aspidosperma series. It is, however, incorporated into the alkaloid system in a rather different fashion and it is best
to discuss it separately. The best known alkaloids in this family are probably
eburnamine (415) and vincamine (416) and they are cited as examples.

The ring system in these alkaloids was initially prepared in connection
with synthetic investigations in the curare alkaloid series[122]. Most of the
subsequent approaches have involved appropriate carbonyl-amine condensations in which tryptamine is the indole template and a suitably functionalised di- or tri-carbonyl system is employed for the 'non-tryptophan'
unit.

Bartlett and Taylor[123] in their structural studies in this series provided
simple seven-step synthesis of ($\pm$)-eburnamonine (419). The key step involves
condensation of β-ethyl-β-formyladipic acid (417) with tryptamine with the
resultant formation of ($\pm$)-eburnamonine lactam (418). The latter substance
is then converted to (419) in a straightforward manner.

Kuehne[124, 125] has provided a synthesis of ($\pm$)-vincamine (416) by condensing tryptamine with 4-ethyl-4-formylpimelate (420) in the initial phases
of the scheme. The resultant product (422) was then converted to the thio-
lactam (423) with the objective of removing the amide carbonyl but retaining
the ester function (424). This latter product, a mixture of isomers, was
equilibrated through the immonium intermediate (425) to the two amino

esters to which the relative configuration shown in (424a) and (424b) could be assigned. One of these (424a) is then converted in low yield (3%) to ($\pm$)-vincamine by the sequence shown.

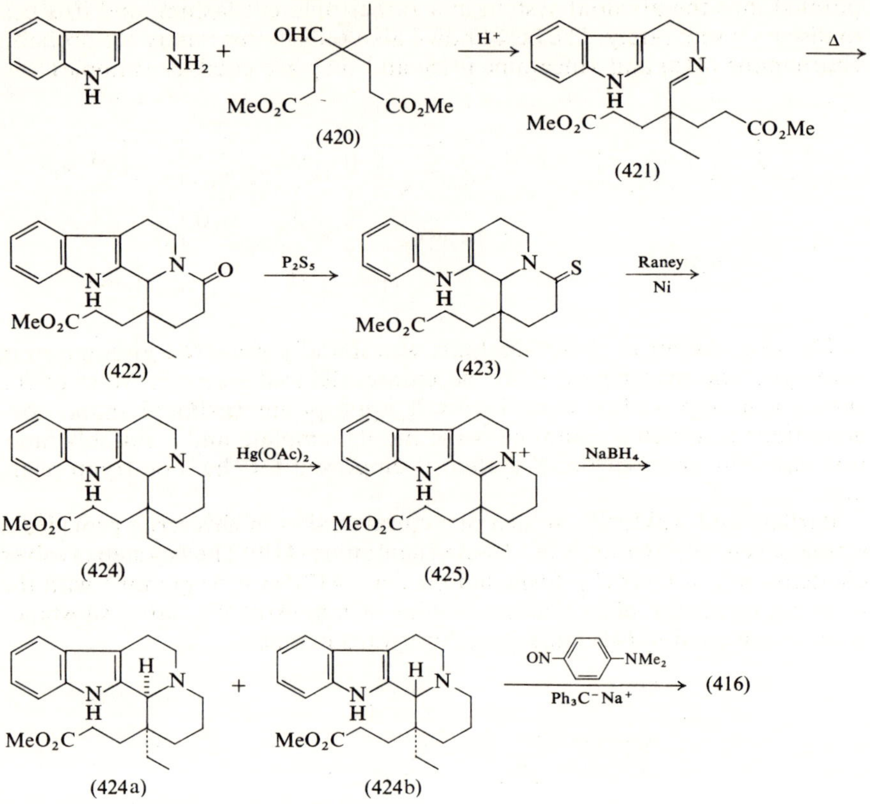

(420)

(421)

(422) $\xrightarrow{P_2S_5}$ (423) $\xrightarrow[\text{Ni}]{\text{Raney}}$

(424) $\xrightarrow{\text{Hg(OAc)}_2}$ (425) $\xrightarrow{\text{NaBH}_4}$

(424a) + (424b) $\xrightarrow[\text{Ph}_3\text{C}^-\text{Na}^+]{\text{ON}-\langle\rangle-\text{NMe}_2}$ (416)

Harley–Mason[126, 127] has employed a similar approach in the condensation of (426) with tryptamine and subsequent elaboration of the lactam (427) to ($\pm$)-eburnamine (415).

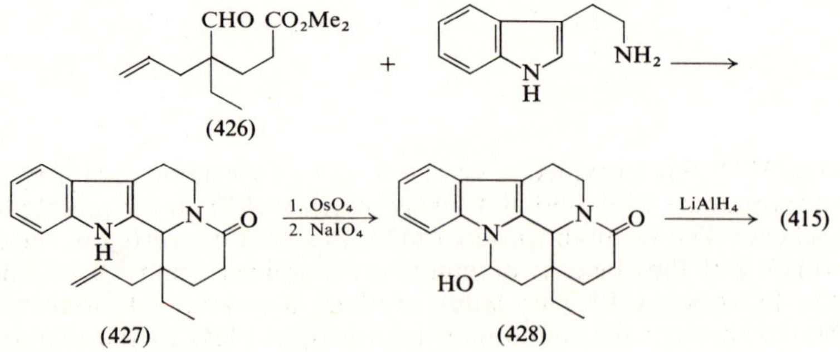

(426)

(427) $\xrightarrow[\text{2. NaIO}_4]{\text{1. OsO}_4}$ (428) $\xrightarrow{\text{LiAlH}_4}$ (415)

Subsequently[128] it was found that the lactam (427) was not homogeneous and in the reaction with osmium tetroxide yielded a mixture of diols which could be separated into the four possible racemates. Either of the *cis* diols (429) on oxidation, cyclisation and reduction provides ($\pm$)-vincaminol (430).

Another synthesis of ($\pm$)-eburnamine has been reported more recently by Saxton[129]. In this scheme the ester diol (431) is the starting material.

An extension of the above provided a new synthesis of ($\pm$)-vincamine[130]. ($\pm$)-Homoeburnamenine (438) available from the above sequence is converted in the manner shown to the alkaloid in racemic form.

(438)

(439)

1. OsO₄
2. DMSO:NEt₃: pyridine:SO₃:H₂O

1. OH⁻
2. CH₂N₂
3. DMSO:NEt₃: pyridine:SO₃:H₂O

416

Wenkert in his general study on the utilisation of piperideine intermediates in indole alkaloid syntheses has published a completely different scheme to ($\pm$)-eburnamonine (419)[131]. The crucial conversion involves the hydrogenation of a pyridinium salt (440) to a Δ^2-piperideine (441) and the latter is then exposed to a Pictet–Spengler cyclisation. The resulting product (442) is reduced and converted to an immonium salt (444) which then undergoes the expected base-catalysed isomerisation to the enamine (445). The latter is then alkylated to (446) which after reduction and treatment with base affords ($\pm$)-eburnamonine (419) and ($\pm$)-epieburnamonine (447), a compound isomeric with the natural alkaloid.

(440)

(441)

(442)

(443)

(444)

(445)

(446)

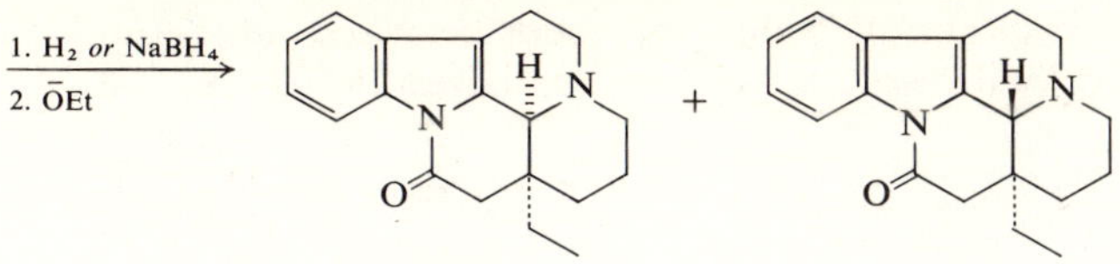

(419)   (447)

2.11 THE IBOGA FAMILY

The Iboga family has been extensively pursued from a structural as well as a synthetic standpoint. Its members also reveal within the structural framework a different variant of the C_{10} non-tryptophan unit (448) from that portrayed in the two previous classes. Since a number of these alkaloids occur along with the Aspidosperma members in the same plant species (for example, *Vinca rosea* Linn) they have also received a great deal of attention in biosynthetic considerations and in medicinal applications. The dimeric Vinca alkaloids of which vinblastine (449) and vincristine (450) may be cited as examples, are important in the clinical treatment of cancer in humans.

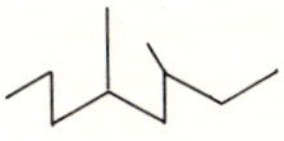

(448)

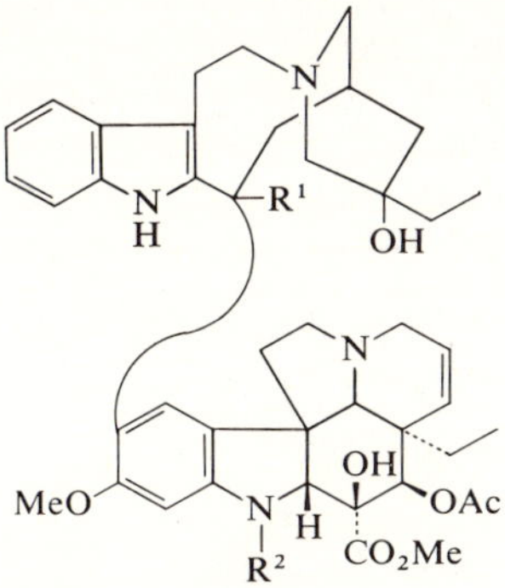

(449) R^1 = CO_2Me; R^2 = Me
(450) R^1 = CO_2Me; R^2 = CHO

These dimers portray an interesting combination of the Iboga and Aspidosperma systems. Thus the dihydroindole unit represents the highly oxygenated aspidosperma alkaloid, vindoline, already mentioned in Section 2.9 while the indole unit reveals one possible natural variant of the Iboga family. This unit, with its nine-membered ring system, has been assigned the name velbanamine (451) and it can be obtained during acidic cleavage of the dimers while the parent system (452) also available from (451) or the dimers has been given the name cleavamine. The cleavamine–velbanamine series has not been isolated from a natural source other than in the dimeric family exemplified above.

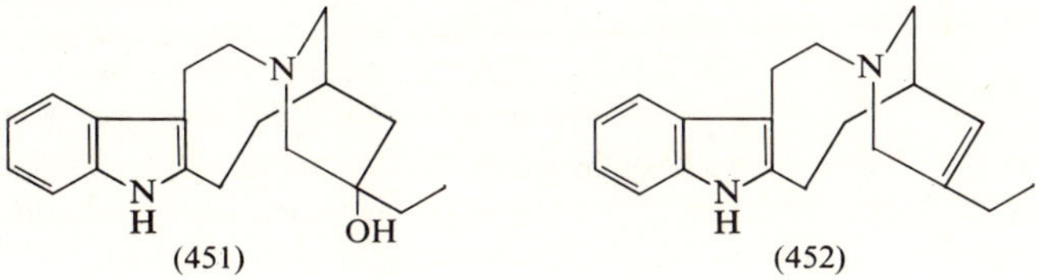

(451)   (452)

The more common Iboga members bear the more rigid cyclic system exemplified by ibogamine (453) and catharanthine (454).

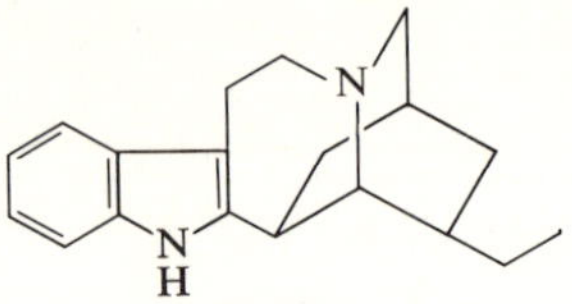

(453) 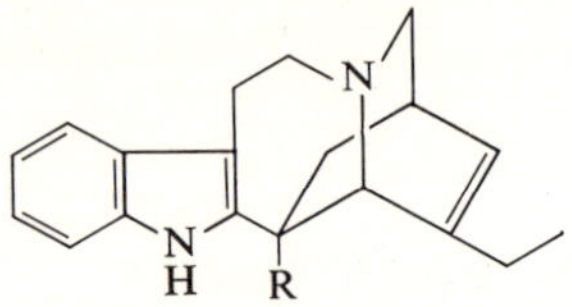   (454) R = CO$_2$Me

The first total synthesis of ibogamine (453) and epi-ibogamine (466) was reported by Buchi[132]. The approach involved first the synthesis of the iso-quinuclidine ring characteristic of the Iboga alkaloids and then subsequent attachment of the indole portion to this ring system. Preparation of the functionalised isoquinuclidine intermediate (456) required for the synthesis proceeded via a Diels–Alder condensation of the dihydropyridine (455) with methyl vinyl ketone. Subsequent elaboration of (456) to the key intermediate (459) followed the pathway indicated. The most interesting conversion of note in proceeding from (456) to the isoquinuclidone (459) is the Hofmann reaction which allows isolation of the intermediate urethan (457). The authors feel that formation of the latter proceeds through an initially formed vinyl isocyanate which then rearranges to the enamine (467) or the conjugated imine (468).

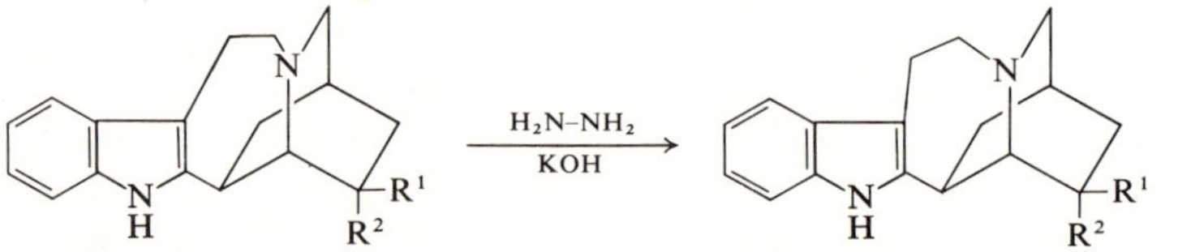

H$_2$N–NH$_2$ / KOH

(464) R^2 = H; R^1 = COMe   (453) R^2 = H; R^1 = CH$_2$Me
(465) R^1 = H; R^2 = COMe   (466) R^2 = CH$_2$Me; R^1 = H

Reaction of (459) with β-indolylacetyl chloride proceeded in the expected fashion to provide (460). However, the subsequent reactions in the scheme involve a series of interesting rearrangements and some comments are therefore appropriate.

In the initial publication in 1965 the authors postulated that in the acid-catalysed cyclisation of (460), the cyclic diol monoacetate thus obtained possesses the structure (469). This would be the expected structure if a normal cyclisation process were to prevail. However, further investigations revealed that the structure of this product must be altered to (461) and it is interesting to speculate on the formation of the latter. It is tempting to propose that formation of (461) proceeds via an aziridine intermediate (470) and as will be seen later in some of Nagata's approaches to the Iboga series, such intermediates can be effectively employed for this purpose. The aziridine system does rationalise the isolation of (461) but as the authors point out, it should be noted that cyclisation of (460) gives *mainly* the monoacetate (461), R = H)

rather than the diacetate ((461), R = Ac) although the reaction is performed
in acetic acid. This product therefore seems to result from internal return
within ion pairs and demands that leaving and entering hydroxyl groups be
located on the same face of the molecule. The authors propose a cation (471)
for this purpose. Under more strenuous conditions the cation does apparently combine with external ions since when (460) is reacted with toluene
p-sulphonic acid in boiling chlorobenzene the product is the acetoxytosylate
((461), R = Tos).

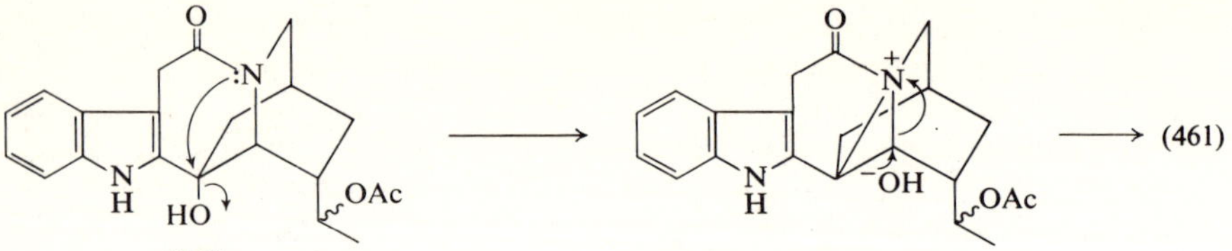

(469)

(470)

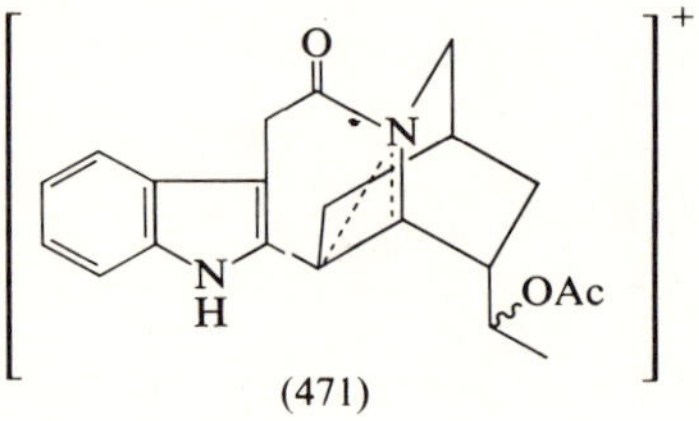

⟶ (461)

(471)

Another reaction of interest in the later stages of the pathway concerns the rearrangement of the azabicyclo[1,2,3]octane system portrayed in (463) to the isoquinuclidine skeleton shown in (464) and (465). This rearrangement prevailed during zinc–acetic acid reduction of the α,β-unsaturated ketone (463) and the mechanistic rationale given by the authors is portrayed in the scheme, (463) → (472) → (464), (465).

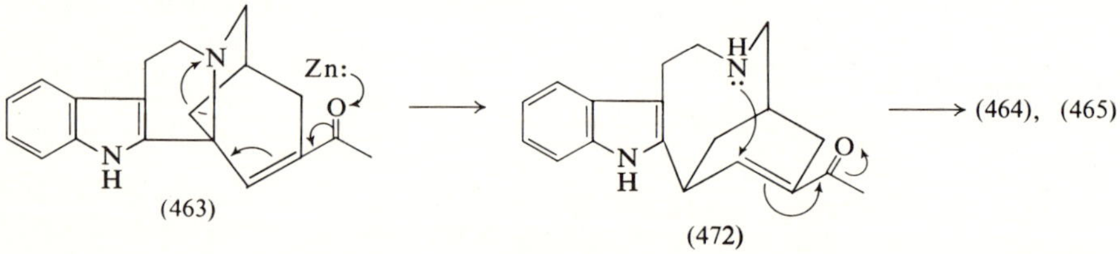

(463)

(472)

⟶ (464), (465)

Extension of the pathway developed by Buchi allowed the synthesis of ibogaine (474) and epi-ibogaine (475). Thus condensation of 3-(5-methoxy-indolyl)acetyl chloride with the amine (459) provided (473) which was then elaborated to the alkaloid in the manner already described.

Investigations in our laboratory[85, 100, 133] provided a general entry into the Iboga family. The approach employed is totally different from that described above and is based on the transannular cyclisation of appropriate nine-membered ring intermediates in the cleavamine series for the final elaboration of the rigid cyclic systems revealed in ibogamine (453), catharanthine (454), etc.

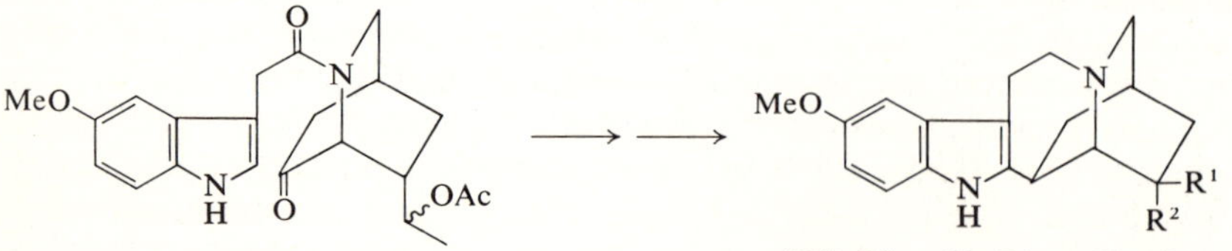

(473)

(474) R^2 = H; R^1 = CH$_2$Me
(475) R^2 = CH$_2$Me; R^1 = H

Initial investigations published in 1964[85] provided the approach involving the cleavamine derivatives. Thus carbomethoxydihydrocleavamine (476) upon reaction with mercuric acetate generates as one of the intermediates an iminium system, which cyclises spontaneously to the Iboga structures, coronaridine (481) and di-hydrocatharanthine (480). The isolation of both of the latter compounds demands that the iminium species initially formed (477) undergoes equilibration via the enamine (478) prior to transannular cyclisation.

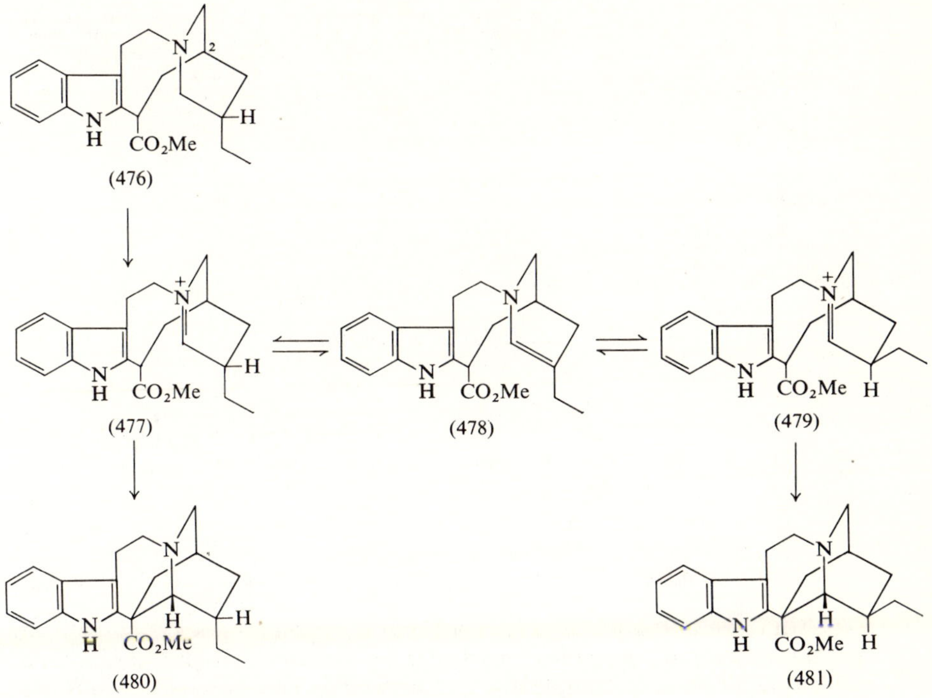

(476)

(477)

(478)

(479)

(480)

(481)

On this basis it was now clear that the synthesis of the requisite cleavamine derivatives would complete the total synthesis of the above alkaloids. The pathway selected[133, 100] paralleled one already discussed under the Aspidosperma alkaloid section. The key step involved the reductive cleavage of the appropriate quaternary system to the nine-membered ring in the final stage (for example (482) → (483)).

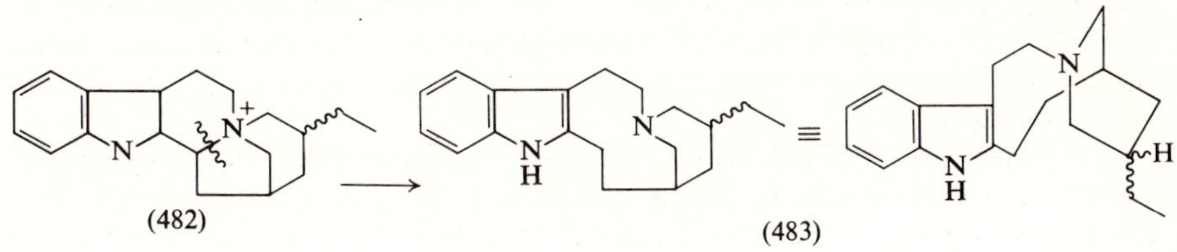

(482)

(483)

The successful sequence which allows the synthesis of (482) and, in turn, the 4α- and 4β-dihydrocleavamines ((489) and (490) respectively) now follows.

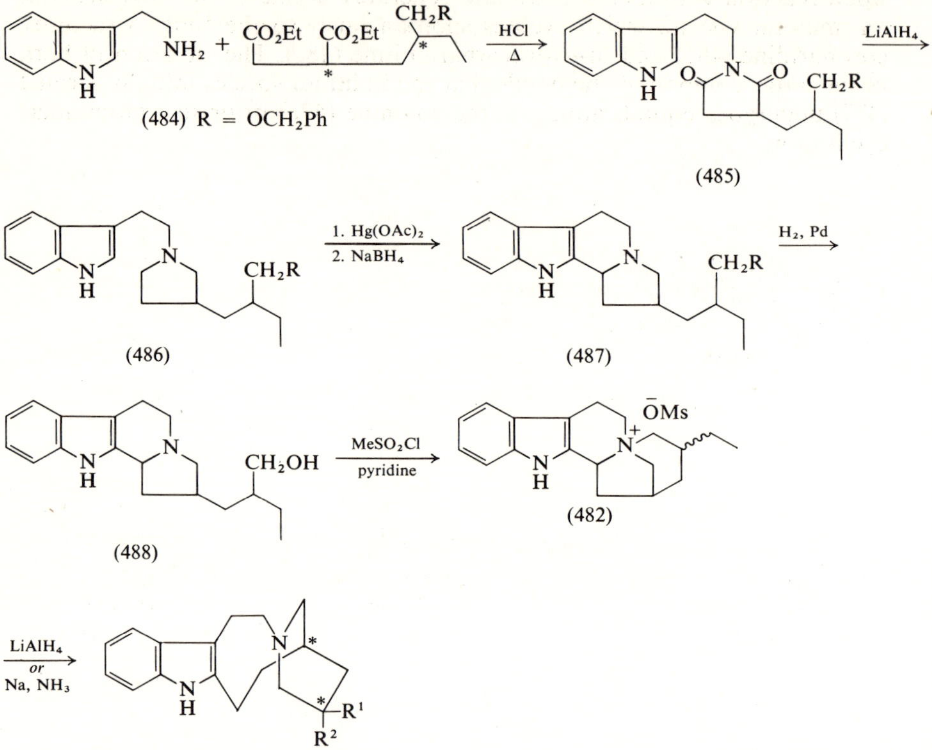

(484) R = OCH$_2$Ph

(485)

(486)

(487)

(488)

(482)

(489) R^2 = H; R^1 = CH$_2$Me
(490) R^2 = CH$_2$Me; R^1 = H

It should be noted that the stereochemical problems associated with the above synthesis are simplified markedly by the fact that the transannular cyclisation process is completely stereospecific. The previous results had already indicated that the asymmetric centre at C-2 (see (476)) in the cleavamine series completely controls the steric course of this cyclisation. Thus it was not necessary to give serious consideration to stereochemistry at the various stages of the pathway. Since the absolute configuration of the 4α- and 4β-dihydrocleavamines ((489) and (490)) is known with certainty from our previous x-ray work, it becomes clear from the above sequence that C-2 and C-4 (asterisked centres in (489), (490)), the two asymmetric centres are derived directly from the two relevant centres in the succinate ester (see asterisked centres in (484)). Therefore the appropriate stereochemistry in (484), as well as in (485) and (486), is completely defined. An extension of this argument allows stereochemical assignments to the remaining intermediates (487), (488) and (482).

The completion of the Iboga alkaloid synthesis made it necessary to introduce a carbomethoxy group into the C-18 position of the dihydrocleavamine molecule. For this purpose the chloroindolenine of 4β-dihydrocleavamine (491) was prepared and an extensive series of investigations were conducted to optimise conditions for introduction of the ester group. It was found in the initial experiments that reaction of (491) with potassium cyanide in a mixture of methanol–water–ether provided a complex mixture from which an 18-cyano-4β-dihydrocleavamine (492) and 18α-methoxy-4β-dihydro-cleavamine were isolated in low yield. However, when (491) was reacted with anhydrous sodium acetate in glacial acetic acid, an unstable 18-acetoxy-4β-dihydrocleavamine (493) was formed, which in time converted to a quaternary species (494) and the latter, without isolation, provided 18β-cyano-4β-dihydrocleavamine (495) upon reaction with potassium cyanide. Hydrolysis of the latter affords 18β-carbomethoxy-4β-dihydrocleavamine (496). The conversion, (491) → (493) → (494) → (495), could be performed without isolation of intermediates and in optimum yields.

Conversion of (496) to dihydrocatharanthine (480) and coronaridine (481) has already been discussed above. Since the conversion of coronoradine to ibogamine has also been reported[134], this sequence also completes a total synthesis of ibogamine (453).

An approach similar to the one described above for the cleavamine series has been investigated by Harley–Mason[135–138]. The triester acetal (497) on reaction with tryptamine afforded the amide ester (498). Conversion of the latter to the quaternary salt (500) and reductive cleavage to (489) and (490) followed the procedures already mentioned in our sequence.

Similarly the conversion of (500) to the 18-cyanocleavamine series (see (492)) and subsequently to the 18-carbomethoxy compounds (see (496)) followed a route described above.

An extension of the above investigation provided a preparation of the 18-hydroxycleavamine derivatives ((503), two isomers at ethyl group).

The most recent publication[138] describes an improvement in the synthesis of (499). In this modification a procedure developed elsewhere[139] and employing sodium cyanide in dimethylsulphoxide to remove an ester group was utilised (497) → (504). The latter on reaction with tryptamine affords (499) in 85% yield.

A fundamentally different approach to the Iboga alkaloid skeleton was developed by Nagata and his group[140, 141]. The crucial reaction involves a new method for isoquinuclidine synthesis in which aziridine intermediates are employed. Thus when the bridged aziridines (505a–c) are treated with an acylating agent (acyl halide or acid anhydride) in an appropriate solvent, the isoquinuclidine system (506) is generated in excellent yield. This reaction initially allowed a synthesis of desethylibogamine[140] and subsequently[141] provided a total synthesis of ibogamine (453) and epi-ibogamine (466). The sequence leading to these latter systems is now described.

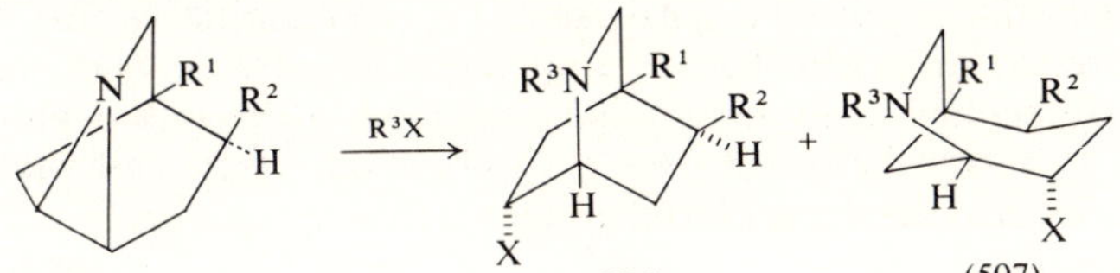

(a) $R^1 = R^2 = H$
(b) $R^1 = Me; R^2 = H$
(c) $R^1, R^2 = -CH_2(CH_2)_2CH_2-$

(505)

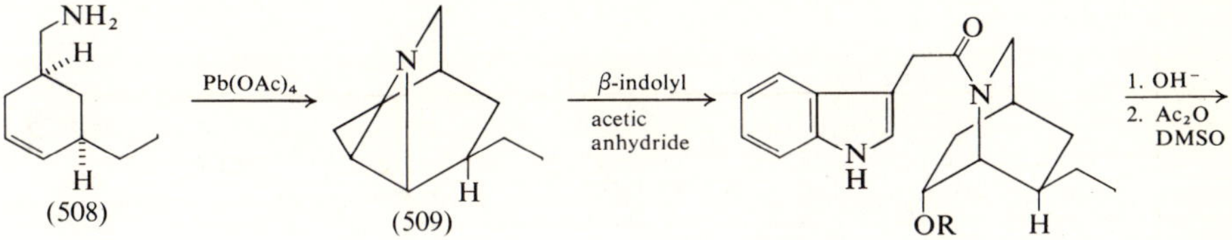

(506)　　(507)

(508)　　(509)

(510) R = β-indolyl-acetyl

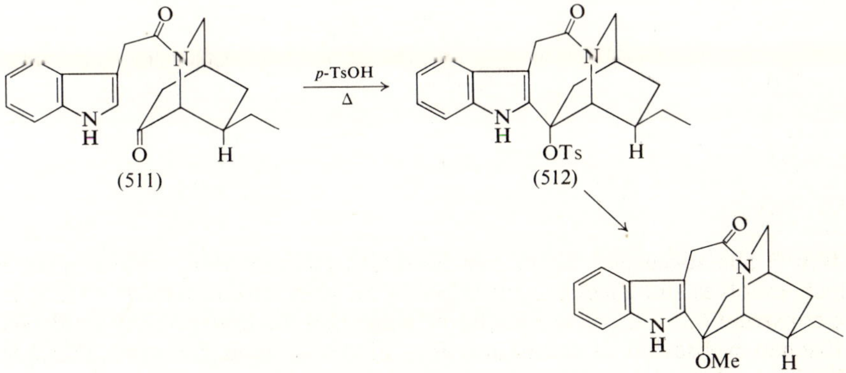

(511)　　(512)

(513)

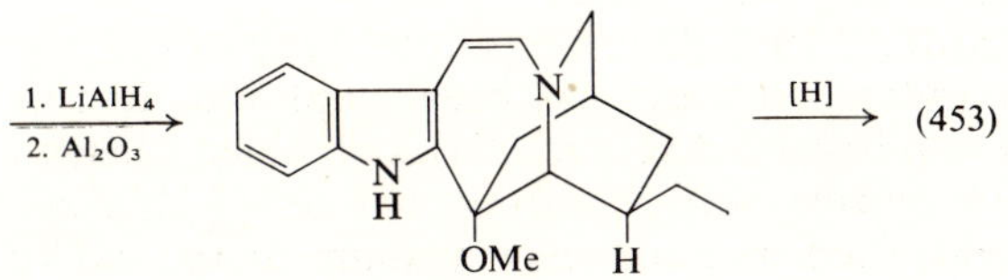

(514)　　(453)

It is of interest to note that in the cyclisation, (511) → (512), no rearrangement of the type observed by Buchi (compare (460) → (461)) was noted.

A parallel series of experiments employing the isomer (515) led to epi-ibogamine (466).

$$\text{(515)} \longrightarrow \longrightarrow \text{(466)}$$

(515)

Sallay[142] has also reported a synthesis of ($\pm$)-ibogamine. In this study the requisite stereochemistry in the isoquinuclidine ring system is developed by employing stereochemically controlled reactions in the initial phases. The final stage of the synthesis involves a Fischer indolisation of the tricyclic ketone (522) to provide the alkaloid system.

(516) $\xrightarrow[\text{pyridine}]{\text{Me} - \text{C}_6\text{H}_4 - \text{SO}_2\text{Cl}}$ (517) $\xrightarrow{\text{PhCO}_3\text{H}}$ (518)

(516)   (517)   (518)

(518) $\xrightarrow[\text{3. Wittig}]{\text{1. LiAlH}_4 \quad \text{2. CrO}_3}$ (519) $\xrightarrow[\text{4. } p\text{-TsCl}]{\text{1. B}_2\text{H}_6 \quad \text{2. LiAlH}_4 \quad \text{3. PhCH}_2\text{OCOCl}}$ (520) $\xrightarrow[\text{HOAc}]{\text{HBr}}$

(519)   (520)

(521) $\xrightarrow[\Delta]{\text{isoamyl alcohol}}$ (522) $\xrightarrow[\text{indole}]{\text{Fischer}}$ (453)

(521)   (522)

Ban[143] has succeeded in the synthesis of ($\pm$)-ibogamine via a sequence which involves the Ziegler cyclisation of an appropriate trinitrile (523) in the construction of the isoquinuclidine ring (524). As in the Sallay synthesis, a Fischer cyclisation of the hydrazones of the isomeric ketones (528) and (529) forms the indole nucleus in the final stages.

Another group of investigators[144] have described a total synthesis of epi-ibogamine in which a new indole synthesis has been employed. In this approach a good yield of the required 1,3-diketone (531) is obtained when an enamine of a ketone (530) is reacted with O-nitrophenyl acetyl chloride. The product (531) is then converted in a reducing medium to the indole system (532). The remaining steps in the sequence are straightforward.

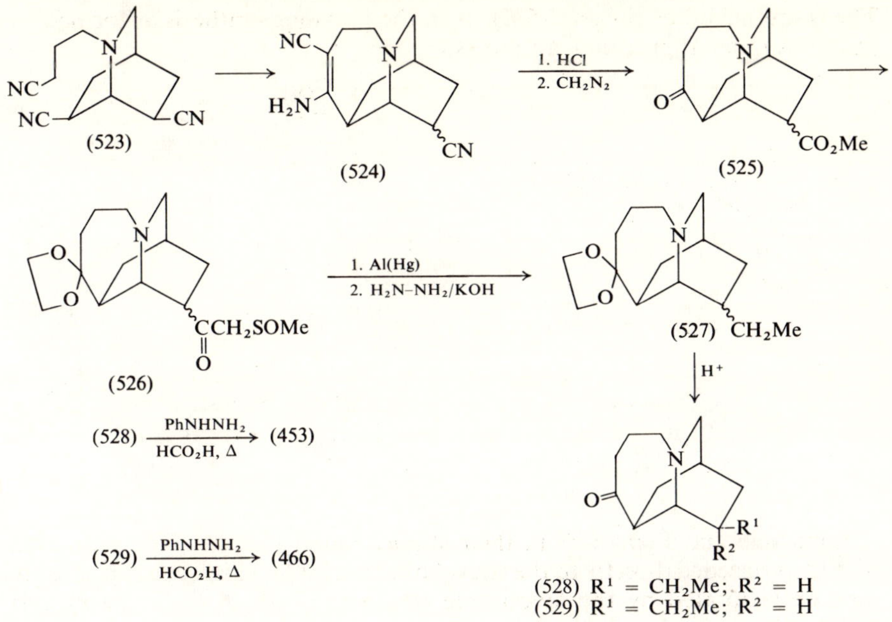

(523)

(524)

1. HCl
2. CH_2N_2

(525) CO_2Me

(526) $\overset{||}{\underset{O}{C}}CH_2SOMe$

1. Al(Hg)
2. H_2N-NH_2/KOH

(527) CH_2Me

H^+

(528) $\xrightarrow[HCO_2H, \Delta]{PhNHNH_2}$ (453)

(529) $\xrightarrow[HCO_2H, \Delta]{PhNHNH_2}$ (466)

(528) $R^1 = CH_2Me$; $R^2 = H$
(529) $R^1 = CH_2Me$; $R^2 = H$

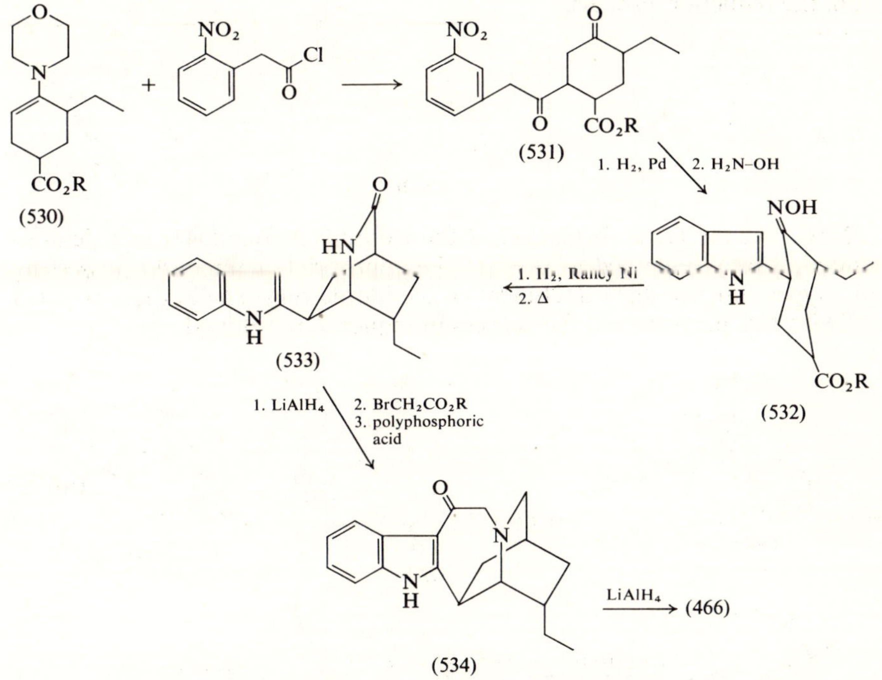

(530)

$+$

(531)

1. H_2, Pd   2. H_2N-OH

(532)

(533)

1. H_2, Raney Ni
2. Δ

1. $LiAlH_4$ 2. $BrCH_2CO_2R$
3. polyphosphoric acid

(534)

$\xrightarrow{LiAlH_4}$ (466)

A synthesis of desethylibogamine (538) in several steps from the epoxide of methyl 3-cyclohexene-1-carboxylate (535) has been described by Huffman[145].

The isoquinuclidine nucleus (537), as in the previous synthesis, is formed by pyrolysis of an appropriate amino ester (536).

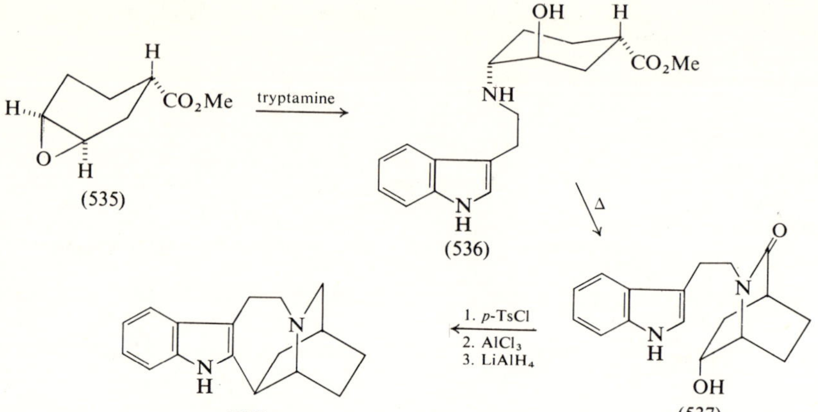

(535) (536) (537) (538)

Augustine and Pierson[146] in their studies found that catalytic reduction of (539) proceeded directly to the tricyclic lactam (541). This facile ring closure may be facilitated by the favourable stereochemistry of the carboxy and amino groups in the *cis*-fused intermediate (540) which would be expected from the reduction process.

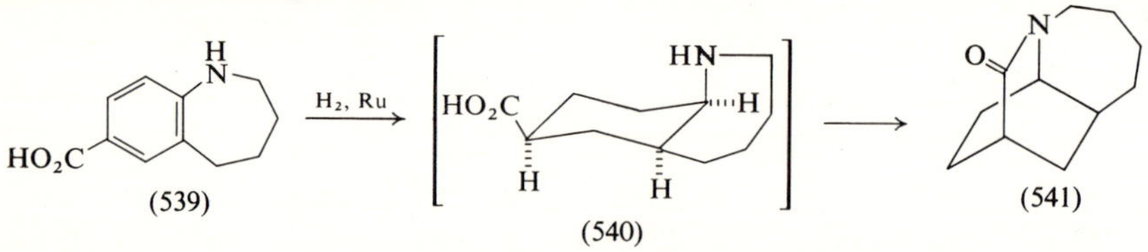

(539) (540) (541)

Now that the facile formation of the isoquinuclidone (541) was demonstrated, it only remained to prepare the appropriate compound possessing an oxygen function for attachment of the indole ring. The authors selected (543) for this purpose and the successful sequence is outlined.

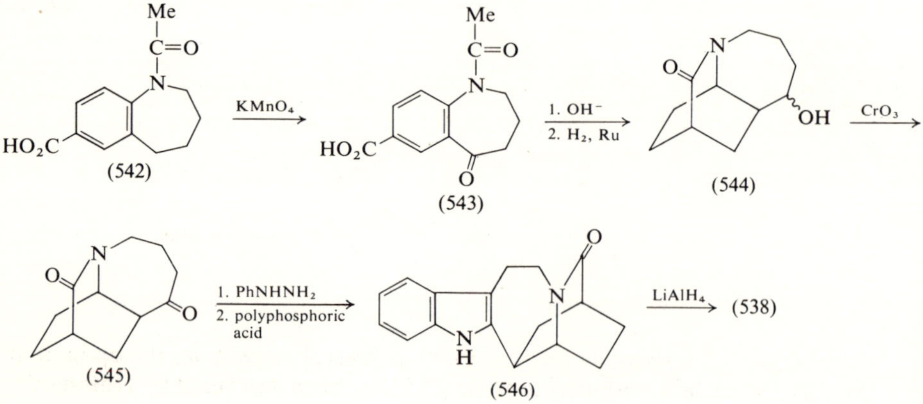

(542) (543) (544) (545) (546)

In some of the most recent studies three groups have reported successful syntheses of velbanamine (451), the monomeric indole unit of the oncolytic Vinca alkaloids, vinblastine (449) and vincristine (450).

Buchi[147, 148] reported the first synthesis of velbanamine employing as starting material the isoquinuclidine (456) previously utilised in the Iboga alkaloid synthesis. Conversion of (456) to the appropriately functionalised isoquinuclidone (550) proceeded according to the sequence of reactions shown. This intermediate was then condensed with sodium indole acetate to provide the amide (551) which is finally elaborated to velbanamine. One of the interesting reactions of note in the final stages is the retroaldolisation process revealed in the conversion, (553) → (554).

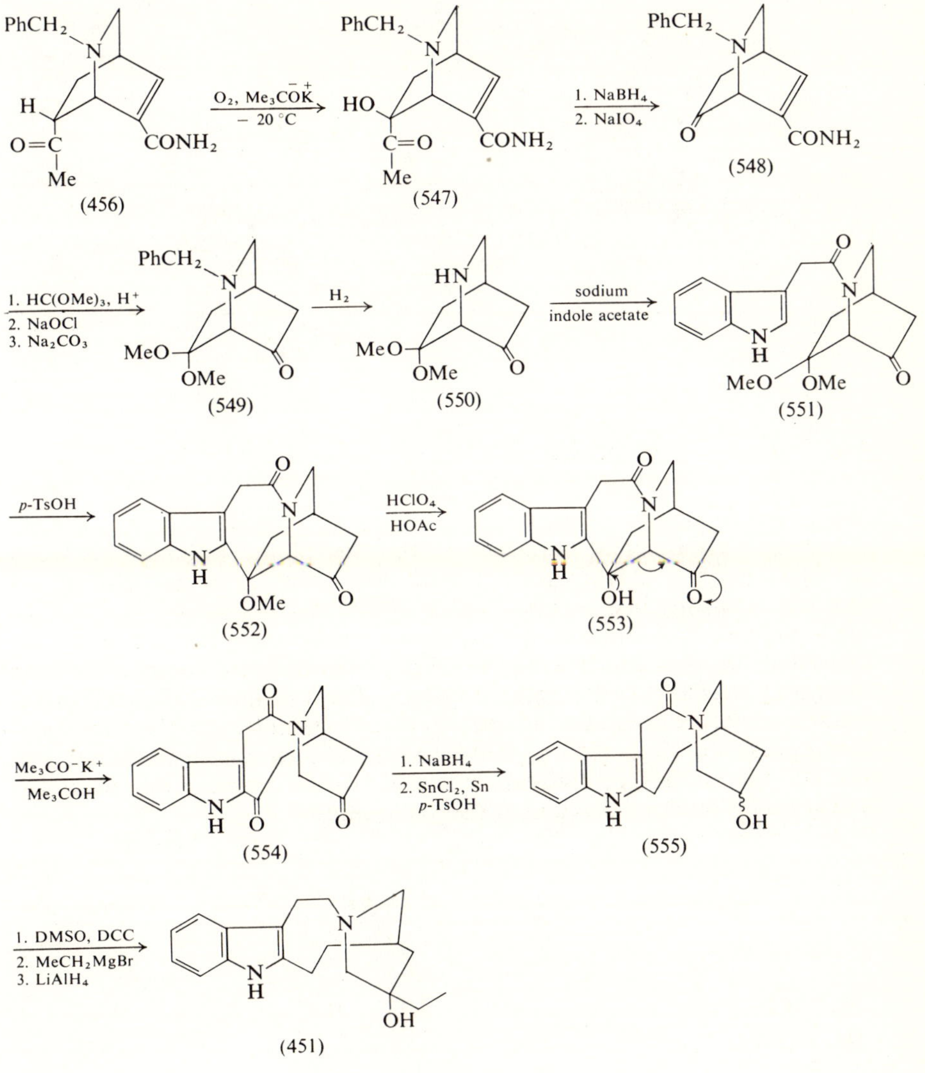

An extension of the above work led to the total synthesis of catharanthine
(454). The starting material for this synthesis was the cyclic ketone (552)
which on exposure to a Grignard reaction with vinylmagnesium bromide
afforded the vinylcarbinol (556). This latter substance could be readily converted to (557) which now bears the ethyl side chain necessary for the catharanthine system. The chloroindolenine method for introducing the ester
function ((558) → (454)) had already been known from previous studies.

A completely different approach to the cleavamine–velbanamine series
was developed in our laboratory[149]. With the view of trying to obtain generality and versatility in a pathway which would lead to a variety of C-3, C-4-
functionalised derivatives in this series, we turned our attention to a fragmentation reaction which would hopefully provide these compounds from
the readily available catharanthine system. It was proposed that the catharanthine molecule possesses the necessary stereochemistry for the fragmentation process shown in (560) → (561) → (562). If such a conversion were to
succeed, the tetracyclic system generated would possess the necessary
activation for functionalisation at the 3,4 positions.

The sequence shown below reveals the successful achievement of this goal.

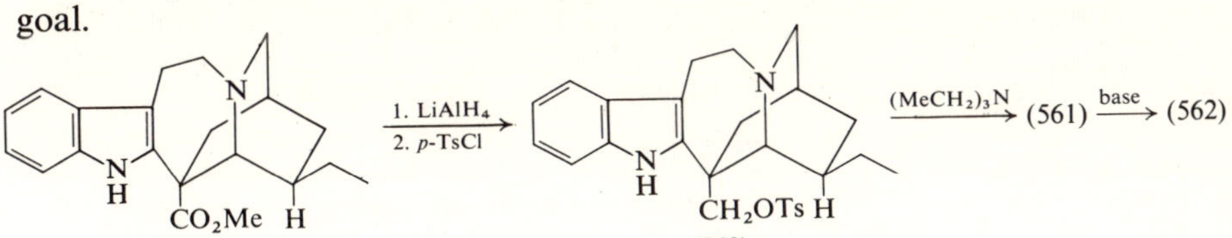

(460) → [1. LiAlH$_4$; 2. p-TsCl] → (563) → [(MeCH$_2$)$_3$N] → (561) → [base] → (562)

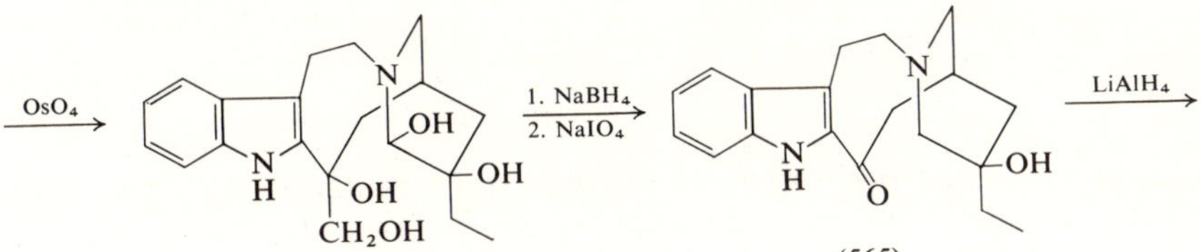

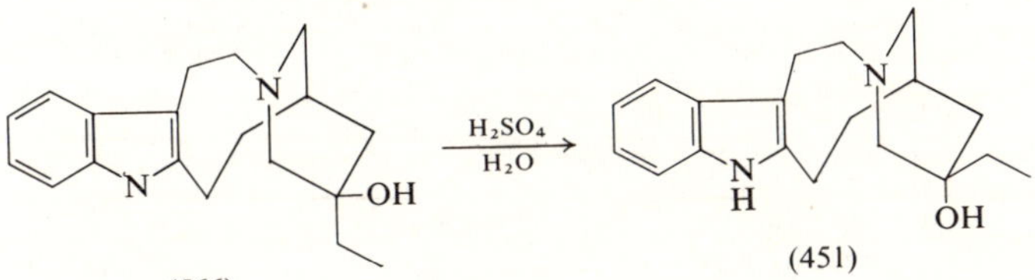

(563) → [OsO$_4$] → (564) → [1. NaBH$_4$; 2. NaIO$_4$] → (565) → [LiAlH$_4$] → (566)

(566) → [H$_2$SO$_4$, H$_2$O] → (451)

The synthetic compound (566) was found to be isomeric with natural vel-banamine and is therefore called isovelbanamine. This latter intermediate not only provided the desired C-4 epimer for further synthetic studies but it represented the crucial compound which allowed the completion of the total synthesis of velbanamine (451), cleavamine (452), and 18β-carbomethoxy-cleavamine (571). It was felt that (566), under acidic conditions, would generate the carbonium ion (567) which would be expected to eliminate a proton to provide cleavamine or hydrate to generate the thermodynamically more favourable velbanamine system. Indeed when isovelbanamine was treated with concentrated sulphuric acid the product was cleavamine (452) but when aqueous sulphuric acid was utilised velbanamine (451) was isolated.

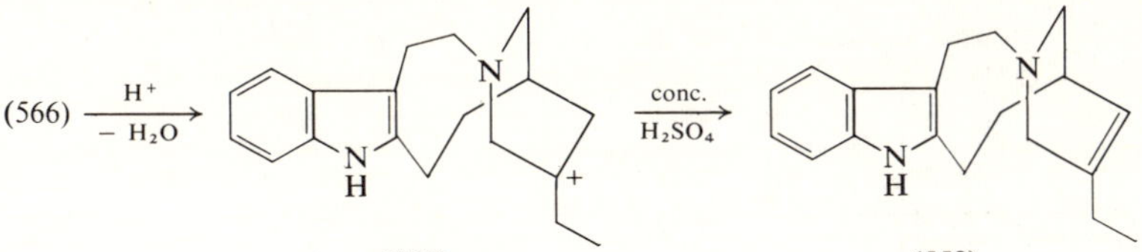

(566) → [H$^+$, $-$ H$_2$O] → (567) → [conc. H$_2$SO$_4$] → (552)

The conversion of cleavamine to the 18β-carbomethoxy derivative (571) was accomplished by the chloroindolenine approach already described earlier.

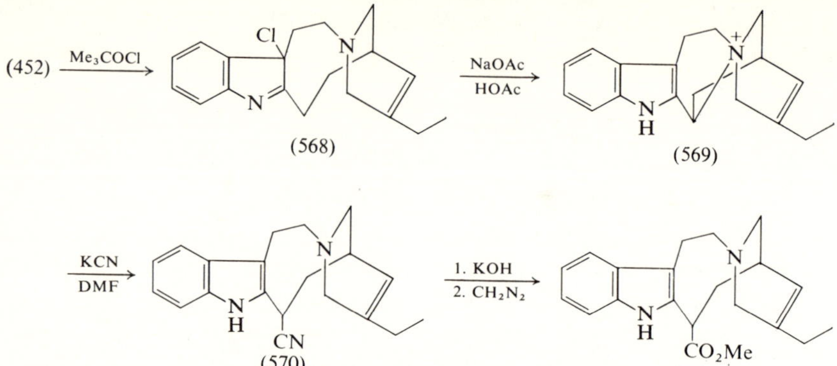

(452) $\xrightarrow{\text{Me}_3\text{COCl}}$ (568) $\xrightarrow[\text{HOAc}]{\text{NaOAc}}$ (569)

$\xrightarrow[\text{DMF}]{\text{KCN}}$ (570) $\xrightarrow[\text{2. CH}_2\text{N}_2]{\text{1. KOH}}$ (571)

The transannular cyclisation of (571) to catharanthine (454) was also performed according to the methods previously developed in our laboratory and discussed earlier in this section. This sequence therefore completed the total synthesis of this Vinca alkaloid.

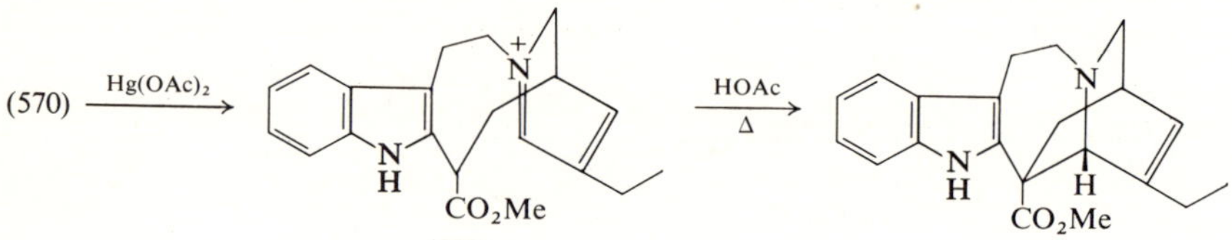

(570) $\xrightarrow{\text{Hg(OAc)}_2}$ (572) $\xrightarrow[\Delta]{\text{HOAc}}$ (454)

Nagata[150] has achieved a synthesis of velbanamine (451) and isovelbanamine (566) utilising an oxidative fragmentation in the key step of the synthetic pathway. The hydroxylactam (573), available from previous studies in the Iboga series on reaction with lead tetra-acetate undergoes a fragmentation to the tetracyclic skeleton shown in (574) and this compound is then converted to the final systems by the sequence of reactions shown.

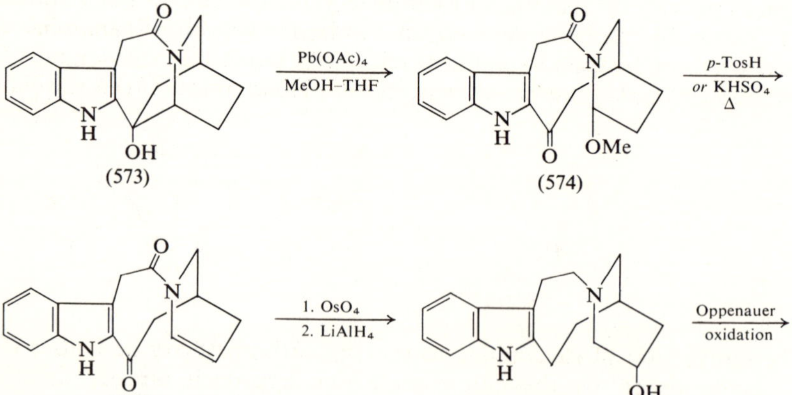

(573) $\xrightarrow[\text{MeOH–THF}]{\text{Pb(OAc)}_4}$ (574) $\xrightarrow[\Delta]{p\text{-TosH} \text{ or } \text{KHSO}_4}$

(575) $\xrightarrow[\text{2. LiAlH}_4]{\text{1. OsO}_4}$ (576) $\xrightarrow{\text{Oppenauer oxidation}}$

(577) $\xrightarrow{\text{MeCH}_2\text{MgBr}}$ (451) + (566)

References

1. Gorman, A. A., Hesse, M., Schmid, H. and Hopff, W. H. (1970). *The Alkaloids, Specialist Periodical Reports,* 201, (London: The Chemical Society)
2. Battersby, A. R. (1967). *Pure and Applied Chem.,* **14,** 117
3. Battersby, A. R. (1970). *The Alkaloids, Specialist Periodical Reports,* 31, (London: The Chemical Society)
4. Scott, A. I. (1970). *Accounts Chem. Res.,* **3,** 151
5. Scott, A. I., Reichardt, P. B., Slaytor, M. B. and Sweeny, J. G. (1971). *Bioorgánic Chem.,* **1,** 157
6. Kutney, J. P., Beck, J. F., Ehret, C., Poulton, G., Sood, R. S. and Westcott, N. D. (1971). *Bioorganic Chem.,* **1,** 194
7. van Tamelen, E. E. and Hester, J. B. (1959). *J. Amer. Chem. Soc.,* **81,** 3805
8. van Tamelen, E. E. and Hester, J. B. (1969). *J. Amer. Chem. Soc.,* **91,** 7342
9. Evstigneeva, R. P., Livshits, R. S., Bainova, M. S., Zakharkin, L. T. and Preobrazhenskii, N. A. (1952). *J. Gen. Chem.,* **22,** 1467
10. van Tamelen, E. E., Aldrich, P. E. and Katz, T. J. (1957). *J. Amer. Chem. Soc.,* **79,** 6426
11. van Tamelen, E. E. and Wright, I. G. (1964). *Tetrahedron Lett.,* 295
12. van Tamelen, E. E. and Wright, I. G. (1969). *J. Amer. Chem. Soc.,* **91,** 7349
13. Powell, J. W. and Whiting, M. C. (1961). *Tetrahedron,* **12,** 168
14. Autrey, R. L. and Scullard, P. W. (1968). *J. Amer. Chem. Soc.,* **90,** 4917
15. Swan, G. A. (1950). *J. Chem. Soc.,* 1534
16. Weisbach, J. A., Kirkpatrick, J. L., Williams, K. R., Anderson, E. L., Yim, N. C. and Douglas, B. (1965). *Tetrahedron Lett.,* 3457
17. Szantay, Cs. and Barczai-Beke, M. (1968). *Tetrahedron Lett.,* 1405
18. Szantay, Cs. and Barczai-Beke, M. (1969). *Chem. Ber.,* **102,** 3963
19. Wenkert, E., Dave, K. G., Lewis, R. G. and Sprague, P. W. (1967). *J. Amer. Chem. Soc.,* **89,** 6741
20. Ziegler, F. E. and Sweeny, J. G. (1969). *Tetrahedron Lett.,* 1097
21. Wick, A. E., Felix, D., Steen, K. and Eschenmoser, A. (1964). *Helv. Chim. Acta,* **47,** 2425
22. Sawa, Y. K. and Matsumura, H. (1968). *Chem. Commun.,* 679; (1969). *Tetrahedron,* **25,** 5319, 5329; (1970). *Tetrahedron,* **26,** 2931
23. van Tamelen, E. E. and Placeway, C. (1961). *J. Amer. Chem. Soc.,* **83,** 2594
24. van Tamelen, E. E., Placeway, C., Schiemenz, G. P. and Wright, I. G. (1969). *J. Amer. Chem. Soc.,* **91,** 7359
25. Korte, F. and Machleidt, H. (1955). *Chem. Ber.,* **88,** 136
26. Taylor, W. I. (1965). *The Alkaloids,* Manske, R. H. F. Ed., Vol. VIII, 789 (New York: Academic Press)
27. Taylor, W. I. (1968). *The Alkaloids,* Manske, R. H. F. Ed., Vol. XI, 41 (New York: Academic Press)

28. Masamune, S., Ang., S. K., Egli, C., Nakatsuka, N., Sarkar, S. K. and Yasunari, Y. (1967). *J. Amer. Chem. Soc.*, **89**, 2506
29. Anet, F. A. L., Chakravarti, D., Robinson, R. and Schlittler, E. (1954). *J. Chem. Soc.*, 1242
30. van Tamelen, E. E. and Oliver, L. K. (1970). *J. Amer. Chem. Soc.*, **92**, 2136
31. van Tamelen, E. E., Shamma, M., Burgstahler, A. W., Wolinsky, J., Tamm, R. and Aldrich, P. E. (1958). *J. Amer. Chem. Soc.*, **80**, 5006; (1969). *J. Amer. Chem. Soc.*, **91**, 7315
32. Bartlett, M. F., Lambert, B. F., Werblood, H. H. and Taylor, W. I. (1963). *J. Amer. Chem. Soc.*, **85**, 475
33. Hobson, J. D. and McCluskey, J. G. (1967). *J. Chem. Soc.*, 2015
34. Mashimo, K. and Sato, Y. (1969). *Tetrahedron Lett.*, 901; (1970). *Tetrahedron*, **26**, 803
35. Yoneda, N. (1965). *Chem. Pharm. Bull. Japan*, **13**, 1231
36. Mashimo, K. and Sato, Y. (1969). *Tetrahedron Lett.*, 905
37. Weisbach, J. A. and Douglas, B. (1965). *Chem. Ind. (London)*, 623; (1966). 233
38. Shiori, T. and Yamada, S. (1968). *Tetrahedron*, **24**, 4159
39. Woodward, R. B., Bader, F. E., Bickel, H., Frey, A. J. and Kierstead, R. W. (1956). *J. Amer. Chem. Soc.*, **78**, 2023, 2657
40. Woodward, R. B., Bader, F. E., Bickel, H., Frey, A. J. and Kierstead, R. W. (1958). *Tetrahedron*, **2**, 1
41. Ernest, I. and Kakac, B. (1965). *Chem. Ind. (London)*, 513
42. Morrinson, G. C., Cetenko, W. A. and Shavel, Jr., J. (1966). *J. Org. Chem.*, **31**, 2695
43. Potts, K. T. and Robinson, R. (1955). *J. Chem. Soc.*, 2675. See also Julian, P. L. and Magnani, A. (1949). *J. Amer. Chem. Soc.*, **71**, 3207; Belleau, B. (1955). *Chem. Ind. (London)*, 229
44. Wenkert, E., Massy-Westropp, R. A. and Lewis, R. G. (1962). *J. Amer. Chem. Soc.*, **84**, 3732
45. Wenkert, E. and Wickberg, B. (1962). *J. Amer. Chem. Soc.*, **84**, 4914
46. Morrinson, G. C., Cetenko, W. and Shavel, Jr., J. (1967). *J. Org. Chem.*, **32**, 4089
47. Beisler, J. A. (1970). *Tetrahedron*, **26**, 1961
48. Fry, E. M. (1963). *J. Org. Chem.*, **28**, 1869; (1964). **29**, 1647
49. Wenkert, E., Dave, K. G., Gnewuch, C. T. and Sprague, P. W. (1968). *J. Amer. Chem. Soc.*, **90**, 5251
50. Beisler, J. A. (1970). *Chem. Ber.*, **103**, 3360
51. Toke, L., Honty, K. and Szanty, Cs. (1969). *Chem. Ber.*, **102**, 3248
52. Szanty, Cs., Toke, L., Honty, K. and Kalaus, Gy. (1967). *J. Org. Chem.*, **32**, 423
53. Szanty, Cs., Honty, K., Toke, L., Buzas, A. and Jacquet, J. P. (1971). *Tetrahedron Lett.*, 4871
54. Meise, W. and Zymalkowski, F. (1969). *Ang. Chem. Int. Ed.*, **8**, 445
55. Ziegler, F. E. and Sweeny, J. G. (1969). *J. Org. Chem.*, **34**, 3545
56. Winterfeldt, E. and Riesner. (1970). *Synthesis*, 261
57. Brutcher Jr., F. V., Vanderwerff, W. D. and Dreikorn, B. (1972). *J. Org. Chem.*, **37**, 297
58. Finch, N. and Taylor, W. I. (1962). *J. Amer. Chem. Soc.*, **84**, 1318, 3871
59. Shavel, J. and Zinnes, H. (1962). *J. Amer. Chem. Soc.*, **84**, 1320
60. Ban, Y. and Oishi, T. (1963). *Chem. Pharm. Bull. Japan*, **11**, 451
61. Hendrickson, J. B. and Silva, R. A. (1962). *J. Amer. Chem. Soc.*, **84**, 643
62. Castedo, L., Harley-Mason, J. and Kaplan, M. (1969). *Chem. Commun.*, 1444
63. van Tamelen, E. E., Yardley, J. P. and Miyano, M. (1963). *Tetrahedron Lett.*, 1011
64. van Tamelen, E. E., Yardley, J. P., Miyano, M. and Hinshaw, W. B. (1969). *J. Amer. Chem. Soc.*, **91**, 7333
65. Woodward, R. B., Cava, M. P., Ollis, W. D., Hunger, A., Daeniker, H. U. and Schenker, K. (1954). *J. Amer. Chem. Soc.*, **76**, 4749
66. Woodward, R. B., Cava, M. P., Ollis, W. D., Hunger, A., Daeniker, H. U. and Schenker, K. (1963). *Tetrahedron*, **19**, 247
67. Dolby, L. J. and Sakai, S. (1964). *J. Amer. Chem. Soc.*, **86**, 1890
68. Freter, K., Hubner, H. H., Merz, H., Schroeder, H. D. and Zeile, K. (1965). *Annalen*, **684**, 159
69. Freter, K. and Zeile, K. (1967). *Chem. Commun.*, 416
70. Dadson, B. A., Harley-Mason, J. and Foster, G. H. (1968). *Chem. Commun.*, 1233
71. Dadson, B. A. and Harley-Mason, J. (1969). *Chem. Commun.*, 665
72. Schumann, D. and Schmid, H. (1963). *Helv. Chim. Acta*, **46**, 1966
73. Dadson, B. A. and Harley-Mason, J. (1969). *Chem. Commun.*, 665

74. Harley-Mason, J. and Taylor, C. G. (1970). *Chem. Commun.*, 812
75. Crawley, G. C. and Harley-Mason, J. (1971). *Chem. Commun.*, 685
76. Hymon, J. R. and Schmid, H. (1966). *Helv. Chim. Acta*, **49**, 2067
77. Wenkert, E. and Sklar, R. (1966). *J. Org. Chem.*, 2689
78. Stork, G. and Dolfini, J. E. (1963). *J. Amer. Chem. Soc.*, **85**, 2872
79. Smith, G. F. and Wrobel, J. T. (1960). *J. Chem. Soc.*, 792
80. Bieman, K. and Spiteller, G. (1961). *Tetrahedron Lett.*, 299
81. Ban, Y., Sato, Y., Inove, I., Nagai, (nee Seo), M., Oishi, T., Terashima, M., Yonemitsu, O. and Kanaoka, Y. (1965). *Tetrahedron Lett.*, 2261
82. Ban, Y. and Iijima, I. (1969). *Tetrahedron Lett.*, 2523
83. Kuehne, M. E. and Bayha, C. (1966). *Tetrahedron Lett.*, 1311
84. Kutney, J. P. and Piers, E. (1964). *J. Amer. Chem. Soc.*, **86**, 953
85. Kutney, J. P., Brown, R. T. and Piers, E. (1969). *J. Amer. Chem. Soc.*, **86**, 2286, 2287
86. Camerman, A., Camerman, N., Kutney, J. P., Piers, E. and Trotter, J. (1965). *Tetrahedron Lett.*, 637
87. Kutney, J. P., Brown, R. T. and Piers, E. (1966). *Can. J. Chem.*, **44**, 637
88. Kutney, J. P., Piers, E. and Brown, R. T. (1970). *J. Amer. Chem. Soc.*, **92**, 1700
89. Kutney, J. P., Abdurahman, N., Le Quesne, P., Piers, E. and Vlattas, I. (1966). *J. Amer. Chem. Soc.*, **88**, 3656
90. Kutney, J. P., Abdurahman, N., Gletsos, C., Le Quesne, P., Piers, E. and Vlattas, I. (1970). *J. Amer. Chem. Soc.*, **92**, 1727
91. Kutney, J. P., Chan, K. K., Failli, A., Fromson, J. H., Gletsos, C. and Nelson, V. R. (1968). *J. Amer. Chem. Soc.*, **90**, 3891
92. Mokry, J. and Kompis, I. (1964). *Lloydia*, **27**, 428
93. Barton, J. E. D. and Harley-Mason, J. (1965). *Chem. Commun.*, 298
94. Harley-Mason, J. and Kaplan, M. (1967). *Chem. Commun.*, 915
95. Ziegler, F. E., Kloek, J. A. and Zoretic, P. A. (1969). *J. Amer. Chem. Soc.*, **91**, 2342
96. Ziegler, F. E. and Spitzner, E. B. (1970). *J. Amer. Chem. Soc.*, **92**, 3492
97. Ziegler, F. E. and Bennett, G. B. (1970). *Tetrahedron Lett.*, 2545
98. Husson, H. P., Thal, C., Potier, P. and Wenkert, E. (1970). *Chem. Commun.*, 480
99. Ziegler, F. E. and Benett, G. B. (1971). *J. Amer. Chem. Soc.*, **93**, 5930
100. Kutney, J. P., Cretney, W. J., Le Quesne, P., McKauge, B. and Piers, E. (1970). *J. Amer. Chem. Soc.*, **92**, 1712
101. Stevens, R. V., Fitzpatrick, J. M., Kaplan, M. and Zimmerman, R. L. (1971). *Chem. Commun.*, 857
102. Inoue, I. and Ban, Y. (1970). *J. Chem. Soc. C*, 602
103. Buchi, G., Matsumoto, K. E. and Nishimura, H. (1971). *J. Amer. Chem. Soc.*, **93**, 3299
104. Harley-Mason, J. and Waterfield, W. R. (1963). *Tetrahedron*, **19**, 65
105. Jackson, A., Gaskell, A. J., Wilson, N. D. V. and Joule, J. A. (1968). *Chem. Commun.*, 364
106. Wilson, N. D. V., Jackson, A., Gaskell, A. J. and Joule, J. A. (1968). *Chem. Commun.*, 584
107. Jackson, A., Wilson, N. D. V., Gaskell, A. J. and Joule, J. A. (1969). *J. Chem. Soc. C*, 2738
108. Dolby, L. J. and Biere, H. (1968). *J. Amer. Chem. Soc.*, **90**, 2699
109. Dolby, L. J. and Biere, H. (1970). *J. Org. Chem.*, **35**, 3843
110. Buchi, G., Gould, S. J. and Naf, F. (1971). *J. Amer. Chem. Soc.*, **93**, 2492
111. Kametani, T. and Suzuki, T. (1971). *J. Chem. Soc. C*, 1053
112. Kametani, T. and Suzuki, T. (1971). *J. Org. Chem.*, **36**, 1291
113. Kametani, T. and Suzuki, T. (1971). *Chem. Pharm. Bull. Japan*, **19**, 1424
114. Woodward, R. B., Iacobucci, G. A. and Hochstein, F. A. (1959). *J. Amer. Chem. Soc.*, **81**, 4434
115. Buchi, G., Mayo, D. W. and Hochstein, F. A. (1961). *Tetrahedron*, **15**, 167
116. Cranwell, P. A. and Saxton, J. E. (1962). *J. Chem. Soc.*, 3482
117. Govindachari, T. R., Rajappa, S. and Sudarsan, V. (1963). *Indian J. Chem.*, **1**, 247
118. Dalton, L. K., Demarac, S., Elmes, B. C., Loder, J. W., Swan, J. M. and Teitei, T. (1967). *Aust. J. Chem.*, **20**, 2715
119. Schmutz, J. and Wittwer, H. (1960). *Helv. Chim. Acta*, **43**, 793
120. Wenkert, E. and Dave, K. G. (1962). *J. Amer. Chem. Soc.*, **84**, 94
121. Mosher, C. W., Crews, O. P., Acton, E. M. and Goodman, L. (1966). *J. Med. Chem.*, **9**, 237
122. Wieland, T. and Neeb, E. (1956). *Annalen*, **600**, 161

123. Bartlett, M. F. and Taylor, W. I. (1960). *J. Amer. Chem. Soc.,* **82,** 5491
124. Kuehne, M. E. (1964). *J. Amer. Chem. Soc.,* **86,** 2946
125. Kuehne, M. E. (1964). *Lloydia,* **27,** 435
126. Barton, J. E. D. and Harley-Mason, J. (1965). *Chem. Commun.,* 298
127. Barton, J. E. D., Harley-Mason, J. and Yates, K. C. (1965). *Tetrahedron Letters,* 3669
128. Castedo, L., Harley-Mason, J. and Leeny, T. J. (1968). *Chem. Commun.,* 1186
129. Gibson, K. H. and Saxton, J. E. (1969). *Chem. Commun.,* 799
130. Gibson, K. H. and Saxton, J. E. (1969). *Chem. Commun.,* 1490
131. Wenkert, E. and Wickberg, B. (1965). *J. Amer. Chem. Soc.,* **87,** 1580
132. Buchi, G., Coffen, D. L., Kocsis, K., Sonnet, P. E. and Ziegler, F. E. (1966). *J. Amer. Chem. Soc.,* **87,** 2073; (1966). **88,** 3099
133. Kutney, J. P., Brown, R. T., Piers, E. and Hadfield, J. R. (1970). *J. Amer. Chem. Soc.,* **92,** 1708
134. Gorman, M., Neuss, N., Cone, N. J. and Deyrup, J. A. (1960). *J. Amer. Chem. Soc.,* **82,** 1142
135. Harley-Mason, J., Rahman, A. and Beisler, J. A. (1966). *Chem. Commun.,* 743
136. Harley-Mason, J. and Rahman, A. (1967). *Chem. Commun.,* 208
137. Harley-Mason, J. and Rahman, A. (1967). *Chem. Commun.,* 1048
138. Harley-Mason, J. and Rahman, A. (1968). *Chem. Ind. (London),* 1845
139. Krapcho, A. P., Glynn, G. A. and Grenon, B. J. (1967). *Tetrahedron Lett.,* 215
140. Nagata, W., Hirai, S., Kawata, K. and Aoki, T. (1967). *J. Amer. Chem. Soc.,* **89,** 5045
141. Nagata, W., Hirai, S., Okumura, T. and Kawata, K. (1968). *J. Amer. Chem. Soc.,* **90,** 1650
142. Sallay, S. I. (1967). *J. Amer. Chem. Soc.,* **89,** 6762
143. Ikezaki, M., Wakamatsu, T. and Ban, Y. (1969). *Chem. Commun.,* 88
144. Rosenmund, P., Haase, W. H. and Bauer, J. (1969). *Tetrahedron Lett.,* 4121
145. Huffman, J. W., Rao, C. B. S. and Kamiya, T. (1967). *J. Org. Chem.,* **32,** 697
146. Augustine, R. L. and Pierson, W. G. (1969). *J. Org. Chem.,* **34,** 1070
147. Buchi, G., Kulsa, P. and Rosati, R. L. (1968). *J. Amer. Chem. Soc.,* **90,** 2448
148. Buchi, G., Kulsa, P., Ogasawara, K. and Rosati, R. L. (1970). *J. Amer. Chem. Soc.,* **92,** 999
149. Kutney, J. P. and Bylsma, F. (1970). *J. Amer. Chem. Soc.,* **92,** 6090
150. Narisada, M., Watanabe, F. and Nagata, W. (1971). *Tetrahedron Lett.,* 3681

3
Biosynthesis of Indole Alkaloids

A. I. SCOTT
Yale University

3.1 SIMPLER DERIVATIVES OF TRYPTOPHAN

Hydroxylation of the aromatic ring of tryptamine usually (but not exclusively) takes place at the 5-position. Thus hydroxylation of L-tryptophan (1) affords L-5-hydroxytryptophan (2), then serotonin (3) by decarboxylation[1]. Conversions of serotonin to 5-hydroxyindole-3-acetic acid (4) has been demonstrated in mammalian systems[2].

Of considerable theoretical interest is the conversion of tryptophan in *Psilocybe semperiva* to psilocibin (5), the phosphate of *N,N*-dimethyl-4-hydroxytryptamine via hydroxylation at the 4 position[3]. Many indole alkaloids show evidence of 5-, 6- and 7-hydroxylation but the stage at which oxidation of the aromatic ring occurs is unknown. We shall return to an

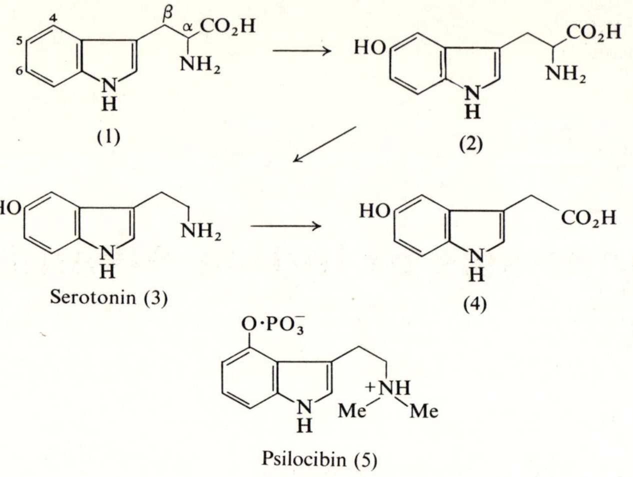

Serotonin (3)

Psilocibin (5)

Figure 3.1 Serotonin and Psilocibin

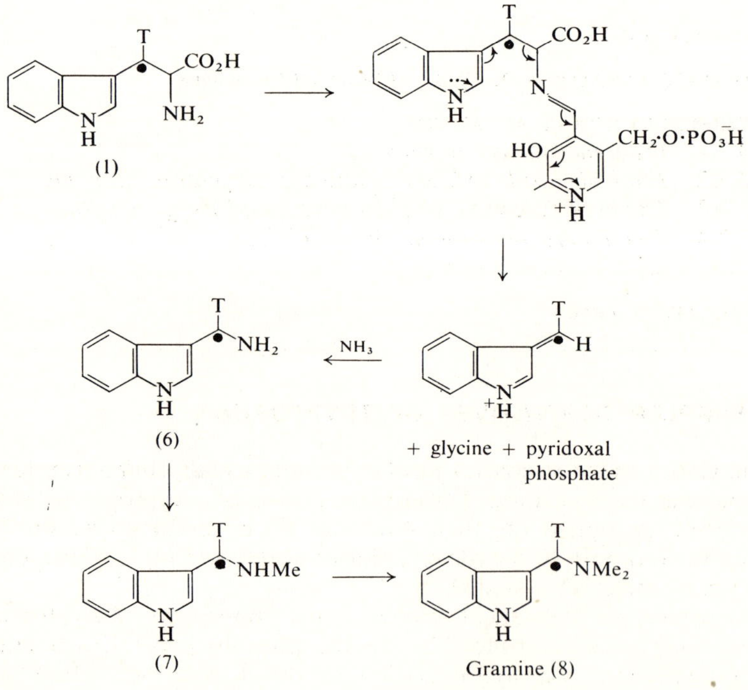

+ glycine + pyridoxal phosphate

Gramine (8)

Figure 3.2 Biosynthesis of gramine

example of electrophilic attack at the 4-position in the formation of the ergot alkaloids. An *in vitro* model of 4-hydroxylation has been recorded[120].

Although by now a classic example of the use of [14]C- and ^{3}H in labelling experiments using higher plant material, this section would not be complete without reference to the biosynthesis of gramine[4] since several of the steps will certainly provide analogy for future studies with more complex alkaloids in which the side chain of tryptophan also suffers contraction. Many such structures are known and mechanisms similar to those implied in Figure 3.2 may well be involved.

Conclusive evidence that the methylene group (·) remains intact during the conversion of tryptophan to gramine (8) was provided by feeding [β-^{14}C] tryptophan and [β-^{3}H]tryptophan and showing that the ^{3}H/^{14}C ratio was maintained in the experiment[5]. Compounds (6) and (7) do in fact occur in barley and form gramine when incubated with *S*-adenosylmethionine and a barley extract[6]. One of the few successful cell free experiments in alkaloid biosynthesis has been carried out with a barley shoot homogenate[7].

3.2 β-CARBOLINE ALKALOIDS

Structural inspection of the simpler β-carboline alkaloids reveals in addition to a tryptophan derived portion, a segment containing one, two, three or five carbon atoms [A:R = H, Me, Et or Bu^i]. There are some 20 alkaloids of the 'C_2' series (R = Me) representing several different oxidation states[8,9].

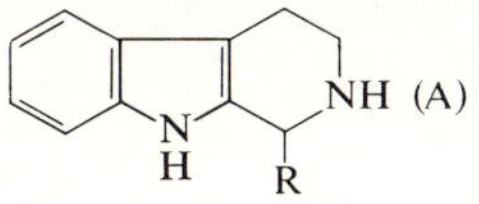

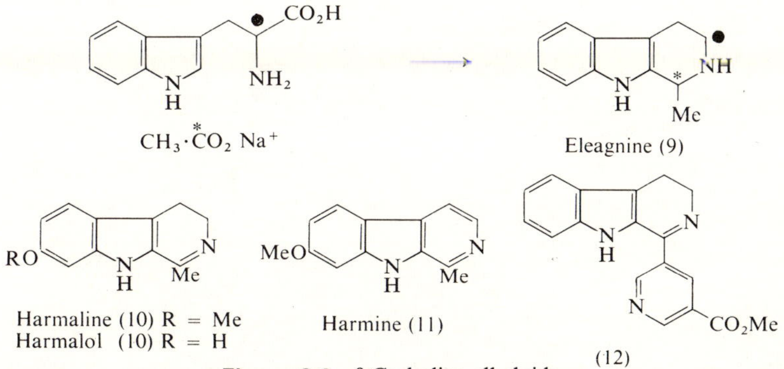

Figure 3.3 β-Carboline alkaloids

Of these only eleagnine (9) in *Eleagnus angustifolia* and harmaline (10; R = OMe), harmalol (10), and harmine (11) in *Peganum harmala* have been studied in biosynthetic detail. A recent addition to the family is (12) obtained from the African species *Nauclea diderrichii*[10].

The theory that the tetrahydro alkaloids are derived by Mannich condensation of tryptophan or tryptamine with an aldehyde, e.g. acetaldehyde has come to be accepted as biogenetic dogma thanks to the efficiency of this reaction in the laboratory. That the temptation to carry this type of analogy into biochemistry has to be resisted – at least in its complete form – is shown in the sequel.

Administration of $[\alpha\text{-}^{14}C]$-DL-tryptophan to *E. angustifolia* followed by careful degradation showed[14] (Figure 3.3) that only the expected carbon atom ($\bullet$C) in (9) was labelled (0.01 % incorporation). Sodium $[1\text{-}^{14}C]$acetate

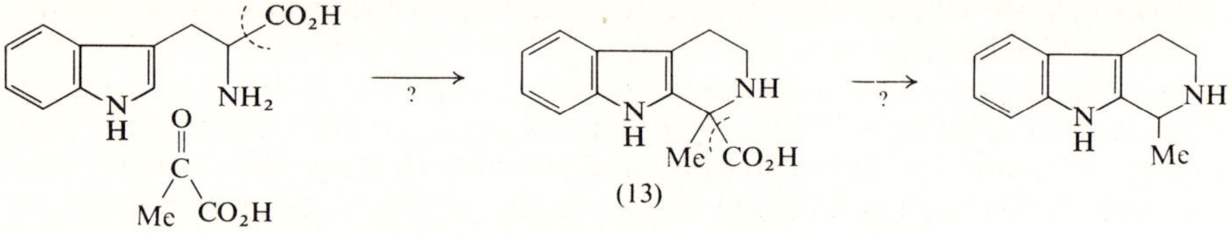

Figure 3.4 Pyruvate mechanism for β-carbolines

incorporation[14] (0.003 %) and degradation next established the expected specific activity at C-1 ($\bullet$C).

In *P. harmala* however, it was reported that $[\beta\text{-}^{14}C\text{-}^{15}N_\beta]$tryptamine was incorporated (0.43 %) into harmine with retention of the isotopic ratio

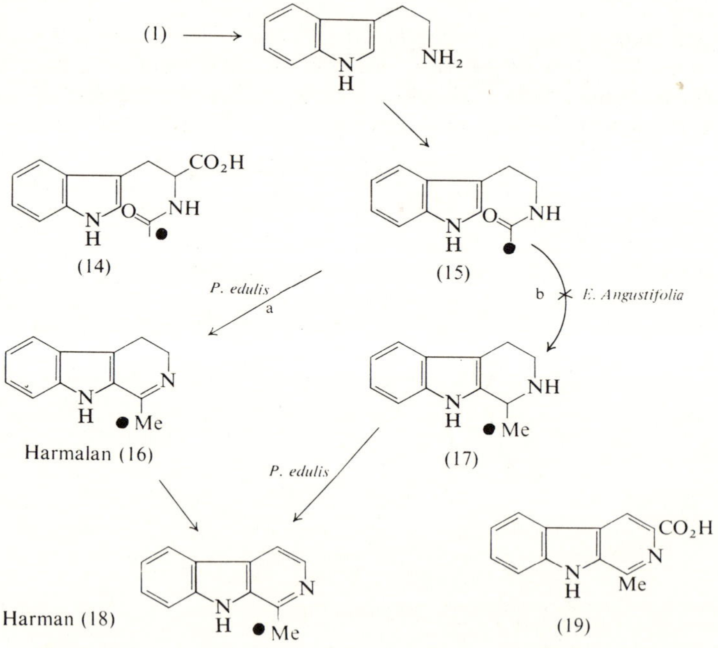

Figure 3.5 Biosynthesis of harman in passion fruit (*Passiéora edulis*)

suggesting that decarboxylation takes place before condensation with the C_2 moiety. Further it was found that sodium $[1\text{-}^{14}C]$acetate gave rise to harmine containing only 7 % of the radioactivity in the C_2 unit and that the majority of this label was at C_1. Since [2- and 3-^{14}C]pyruvate were also incorporated into harmine (*c.* 0.5 %), Stolle and Gröger[15] concluded that the species (13) was involved (Figure 3.4). However, well-known interconversions involving acetate and pyruvate, together with the spontaneous non-enzymic formation of tetrahydro-β-carbolines in the presence of aceteldehyde, render interpretation of the above results quite problematical[16].

An interesting experiment designed to overcome these difficulties was recently carried out[17]. N-[2-^{14}C]acetyltryptophan (14), N-[2-^{14}C]acetyl tryptamine (15), [9-^{14}C]harmalan (16) and [9-^{14}C]tetrahydroharman (17) were fed to *Passiflora edulis* and the resultant harman (18) analysed. The results of these feeding (and trapping) experiments support the view that the pathway tryptophan → tryptamine → N-acetyltryptamine (14) → harmalan (16) → harman (18) is followed (Figure 3.5).

In spite of the occurrence of compounds such as harman-3-carboxylic acid (19) or cordifoline ((63), p.120) decarboxylation to tryptamine is probably the first event in *most* biosynthetic operations leading to indole alkaloids.

However, although tryptamine served as a source of the indole unit of eleagnine (9) in *E. angustifolia*, the incorporation of N-acetyltryptamine into (9) could not be demonstrated[18]. This result is interpreted in terms of two different pathways — one to the fully aromatic harman via N-acetyltryptamine and harmalan (16) (Figure 3.5a), the other to the tetrahydro-β-carbolines, perhaps via pyruvate (Figure 3.5b). However, the reported conversion of (17) → (18) seems anomalous and perhaps is not a specific reaction for harman biosynthesis.

It is clear that much work remains to be done in this comparatively simple series, e.g. feeding of precursors labelled in *both* the tryptamine and C_2 or C_3 fragment.

3.3   THE ERGOT ALKALOIDS

3.3.1   Formation of the tetracyclic ergoline system

In contrast to the 'C$_9$–C$_{10}$' indole alkaloids it was recognised comparatively early in their biogenetic history that the ergot alkaloids of *Claviceps spp.* were composed of tryptophyl and mevalonate segments[19]. Thus L-tryptophan is incorporated into the tetracyclic ergoline system (Figure 3.6) of elymoclavine (20) with loss of the carboxyl group, whilst C-2 of mevalonate becomes C-7 of the ergolines and the N-methyl group is methionine derived. Mevalonic acid is incorporated via γ,γ-dimethylallylpyrophosphate[20] (dimethylallyl [1-^{3}H]-*alcohol* is *not* specifically incorporated)[21] and in fact the first encounter of the C_5 unit must be an electrophilic substitution to give 4(γ,γ-dimethylallyl) tryptophan[22] (22) (Figure 3.7), a known metabolite of ergot fungus under conditions of inhibition of alkaloid synthesis[23–25]. Inhibition also produces the interesting structure clavicipitic acid (Figure 3.6) which however is not a precursor[23]. Intact incorporation of (22) into ergot alkaloids has been demonstrated[20], but there is still some question as to whether 4-dimethylallyl-tryptamine (22; $CO_2H = H$) is a true precursor[20, 25].

Figure 3.7 summarises current evidence of the course of biosynthesis of the ergoline system. The occurrence of chanoclavines-I (25) and -II (27) and of isochanoclavine-I (26) is suggestive that ring C is formed before ring D. The history of the experiments leading to current theories of ring C formation make fascinating reading and the original references should be consulted[26-34]. In summary, it was shown that deoxychanoclavine-I (28) and nordeoxy-I (29) are *not* precursors[27, 28]. Thus one of the methyls of the dimethylallyl group must be hydroxylated prior to ring C closure. Inspection of the stereochemistry of iso-I (5*R*, 10*R*) (26) which carries the hydroxymethyl group *cis* to the C_9 substituent, would seem to offer a perfectly poised candidate for

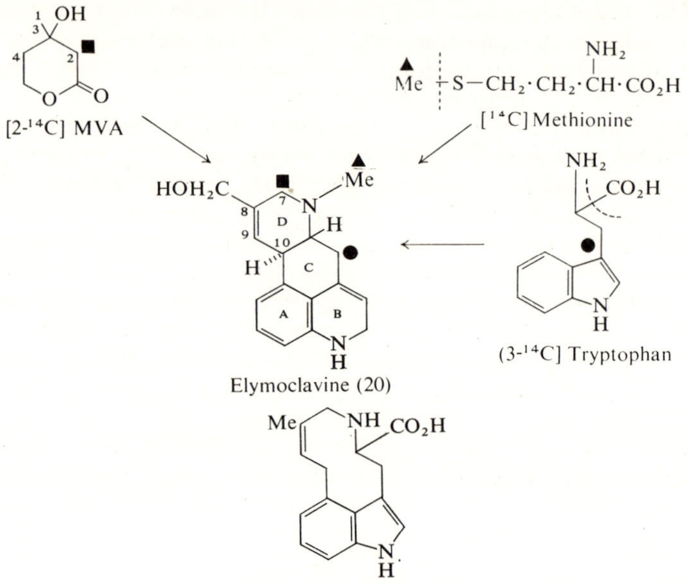

Figure 3.6 Biosynthesis of elymoclavine

ring closure. In support of this biogenetic reasoning it was found that all of the isomeric chanoclavines carried the label from C-2 of mevalonate in the C methyl group (■) (Figure 3.7)[27, 28]. That Nature had another method for the cyclisation was unambiguously confirmed by three independent studies on the positive incorporation of chanoclavine-I into the ergolines (as 30). Iso-I (26) and chano-II (27) are *not* incorporated[28-31] into agroclavine (30). In other words, cyclisation involves *cis–trans* isomerisation at the allylic double bond. To add a further biochemical twist to this mechanism it was shown that tritium from 4*R*, 4^{3}H-mevalonate (but not from the 4*S*, 4^{3}H-isomer) is incorporated into chanoclavine-I (25) and elymoclavine (20) indicating that the original dimethylallyl residue carries the label from C-2 of MVA in the *trans*-methyl group[31, 32]. Thus the initial hydroxylation must take place at the *cis* methyl group, which later becomes the *trans*-hydroxymethyl group of chanoclavine-I. The biosynthetic operation involves *two cis–trans* isomerisations!

In a double-labelling experiment it could be shown that chanoclavine-I (25) cyclised[33] with complete retention[31] of the C-10 proton (*). However, [7-^{14}C-9-^{3}H]chanoclavine-I lost between 30 and 70% of the [^{3}H] label depending on the experimental conditions. These experiments and companion studies with 3′, 4-d_5-MVA indicate an *inter*molecular transfer of the

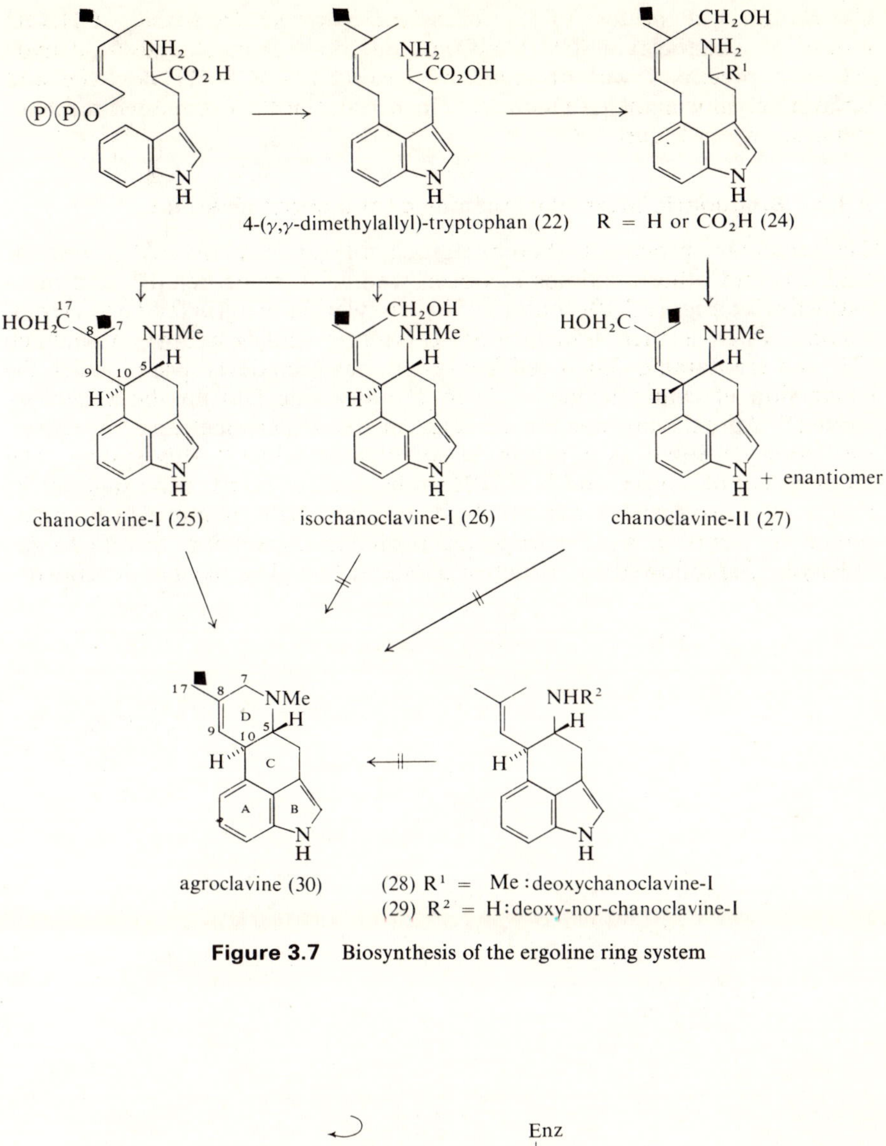
4-(γ,γ-dimethylallyl)-tryptophan (22)
R = H or CO₂H (24)
chanoclavine-I (25)
isochanoclavine-I (26)
chanoclavine-II (27)
+ enantiomer
agroclavine (30)
(28) R¹ = Me: deoxychanoclavine-I
(29) R² = H: deoxy-nor-chanoclavine-I

Figure 3.7 Biosynthesis of the ergoline ring system

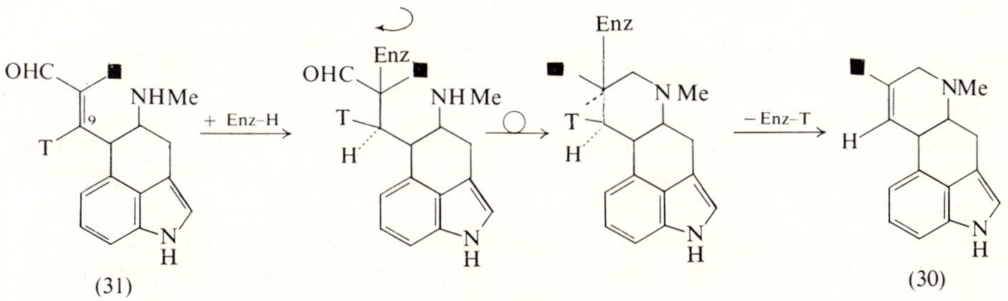
+ Enz–H
– Enz–T
(31)
(30)

Figure 3.8 Mechanism of inversion for chanoclavine-I aldehyde

C-9 proton and the loss of *one* methylene hydrogen from the —CH$_2$OH group of chanoclavine-I[34]. Confirmation that chanoclavine-I-aldehyde (31) is a precursor was obtained by feeding [7-^{14}C-9-^{3}H]aldehyde and isolating elymoclavine (20) with 40% incorporation[36]. A suggested mechanism is shown in Figure 3.8.

3.3.2   Biogenetic interrelationships of the ergot alkaloids

The secondary processes whereby agroclavine (30) is converted to lysergic acid (21) and other alkaloids has been studied in some detail[35]. The main pathway (see Figure 3.10) consists of hydroxylation of agroclavine to elymoclavine which is then transformed to lysergic acid, α-hydroxyethylamide (32), and ergotamine (35). A cell free system has been developed in which the conversion of chanoclavine-I (25) to elymoclavine (20) has been demonstrated[37]. Agroclavine was *not* detected in this experiment and, moreover, was poorly utilised as a precursor for elymoclavine in this system. The requirement of oxygen and NADPH in the cell free preparation suggested[38] that a mono-oxygenase was involved but the results might also be interpreted in terms of a dehydrogenase-mediated conversion of (25) to the aldehyde (31) followed by reductive condensation. The oxygen dependence

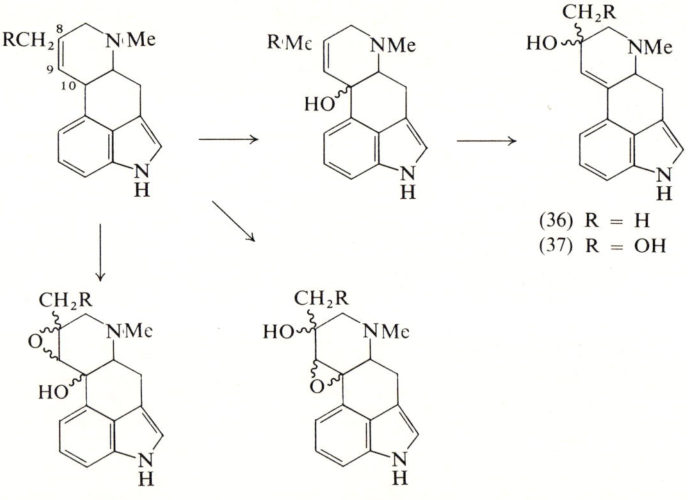

(36) R = H
(37) R = OH

(38)   (39)

Figure 3.9   Relationships of seto- and penniclavines: R = H (seto-series); R = OH (penni-series)

would need confirmation, however, for this to be valid. The loss of tritium from C-9 is at present explained by the exchange mechanism[31, 34] summarised in Figure 3.8.

   A number of hydroxylations catalysed by peroxidase have been discovered. In particular, 8-hydroxylation of agroclavine and elymoclavine leads to the diastereomeric setoclavines and penniclavines respectively[39, 40]. The initial attack (which can be brought about by a number of organisms) takes place at C-10 and thence by allylic rearrangement to the 8-hydroxy compound[41, 42]. The corresponding epoxides (38, 39) have also been isolated as shown in Figure 3.9.

The conversion of elymoclavine to lysergic acid derivatives has been established only for the case of the α-hydroxyethylamide (32) and ergotamine (35)[34, 43]. The attractive possibility that lysergylalanine (33) (Figure 3.10)

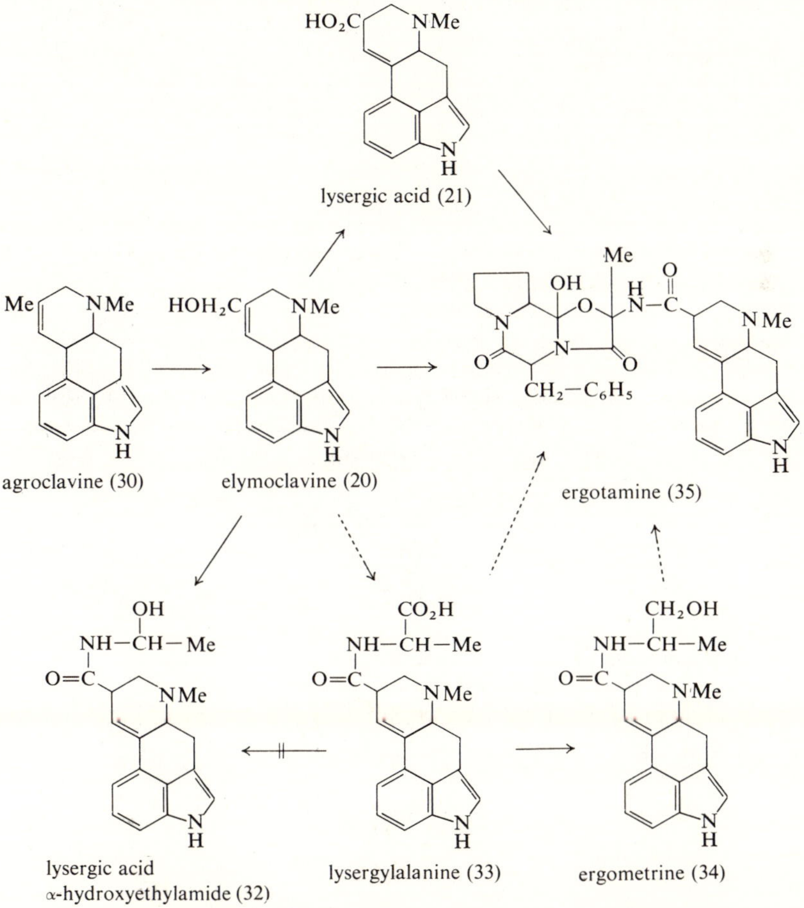

Figure 3.10   Biosynthetic connections of the ergot alkaloids

plays a central role in these interconversions has been confirmed only for its biotransformation to ergometrine (34) which in turn does not seem to be involved in ergotamine synthesis[44, 45]. No incorporation of lysergylalanine into the α-hydroxyethylamide (32) could be demonstrated[44, 45]. While lysergic acid (21) was incorporated into ergotamine (35), the corresponding amide does not serve as a precursor[45]. Some confusion exists as to the comparative roles of L-alanine, pyruvate and L-alaninol in the metabolism[45, 46] and the status of the biosynthesis of the amide-type alkaloids is obviously an area of future definition[34].

3.4 CYCLOPIAZONIC ACID

The toxic principle of *Penicillium cyclopium* was shown to have the structure (36) (Figure 3.11) and more recently the compound (37) was isolated from the same organism[47]. It could be shown by feeding and degradative experiments that, by analogy with the ergoline system, (36) was formed from

Cyclopiazonic acid (36)   β-cyclopiazonic acid (37)

Figure 3.11

mevalonate, tryptophan and acetate. It was also found that (37) served as a precursor for cyclopiazonic acid[48].

3.5 INDOLMYCIN

The incorporation of the naturally occurring stereoisomer (38) of indolmycenic acid into the tryptophan derived antibiotic indolmycin (39) in *Streptomyces griseus* has been demonstrated[49]. Tryptophan supplies the C_3 side chain

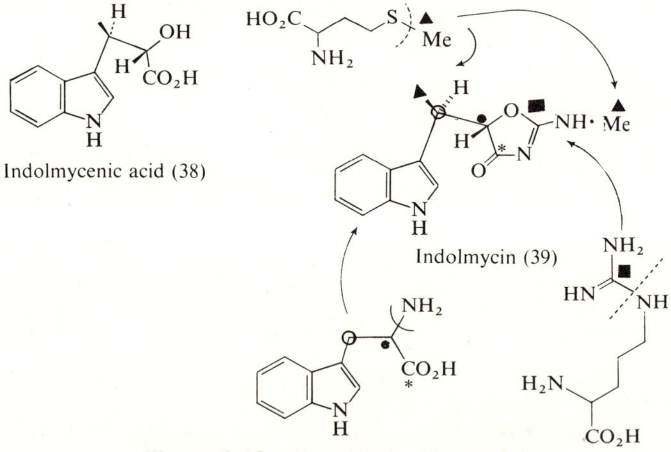

Indolmycenic acid (38)

Indolmycin (39)

Figure 3.12   Biosynthesis of indolmycin

intact, the two 'extra' methyl groups (▲) are methionine derived and the guanido group of arginine labels C■ of the oxazoline ring as summarised in Figure 3.12 [49].

3.6 DIMERIC ALKALOIDS OF THE CALYCANTHINE SERIES

$[\beta\text{-}^{14}C, 2\text{-}^{3}H]$tryptophan (40) is incorporated without change in the $^{3}H/^{14}C$ ratio into chimonanthine[50] (41) showing that oxidative dimerisation most

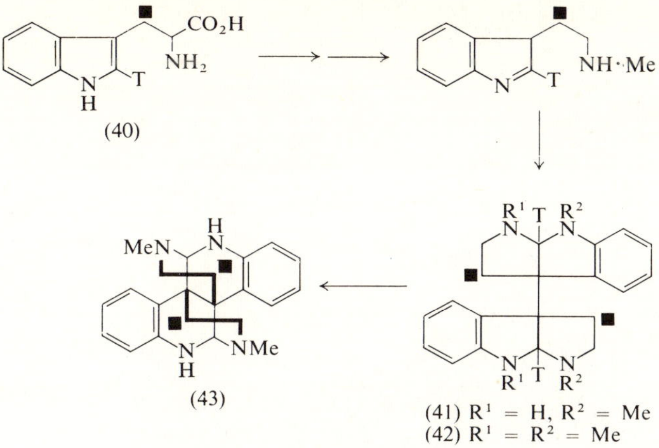

Figure 3.13   Calycanthine biosynthesis

probably occurs by the coupling reaction shown in Figure 3.13 for which there is good laboratory analogy[51]. Tryptophan is also incorporated specifically into folicanthine (42) in *C. floridus*[52]. The related calycanthine (43) is probably formed as indicated[51] in Figure 3.13.

3.7 TERPENOID INDOLE ALKALOIDS

In spite of much speculation, until 1965 little was known about the non-tryptophan derived segment of these alkaloids represented by the *Corynanthé-Strychnos* (44), *Aspidosperma* (45) and *Iboga* (46) families. Tryptophan serves as a good precursor for the 'tryptamine' segment and recently it has been observed that labelled tryptamine[53–56] also finds its way specifically into the major alkaloids of *Vinca* and other species. In Figure 3.14 are shown the principal members of each family, ajmalicine (47), akuammicine (48), vindoline (49) and catharanthine (50) which occur in *V. rosea*. The initial observation of $[2\text{-}^{14}C]$mevalonate incorporation into vindoline (49) in *V. rosea* plants set the stage for vigorous efforts in several laboratories to define the entire pathway of biosynthesis of these complex alkaloids. We shall pass rapidly over the historical development, since this has been reviewed extensively in several places[57–62], and concentrate on more recent experimental evidence for each state of the proposed pathway. The sequence is conveniently divided into three eras. The early stages cover the development of a highly oxygenated and reactive glucoside, secologanin (64) and its conversion to vincoside (69). Next is defined the transformation of vincoside to the *Corynanthé* (44) and *Strychnos* (52) alkaloids at their lowest oxidation level. Finally, the *Aspidosperma*, *Iboga* and *Eburna* families are developed from a key member by various oxidative and rearrangement processes.

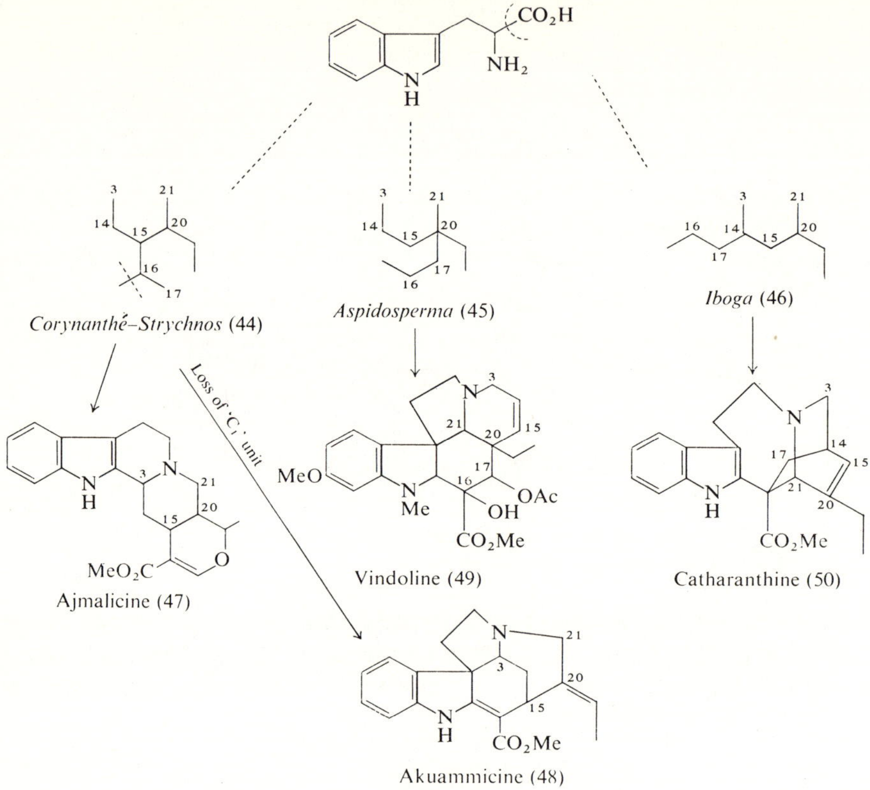

CO_2H
N
H
NH_2
3
21
14
15
20
16
17
Corynanthé–Strychnos (44)
Aspidosperma (45)
Iboga (46)
Loss of 'C₁' unit
MeO
N
Me
OAc
OH
CO_2Me
Vindoline (49)
Catharanthine (50)
MeO_2C
O
Ajmalicine (47)
Akuammicine (48)

Figure 3.14

3.7.1 From mevalonate to vincoside

It was proposed by Thomas[63] and Wenkert[64] more than a decade ago that the indole alkaloids were rather elaborate monoterpenes and in fact stemmed from the encounter of tryptamine with a cleaved or seco-version of a cyclopentanoid monoterpene of the general class (51) (Figure 3.15). The connection with the *Yohimbè* family (53) is also shown. As long ago as 1964, Thomas[65] also suggested that the particular monoterpene might be the glucoside loganin (56) which undergoes cleavage to secologanin (64) as part of the intermediary metabolism. All of these speculations have been vindicated by a prodigious effort involving several laboratories since 1965. Thus 2-^{14}C (●)[57,58] and 3′-^{14}C (○)[66] mevalonates gave a distribution of radioactivity in the alkaloids shown for vindoline (49) in Figure 3.16, a result which is in accord with the loss of identity of the two carbon atoms. These points have been confirmed and amplified with experimental detail[67–70] summarised in Figure 3.16. The role of loganin in the biosynthetic pathway was revealed by transformation (in *Menyanthes trifoliata*) of [2-^{14}C]- and [4-^{14}C]-geraniol (54) to appropriately labelled loganin[68] samples followed by separate feeding of these to *V. rosea* shoots[71,72]. The labelling patterns in the alkaloids ajmalicine

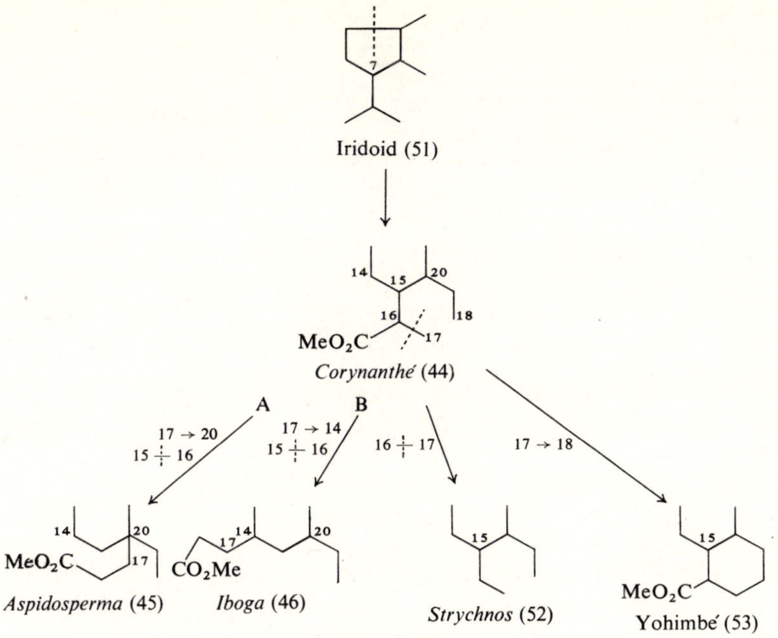

Figure 3.15

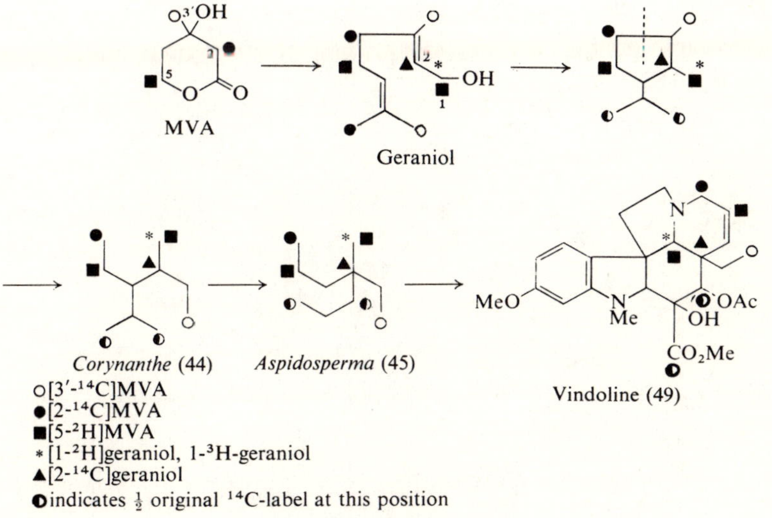

○[3'-¹⁴C]MVA
●[2-¹⁴C]MVA
■[5-²H]MVA
∗[1-²H]geraniol, 1-³H-geraniol
▲[2-¹⁴C]geraniol
◐indicates ½ original ¹⁴C-label at this position

Figure 3.16

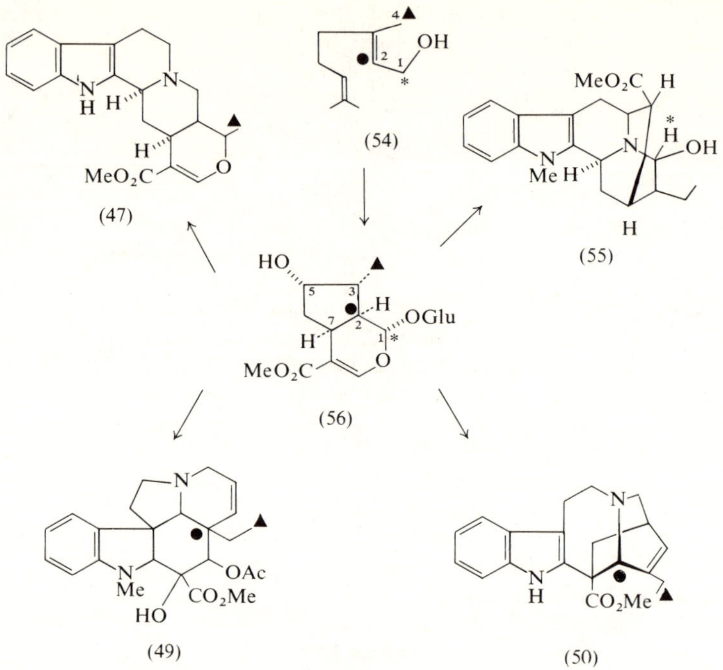

(47)

(54)

(55)

(56)

(49)

(50)

Figure 3.17

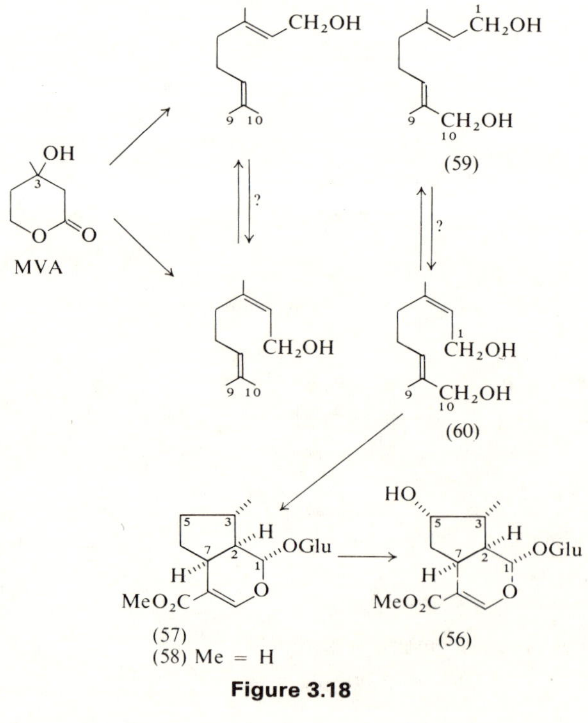

MVA

(59)

(60)

(57)
(58) Me = H

(56)

Figure 3.18

(47), vindoline (49), and catharanthine (50) are shown in Figure 3.17. Similarly, $[1-^3H_2]$geraniol gave $[1-^3H]$loganin which was incorporated into ajmalicine (53) in *Rauwolfia serpentina*[71]. Loganin (56) was found in *V. rosea* by radio-chemical dilution[68, 72] and by isolation[73].

Working back in the sequence from loganin suggests deoxyloganin (57) as a likely precursor and this idea has been confirmed experimentally by partial synthesis of $[O\text{-methyl-}^3H]$deoxyloganin (57) from loganin and demonstration of its incorporation into loganin (Figure 3.18) and the three main classes of alkaloid in *V. rosea*. Deoxyloganin was also found in *Strychnos nux-vomica*, *Menyanthes trifoliata*, and *Vinca rosea*.

A further set of interconnecting data relates deoxyloganic acid and loganin. The free acid of loganin has been isolated from *Swertia caroliniensis* and shown to be derived from $[2-^{14}C]$mevalonate and $[1-^{14}C]$geranyl pyrophosphate[74].

The gap between geraniol and loganin has almost been filled in. Thus both geraniol and nerol were converted to the corresponding 10-hydroxy derivatives (59) and (60) with a $9-^{14}C$ label. When these were administered to *V. rosea* plants both were incorporated into loganin and the main alkaloids. It was found that the *cis*-compound (60) was somewhat more efficient as a precursor than the *trans* and also that in accord with earlier experiments with $[2-^{14}C]$mevalonate, the methoxy-carbonyl groups of the isolates contained *c.* 40–48% of the total activity[75]. In a related study $[1-^3H_2]$-10-hydroxy geraniol (59) and $[1-^3H_2]$-10-hydroxy nerol (60) were administered together. Again good incorporations were recorded in loganin and the main alka-loids[76]. The foregoing evidence together with the results of non-incorporation of a number of structural variants of the hydroxy geraniol system[75] and of iridodial[77] culminates in the scheme[75] shown in Figure 3.19 for the cyclisation

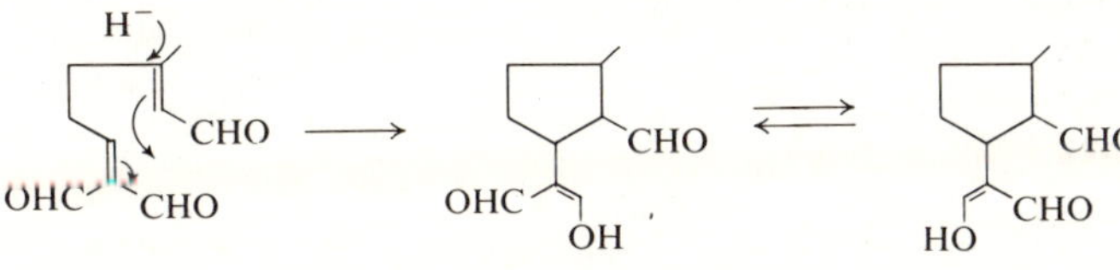

Figure 3.19

process which also rationalises the equipartition of label at the terminal aldehyde carbons.

The details of the segment of the pathway between loganin (56) and vinco-side (69) have been uncovered thanks to the isolation and structural studies[78, 79] on two new glucosides of *Menyanthes trifoliata* namely folia-menthin (61) and menthiafolin (62) (Figure 3.20). Hydrolysis of (62) to (63), ring opening and methylation furnished secologanin (64) which in its $[O\text{-methyl-}^3H]$-labelled form proved a good precursor for the three families of indole alkaloid[80, 81]. It also satisfied a second criterion of intermediacy since it could be isolated from *V. rosea*[80, 81] and from *Lonicera morrowii*[82].

Secologanin serves as a precursor of the glucosides (61) and (62) and of sweroside (65) in *M. trifoliata*[83]. In *V. rosea* and *M. trifoliata* $[O\text{-methyl-}^3H]$-loganin is cleaved to secologanin (5–6% incorporation) by a mechanism

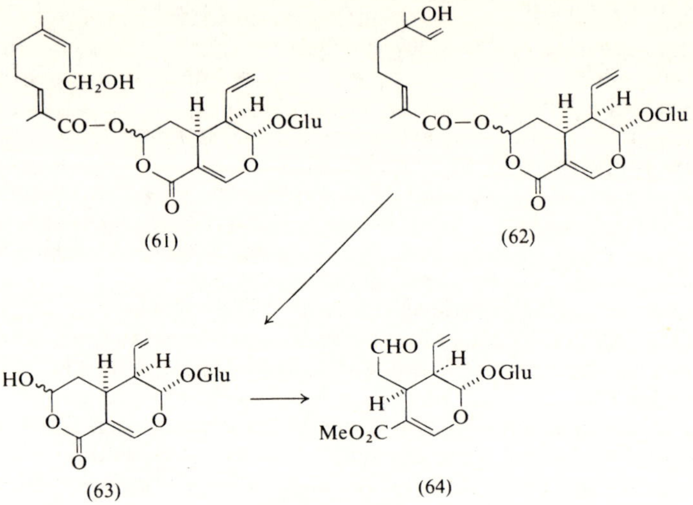

(61) (62) (63) (64)

Figure 3.20

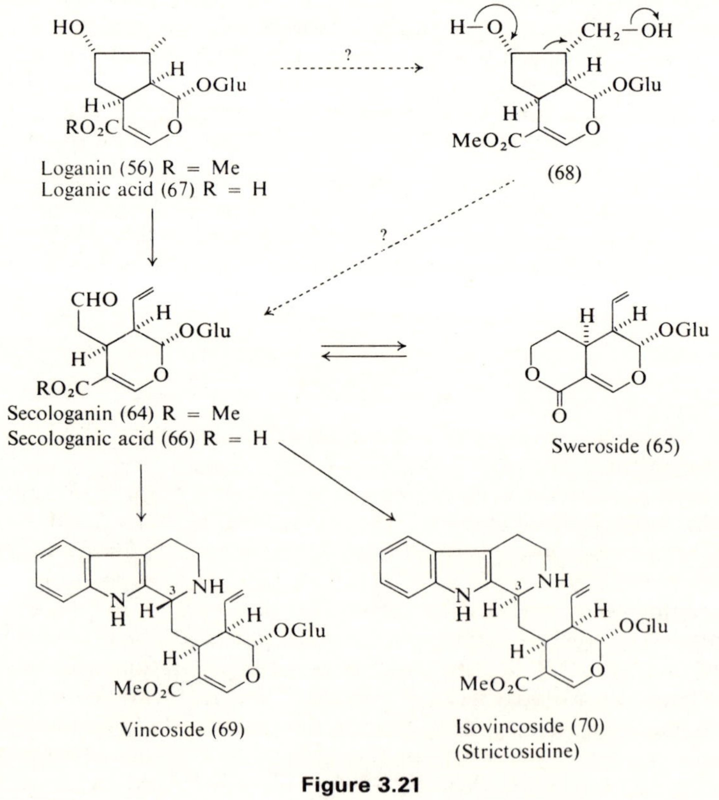

Loganin (56) R = Me
Loganic acid (67) R = H

(68)

Secologanin (64) R = Me
Secologanic acid (66) R = H

Sweroside (65)

Vincoside (69)

Isovincoside (70)
(Strictosidine)

Figure 3.21

which may involve the hitherto undiscovered hydroxyloganin (68) in a 1,3-fragmentation reaction[81] (Figure 3.21).

Since the lactone sweroside labelled with ^{14}C at C-4 is converted into vindoline[84] the interconversion of secologanin (64) to sweroside (65) was assumed and this was confirmed as above and also in the reverse direction by feeding [6-$^{3}H_2$]sweroside to *V. rosea* to afford [6-$^{3}H_2$]secologanin in a specific process[83]. It has recently been found that the metabolism of sweroside· proceeds via secologanic acid (66)[85, 86]. Purification of a methyl transferase from *V. rosea* which methylates both loganic acid (67) and secologanic acid (66) at comparable rates[87] together with evidence[85-89] that loganin is converted to secologanic acid and that loganic acid (67) affords loganin (56), secologanic acid (66) and secologanin (64) requires the operation of a dual pathway or metabolic grid summarised in Figure 3.21. The final compound in the sequence, secologanin (64) condenses with tryptamine *in vitro* to form a separable mixture of vincoside[90-92] (69) and the 3-epimer, isovincoside (70). The latter compound is identical with strictosidine isolated from *Rhazya stricta*[83] while radiochemical dilution analysis and isolation from *V. rosea* proved the presence of vincoside (69), isovincoside (70), and *N*-acetylvincoside[90-92].

Thus the pathway from mevalonate to the first alkaloidal material, vincoside (a proved precursor of the three main families), is virtually defined. The major remaining problems concern the cleavage mechanism (56) → (61) and the cyclisation details proposed in Figure 3.19.

3.7.2   From vincoside to corynanthé and strychnos alkaloids

Formation of the *Corynanthé* skeleton from vincoside *formally* requires enzymatic removal of glucose and cyclisation of the resultant dialdehyde ester (71). The reactive species thus produced can cyclise to the salt from which geissoschizine (74), corynantheine (72), its aldehyde (73), and the pentacyclic ajmalicine (47) can be derived. All of these *Corynanthé* alkaloids have been discovered in *V. rosea*[93]. The key to successful isolation of the rarer members lies in the utilisation of very young (1–2 day-old) seedlings[93, 94]. The recent definition[92] of the stereochemistry of vincoside as in (69) presents a problem in mechanism since experiments with [5-^{3}H]loganin reveal that ^{3}H at C-3 in vincoside marked Ⓗ passes into the later alkaloids unscathed. Any mechanism converting vincoside to geissoschizine and the later alkaloids must therefore explain inversion without loss of tritium. One such possibility is offered in Figure 3.23. The stage at which this inversion occurs must be after vincoside since isovincoside is not incorporated into the later alkaloids[90-92], i.e. R^1 and R^2 in Figure 3.23 are still unknown.

Four principal methods have been brought to bear on the intricate problem of mechanism connecting the veritable maze of structures present in *V. rosea* (at least 100 alkaloids have been characterised from the mature plant) and may be summarised as follows:

(i) The combined techniques of isolation and dilution analysis have been used in sequence studies to gain some insight into the order of appearance of the alkaloid families in *V. rosea*.

(ii) Enlightened speculation is a second method whereby a reasonable guess as to structure and oxidation level has to be followed up by synthesis of a labelled precursor.

(iii) By double labelling of the smaller building blocks (e.g. mevalonate, loganin), the pattern of the label in the final alkaloid can be used to infer

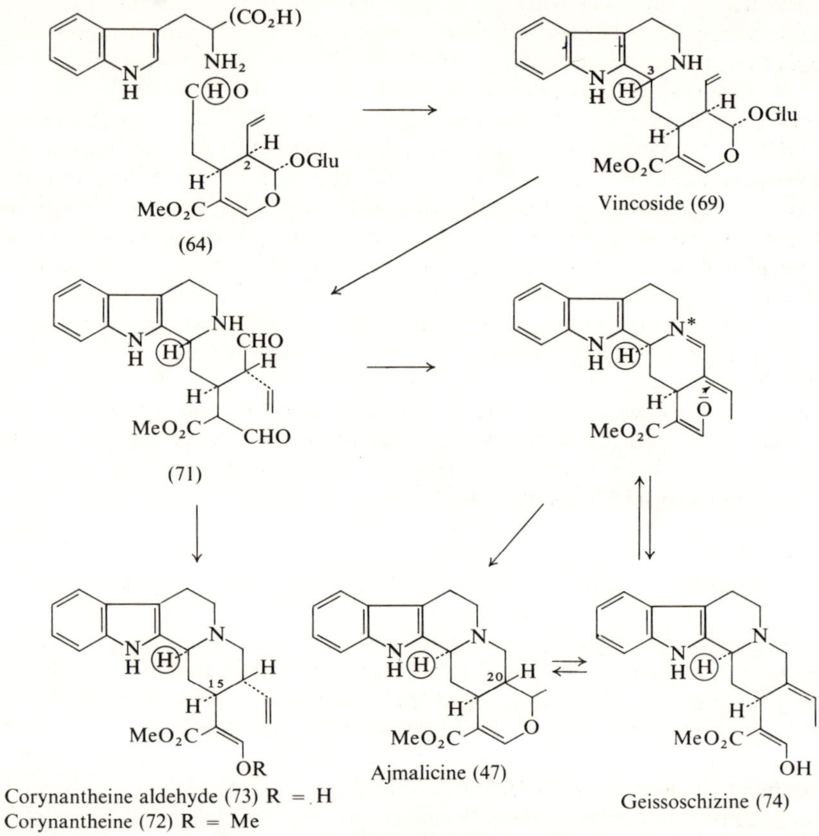

(64)

Vincoside (69)

(71)

Corynantheine aldehyde (73) R = H
Corynantheine (72) R = Me

Ajmalicine (47)

Geissoschizine (74)

Figure 3.22

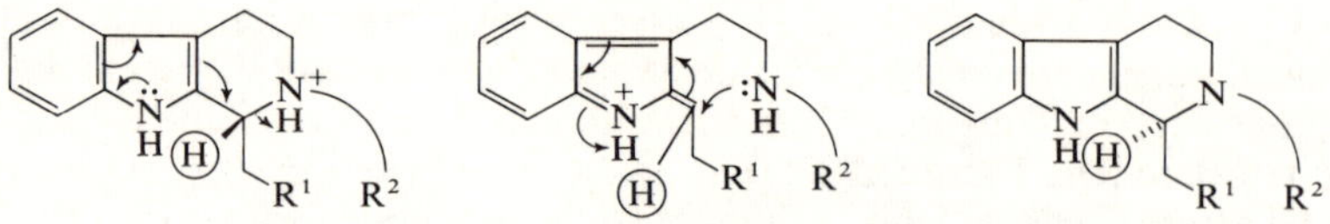

Figure 3.23

mechanisms and to suggest and exclude intermediate structures which may or may not accord with the loss or retention of ^{3}H.

(iv) Pulse or steady state labelling with tryptophan and sequential auto-radiographic assay of those compounds whose 'radio profile' is sufficiently dynamic to suggest that a true intermediate is involved.

All of these methods have been combined to solve the complex problem of rearrangements and interconnections among the alkaloidal classes almost exclusively in *V. rosea.*

Feeding experiments with corynantheine aldehyde[93, 94] (73) showed little or no incorporation into later alkaloids, with the exception of the methyl ether (72). However, [*O*-methyl-^{3}H, Ar-^{3}H]geissoschizine (74) was incorporated[95] into ajmalicine (47), serpentine (75) and most significantly, akuammicine (48), vindoline (49) and catharanthine (50). Mass spectrometry was

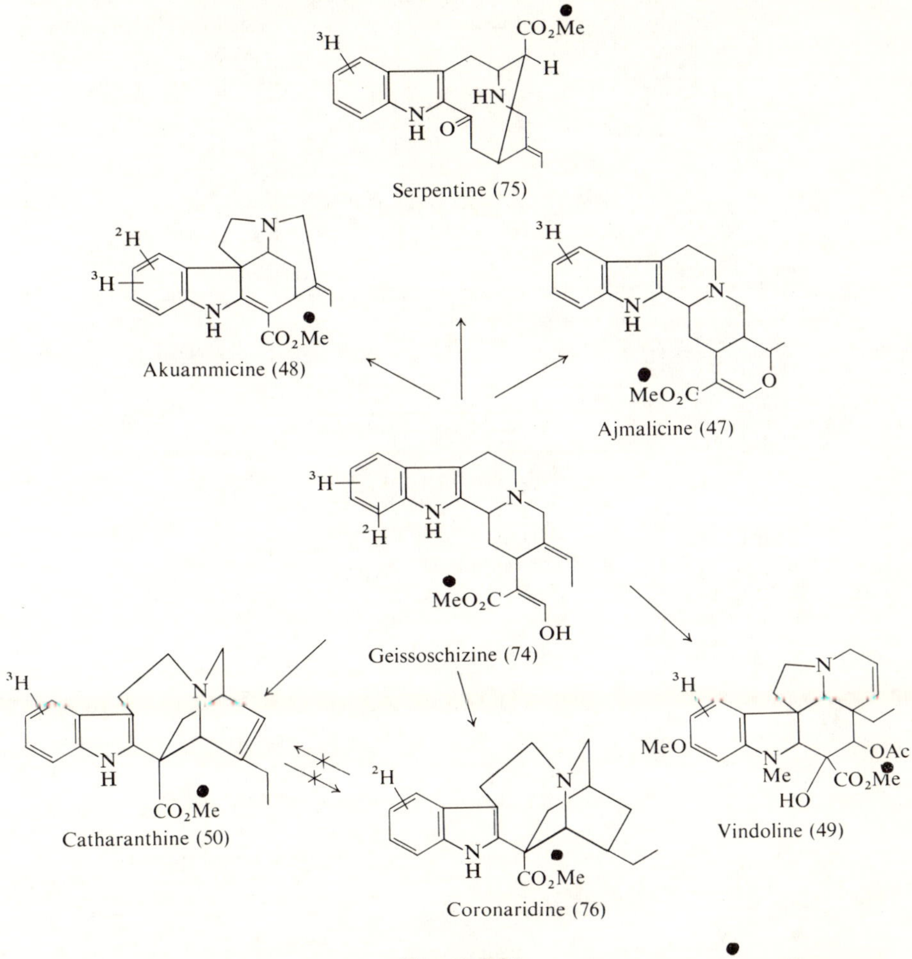

Serpentine (75)

Akuammicine (48)

Ajmalicine (47)

Geissoschizine (74)

Catharanthine (50)

Coronaridine (76)

Vindoline (49)

Figure 3.24   Me = [^{3}H—CH$_3$]

used[96] to follow the fate of [Ar-^{2}H$_4$]geissoschizine into akuammicine (48) and coronaridine (76). Thus geissoschizine takes its place as the *Corynanthé* alkaloid whose reactivity and oxidation level is that required to forge the connection with the other families. In further support of this premise, the radioactivity of geissoschizine following administration of DL-[2-^{14}C] tryptophan to five day-old seedlings of *V. rosea* was found to reach a maximum after *c.* 1.5–2 h, at which time some 4% of the total activity of the alkaloid

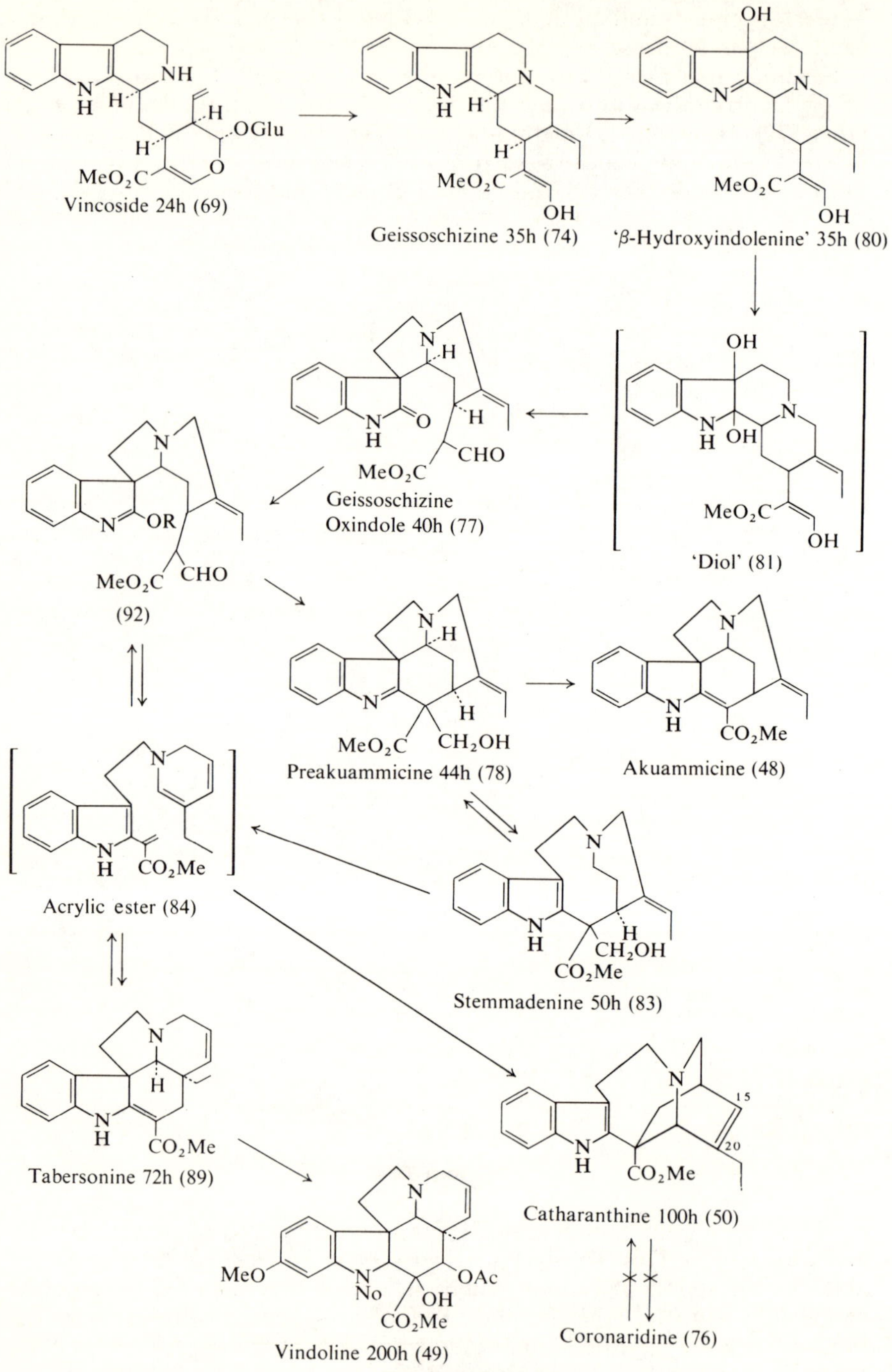

124

Vincoside 24h (69)

Geissoschizine 35h (74)

'β-Hydroxyindolenine' 35h (80)

Geissoschizine
Oxindole 40h (77)

'Diol' (81)

(92)

Preakuammicine 44h (78)

Akuammicine (48)

Acrylic ester (84)

Stemmadenine 50h (83)

Tabersonine 72h (89)

Catharanthine 100h (50)

Vindoline 200h (49)

Coronaridine (76)

Figure 3.25

fraction resided in geissoschizine[97]. This rapid rise in activity over the first 60–90 min after feeding tryptophan is followed by a slow decline to 1 % over the 8-day experiment. In sharp contrast, the profile for ajmalicine showed no such effect and the counts from tryptophan remained at c. 1 % throughout after a slow climb over the first 2 days[97]. The demonstration that geissoschizine is bioconverted to akuammicine with loss of one of the original carbons of the 'C_{10}' unit led to the search and discovery of a new alkaloid preakuammicine (78). Part of the structure proof for this compound, which is isomeric with precondylocarpine (79), is the ready loss of formaldehyde to give akuammicine (48), the *Strychnos* representative of *V. rosea*. The details of the mechanism connecting geissoschizine with akuammicine have not been filled in rigorously, but several interesting preliminary results have been reported in which the sequence shown in Figure 3.25 is suggested. Geissoschizine oxindole (77) has been isolated from *V. rosea* and a labelled specimen incorporated into akuammicine and vindoline[98]. In accord with the postulated mechanism[96] (Figure 3.25) an indole alkaloid with imino-ether functionality has recently been isolated[99]. The successful transformation of geissoschizine to the *Strychnos* alkaloids in *Vinca* has been more difficult to demonstrate in *Strychnos nuxvomica*. By feeding very young *S. nux vomica* seedlings over a period of months, an incorporation of 0.23% from Ar- H geissoschizine to strychnine has been observed. Degradations are currently in progress[100].

3.7.3   The pre-Strychnos, Aspidosperma, and Iboga alkaloids

The later and perhaps most exciting rearrangements which must be wrought upon alkaloids closely related to preakuammicine have the effect of opening up the pre-*Strychnos* framework, thereby passing through *three* asymmetric centres, then re-forming the *Aspidosperma* and *Iboga* alkaloids. The pivotal compound which could mediate in these remarkable transformations is postulated[104] to be an acrylic ester (84). The original ideas of Wenkert[101] concerning this type of transformation are shown in Figure 3.26 since they have played a large part in the rationalisation of the various rearrangements and have prompted searches for certain classes of intermediate. A later version[104] of the 'seco' concept was evolved to accommodate the oxidation level of the isolates of *V. rosea* and operates on a rather different mechanistic rationale (Figures 3.25 and 3.28).

Recent feeding and isolation work has confirmed the essential correctness of the seco-mechanism. Based on the oxidation chemistry[58] of the rare alkaloid stemmadenine (83) both preakuammicine (78) and precondylocarpine (79) have been prepared and are now available in sufficient quantity for study of model reactions[103] (Figure 3.27). Following up preliminary observations[104] on the rearrangement of stemmadenine (83) as a biogenetic model in which the secodine system (84) was first suggested, it has become evident that the collapse of the pre-*Strychnos* system[104, 105] as in Figure 3.28 is even more facile than indicated earlier. In sharp contrast to the unfounded criticism of certain investigators who have left an incorrect impression[106, 107] in the literature that the bond marked (– – –) in stemmadenine (83) (Figure

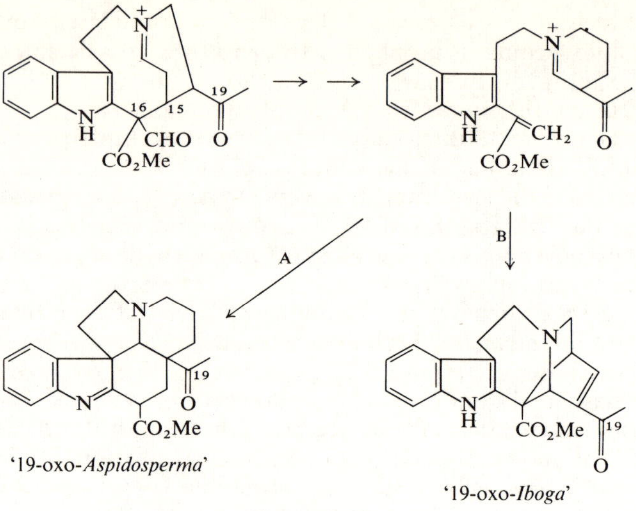

'19-oxo-*Aspidosperma*'

'19-oxo-*Iboga*'

Figure 3.26

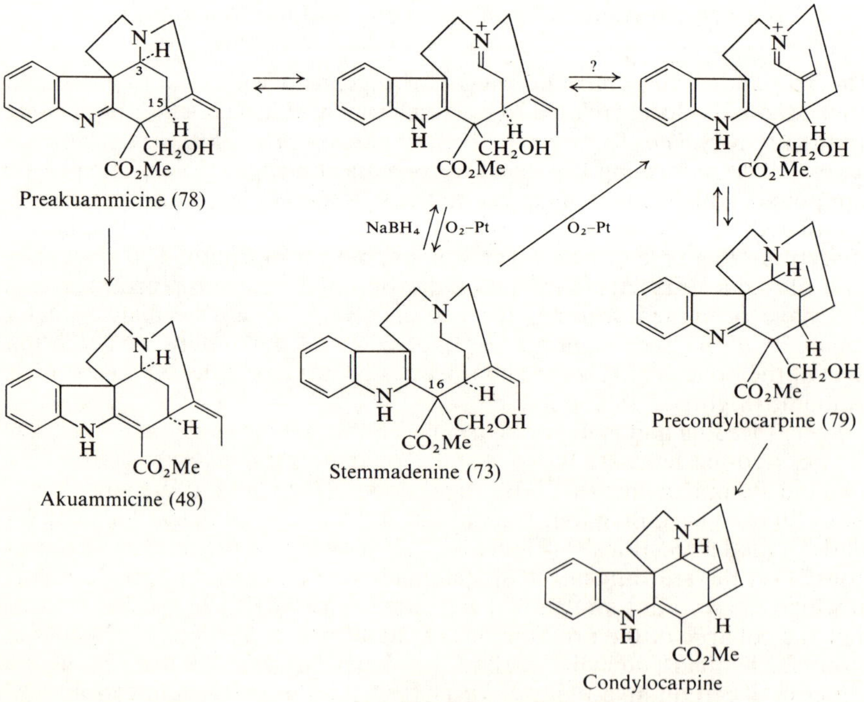

Preakuammicine (78)

Akuammicine (48)

NaBH$_4$ / O$_2$–Pt

O$_2$–Pt

Stemmadenine (73)

Precondylocarpine (79)

Condylocarpine

Figure 3.27

3.29) is difficult if not impossible to sever, the partial reactions of a previous brief report have now been studied and confirmed. In fact it is possible to cleave bond (– – –) in (83) at room temperature[103] under conditions which probably represent the first stages of the overall rearrangement reported earlier (Figure 3.29). In a second mild reaction precondylocarpine has been

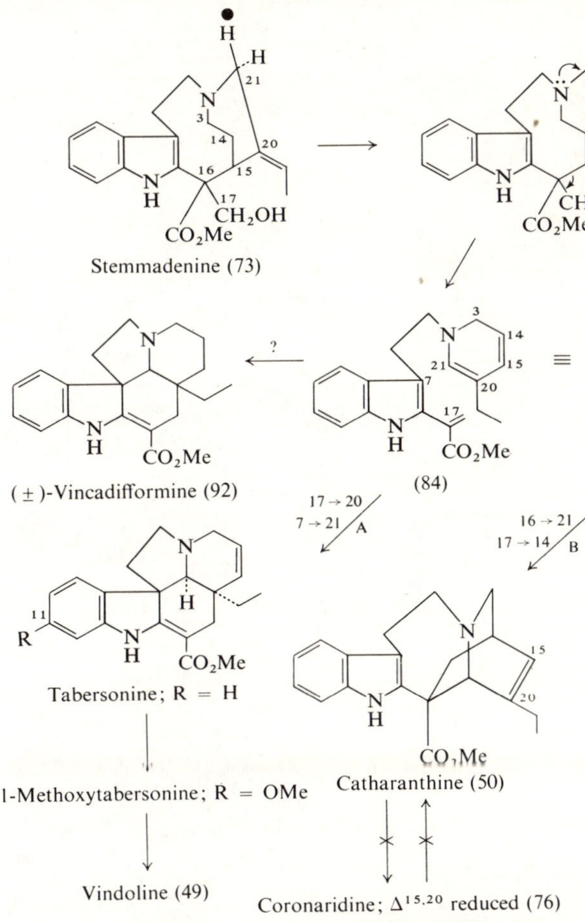

Stemmadenine (73)

(±)-Vincadifformine (92)

(84)

(84)

$17 \rightarrow 20$
$7 \rightarrow 21$ A

$16 \rightarrow 21$
$17 \rightarrow 14$ B

C $17 \rightarrow 20$

Tabersonine; R = H

Catharanthine (50)

Vincadine (91)

11-Methoxytabersonine; R = OMe

Vindoline (49)

Coronaridine; $\Delta^{15,20}$ reduced (76)

Figure 3.28

rearranged to (85) (Figure 3.29) which provides[103] even deeper detail for a second reaction sequence which was also challenged[106].

This review does not permit a detailed analysis of the conflicting viewpoints. It is sufficient to note at this juncture that all of the criticisms levelled by Smith *et al.*[106] are based on correlation (or lack of correlation) with an amorphous *optically active*[107] version of pseudo-catharanthine* which in fact may not have structure (86). All of the work reported by Scott has been related to pure crystalline pseudocatharanthine (86), $[\alpha]_{300-600}$ 0 degrees, m.p.

---

*In fact it is obvious from their statements[106] that Smith *et al.* are not aware of the optical purity of many of their preparations.

116–118 °C (identical with material supplied by Lilly Laboratories) obtained by generation of (90) which of course confers optical inactivity on *all* subsequent products.

In this critical account of the field the reviewer feels impelled to point out that in the original concept[102, 104] many of the partial reactions and

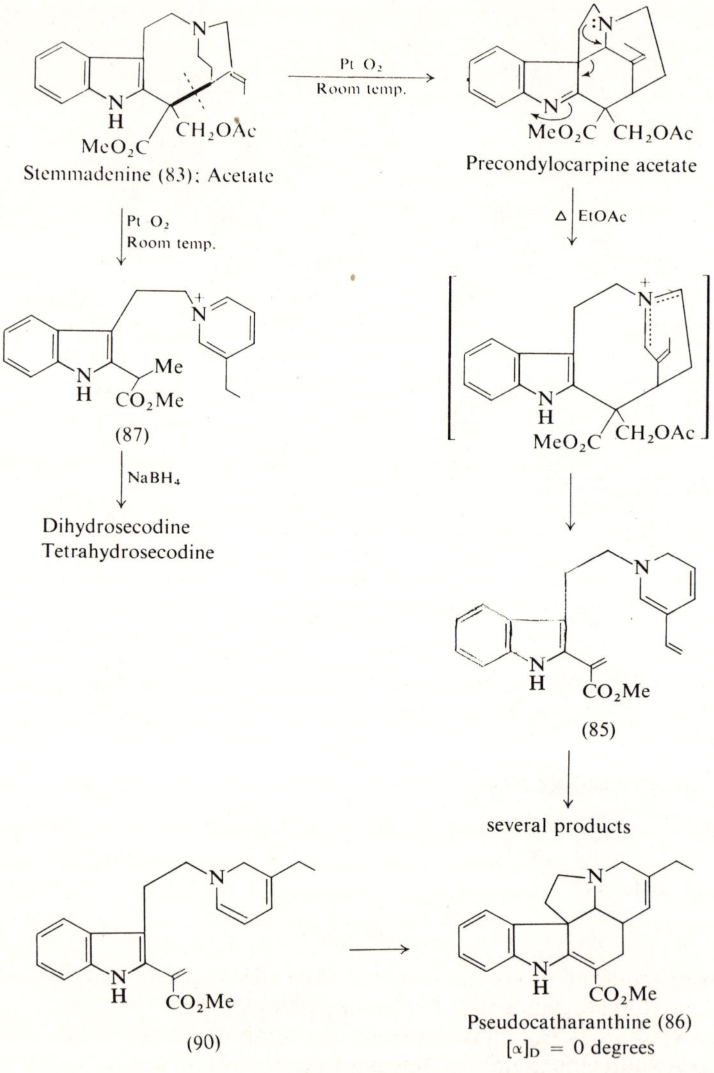
Pt O_2
Room temp.
MeO_2C CH_2OAc
Stemmadenine (83); Acetate
Precondylocarpine acetate
Pt O_2
Room temp.
Δ EtOAc
Me
CO_2Me
(87)
$NaBH_4$
Dihydrosecodine
Tetrahydrosecodine
MeO_2C CH_2OAc
CO_2Me
(85)
several products
CO_2Me
(90)
CO_2Me
Pseudocatharanthine (86)
$[\alpha]_D$ = 0 degrees

Figure 3.29

almost all of the subsequent chemistry and biochemistry of the secodine system (84, 90) were predicted on the basis of stemmadenine chemistry. In fact a very successful total synthesis of *Aspidosperma* alkaloids uses the recyclisation of a secodine system as a model[108]. Thus we would suggest that taken together with the obvious errors of fact and logic inherent in Smith's challenging paper[106], the subsequent chemistry and biochemistry has vindicated and indeed expanded successfully (in our hands and those of others) our original premise. At this juncture it can only be assumed that the two laboratories concerned were operating under different experimental conditions, albeit difficult to define, but by no means a unique situation in our own experience of successful biogenetic type synthesis!

Cleavage of both *a* or *b* bonds in catharanthine (50) and tabersonine (89) has also been used in the laboratory to prepare the secodine system as the salt (87) in high yield (Figure 3.30). These processes in fact constitute an ex-

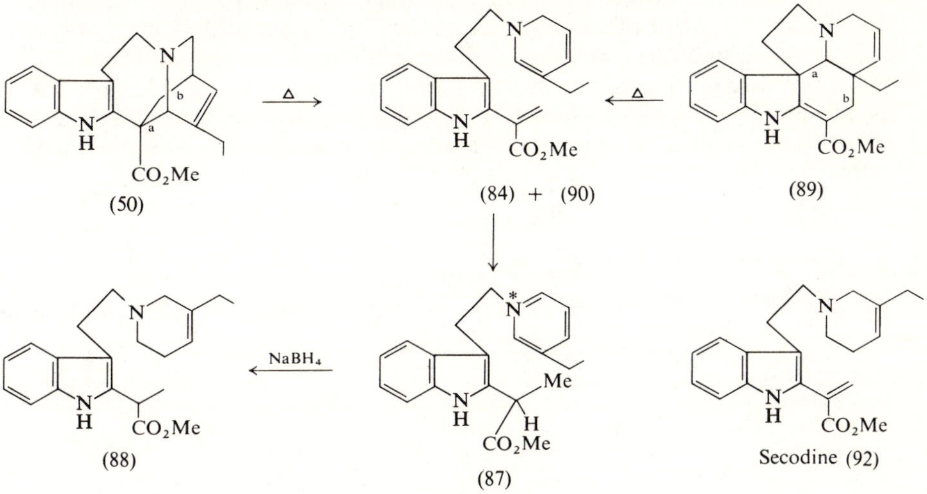

$$\underset{(50)}{} \xrightarrow{\Delta} \underset{(84) + (90)}{} \xleftarrow{\Delta} \underset{(89)}{}$$

$$\underset{(88)}{} \xleftarrow{NaBH_4} \underset{(87)}{} \qquad \underset{Secodine\ (92)}{}$$

Figure 3.30

cellent source of rare seco-alkaloids, e.g. dihydrosecodine (88) which is now available as a crystalline substance identical with natural material (Figure 3.30). Thus the acrylic ester system has been generated from *Strychnos*, *Iboga*, and *Aspidosperma* alkaloids of the correct oxidation level and frequently in good yield.

The observed conversion of stemmadenine (83) to vindoline (49) (*Aspidosperma*) and catharanthine (50) (*Iboga*) in *V. rosea* can be rationalised by invoking the acrylic ester (84), i.e. dehydrosecodine (Figures 3.25 and 3.28). Moreover, this type of intermediate (84) must surely play a part in the conversion of [O-methyl-^{3}H,11-^{14}C]- and [Ar-^{3}H]-tabersonine (89) to vindoline (49) and most remarkably to catharanthine (50) in *V. rosea*. The dynamic role of tabersonine (89) was also revealed by a short term incubation experiment in which [2-^{14}C]tryptophan gave a specific incorporation of 30% into tabersonine at 9 h [97]. At this time tabersonine contained *c.* 14% of all the radioactivity in the alkaloids. The counts in this key representative of the *Aspidosperma* family fell away rapidly and re-feeding (−)-tabersonine confirmed its conversion into the *Iboga* series, this time into coronaridine (77), the naturally occurring epimer of dihydro-catharanthine[97].

Returning to the key progenitor, stemmadenine (83), a radio profile of this compound proved to be unremarkable and a steady value of *c.* 1%

of the radioactivity from tryptophan was recorded over an 8-day experiment. An explanation of this behaviour is to suggest that administered stemmadenine finds its way to the precondylocarpine–preakuammicine set and that the secoalkaloids and hence the more complex families are derived by the process shown in Figure 3.28 which draws heavily on the model reactions developed in Figures 3.27 and 3.29.

An alternative transannular ring closure of vincadine (91) to vincadifformine (92) derivatives (see Figure 3.28), which provides a very attractive route for total synthesis in this series, was considered to be invalid[109, 110] as a biochemical route since the blank experiments gave higher incorporations than those with plant material present. However, in order to reach alkaloids such as vincadine (91) only one reactive terminus of the system is closed so that a stepwise mechanism cannot be entirely ruled out even at this stage. The isolation and synthesis of compounds related to secodine make this type of structure most attractive and recent feeding experiments indicate (albeit with rather low incorporation) that secodine (92) is transformed to vindoline in

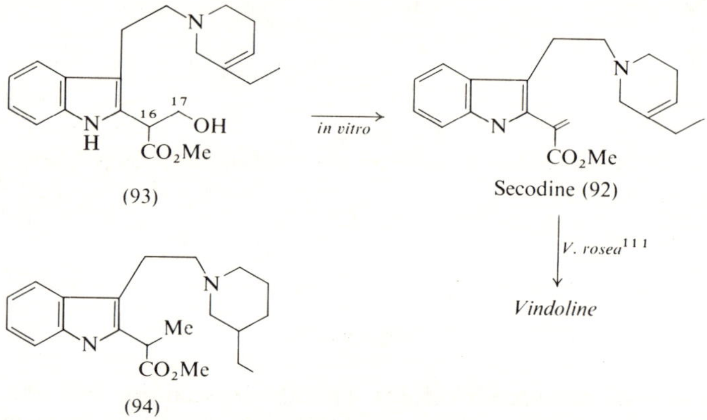

in vitro

(93)   Secodine (92)

V. rosea[111]

Vindoline

(94)

Figure 3.31

V. rosea[111]. The corresponding hydration product, dihydrosecodin-17-ol[112] (93) and the tetrahydro compound (94) as well as some dimers related to (92) have been found in **R. stricta** and **R. orientalis**[113] (Figure 3.31).

The experiments described above represent three of the four methods employed namely isolation, feeding of likely intermediates, and short-term labelling. The final method which ultimately sheds light on the fate of the protons of all the intermediates postulated so far and which constitutes an exacting probe for these concepts and their stereochemical consequences is the use of doubly labelled specimens of mevalonate, geraniol and loganin. Some recent data for this type of experiment is summarised in Table 3.1[83].

The following conclusions can be drawn from the data of Table 3.1. All of these conclusions accord with the schemes delineated in Figures 3.25 and 3.28.

(i) The formation of geraniol takes place with the same stereochemical control in *V. rosea* as in other terpenoid producing systems.

(ii) The conversions shown in Figure 3.28 to explain the incorporation of

Table 3.1 Tritium retention in MVA, geraniol and loganin metabolism in V. rosea

Precursor	% Retained ^{3}H*			
	Loganin	Corynanthe	Aspidosperma	Iboga
$[1\text{-}^3\text{H}_2, 2\text{-}^{14}\text{C}]$geraniol	45	45	47	48
$[2\text{-}^3\text{H}, 2\text{-}^{14}\text{C}]$geraniol	95	<5	<5	<5
$[4\text{-}R\text{-}^3\text{H}, 2\text{-}^{14}\text{C}]$MVA†	103	49	52	49
$[4\text{-}S\text{-}^3\text{H}, 2\text{-}^{14}\text{C}]$MVA‡	10±5	<5	<5	<5
$[5\text{-}RS\text{-}^3\text{H}_2, 2\text{-}^{14}\text{C}]$MVA	73	—	45	60
$[5\text{-}S\text{-}^3\text{H}, 5\text{-}^{14}\text{C}]$MVA	—	—	12	40
$[5\text{-}^3\text{H}, O\text{-}methyl\text{-}^3\text{H}]$loganin	—	96	99	102

*Several of the values are averaged.
†Fed as $(3R, 4R)+(3S, 4S)[4\text{-}^3\text{H}, 2\text{-}^{14}\text{C}]$MVA to test the $4R$ label ($3S$ not incorporated).
‡Fed as $(3R, 4S)+(3S, 4R)[4\text{-}^3\text{H}, 2\text{-}^{14}\text{C}]$MVA to test the $4S$ label ($3S$ not incorporated).

stemmadenine into the later alkaloids must involve a stereospecific retention of the C-21 proton marked (•) in Figure 3.28. The proton at C-21 which survives in vindoline and catharanthine is that derived from C-1 of loganin and geraniol and in fact must correspond to the 5-pro-R-proton of mevalonate.

(iii) The ^{3}H label at C-7 of loganin becomes C-15 in the *Corynanthé* alkaloids and survives all subsequent rearrangement. This implies that the stereochemistry at C-17 of loganin determines the configuration of the later alkaloids.

(iv) Conclusion (iii) has interesting implications for the biogenesis of ($\pm$)-akuammicine, a *racemic* alkaloid of the *Strychnos* family. One might be tempted to invoke an antipodal loganin to explain this compound. More likely is the unravelling and recombination of preakuammicine to give the antipodal and/or racemic series.

(v) ^{3}H at C-5 of loganin appears in all three classes of alkaloid at the expected position [C-3 in (47), (48) and (49)].

(vi) The experiment with 5R, S and 5S-^{3}H-MVA[114] shows stereoselectivity (not quite complete) for one of the protons in the steps from geissoschizine to vindoline (H_R, H_R) and catharanthine (H_R, H_S).

(vii) The conclusion (vi) implies that ($-$)-tabersonine (H_R, H_R) (89) cannot serve as a true precursor for catharanthine (H_R, H_S). In order to resolve this paradox it has been suggested[121] either that ($-$)-tabersonine is bioconverted in *Vinca* to catharanthine by a minor pathway[97, 114], or simply that by analogy with the variable rotations of vincadifformine (92) [($+$)-, ($-$)- or ($\pm$)-dihydrotabersonine] the optical purity of tabersonine isolated and used in all of the experiments is c. 80–90%. Thus it is the ($+$)-isomer (H_R, H_S) which serves as the precursor of catharanthine according to Figure 3.32 whilst ($-$)-tabersonine labels vindoline in the H_R, H_R mode.

Almost all of the above experiments were carried out with *V. rosea* according to the summary given in Figure 3.25. Mention must now be made of recent work with related species. The alkaloid vincamine (95) has been shown to be derived from geissoschizine, stemmadenine, secodine, and tabersonine in *V. minor* and the suggested route is shown in Figure 3.33[115].

Apparicine (96) which has apparently lost one of the carbons of the tryptamine C_2 side chain, is in fact derived from tryptophan by loss of C_2

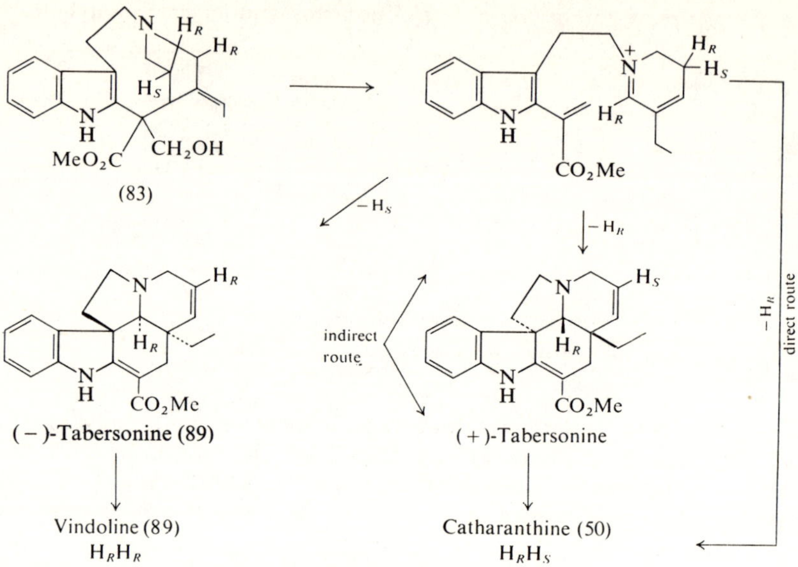

Figure 3.32

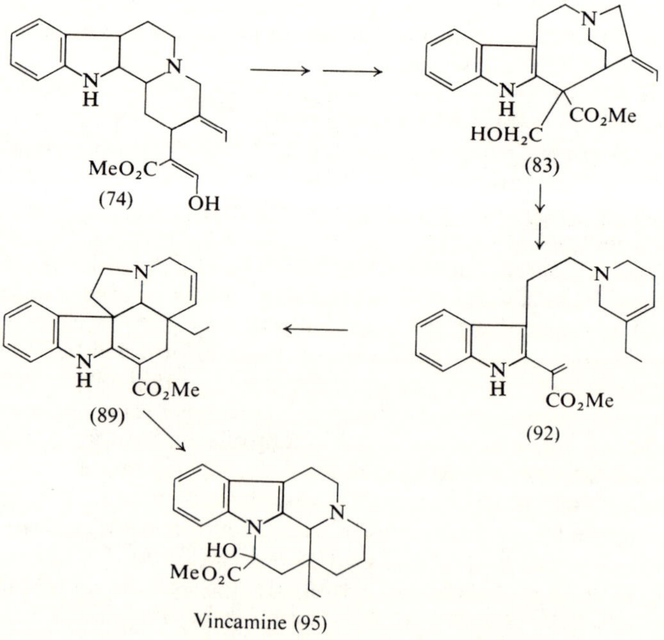

Figure 3.33

from the side chain as shown by feeding [2-^{14}C, Ar-^{3}H]- and [3-^{14}C, Ar-^{3}H]-tryptophan to *Aspidosperma pyricollum*[116]. The same plant incorporates [Ar-^{3}H]stemmadenine (83) rather than vallesamine (97) into apparicine indicating a late loss of the side chain carbon. Interestingly [2-^{14}C, Ar-^{3}H] tryptophan is not incorporated into uleine[116] (98) (Figure 3.34).

The role of glycine in indole alkaloid biosynthesis has been evaluated[117]. Apart from labelling the tryptophan side chain (via glycine–serine, *etc.*)

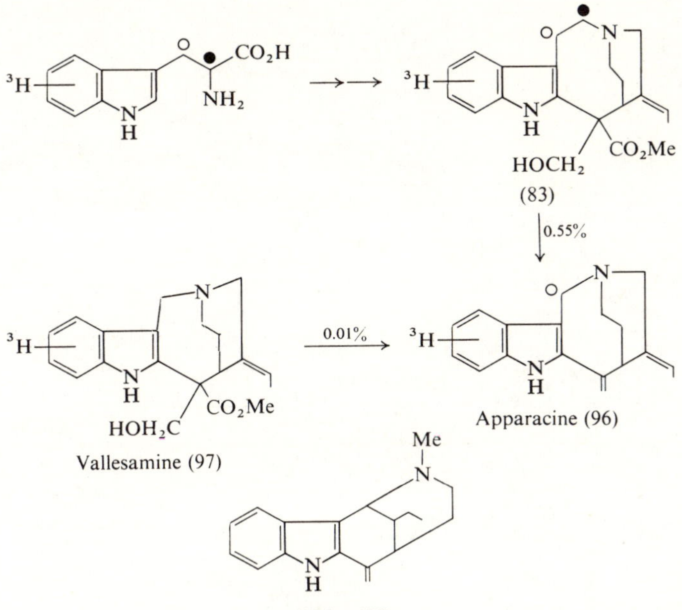

Vallesamine (97)

Apparacine (96)

Uleine (98)

Figure 3.34

and a very slight diffusion into the 'C$_1$' pool to give label in —OMe groups, it appears that the role of glycine in the plants studied so far is at best a tertiary one.

In *Strychnos nux-vomica* strychnine (99) appears to be formed from tryptophan and geraniol with addition of the C$_2$ unit (Figure 3.35) from acetate[122]. *Strychnos* biosynthesis has proved much more difficult to evaluate in *S. nux-vomica* and *S. lucida* (strychnine, brucine) than in *V. rosea* (akuammicine), perhaps due to permeability problems. For example the Wieland Gumlich (W–G) aldehyde[100, 122] (100), diaboline[122] (101) are within experimental error (<0.001%) not incorporated into strychnine. By autoradiography and isotope dilution analysis however, the W–G aldehyde, geissoschizine, and more importantly, decarbomethoxy geissoschizine (but not diaboline) have been found in very young *S. nux vomica* seedlings[100].

In isolation studies, the following compounds which are believed to be potential precursors of the main families have been obtained from 'early' *V. rosea* preparations (24–100 h after germination) and have not previously been found in the mature plant: geissoschizine hydroxyindolenine (80),

the diol (81), geissoschizine oxindole (77), preakuammicine (78), stemmadenine (83), and tabersonine (89) (see Figure 3.25). Short-term incubation
has revealed an unknown alkaloid on the time scale between vincoside and
geissoschizine which after 5 min carries *c*. 35% of the total radioactivity
and which contains almost no counts after 1 h[97].

The isolation of coronaridine (77) could be significant since it appears at
45 h., i.e. much earlier than catharanthine. Very recently it has been shown[118]
that there is no incorporation of [*O*-methyl-³H]- or [Ar-³H]-coronaridine

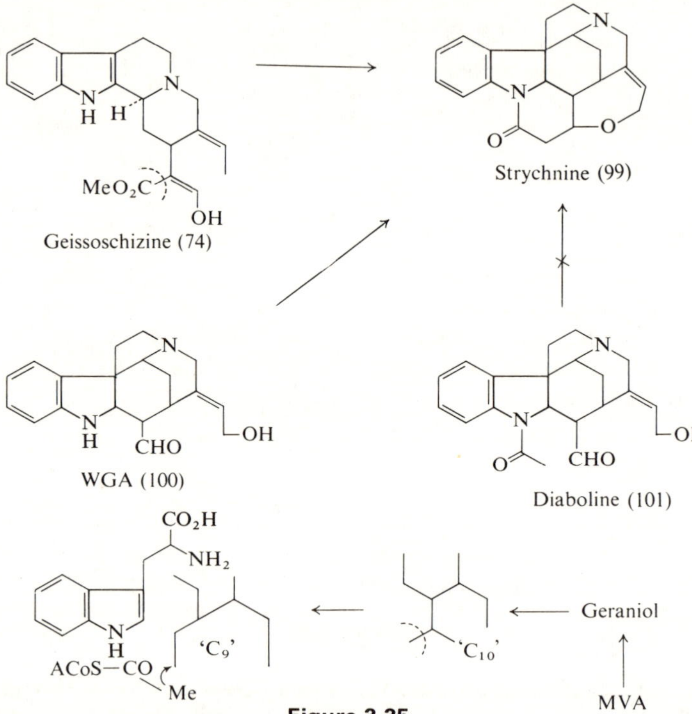

Geissoschizine (74)

Strychnine (99)

WGA (100)

Diaboline (101)

Geraniol

MVA

Figure 3.35

(77) into catharanthine (50) and very surprisingly that [Ar-³H]-catharan-
thine did not produce [Ar-³H]coronaridine in *V. rosea* even after several
weeks. This suggests that the common precursor of *Iboga* alkaloids might
be (102) and in turn this could be formed by a dehydrogenation of stemmadenine (or of the acrylic ester) as suggested in Figure 3.36 (for a laboratory
analogy, see Figure 3.29). Further work on this aspect is required, since
non-permeability could always be an alternative explanation. An outcome of this proposal is that tabersonine (89) would be reached via (103)
and (104).

3.7.4 The dimeric antitumour alkaloid VLB

The structure of vinblastine (VLB) (105) is formally composed of vindoline
and a modified hydroxylated catharanthine unit. Several schemes for the

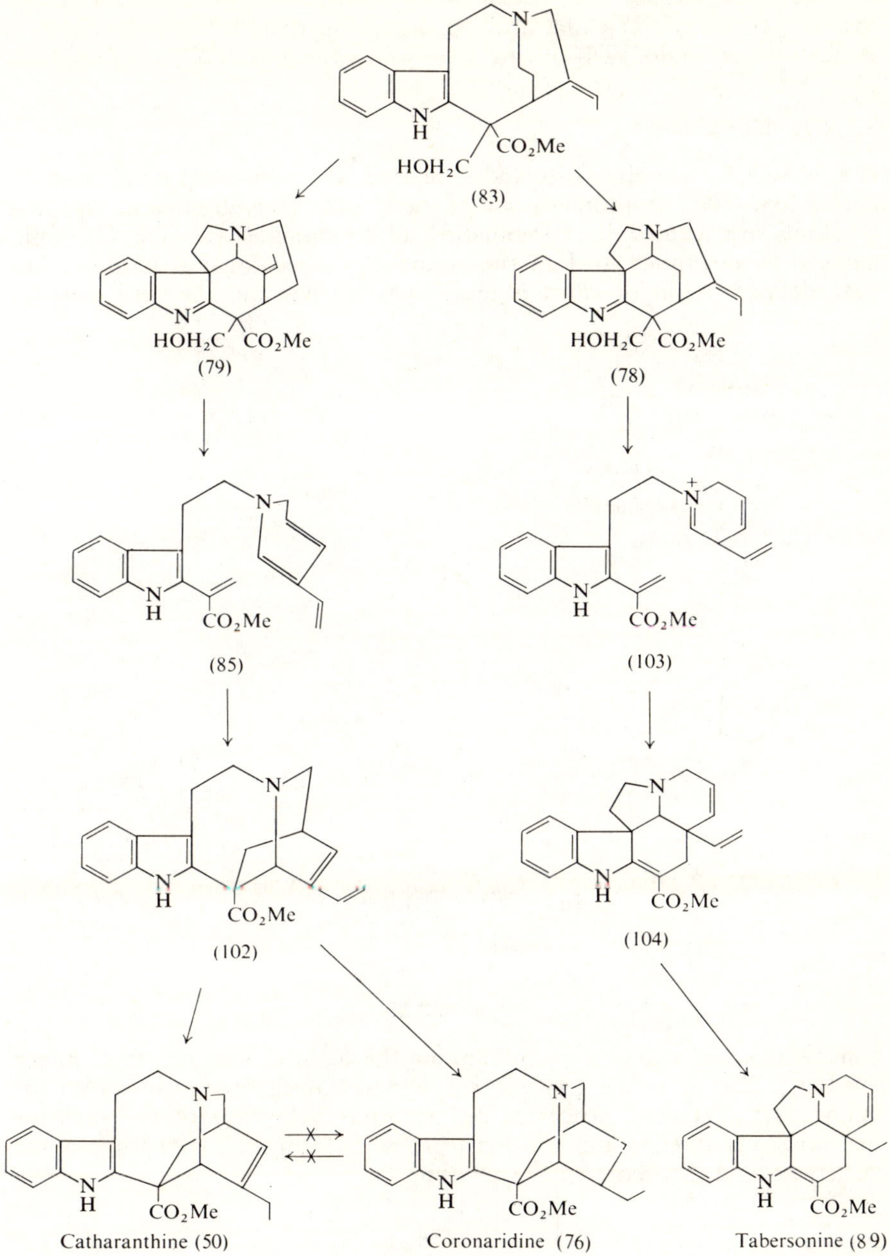

Figure 3.36

connection of *Aspidosperma* and *Iboga* or *Cleavamine* type alkaloids can be envisaged. So far the only biochemical evidence shows that in *V. rosea* neither [*O*-methyl-^{3}H]catharanthine nor [*O*-methyl-^{3}H]coronaridine (76) is incorporated into VLB during a 5-week feeding period[119] (Figure 3.37).

3.7.4.1 Conclusion

It now seems clear that with the exception of a few compounds such as cordifoline (106), tryptophan suffers early decarboxylation and the first alkaloids are formed by combination of tryptamine with the C_{10} unit, namely secologanin. Most of the geraniol ⟶ secologanin pathway has been defined. A major effort is under way to map out the very complex

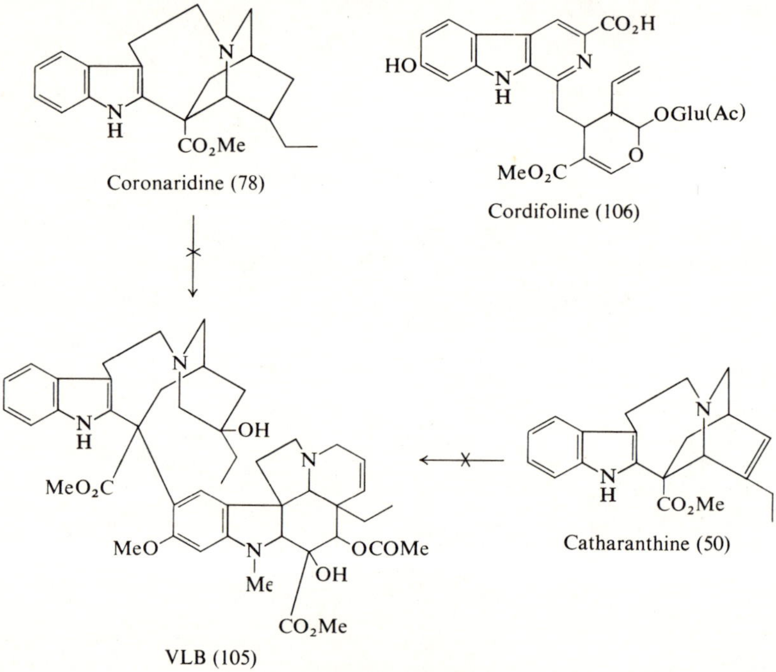

Coronaridine (78)

Cordifoline (106)

VLB (105)

Catharanthine (50)

Figure 3.37

connections and rearrangements linking the main classes and their minor offshoots. The 'secodine concept' serves as a satisfactory rationale of the major rearrangement processes but rigorous definition of the intricate mechanism is awaited and will most probably form the major topic of the next review on this series in 2 years' time.

Note added in proof

Addenda (Section 3.7.3)
1. Recent experiments have shown[103] that (+) catharanthine is transformed *in vitro* to (+) coronaridine. This correlation sets catharanthine

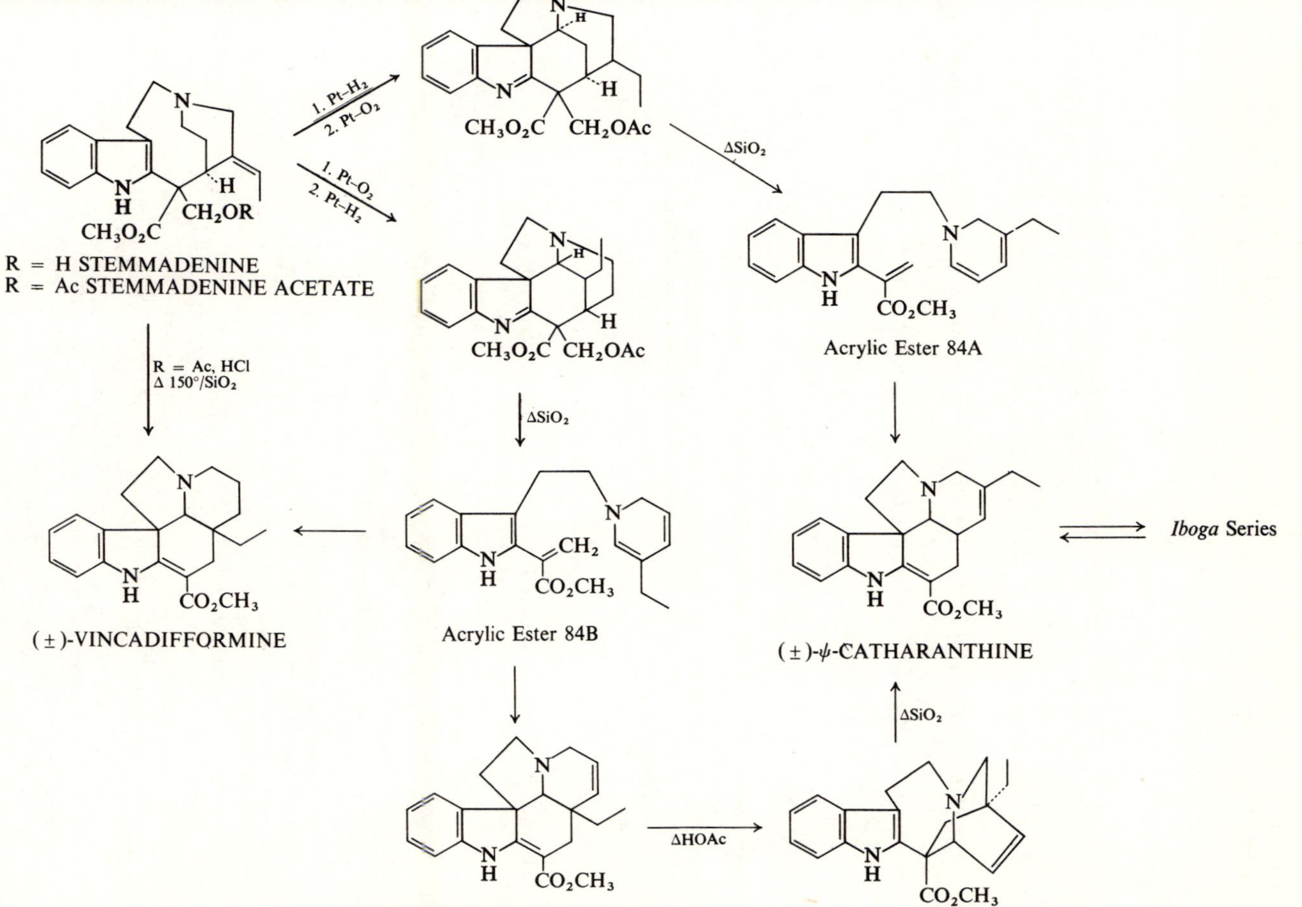

Figure 3.38 Biogenetic-type synthesis of *Iboga* and *Aspidosperma* alkaloids

apart as the only member of the *Iboga* series (except for the dimeric alkaloids VLB, etc.) with the absolute stereochemistry depicted in this review and throughout the current literature. All of the other members, e.g. Ibogamine and Coronaridine must now be represented by the absolute configuration (107).

$$CO_2Me$$

(107)

2. The mechanisms implicit in the biogenetic schemes suggested in Figures 3.28, 3.31, 3.32, 3.33 and 3.36 have received remarkable and extremely powerful confirmation from the *in vitro* experiments[123] summarised in Figure 3.38.

References

1. Meister, A. (1965). *Biochemistry of the Amino Acids,* Vol. 2, 2nd Ed., Chapter 6. (New York: Academic Press)
2. Daly, J. W. and Witkop, B. (1963). *Angew Chem. Internat. Ed.,* **2,** 421
3. Brack, A., Hofmann, A., Kalberer, E., Kobel, H. and Rutschmann, J. (1961). *Arch. Pharm.,* **294,** 230
4. Spenser, I. D. (1968). *Comprehensive Biochemistry,* Vol. 20, Chapter 6. (M. Florkin and E. H. Stotz, editors). (New York: Elsevier)
5. O'Donovan, D. and Leete, E. (1963). *J. Amer. Chem. Soc.,* **85,** 461
6. Mudd, S. H. (1961). *Nature (London),* **189,** 489
7. Breccia, A., Crespi, A. M. and Rampi, M. (1966). *Z. Naturforsch,* **21B,** 1243
8. Hesse, M. (1968). *Indole Alkaloide in Tabellen.* (Vienna, Springer-Verlag); Snieckus, V. (1968). *The Alkaloids.* (R. H. F. Manske, editor). (New York: Academic Press); Joule, J. A. (1971). *Spec. Per. Rep., The Alkaloids,* I, Chapter 13. (London: Chemical Society)
9. Spenser, I. D. (1968). *Comprehensive Biochemistry,* 20, Chapter VI. (New York: Elsevier)
10. McLean, S. and Murray, D. G. (1970). *Can. J. Chem.,* **48,** 867
11. Hahn, G. and Rumpf, F. (1938). *Chem. Ber.,* **71,** 2141
12. Schopf, C. (1937). *Angew Chem.,* **50,** 779
13. Hahn, G. and Stiehl, K. (1936). *Chem. Ber.,* **69,** 2627
14. O'Donovan, D. G. and Kenneally, M. F. (1967). *J. Chem. Soc. C,* 1109
15. Stolle, K. and Gröger, D. (1968). *Arch. Pharm.,* **301,** 561
16. Robinson, (Sir) R. (1936). *J. Chem. Soc.,* 1079
17. Slaytor, M. and McFarlane, I. J. (1968). *Phytochemistry,* **7,** 605
18. McFarland, I. J. and Slaytor, M. (1971). *Phytochemistry,* **10**
19. Weygand, F. and Floss, H. G. (1963). *Angew. Chem.,* **75,** 783
20. Plieninger, H., Immel, H. and Völkl, L. (1967). *Liebig's Ann. Chem.,* **706,** 223
21. Günther, H. F. and Floss, H. G. (1971). Private communication
22. Plieninger, H., Fischer, R. and Liede, V. (1964). *Liebig's Ann. Chem.,* **672,** 223
23. Robbers, J. E. and Floss, H. G. (1969). *Tetrahedron Letters,* 1857
24. Aguvell, S. and Lindgren, J. E. (1968). *Tetrahedron Letters,* 5127
25. Robbers, J. E. and Floss, H. G. (1968). *Arch. Bioch. Biophys.,* **126,** 967
26. Stauffacher, D. and Tscherter, H. (1964). *Helv. Chim. Acta,* **47,** 2186
27. Acklin, W., Fehr, T. and Arigoni, D. (1966). *Chem. Commun.,* 799
28. Fehr, T., Acklin, W. and Arigoni, D. (1966). *Chem. Commun.,* 801
29. Groger, D., Erge, D. and Floss, H. G. (1966). *Z. Naturforsche,* **21b,** 827
30. Voigt, R., Bornschein, M. and Rabitzsch, G. (1967). *Pharmazie,* **22,** 326

31. Floss, H. G., Homemann, V., Schilling, N., Gröger, D. and Erge, D. (1968). *J. Amer. Chem. Soc.*, **90**, 6500
32. Floss, H. G. (1967). *Chem. Commun.*, 804
33. Baxter, R. M., Kandel, S. I., Okany, A. and Tam, K. L. (1962). *J. Amer. Chem. Soc.*, **84**, 4350
34. Floss, H. G. (1969). *4th International Symposium: Biochemistry and Physiology of Alkaloids*, Halle. (Berlin: Abhandlagen de Deutsche Akademie der Wissenschaften)
35. Agurell, S. and Ramstad, E. (1962). *Arch. Biochem. Biophys.*, **98**, 457
36. Naidoo, B., Cassady, J. M., Blair, G. E. and Floss, A. G. (1970). *Chem. Commun.*, 671
37. Ogunlana, E. O., Wilson, B. J., Tyler, V. E. and Ramstad, E. (1970). *Chem. Commun.*, 775
38. Ogunlana, E., Ramstad, E. and Tyler, V. E. (1969). *J. Pharm. Sci.*, **58**, 143
39. Belivean, J. and Ramstad, E. (1966). *Lloydia*, **29**, 234
40. Taylor, E. H., Goldner, K. J., Pong, S. F. and Shough, H. R. (1966). *Lloydia*, **29**, 239
41. Taylor, E. H. and Shough, H. R. (1967). *Lloydia*, **30**, 197
42. Chan Lin, W. N., Ramstad, E. and Taylor, E. H. (1967). *Lloydia*, **30**, 202
43. Agurell, S. (1966). *Acta Pharm. Suecica*, **3**, 71
44. Basmadjian, G., Floss, H. G., Gröger, D. and Erge, D. (1969). *Chem. Commun.*, 418
45. Minghetti, A. and Arcamone, F. (1969). *Experientia*, **25**, 926
46. Nelson, V. and Agurell, S. (1969). *Acta Chem. Scand.*, **23**, 3393
47. Holzapfel, C. W. (1968). *Tetrahedron*, **24**, 2101
48. Holzapfel, C. W. and Wilkins, D. C. (1968). *5th International Symposium on the Chemistry of Natural Products*, Abstract C65. (London)
49. Homemann, U., Hurley, L. H., Speedle, M. K. and Floss, H. G. (1969). *Chem. Commun.*, 245; (1970). *Tetrahedron Letters*, 2255; (1971). *J. Amer. Chem. Soc.*, **93**, 3028
50. Kirby, G. W., Shah, S. W. and Herbert, E. J. (1969). *J. Chem. Soc. C*, 1916
51. Hall, E. S., McCapra, F. and Scott, A. I. (1967). *Tetrahedron*, **23**, 4131
52. O'Donovan, D. G. and Keogh, M. F. (1966). *J. Chem. Soc. C*, 1570
53. Battersby, A. R., Burnett, A. R. and Parsons, P. G. (1968). *Chem. Commun.*, 1282
54. Battersby, A. R., Burnett, A. R. and Parsons, P. G. (1969). *J. Chem. Soc. C*, 1193
55. Kutney, J. P., Cretney, W. J., Hadfield, J. R., Hall, E. S., Nelson, V. R. and Wigfield, D. C. (1968). *J. Amer. Chem. Soc.*, **90**, 3566
56. Scott, A. I. and Cherry, P. C., unpublished results
57. Battersby, A. R. (1967). *Pure and Applied Chemistry*, **14**, 117
58. Scott, A. I. (1970). *Accounts Chem. Res.*, **3**, 151
59. Begue, J. P. (1969). *Bull. Soc. Chim. France*, 2545
60. Leete, E. (1969). *Accounts Chem. Res.*, **2**, 59
61. Spenser, I. D. (1968). *Comprehensive Biochemistry*, **20**, 231. (New York: Elsevier)
62. Leete, E. (1967). *Biogenesis of Natural Compounds*, Chapter 17. (P. Bernfeld, editor). (Oxford: Pergamon)
63. Thomas, R. (1961). *Tetrahedron Letters*, 544
64. Wenkert, E. (1962). *J. Amer. Chem. Soc.*, **84**, 98
65. Thomas, R. (1965). *Biogenesis of Antibiotic Substances* (Z. Vanek and Z. Hostalek, editors), (Symposium, June 1964), p. 163. (New York: Academic Press)
66. Escher, P., Loew, P. and Arigoni, D. (1970). *Chem. Commun.*, 823
67. Gröger, D., Stolle, K. and Mothes, K. (1966). *Z. Naturforsch.*, **21b**, 206; (1967). *Arch. Pharm.*, **300**, 393
68. Battersby, A. R., Brown, R. T., Kapil, R. S., Knight, J. A., Martin, J. A. and Plunkett, A. O. (1966). *Chem. Commun.*, 888, 890
69. Leete, E. and Veda, S. (1966). *Tetrahedron Letters*, 4915
70. Money, T., Wright, I. G., McCapra, F., Hall, E. S. and Scott, A. I. (1968). *J. Amer. Chem. Soc.*, **90**, 4144
71. Battersby, A. R., Kapil, R. S., Martin, J. A. and Mo, L. (1968). *Chem. Commun.*, 133
72. Arigoni, D. and Loew, P. (1968). *Chem. Commun.*, 137
73. Battersby, A. R., Burnett, A. R., Hall, E. S. and Parsons, P. B. (1968). *Chem. Commun.*, 1582
74. Coscia, C. J. and Guarnaccia, R. (1968). *Chem. Commun.*, 138
75. Escher, S., Loew, P. and Arigoni, D. (1970). *Chem. Commun.*, 823
76. Battersby, A. R., Brown, S. H. and Payne, T. G. (1970). *Chem. Commun.*, 827
77. Leete, E. and Bowman, R. M. (1969). *Phytochemistry*, **8**, 1003
78. Loew, P., von Szczepanski, C., Coscia, C. J. and Arigoni, D. (1968). *Chem. Commun.*, 1276

79. Battersby, A. R., Burnett, A. R., Knowles, G. D. and Parsons, P. G. (1968). *Chem. Commun.*, 1277

80. Battersby, A. R., Burnett, A. R. and Parsons, P. G. (1968). *Chem. Commun.*, 1280

81. Battersby, A. R., Burnett, A. R. and Parsons, P. G. (1969). *J. Chem. Soc. C*, 1187

82. Souzu, I. and Mitsuhashi, H. M. (1970). *Tetrahedron Letters*, 191

83. Battersby, A. R. (1971). *Specialist Periodical Report, The Alkaloids*, Vol. I, p. 39–40. (London: Chemical Society)

84. Inoue, H., Veda, S. and Takeda, Y. (1968). *Tetrahedron Letters*, 3453

85. Guarnaccia, R., Botta, L. and Coscia, C. J. (1970). *J. Amer. Chem. Soc.*, **92**, 6098

86. Guarnaccia, R. and Coscia, C. J. (1971). *J. Amer. Chem. Soc.*, **93**, 6320

87. Coscia, C. J., Madyastha, K. M. and Guarnaccia, R. (1971). *Fed. Proc., Fed. Amer. Soc. Exp. Biol.*, **30**, 1472

88. Coscia, C. J., Guarnaccia, R. and Botta, L. (1969). *Biochemistry*, **8**, 5236

89. Coscia, C. J., Botta, L. and Guarnaccia, R. (1970). *Arch. Biochem. Biophys.*, **136**, 498

90. Battersby, A. R., Burnett, A. R. and Parsons, P. G. (1968). *Chem. Commun.*, 1282

91. Battersby, A. R., Burnett, A. R. and Parsons, P. B. (1969). *J. Chem. Soc. C*, 1193

92. Battersby, A. R. and Parry, R. J. (1971). *Chem. Commun.*, 902 and refs. therein

93. Qureshi, A. A. and Scott, A. I. (1968). *Chem. Commun.*, 948

94. Battersby, A. R., Byrne, J. C., Kapil, R. S., Martin, J. A., Payne, T. G., Arigoni, D. and Loew, P. (1968). *Chem. Commun.*, 951

95. Battersby, A. R. and Hall, E. S. (1969). *Chem. Commun.*, 793

96. Scott, A. I., Cherry, P. C. and Qureshi, A. A. (1969). *J. Amer. Chem. Soc.*, **91**, 4932

97. Scott, A. I., Slaytor, M. B., Reichardt, P. B. and Sweeny, J. G. (1971). *Bio-organic Chemistry*, **1**, 157

98. Scott, A. I. and Bennett, C. R., unpublished observations

99. Sakai, S., Aimi, N., Kubo, A., Kitagawa, M., Shiratori, M. and Haginiwa, J. (1971). *Tetrahedron Letters*, 2057; Aimi, N., Sakai, S., Iitaka, Y. and Itai, A. (1971). *Tetrahedron Letters*, 2061

100. Scott, A. I. and Heimberger, S., unpublished work

101. Wenkert, E. and Wickberg, B. (1965). *J. Amer. Chem. Soc.*, **87**, 1580

102. Scott, A. I. (1968). *Chimia*, **22**, 311; Scott, A. I. (January, 1968). *2nd Natural Products Symposium*, Jamaica

103. Scott, A. I. and Wei, C. C., unpublished work

104. Qureshi, A. A. and Scott, A. I. (1968). *Chem. Commun.*, 945, 947

105. Cf. Battersby, A. R. (1968). *Chimia*, **22**, 313

106. Brown, R. T., Hill, J. S., Smith, G. F., Stapleford, K. S. T., Porsson, J., Muquet, M. and Kunesch, N. (1969). *Chem. Commun.*, 1475

107. Brown, R. T., Hill, J. S., Smith, G. F. and Stapleford, K. S. J. (1971). *Tetrahedron*, **27**, 5217

108. Ziegler, F. E. and Spitzner, E. B. (1970). *J. Amer. Chem. Soc.*, **92**, 3492

109. Kutney, J. P., Beck, J. F., Ehret, C., Poulton, G., Sood, R. S. and Westcott, N. D. (1971). *Bio-organic Chemistry*, **1**, 194

110. Kutney, J. P., Ehret, C., Nelson, U. R. and Wigfield, D. C. (1968). *J. Amer. Chem. Soc.*, **90**, 5929

111. Kutney, J. P., Beck, J. F., Eggers, N. J., Hanssen, H. W., Sood, R. S. and Westcott, N. D. (1971). *J. Amer. Chem. Soc.*, **93**, 7322

112. Battersby, A. R. and Bhatnagar, A. K. (1970). *Chem. Commun.*, 193

113. For review see Joule, J. A. (1971). *Specialist Periodical Report, The Alkaloids*, Vol. I, Chapter 13. (London: Chemical Society)

114. Scott, A. I., Reichardt, P. B. and Sweeny, J. G., submitted for publication

115. Kutney, J. P., Beck, J. F., Nelson, V. R. and Sood, R. S. (1971). *J. Amer. Chem. Soc.*, **93**, 255

116. Kutney, J. P., Nelson, V. R. and Wigfield, D. C. (1969). *J. Amer. Chem. Soc.*, **91**, 4278, 4279

117. Kutney, J. P., Beck, J. F., Nelson, U. K., Stuart, K. L. and Bose, A. K. (1970). *J. Amer. Chem. Soc.*, **92**, 2174

118. Scott, A. I., Reichardt, P. B. and Michael, J., unpublished work

119. Michael, J., unpublished work

120. Julia, M. and Ricaleus, F. (1969). *Compt. Rend.*, **269C**, 51

121. Scott, A. I. (1972). 4th Natural Products Symposium, Jamaica

122. Schlatter, Ch., Waldner, E. E., Schmid, H., Maier, W. and Gröger, D. (1969). *Helv. Chim. Acta,* **52,** 776
123. Scott, A. I. and Wei, C. C. (1972). *J. Amer. Chem. Soc.* (Nov. 15). (In press)

4
Tuberostemonine and Related Compounds: The Chemistry of the *Stemona* Alkaloids

M. GÖTZ
Ayerst Research Laboratories, Montreal, Quebec

G. M. STRUNZ
Canadian Forestry Service, Fredericton, New Brunswick

4.1 INTRODUCTION

The physiological or medicinal activity associated with a plant species has, in many cases, provided the initial stimulus for research on its metabolic products. Thus, in prefacing a report on his early studies on tuberostemonine, isolated from *Stemona sessilifolia*, Schild[1] noted that extracts of *Stemona* species have been used in their countries of origin, China and Japan, as drugs for the treatment of respiratory diseases and as anthelmintics.

A recent investigation of the biological properties of the major *Stemona* alkaloid, tuberostemonine, indicates that this substance does not contribute to the therapeutic activity attributed to *Stemona* extracts. Tuberostemonine shows no effect against gram-positive or -negative bacteria *in vitro*, and is inactive against common fungi of human pathogenesis[2]. The alkaloid had no effect against histamine aerosol induced bronchospasm in conscious guinea-pigs[3] and was inactive when tested as an anthelmintic against *Hymenolepis nana* and *Nematospiroides dubius*[4].

Notwithstanding the questionable utility of *Stemona* species for therapeutic purposes, the chemical studies have yielded a handsome dividend in terms of the discovery of a novel structural class of alkaloids, apparently unique to the genus. After elucidation of the structure of tuberostemonine, the prototype of this class, structures were advanced for several closely related minor alkaloids within a relatively short interval.

This review describes studies directed towards structure determination of *Stemona* alkaloids, and also attempts to shed some light on the confusing literature reports of minor alkaloids of, as yet, undefined structure.

4.2 ALKALOIDS OF DEFINED STRUCTURE

4.2.1 Tuberostemonine

Tuberostemonine is obtained from two Roxburghiaceae, *Stemona tuberosa* Loureiro (Tamabyakubu)[5] of Formosan origin, and *Stemona sessilifolia* Miq.[1].

4.2.1.1 Early studies

The correct elemental composition of the alkaloid, $C_{22}H_{33}NO_4$, was established by Schild[1], who also recognised the tertiary nature of the nitrogen, the presence of lactone functionality and the oxidation to a pyrrole under mild conditions. On permanangate oxidation, Schild obtained 1-methyl-succinic acid and a compound with analytical data corresponding to $C_{17}H_{25}NO_3$. Kondo *et al.*[6] reported, in addition, the absence of *N*-methyl and *O*-methyl groups. The Hofmann degradation[7], cyanogen bromide reaction[8] and further oxidation studies[9, 10] were inconclusive. Hydrogenation experiments demonstrated the absence of carbon–carbon double bonds[11].

4.2.1.2 Structure determination

The initiation of the work on tuberostemonine at the University of New Brunswick in the autumn of 1960, was favoured by the ready availability of substantial quantities of the pure alkaloid, obtained from *Stemona tuberosa*[12] using the procedure developed at the ITSUU laboratory[13]. The late 1950s had seen exciting developments in the application of n.m.r. to problems in structural organic chemistry. This powerful new technique, in combination with other physical methods, was of central importance during the course of the structural studies described below. Spectroscopic examination of tuberostemonine confirmed some of the earlier observations. Thus, the infrared spectrum demonstrated the absence of hydroxyl or N–H groups, and the presence of γ-lactone functionality was substantiated by absorption at 1765 cm^{-1}. Saponification experiments provided evidence for the existence of two lactone groups[14]. The absence of a conjugated chromophore was indicated by the ultraviolet spectrum[9]. The n.m.r. spectrum revealed the presence of three C-methyl groups. Two sharp peaks, at τ 8.63 and 8.77 were subsequently shown to comprise superimposed doublets, corresponding to two secondary methyl groups. A third methyl signal, split into a poorly resolved triplet at τ 9.02 indicated the presence of an ethyl group. The latter had been postulated previously, on the basis of the detection of propionic acid when the Karrer procedure for C-alkyl group determination was applied to tuberostemonine[14].

Some of Schild's experiments[1] were repeated in order to establish the arrangement of atoms in the neighbourhood of the nitrogen. The mild oxidation of tuberostemonine to the pyrrole derivative, and the permanganate oxidation, appeared to be particularly promising in this regard. The pyrrole could, in fact, be readily obtained, in good yield, from tuberostemonine mother liquors which had been exposed to air for some time. A single pyrrole β-proton at τ 4.08 in the n.m.r. spectrum, split by 1,3 coupling into a narrow doublet ($J = 2.2 \text{ Hz}$), immediately clarified the substitution pattern as shown (1). In view of the extremely mild conditions under which the pyrrole was formed, a rearrangement in the course of the oxidation seemed unlikely. As an additional bonus, the two methyl doublets, superimposed in the spectrum of tuberostemonine, were now resolved (four peaks) demonstrating that each methyl group was attached to a carbon atom bearing a single hydrogen. Furthermore, comparison of the spectrum of the pyrrole derivative with that of tuberostemonine revealed a downfield shift of $c.$ 1 p.p.m. for one of two protons originally included in a multiplet at τ 5.62. In view of the functionality established for the alkaloid, it is reasonable to assign these low-field signals to the γ-hydrogens of the two lactone rings, one of which becomes significantly more deshielded in the oxidised product due to the proximity of the pyrrole system.

The relationship of the lactone thus affected to the pyrrolidine ring was clarified by examination of the products obtained from permanganate oxidation of tuberostemonine. In agreement with Schild's report[1], 1-methylsuccinic acid, $C_5H_8O_4$ (2), and a non-basic substance, $C_{17}H_{25}NO_3$, were isolated. These two fragments account for all carbons of the alkaloid. The larger fragment showed a band at 1685 cm^{-1} in the i.r. spectrum, indicative of γ-lactam functionality, thus providing further evidence for the presence of a pyrrolidine ring in the starting material. Lactone absorption was still present at 1765 cm^{-1}. The product can be represented by the partial structure (3). The carbon atom of the lactam carbonyl was clearly the point of attachment of the five-carbon moiety, allowing the part-structure of tuberostemonine to be extended to (4). The methyl group was placed α to the carbonyl, since its signal in the n.m.r. spectrum was relatively deshielded. Additional proof for this assignment was obtained later.

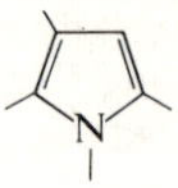

$$MeCH \cdot CO_2H$$
$$CH_2CO_2H$$

(1)    (2) 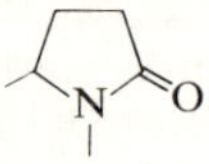   (3)

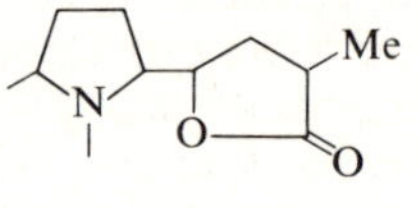

(4)

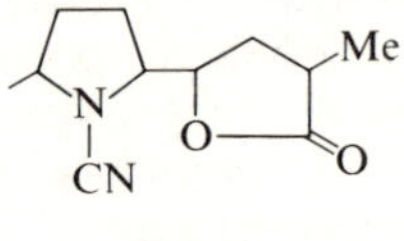

(5a) X = Br
(5b) X = OAc

$$-CH_2X$$

(5)

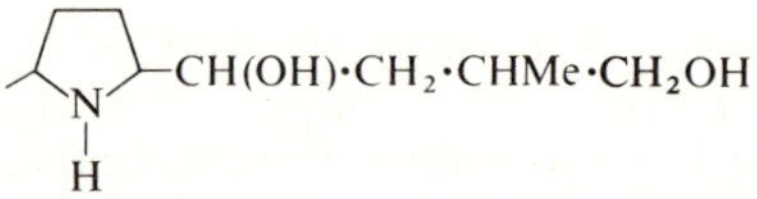

$$-CH(OH)\cdot CH_2 \cdot CHMe \cdot CH_2OH$$

$$-Me$$

(6)

$$-CH(OH)\cdot CH_2 \cdot CHMe \cdot CH_2OH$$

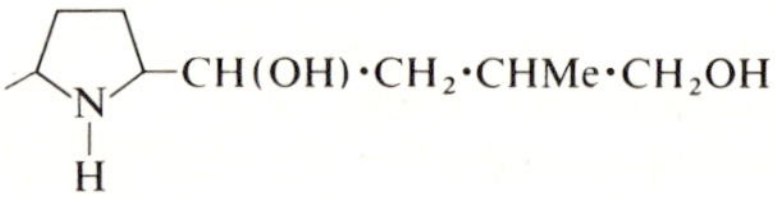

$$-Bu^n$$

(7)

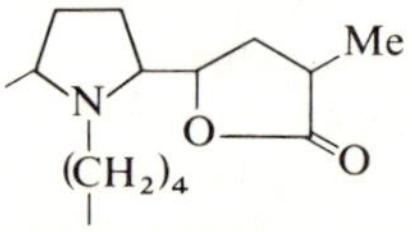

(8) 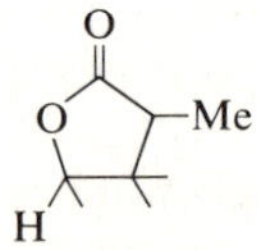 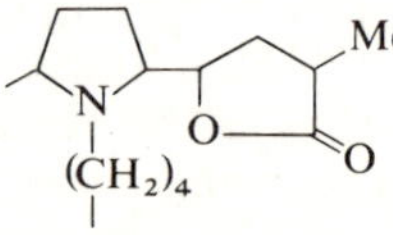   (9)

Further information relating to the structural environment of the nitrogen was derived from the reaction of tuberostemonine with cyanogen bromide. The product obtained showed different properties from that described by Kondo *et al.*[8]. Since Zeisel determination had previously demonstrated the absence of *N*-alkyl functionality in the alkaloid[6], it was reasoned that cyanogen bromide reaction must involve ring opening without loss of carbon. Evidence for the introduction of a cyanide group was provided by a strong band at $2200\ cm^{-1}$ in the i.r. spectrum, which also showed the usual lactone absorption at $1775\ cm^{-1}$. N.M.R. studies demonstrated that the product was a *primary* bromide. Thus, the signals corresponding to two hydrogens, included in a three-proton multiplet at τ 6.66 experienced a shift downfield to τ 6.03 on conversion of the bromide to an acetate, by the action of silver acetate in pyridine. These results are accommodated by partial structure (5).

Reductive removal of the bromine was effected by treatment of (5a) with lithium aluminium hydride, giving rise to (6). The reduced product (6) was subjected to a modified Kuhn–Roth oxidation[15], and the resulting acids were esterified and identified by gas–liquid chromatography. The highest member of the homologous series of acids detected in this manner was valeric acid, indicating the presence of an n-butyl chain. Tuberostemonine under the same conditions yielded only acetic and propionic acids. Structure (6) could consequently be extended to (7), allowing tuberostemonine to be represented as (8) by implication. The environment of the nitrogen, including its relationship to the five-carbon γ-lactone appendage, was thus defined.

The second lactone could be tentatively represented as in the expanded formulation (9). The n.m.r. spectra of tuberostemonine and its bisdehydro (pyrrole) derivative suggested that each lactone bore a single γ-hydrogen, and the location of the methyl group could be inferred from the similarity of the two deshielded methyl signals described above.

Grignard reaction, involving the lactone functionality of tuberostemonine, appeared to be a potentially useful structural probe. Phenylmagnesium bromide was selected, in preference to the methyl Grignard reagent[16], since phenyl groups introduced would be easily recognisable, without obscuring the methyl groups already present in the molecule. In the event, three phenyl groups were incorporated, and the i.r. spectrum of the product showed strong hydroxyl absorption at 3500 cm^{-1}, a weak band at 1605 cm^{-1}, and the absence of a carbonyl group. It was apparent that the lactone which reacted with only one equivalent of Grignard reagent had been converted to a hemiketal. Separated from the main group of signals, the n.m.r. spectrum displayed, in addition to 15 aromatic protons at τ 2.60–2.84, a single one-proton signal at τ 5.82 which was attributed to the hydrogen α to the hemiketal ether oxygen. The change in functionality was accompanied by a shift upfield of the two methyl signals formerly deshielded by the lactone carbonyls: these appeared in the Grignard product as part of a nine-proton multiplet between τ 8.92 and 9.15.

The acidic material resulting from chromium trioxide–pyridine oxidation of the Grignard product was found to contain, *inter alia*, benzoic acid and γ,γ-diphenyl-β-methylbutyrolactone (10). In view of the facile loss of the

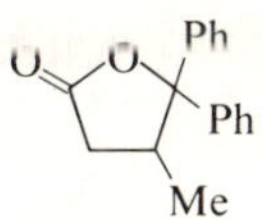

(10)

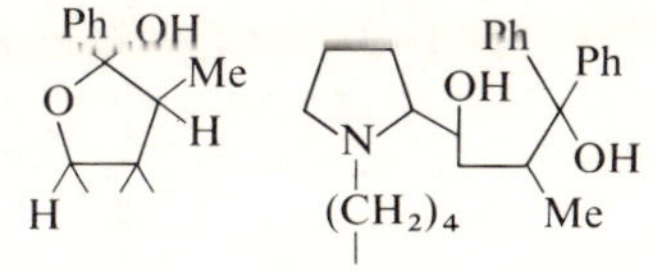

(11)

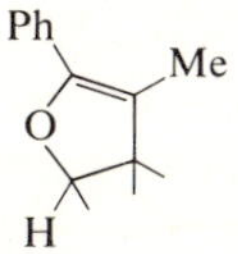

(12) 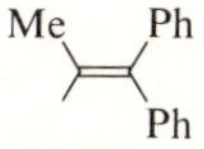   (13) 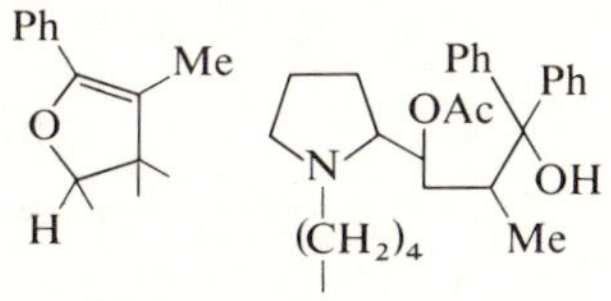   (14)

methyl butyrolactone appendage as methylsuccinic acid on oxidation of tuberostemonine, it could be surmised that (10) was derived from the same moiety. Structure (11) could, therefore, be assigned, as a reasonable working hypothesis, to the Grignard product.

Reaction of (11) with acetic anhydride afforded a crystalline anhydro-monoacetate. Infrared absorption at 1735 cm^{-1} and a three-proton singlet at τ 7.85 in the n.m.r. spectrum established the presence of the *O*-acetyl group, while a one-proton multiplet at τ 4.81 evidenced the secondary nature of the acetylated hydroxyl. One of the three methyl groups of the dehydrated product appeared downfield as a sharp singlet at τ 8.10, indicating its location on the newly-introduced double bond. Ultraviolet absorption at 280 nm (ε 9900) did not distinguish clearly between the possible partial structures (12) and (13). The fact that benzophenone could not be detected as an oxidation product of the anhydro compound under a variety of conditions, however, appeared to eliminate the second possibility. These observations could be accommodated by the structure (14) for the anhydromonoacetate.

On oxidation of (14) with chromium trioxide in acetic acid, the pronounced u.v. chromophore was replaced by maxima at 270 nm (ε 1200) and 281 nm (ε 810) indicating a cleavage of the double bond. The product was very labile, but could be readily characterised on the basis of its spectroscopic features. A plausible product from such a cleavage would possess benzoate and methyl ketone functionality as in (15). Indeed, the n.m.r. spectrum showed the two *ortho* hydrogens of a benzoate at τ 2.08, and the 13 other aromatic protons between τ 2.54 and 2.89. The single hydrogen, α to the benzoate group, is additionally deshielded by the nearby ketone function and gave rise to a signal at τ 4.54. The proton geminal with the acetate at τ 4.91 was little displaced from its original position in (14). Three-proton singlets at τ 7.89 and 7.92 are in accord with the presence of a methyl ketone and an acetate group. The i.r. spectrum showed bands at 3400 and 1723 cm^{-1}, corresponding to hydroxyl and acetate functionality respectively, and a strong band at 1715 cm^{-1} could be attributed to ketone and benzoate carbonyl absorption.

(15)

(16)

(17)

Elution of (15) through a column of basic alumina led to the development of a new chromophore, and the absorption at 225 nm (ε 12 000) was indicative of an α,β-unsaturated ketone. The n.m.r. spectrum of the product showed a one-proton doublet at τ 3.37. This single olefinic signal, representing the β-hydrogen of the conjugated ketone system, demonstrated that the double bond is tri-substituted. Both the absence of acetate absorption in the i.r. spectrum and the corresponding signal in the n.m.r. spectrum showed that the above transformation was accompanied by hydrolysis of the acetate. The formulation (16) is seen to represent the partial structure of the conjugated ketone, and the complete substitution pattern of both lactone groups of tuberostemonine is established as shown in (17). This structure describes the immediate environment of all heteroatoms. The structural problem was thus reduced to that of elucidating the nature of the attachment of the second lactone group, accommodation of the ethyl substituent[14] and defining the nature of two more carbons. Since tuberostemonine analysed for $C_{22}H_{33}NO_4$ and was without carbon–carbon double bonds, the alkaloid clearly contained two more rings than shown in (17).

Refluxing of (16) in methanolic potassium hydroxide in the presence of air, furnished a product which essentially clarified the location of the methyl ketone, and the lactone from which it was derived, with respect to the pyrrolidine system. The facile air oxidation of tuberostemonine to a pyrrole suggested that (16) might undergo a similar transformation. The loss of four hydrogens, indicated by elemental analysis of the product, appeared to be in accord with this expectation. In the n.m.r. spectrum, however, a pyrrole β-hydrogen (displayed by (1) at τ 4.08) was conspicuously absent. Inspection of the u.v. spectrum revealed a chromophore distinctly different from either that of a pyrrole or an α,β-unsaturated ketone. Maxima were recorded at 252, 275 (infl.), and 354 nm (ε 17 800, 8700 and 1900 respectively). Measurement of the spectrum in acidic solution provided a significant clue as to the structure of the dehydrogenation product. Maxima at 252 nm (ε 12 900) and 282 nm (ε 3200) were recognised as being typical for an alkyl substituted acetophenone[17]. This implied that the α,β-unsaturated ketone must have been part of a six-membered carbocyclic ring, converted to an acetophenone on air oxidation. From the dramatic change in the u.v. spectrum on acidification of the ethanol solution of the aromatic compound, it was clear that the nitrogen was attached to the newly formed aromatic ring.

To bring the structural analysis to a successful conclusion, it was necessary to establish the precise location of the nitrogen with respect to the methyl ketone. The u.v. spectrum as described was consistent with either an o- or m-aminoacetophenone[18]. An n.m.r. study allowed a distinction between these two possibilities. The aromatic region of the spectrum revealed eleven hydrogens at τ 2.49–2.77 of which ten are accounted for by the phenyls of Grignard origin, proving that the newly formed aromatic ring is pentasubstituted. Two three-proton signals, one a singlet at τ 7.47 and one a triplet at τ 8.82 ($J = 6.6$ Hz), were appropriate for the methyl hydrogens of an acetophenone and an ethyl benzene respectively. Addition of hydrogen chloride to the solution resulted in a significant change in the aromatic pattern: a one-proton singlet now emerged at τ 2.33, downfield from the multiplet centred at τ 2.78 for the ten remaining aromatic hydrogens. Only a hydrogen ortho to the methyl ketone would be expected to experience significant deshielding in acidic medium. The smaller deshielding effect on a meta or para hydrogen would be compensated by the shielding influence of the alkyl substituents[19]. To determine the relative position of the single aromatic proton with respect to the nitrogen, it was necessary to remove the deshielding influence of the carbonyl group. The desired modification was effected by reduction of the ketone with sodium borohydride. In addition to ten phenyl hydrogens at τ 2.78, the n.m.r. spectrum of the reduced product revealed a one-proton singlet at τ 3.36. This substantial shift upfield could only arise if the hydrogen were *ortho* or *para* to the amino group. Since the proton in question had been shown to be also *ortho* to the methyl ketone, it followed that the ketone function occupied a *meta* position with respect to the nitrogen. Partial structures (18) and (19) are compatible with these findings, but lack two substituents on the penta-substituted aromatic ring. An ethyl group occupies one of the vacant sites, and the other position must be involved in the completion of the hitherto-undefined ring required by the analytical data for tuberostemonine. All of the carbon atoms constituting the original skeleton of tuberostemonine are still present in the aromatic product, with the exception of that lost as the benzoate carbonyl. The aromatic ring is clearly, therefore, the site of attachment of the four-carbon moiety, having the nitrogen atom as its other terminus. Such an attachment is only feasible at the position *ortho* to the nitrogen, allowing the structures (18) or (19) to be expanded to (20) or (21). The complete structure of tuberostemonine, represented accordingly as (22) or (23) was published by Götz *et al.*[20] in 1961.

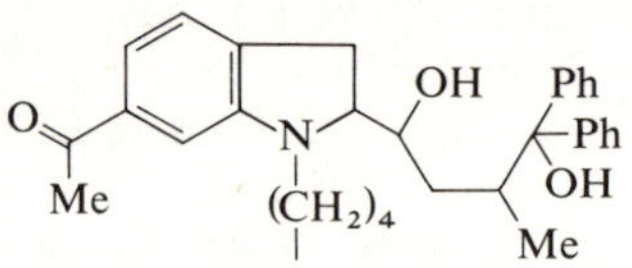

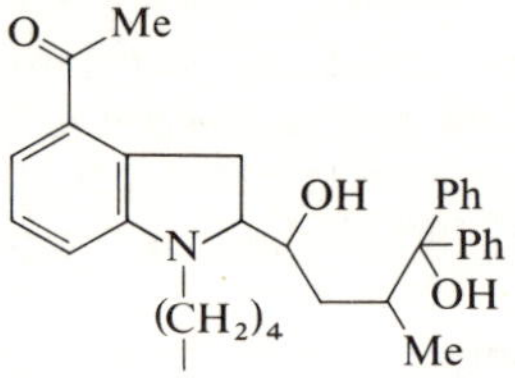

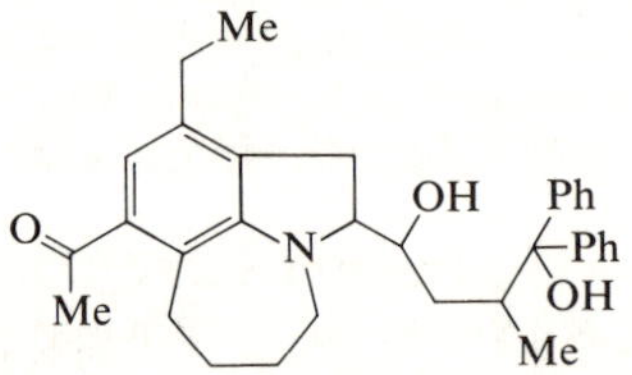

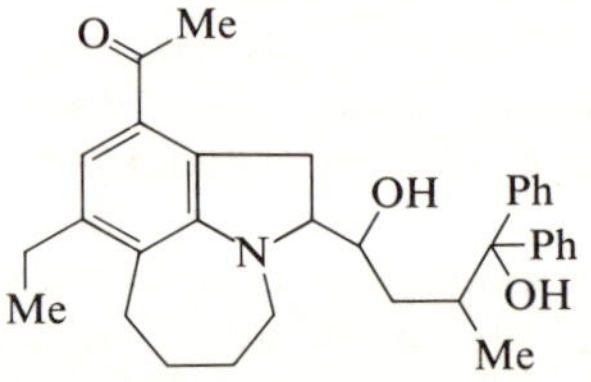

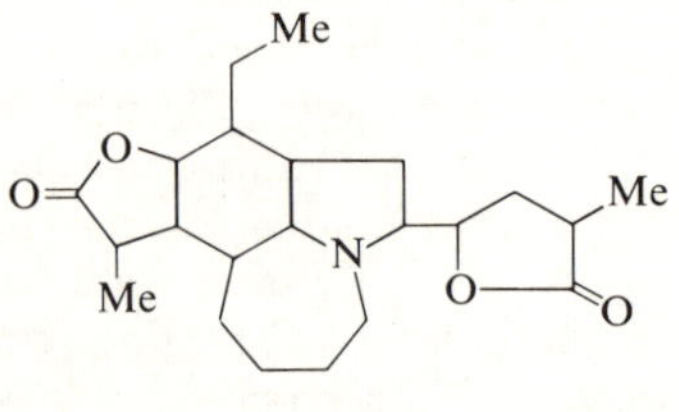

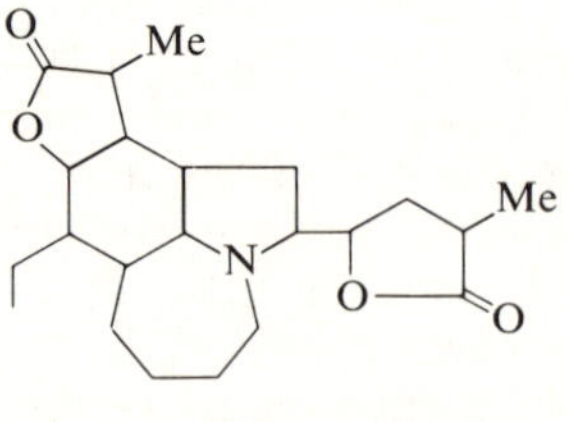

A subsequent report by Edwards *et al.*[21] described independent studies which corroborated the substitution pattern of the pyrrolidine ring. Thus bisdehydrotuberostemonine, depicted there by the partial structure (24), underwent hydrogenolysis under appropriate conditions giving the acid (25). Treatment of the latter with an acetic acid–acetic anhydride mixture resulted in cyclisation of the methyl butyric acid side-chain to the unsubstituted β-position of the pyrrole, yielding the ketone (26).

Edwards and Feniak[22] later advanced spectrometric evidence which allowed a decision to be made between the structures (22) and (23) for tuberostemonine proposed previously by Götz et al.[20]. The n.m.r. spectrum of bisdehydrotuberostemonine (27) or (28) displayed, *inter alia*, triplets at τ 5.45 and 6.9. The former signal could be readily assigned to the hydrogen on C-10 on the basis of its chemical shift and multiplicity. It was surmised that the second triplet corresponded to the hydrogen attached to the neighbouring carbon, C-11, which would be deshielded by the pyrrole ring and nearby oxygen atoms. This being the case, it was argued that the simplicity of the signal at τ 6.9 was not consistent with structure (27), whose C-11 hydrogen should give rise to a multiplet comprising at least four lines. The τ 6.9 triplet collapsed to a doublet when decoupled[23] from the hydrogen on C-10 in accord with the above reasoning. It follows that the pyrrole has structure (28) and that tuberostemonine is correctly represented as (23).

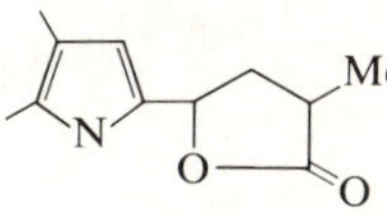

(24) 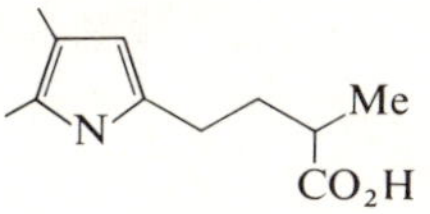   (25) 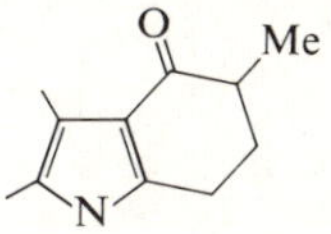   (26)

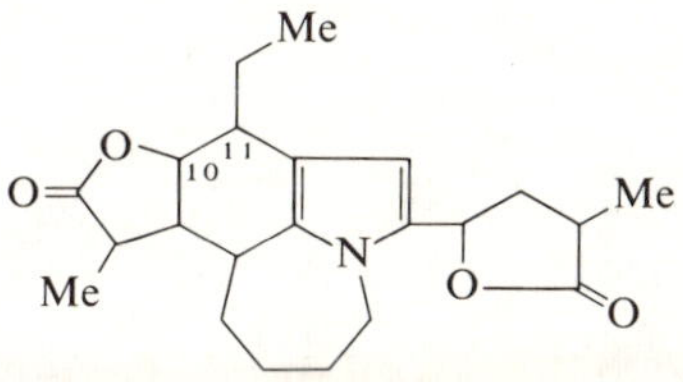

(27) 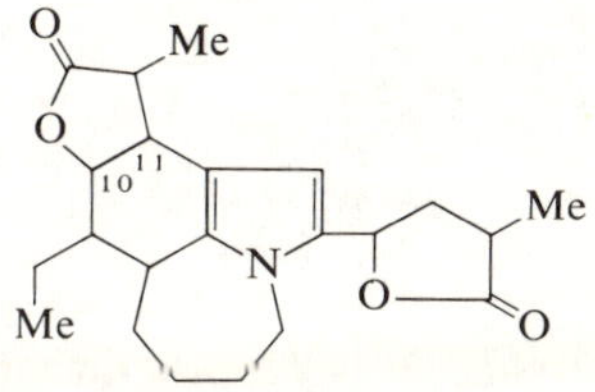   (28)

The correctness of the formulation (23) was substantiated by dehydrogenation experiments. It was soon discovered that dehydrogenation of tuberostemonine, or its derivatives containing oxygen functionality in the side-chains, gave rise to diethyl indoles[24–26]. Palladium catalysed dehydrogenation of tuberostemonine afforded, *inter alia*, the indole (29), designated tuberostemonane[24]. A hydroxy indole, based on the cyclised skeleton (26) and retaining an intact lactone function, was also isolated[26]. The identification of such compounds, while confirming the nature of three of the five rings of tuberostemonine, did nothing to facilitate the distinction between structures (22) and (23). Dehydrogenation of the isopropenyl derivative (33), derived from tuberostemonine via the lactam (3) and the intermediates (30) through (32), did, however, afford the indole (34) which clearly defined the structure of the alkaloid as (23)[28]. The i.r. spectrum of (34) showed a sharp medium-intensity peak at 1165 cm^{-1} not present in the spectrum of the corresponding diethyl indole. U.V. maxima were observed at 227, 276, 286 (infl.) and 297 (infl.) nm (ε 32 600, 9100, 7900 and 5200 respectively). The n.m.r. spectrum was highly diagnostic, showing a pair of doublets at τ 3.30 and 3.76 ($J =$ 2.8 Hz) for one α- and one β-indole hydrogen[27] respectively. A one-proton singlet at τ 3.44 represented the single benzenoid hydrogen. The two protons on C-4, α to the nitrogen, showed as a poorly-defined triplet centred at τ 5.84. A multiplet extending between τ 6.85 and 7.54 accounted for the five benzylic hydrogens, and included a quartet centred at τ 7.34 for the methylene protons of the ethyl group. A nine-proton multiplet at τ 8.62–8.98 represented the three C-methyl groups. The mass spectrum of (34) showed a molecular ion at m/e 241, and a base peak at m/e 226 corresponding to the loss of a methyl group[28].

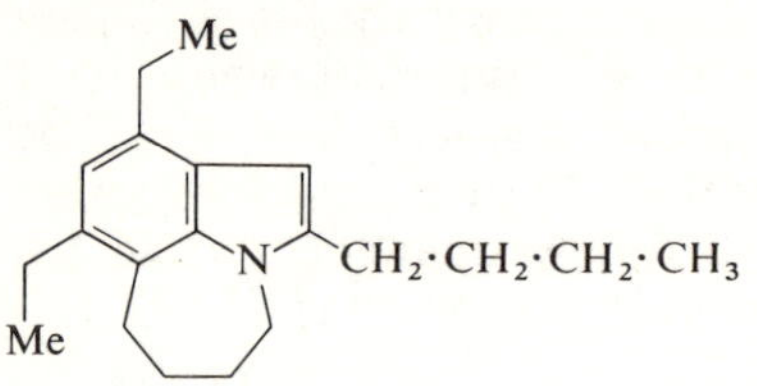

(29) 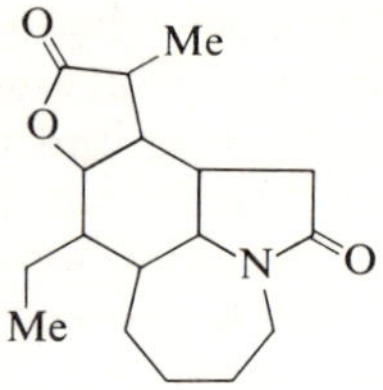   (3)

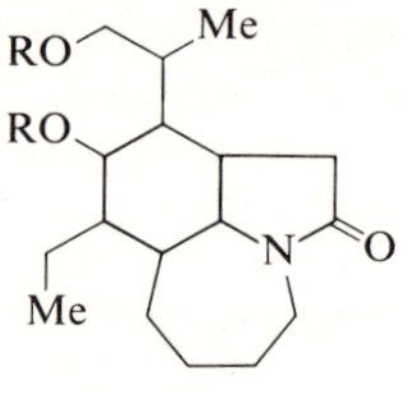

(30) R = H
(31) R = COPh

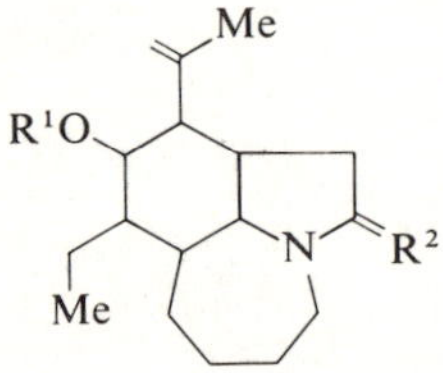

(32) R^1 = COPh, R^2 = O
(33) R^1 = H, R^2 = H$_2$

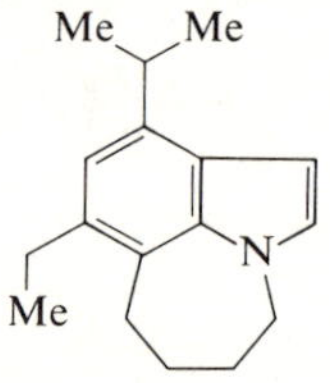

(34)

Comparison with a synthetic sample of (34), prepared by an unambiguous route[29], revealed complete identity of the two, as established by i.r., u.v., n.m.r. and mass spectrometry, and by thin-layer chromatography. The melting point of the indole (34), derived from the natural source, was undepressed on admixture with the synthetic material. These results confirmed the carbon skeleton of the lactam (3), and the structure of tuberostemonine was thus rigorously established as (23).

The evidence in favour of structure (23) had earlier also been well reinforced by the independent researches of Kaneko[30]. Thus, the pyrrolidine diol (35), obtained from lithium aluminium hydride reduction of (3), afforded the bromocyanamide (36) on reaction with cyanogen bromide. The bromine atom of (36) was reductively removed on catalytic hydrogenation, and the product (37), when subjected to lithium aluminium hydride reduction, followed by hydrolysis, yielded (38). Oxidation of (38) with permanganate, under alkaline conditions, gave 1-n-butyl-2-ethyl succinic acid (39), identified by comparison with a synthetic sample. Isolation of this product confirmed the position of the ethyl group with respect to the seven-membered ring, and is clearly compatible only with the structure (23) for tuberostemonine.

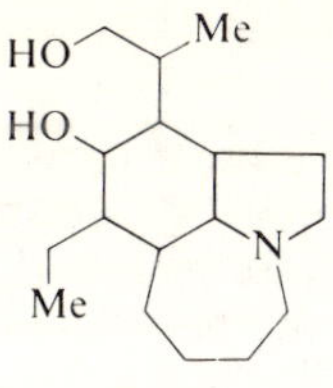

(35)

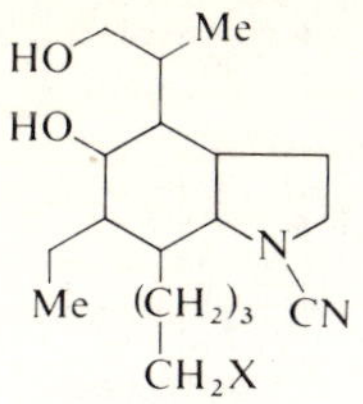

(36) X = Br
(37) X = H

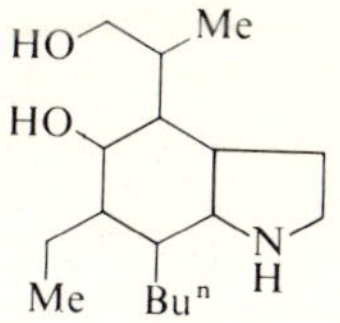

(38)

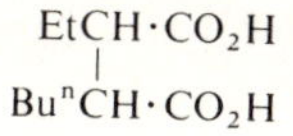

EtCH·CO$_2$H
|
Bu^{n}CH·CO$_2$H

(39)

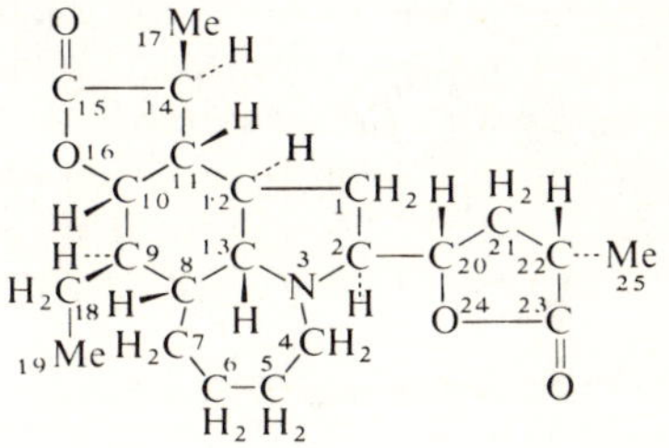

(40)

4.2.1.3 Relative and absolute configuration

Single-crystal x-ray analysis by Harada et al.[31] of tuberostemonine metho-bromide dihydrate (orthorhombic) showed the alkaloid to possess the relative and absolute configuration depicted in (40). This result is in accord with the earlier observation that oxidation of tuberostemonine gives 1-methylsuccinic acid[1], whose absolute configuration has been established[32].

4.2.2 Stenine

4.2.2.1 Structure determination

In 1967, Uyeo et al.[33] reported the isolation of a hitherto unknown alkaloid from the basic extracts of Stemona tuberosa roots. The new base, which was named stenine, could be readily separated from the other alkaloids as a result of its greater solubility in hexane. Elemental analysis and mass spectrometry established the molecular formula as $C_{17}H_{27}NO_2$.

Stenine displayed a number of spectrometric features reminiscent of tuberostemonine (40). The i.r. spectrum exhibited a γ-lactone carbonyl band at $1775\,cm^{-1}$, and showed no absorption corresponding to hydroxyl or N—H functionality. End absorption was observed in the u.v. spectrum, but n.m.r. and i.r. spectrometry indicated the absence of vinyl hydrogens or a carbon–carbon double bond. The n.m.r. spectrum of stenine showed a three-proton triplet at τ 9.09 similar to the signal for the methyl hydrogens of the tuberostemonine ethyl group. The only other C-methyl signal was a doublet at τ 8.73 ($J = 6.5\,Hz$) assigned to the hydrogens of a secondary methyl group, α to the lactone carbonyl. A broad one-proton triplet at τ 5.53

($J = 8\,\text{Hz}$) had a chemical shift appropriate for a hydrogen on the carbon bearing the (secondary) lactonised hydroxyl group.

Mass spectrometry, in conjunction with the above data, demonstrated the close structural relationship between this alkaloid and tuberostemonine. The most abundant ion in the spectrum of tuberostemonine is found at $m/e = 276$ ($M - C_5H_7O_2$). This species results from loss of the γ-lactone appendage at C-2, and has been depicted as (41)[34]. Similarly, the base peak in the spectrum of stenine occurs at $m/e = 276$, corresponding in this case to the $M - 1$ ion. These findings suggested that stenine might possess the same basic structural features as the ion (41); this is tentatively formulated accordingly as (42). Rigorous proof for structure (42) was provided by correlation of stenine with a tuberostemonine derivative whose structure was unequivocally established. Thus, reduction of stenine with lithium aluminium hydride afforded a diol (43), identical in all respects with that previously derived from tuberostemonine by Kaneko (35)[30].

(41)   (42)

(43)   (44)

4.2.2.2   *Relative and absolute configuration*

Since the stereochemistry of (35) can be inferred from the x-ray analysis of tuberostemonine methobromide dihydrate, it follows from the correlation described above, that the relative and absolute configuration of stenine is represented by the formulation (44)[31].

4.2.3   Oxotuberostemonine

The isolation of a new base from *Stemona tuberosa* was reported by Kondo *et al.*[35] in 1954. The alkaloid, designated oxotuberostemonine, could be obtained from tuberostemonine mother liquors which had been set aside for an extended period. Since the same base was subsequently obtained by mercuric acetate oxidation of tuberostemonine[21], it was evident that the structures of the two substances were closely related. The possibility is not excluded that oxotuberostemonine is actually an artifact, formed by air oxidation of tuberostemonine.

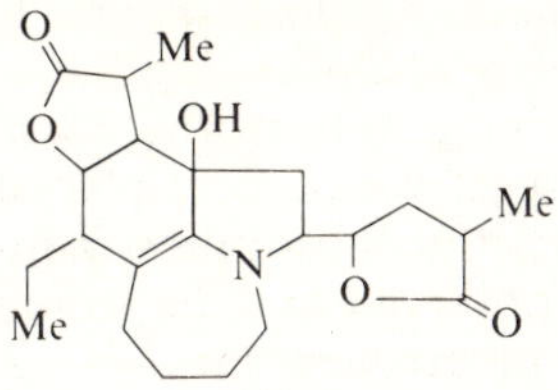

(45) 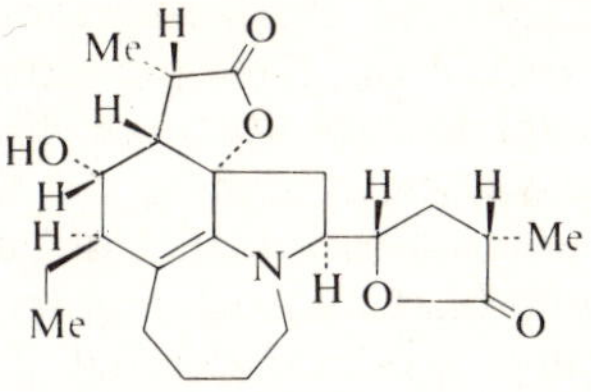   (46)

The molecular formula of the new base, $C_{22}H_{31}NO_5$, demonstrated that oxotuberostemonine differs from tuberostemonine by the presence of an extra oxygen atom and an additional unsaturation. The hydroxylic nature of one of the oxygens was established by acetylation[35]. The double bond was evidently tetrasubstituted since no vinyl hydrogens could be observed in the n.m.r. spectrum. Absorption at 243 nm (ε 8600) in the u.v. spectrum suggested an eneamine chromophore, and i.r. evidence was provided for the presence of two lactone functions[21]. In a review on *Stemona* alkaloids, Edwards[36] tentatively advanced the reasonable structure (45) for oxotuberostemonine on the basis of the evidence available[21]. Subsequent x-ray analysis by Huber *et al.*[37] demonstrated, however, that an unexpected re-lactonisation had occurred in the formation of oxotuberostemonine, and that the base is correctly represented by the structure (46). The x-ray studies[37], besides demonstrating the *cis*-fusion of the new lactone ring, indicated that the configurations of all centres not affected by the oxidation were comparable with their counterparts in tuberostemonine (40).

4.2.4   Stemonine

Stemonine, a metabolic product of *Stemona japonica*, was first isolated and investigated by Suzuki[38, 39]. It was later reported that the alkaloid could also be obtained as a degradation product of protostemonine[40, 41] (*vide infra*).

Spectroscopic examination[41] of stemonine, $C_{17}H_{25}NO_4$, and the corresponding pyrrole (obtained by manganese dioxide oxidation of the base), demonstrated that although it possesses major structural features in common with the *Stemona* alkaloids considered heretofore, stemonine represents a departure from their characteristic perhydroazepinoindole skeletal system. Thus, the i.r. spectrum of stemonine showed the familiar lactone band at 1765 cm^{-1}, and the absence of hydroxyl or N–H absorption, while the spectrum of the bisdehydro product demonstrated that two γ-lactones were present. The u.v. spectrum indicated the absence in stemonine of a conjugated chromophore. In the n.m.r. spectrum of the alkaloid, three-proton doublets at τ 8.78 (J = 6 Hz) and at τ 8.80 (J = 6.5 Hz) are comparable with the signals for the secondary methyl groups adjacent to the lactone carbonyls of tuberostemonine[20]. A two-proton multiplet at τ 5.50–6.02 is similarly in accord with the γ-hydrogens of the lactone groups.

The mass spectrum of stemonine exhibited a base peak at m/e 208 $(M - C_5H_7O_2)$. Such a fragmentation is clearly analogous to that observed in the spectrum of tuberostemonine, involving loss of the five-carbon γ-lactone moiety from the α position of the pyrrolidine ring[33].

A significant feature, distinguishing stemonine from the other alkaloids in the series, was the absence in the n.m.r. spectrum of the triplet corresponding to the ethyl substituent. Furthermore, the substitution pattern of the pyrrole ring in bisdehydrostemonine was observed to differ from that in the corresponding product derived from tuberostemonine[1, 20]. The n.m.r. spectrum of bisdehydrostemonine displayed an AB pattern with doublets $(J = 4\,\text{Hz})$ at τ 3.86 and τ 4.08, consistent with the two β-protons of an otherwise fully-substituted pyrrole ring, as depicted in (47).

(47)

(48)

Single-crystal x-ray analysis[42] of stemonine hydrobromide hemihydrate (orthorhombic) established the complete structure and absolute configuration as represented in (48).

4.2.5 Protostemonine

Isolation of the alkaloid protostemonine from *Stemona* root was first described by Kondo and Satomi[40] in 1947. These authors advanced evidence for the presence in the molecule of three *C*-methyl groups, a methoxyl group, and lactone functionality, and reported its facile degradation to stemonine.

Irie *et al.*[41] established the correct formula of protostemonine, from roots of *Stemona japonica*, as $C_{23}H_{31}NO_6$ $(m/e = 417)$. Re-examination of the reaction sequence by which the alkaloid is converted to stemonine, revealed that on acid treatment, protostemonine first adds a molecule of water, giving a hydrate hydrochloride, $C_{23}H_{33}NO_7 \cdot HCl$ $(m/e = 435)$. The latter, on treatment with potassium carbonate or by pyrolysis *in vacuo*, undergoes cleavage affording stemonine (48) or its hydrochloride respectively, in addition to a neutral oil, identified as 4-methoxy-3-methyl-2(5*H*)-furanone (49).

These results reduce the structural problem to that of determining the precise nature and location of the bonding between the fragment (49) and stemonine (48).

Protostemonine displayed strong absorption in the u.v. spectrum at 305 nm (ε 23 000) indicative of a conjugated dienone system. The latter was shown to be fully substituted by the absence of olefinic protons in the n.m.r. spectrum. Singlets at τ 5.87 and 7.94 were assigned to methyl ether protons and the hydrogens of an allylic methyl group respectively. Doublets corresponding to secondary methyl groups were at τ 8.66 and 8.75. Infrared bands at 1740, 1679 and 1618 cm^{-1} can be attributed to the conjugated lactone system involving (49), while a peak at 1770 cm^{-1} constitutes normal γ-lactone carbonyl absorption. A dramatic change in the u.v. chromophore occurs on hydration of protostemonine. Thus, the spectrum of the hydrate–hydrochloride displayed a maximum at 233 nm (ε 10 500), in accord with the absorption of an α,β-unsaturated lactone. Corresponding changes were also evident in the i.r. spectrum.

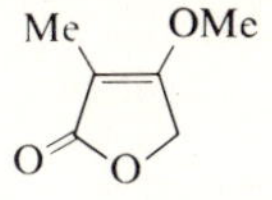

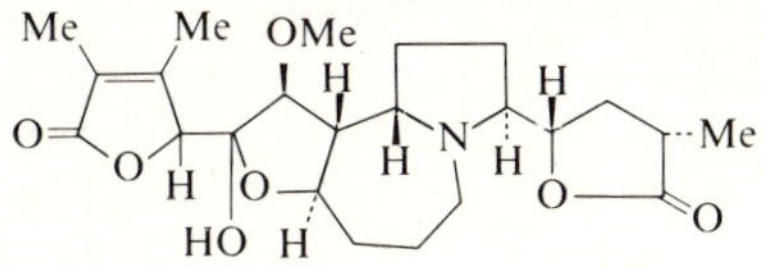

These data suggest that the hydrate hydrochloride is the product of Michael-type addition of water to the dienone system, and that the subsequent degradation to (48) and (49) occurs as a consequence of retro-aldol-type cleavage. Structure (50) was accordingly advanced for the hydrated intermediate, and (51) or (52) were proposed for protostemonine[41]. On the basis of the published evidence, the attachment of (49) to the other lactone group of stemonine does not appear to have been rigorously excluded.

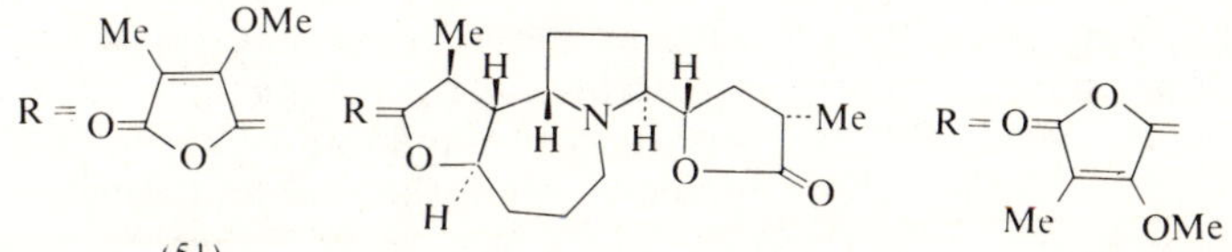

4.2.6   Stemofoline

An interesting modification of the protostemonine skeleton is found in stemofoline, isolated from the stems and leaves of *Stemona japonica* by Irie *et al.*[43]. The alkaloid, $C_{22}H_{29}NO_5$ (m/e = 387) possesses a chromophore similar to that of protostemonine, displaying a maximum in the u.v. spectrum at 296 nm (ε 24 200).

The structure and absolute configuration of stemofoline (53) were determined by x-ray crystallographic analysis of its hydrobromide monohydrate[43]. A close relationship between the novel cage structure of stemofoline (53) and the protostemonine system is readily apparent if one visualises removal of the additional ether bridge and the extra carbon–carbon bond (indicated by arrow in (53)). A further remarkable feature, hitherto unknown in other *Stemona* alkaloids, is the presence of an n-butyl chain α to the pyrrolidine nitrogen, in place of the usual γ-lactone system present in tuberostemonine[20], oxotuberostemonine[37], and protostemonine[41].

(53)   (54)

4.2.7   Tuberostemonine A

Tuberostemonine A, $C_{22}H_{33}NO_4$, isomeric with tuberostemonine, was obtained from rhizomes of *Stemona sessilifolia* by Edwards *et al.*[21]. The base afforded the same pyrrole derivative as tuberostemonine on silver oxide oxidation. Furthermore, treatment with alkaline permanganate gave rise to a lactam, $C_{17}H_{25}NO_3$, identical with that obtained in a like manner from tuberostemonine. These results indicated that the alkaloid differed from tuberostemonine only in the stereochemistry at C-2[21,36]. Tuberostemonine A is thus represented by structure (54).

4.3   ALKALOIDS OF UNKNOWN STRUCTURE

In his comprehensive treatise on alkaloid chemistry published in 1961, Boit[44] included a list of alkaloids isolated from members of the Roxburghiaceae family. At the time the compilation was prepared, the structure of none of the *Stemona* alkaloids had been elucidated. Those alkaloids whose structures still remain obscure at present are grouped below according to their plant source. Preliminary chemical investigations have been carried out on several of these compounds.

4.3.1   *Stemona tuberosa*

In addition to tuberostemonine, this species has been reported to yield three other alkaloids having the composition $C_{22}H_{33}NO_4$:

Stemonine[45,46,48]: m.p. 162 °C, $[\alpha]_D^{20} + 76$ degrees (ethanol).

Isostemonine[46]: m.p. 212–216 °C.

Isotuberostemonine[47]: m.p. 123–125 °C, $[\alpha]_D^{19} - 86$ degrees (methanol).

Two further alkaloids were also attributed to this plant source:

Hypotuberostemonine[47]: $C_{19}H_{21-23}NO_3$, m.p. 183 °C $[\alpha]_D + 63$ degrees (methanol).

Stemotuberin[46]: m.p. 77–82 °C.

4.3.2 *Stemona ovata*

This species may be synonymous with *Stemona japonica*; it has been reported
to yield two alkaloids having the formula $C_{19}H_{31}NO_5$:
  Stemonidine[38, 6] : m.p. 116 °C, $[\alpha]_D$ -8 degrees.
  Isostemonidine[49]: m.p. 137 °C $[\alpha]_D$ -49 degrees (acetone).

4.3.3 *Stemona sessilifolia*

Hodorine[50]: $C_{19}H_{31}NO_5$.
Sessilistemonine[51]: $C_{25}H_{35}NO_7$, m.p. 170 °C, $[\alpha]_D$ $+112$ degrees
       (methanol).

4.3.4 Pai-Pu (*Stemona* species)

Sinostemonine[52]: $C_{21}H_{35-37}NO_5$, m.p. 138 °C, $[\alpha]_D^{25}$ -37 degrees (water).
Paipunine[52]: $C_{24}H_{37}NO_4$, m.p. 106 °C, $[\alpha]_D^{25}$ -54 degrees (acetone).

References

1. Schild, H. (1936). *Ber.*, **69**, 74
2. Vezina, C. and Baker, H. Personal communication
3. Herr, F. and Greenberg, R. Personal communication
4. Rockhold, W. T. and Wetzel, J. Personal communication
5. Suzuki, K. (1934). *J. Pharm. Soc. Japan*, **54**, 96
6. Kondo, H., Suzuki, K. and Satomi, M. (1939). *J. Pharm. Soc. Japan*, **59**, 177
7. Kondo, H., Suzuki, K. and Satomi, M. (1940). *J. Pharm. Soc. Japan*, **60**, 149
8. Kondo, H., Suzuki, K. and Satomi, M. (1941). *J. Pharm. Soc. Japan*, **61**, 111
9. Kondo, H., Satomi, M. and Kaneko, T. (1956). *Ann. Rep. ITSUU Lab. Tokyo*, **7**, 57
10. Kondo, H., Satomi, M. and Kaneko, T. (1958). *Ann. Rep. ITSUU Lab. Tokyo*, **9**, 107
11. Kondo, H., Satomi, M. and Kaneko, T. (1957). *Ann. Rep. ITSUU Lab. Tokyo*, **8**, 51
12. We would like to thank Prof. K. Wiesner for allowing us to select *Stemona tuberosa* from
    his large collection of exotic plant materials, stored in the basement of the Chemistry
    Department of U.N.B., quite reminiscent of a well stocked cellar in a private house
13. Kondo, H., Satomi, M. and Kaneko, T. (1958). *Ann. Rep. ITSUU Lab. Tokyo*, **9**, 99
14. Kaneko, T. (1960). *Ann. Rep. ITSUU Lab. Tokyo*, **11**, 39
15. Bickel, H., Schmid, H. and Karrer, P. (1955). *Helv.*, **38**, 649
16. The reaction of tuberostemonine with methylmagnesium iodide was reported[8] to be
    complete with the uptake of two methyl groups
17. Braude, E. A. and Sondheimer, F. (1955). *J. Chem. Soc.*, 3754
18. Lutskii, A. E. and Dorofeev, V. V. (1957). *Zhur. Obshchei Khim.*, **27**, 1064
19. Jackman, L. M. (1959). *Application of Nuclear Magnetic Resonance Spectroscopy in
    Organic Chemistry* (Pergamon Press: New York)
20. Götz, M., Bögri, T. and Gray, A. H. (1961). *Tetrahedron Letters*, **20**, 707
21. Edwards, O. E., Feniak, G. and Handa, K. L. (1962). *Can. J. Chem.*, **40**, 455
22. Edwards, O. E. and Feniak, G. (1962). *Can. J. Chem.*, **40**, 2416
23. Kaiser, R. (1960). *Rev. Sci. Instrum.*, **31**, 963
24. Shingu, T., Tsuda, Y., Uyeo, S., Yamato, Y. and Harada, H. (1962). *Chem. and Ind.*, 1911
25. Strunz, G. M. (1963). *Ph.D. Thesis*, University of New Brunswick
26. (a) Uyeo, S., Shingu, T., Tsuda, Y. and Yamato, Y. (1964). *J. Pharm. Soc. Japan*, **84**, 548
    (b) Uyeo, S. and Shingu, T. (1964). *J. Pharm. Soc. Japan*, **84**, 552
    (c) Uyeo, S. and Shingu, T. (1964). *J. Pharm. Soc. Japan*, **84**, 555
    (d) Uyeo, S., Shingu, T. and Tsuda, Y. (1964). *J. Pharm. Soc. Japan*, **84**, 663

27. Cohen, L. A., Daley, J. W., Kuy, H. and Witkop, B. (1960). *J. Amer. Chem. Soc.,* **82,** 2184
28. Götz, M., Bögri, T., Gray, A. H. and Strunz, G. M. (1968). *Tetrahedron,* **24,** 2631
29. Strunz, G. M. (1968). *Tetrahedron,* **24,** 2645
30. Kaneko, T. (1965). *Ann. Rep. ITSUU Lab. Tokyo,* **14,** 49
31. Harada, H., Irie, H., Masaki, N., Osaki, K. and Uyeo, S. (1967). *Chem. Commun.,* 460
32. Klyne, W. (1964). *Progress in Stereochemistry,* **1,** 188
33. Uyeo, S., Irie, H. and Harada, H. (1967). *Chem. Pharm. Bull.,* **15,** 768
34. See footnote (5) of Ref. 33
35. Kondo, H., Satomi, M. and Odera, T. (1954). *Ann. Rep. ITSUU Lab. Tokyo,* **5,** 99
36. Edwards, O. E. (1967). *The Stemona Alkaloids, in* Manske, R.H.F., *The Alkaloids* (Academic Press: New York – London)
37. Huber, C. P., Hall, S. R. and Maslen, E. N. (1968). *Tetrahedron Letters,* **38,** 4081
38. Suzuki, K. (1929). *J. Pharm. Soc. Japan,* **49,** 457
39. Suzuki, K. (1931). *J. Pharm. Soc. Japan,* **51,** 419
40. Kondo, H. and Satomi, M. (1947). *J. Pharm. Soc. Japan,* **67,** 182, 185, 188
41. Irie, H., Harada, H., Ohno, K., Mizutani, T. and Uyeo, S. (1970). *Chem. Commun.,* 268
42. Koyama, H. and Oda, K. (1970). *J. Chem. Soc. B,* 1330
43. Irie, H., Masaki, N., Ohno, K., Osaki, K., Taga, T. and Uyeo, S. (1970). *Chem. Commun.,* 1066
44. Boit, H. G. (1961). *Ergebnisse der Alkaloid-Chemie bis 1960,* (Akademie-Verlag: Berlin)
45. Lobstein, J. E. and Grumbach, I. (1932). *Bull. Sci. Pharmacol.,* **39,** 26
46. Pfeifer, S. and Nastewa, W. (1968). *Pharmazie,* 23(6), 342
47. Kondo, H., Satomi, M. and Kaneko, T. (1956). *Ann. Rep. ITSUU Lab. Tokyo,* **7,** 64
48. The reported properties of this alkaloid differ from those of the *Stemona japonica* metabolite, also named stemonine; see Section 4.2.4
49. Suzuki, K. (1934). *J. Pharm. Soc. Japan,* **54,** 100
50. Furuya, T., *Arb. Pharm. Inst. Univ. Berlin,* **9,** 112 [*Chem. Abstr.* (1914). **8,** 550]
51. Chu, I. H. and Young, (1955). *Acta Chim. Sinica,* **21,** 173
52. Lee, H. M. and Chen, K. K. (1940). *J. Amer. Pharm. Assoc.,* **29,** 391

5
Spirobenzylisoquinoline Alkaloids

S. McLEAN
and
J. WHELAN
University of Toronto

5.1 INTRODUCTION

The description 'benzylisoquinoline alkaloids' has been used to define an
extremely large group of alkaloids, the structures of which can be con-

sidered to be based on the 1-benzylisoquinoline skeleton; the degree of elaboration and modification of the skeleton varies considerably, and numerous subdivisions of the class, based partly on structural features and partly on the botanical origins of the alkaloids, have been recognised, and the growing body of biogenetic information provides additional justification in many cases for the earlier classification based on empirical criteria. The spirobenzylisoquinoline alkaloids, up to the present time found only in certain *Corydalis* and *Fumaria* species (Papaveraceae), represent a structural group that gained recognition only in the past few years, and the present review will confine itself to an account of the chemistry of these alkaloids.

Table 5.1

	R_1	R_2	R_3	R_4	R_5	R_6
Ochotensimine (1)	Me	Me	H	H		$=CH_2$
Ochotensine (2)	Me	H	H	H		$=CH_2$
Fumaricine (3)	Me	Me	OH	H	H	H
Fumaritine (4)	H	Me	OH	H	H	H
Fumariline (5)	$-CH_2-$		$=O$		H	H
Sibiricine (7)	$-CH_2-$		OH	H	$=O$	
Ochrobirine (8)	$-CH_2-$		H	OH	OH	H
Parfumine (9)	H	Me	$=O$		H	H
Fumarophycine (10)	H	Me	OAc	H	H	H
Fumarofine (11)	H	Me	$=O$		H	OH
Corydaine (45)	$-CH_2-$		H	OH	$=O$	
Corpaine (46)	H	Me	H	OH	$=O$	

Members of this group have the characteristic structural feature of a spiro union linking a reduced isoquinoline unit and an indane moiety, which is clearly a modified benzyl unit, and can be represented by the generalised structural formula shown in Table 5.1; structural differentiation within the group is most significantly dependent on the nature and position of substituents on the five-membered ring C, a variety of oxygen functions has also been found in ring A (always at positions 2 and 3), but all of the alkaloids described so far carry a methylenedioxy group on ring D*.

5.2 DETERMINATION OF STRUCTURES

5.2.1 Ochotensimine (1) and ochotensine (2)

The alkaloids ochotensimine, $C_{22}H_{23}NO_4$, and ochotensine, $C_{21}H_{21}NO_4$, were described by Manske[1] in 1940 as constituents of *Corydalis ochotensis* Turcz., a member of a genus that commonly affords benzylisoquinoline alkaloids; the presence in the plant of protopine, cryptocavine, aurotensine, and a $C_{19}H_{23}NO_4$ alkaloid was also recorded. The molecular formulae and optical activity of the alkaloids were determined, and the relationship of ochotensimine to ochotensine was established through the conversion of the latter to the former by the methylation of a phenolic hydroxyl group. Apart from the conversion of ochotensimine methiodide to a crystalline product by Emde reduction, standard degradative methods failed to yield products of value in the investigation of the structures of the alkaloids, and it was only with the advent of nuclear magnetic resonance (n.m.r.) spectroscopy that the structural problem was solved[2].

Ochotensine is crystalline and extremely insoluble in common solvents but ochotensimine, which has not been obtained crystalline, is soluble and it proved ideally suited to investigation by n.m.r. spectroscopy. The signals associated with the two *O*-methyl groups, the *N*-methyl group and the methylenedioxy group were readily recognised in the 60 MHz spectrum of ochotensimine, and it was clear that the molecule contained two separate tetra-substituted benzene rings, in one of which the unsubstituted positions were *para*-related since the benzenoid protons were not significantly coupled, and, in the other, the unsubstituted positions were *ortho*-related since the protons at these positions were coupled with a J value of 8.0 Hz. Two one-proton singlets associated with vinyl protons indicated the presence of the exocyclic methylene group, and confirmation of this assignment was obtained by catalytic reduction which transformed the group into a $CH_3CH<$ moiety; the singlet character of the signals of the methylene group showed that the J_{gem} value was zero, in accord with the low value expected for protons with this relationship, but more significantly it also showed that there were no other protons close enough to couple with either of the vinyl protons. The remaining six protons produced a complex spectrum, but at 100 MHz the presence of an AB quartet centred at τ 6.80 with $J = 18.0$ Hz became apparent and it was possible to attribute this to the protons of a $>CH_2$ group adjacent to a benzene ring and part of the five-membered ring of an indane; moreover, there were no other protons close enough to produce

---

*Locants are defined by the arbitrary numbering scheme shown in the generalised formula (see Table 5.1) so that, throughout the review, structures of alkaloids can be related to the generalised formula; the scheme does not necessarily correspond to that used for a particular alkaloid in the original literature, and it may be noted that particular care must be taken with respect to the relationship of substituents on rings C and D.

further appreciable splitting of this quartet. On the assumption that the remaining four protons were part of the expected $ArCH_2CH_2N<$ grouping, structure (1) was assigned to ochotensimine with the proviso that, although the skeleton was secure if the assumption were valid, the data could still be satisfied by several other arrangements of the oxygen functions.

The skeletal arrangement was confirmed through a demonstration that the postulated $ArCH_2CH_2N<$ grouping was in fact present by degrading it through a Hofmann sequence to $ArCH\!=\!CH_2$, and support, but not proof, for the distribution of oxygen functions came from the identification of metahemipinic acid as a product of an oxidative degradation.

Although the final confirmation of the structure came from an x-ray crystallographic analysis[3] of ochotensine methiodide, which immediately established the structures of both ochotensine (1) and ochotensimine (2) (Table 5.1), it is interesting to note in retrospect that a further development of n.m.r. spectroscopy, namely the nuclear Overhauser effect (NOE), could have provided the complete answer to the structural problem.

In the context of proton n.m.r. spectroscopy, an NOE is measured by the increase (or decrease) in the integrated signal area associated with one proton when a frequency is applied that selectively saturates a second proton; the effect may be observed when the two protons have a sufficiently close spatial relationship but a zero or negligible spin-coupling constant (J), and can, therefore, provide a measure of the proximity of the two protons[4]. In the case of ochotensimine (1), saturation of one vinylic proton (H_B) at

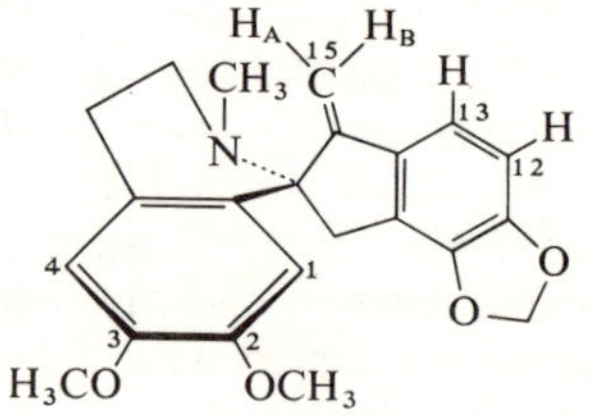

(1)

C-15 produced a positive NOE (24% increase in area) at the proton on C-13 and a smaller negative NOE (7% decrease in area) at the proton at C-12; it followed, therefore, that the three protons had the spatial relationship shown. The proximity of the proton at C-1 to one methoxyl group and of the proton at C-4 to the other methoxyl group was demonstrated in a similar manner. The exhaustive NOE study of ochotensimine[5], besides demonstrating certain theoretically interesting features associated with negative NOE values and the NOE between the geminal protons at C-15, also revealed further spatial relationships within the molecule that would have helped to confirm the structural assignments made previously, but it may be seen that the NOE experiments which demonstrated the nature and relationship of substituents at positions 1, 2, 3 and 4 and those at positions 12, 13 and 15 were sufficient to remove the reservations applied to the structural assignments based on the original n.m.r. study described above, and would, if they had been available earlier, have been sufficient to allow the determination of the structure of ochotensimine to have been based entirely on n.m.r. spectroscopy.

5.2.2 Fumaricine (3), fumaritine (4), and fumariline (5)

The validity of these arguments and the power of the spectroscopic techniques they depend on were soon demonstrated when the techniques were applied to an investigation[6] of the structures of three of the alkaloids that had been isolated from *Fumaria officinalis*[7], but which had resisted structure elucidation by traditional methods. These alkaloids, fumaricine (3), fumaritine (4), and fumariline (5), proved to be spirobenzylisoquinoline alkaloids bearing an oxygen function in ring C instead of the exocyclic methylene group of ochotensine and ochotensimine. The n.m.r. spectra showed that each alkaloid had an *N*-methyl group and two benzene rings, one 1, 2, 4, 5-tetrasubstituted and one 1,2,3,4-tetrasubstituted, and each ring carried two oxygen functions.

Fumariline (5) (Table 5.1) proved to be a non-enolisable 1-indanone derivative (5.85 μm peak in its i.r. spectrum) and the oxygen functions on

(6)

both benzene rings were shown to be methylenedioxy groups. The spectroscopic characteristics associated with rings B, C and D were similar to those of an intermediate (6) in the synthesis of ochotensimine (to be discussed later) but there were enough discrepancies to cause a search to be made for the reason for the differences. The appearance of the n.m.r. signals associated with the CH_2 protons of the five-membered ring provided the first clue; the anticipated AB quartet with $J_{AB} = 18.0$ Hz was observed, but each line showed further fine splitting which was attributed to long-range coupling between these protons and near-by aromatic protons. Because of the magnitude of this coupling it appeared that in fumariline, in contrast to the situation in ochotensimine, the aromatic position *ortho* to the benzylic methylene group bore a hydrogen atom rather than an oxygen function; decoupling experiments then confirmed that the protons at the benzylic position (C-14) were adjacent to aromatic protons at C-13 and C-12, and that fumariline had the alternative substitution pattern about the indanoid system, as shown in (5) (Table 5.1). The application of the NOE technique, particularly when compared with the effects observed with the synthetic model (6), provided further confirmation of these and other internuclear relationships and established structure (5) for fumariline.

Fumaricine (3) (Table 5.1) was shown to be the methyl ether of the phenol fumaritine (4) (Table 5.1) by converting (4) to (3) by the action of diazomethane, and the skeletal features of (3) (Table 5.1) were established by methods analogous to those used for fumariline. The presence of a secondary alcohol function at C-9, leading to a peak at 2.81 μm (carbon disulphide solution) in the i.r. spectrum, introduced a stereochemical question not present in the ketone, fumariline (5) (Table 5.1). Because of the position of the OH-stretch in the i.r. spectrum, it was suspected that the OH group was hydrogen-bonded to the nitrogen of the *N*-methyl group and that these groups had a *cis*-relationship on ring C. However, the demonstration of an observable NOE between the *N*-methyl group and the proton bonded to the carbon (C-9) bearing the hydroxyl group, required the reverse stereochemical assignment, as shown in (3), with the C—H bond proximal to the *N*-methyl group, and another explanation, π-bonding of the OH group to an aromatic ring, had to be invoked to account for the position of the OH peak in the i.r. spectrum. Since the alkaloids had already been interrelated, all of the structural and stereochemical features established for fumaricine (3) were equally valid for fumaritine (4), but the problem remained of deciding which of the oxygen functions on ring A of fumaritine was OMe and which OH. The structure (4) (Table 5.1) was established by demonstrating through an NOE experiment that the methoxyl group was adjacent to the aromatic proton (at C-4) which was also adjacent to a benzylic methylene group (at C-5).

5.2.3 Sibiricine (7) and ochrobirine (8)

Sibiricine[8] and ochrobirine[9, 10] have been isolated from *Corydalis sibirica* and the latter alkaloid has also been identified in other species of the same genus[11]. Their n.m.r. spectra[8, 10] showed that each alkaloid had an *N*-methyl group and two methylenedioxy groups on a spirobenzylisoquinoline skeleton. They differed from previous members of this class, however, by having no intact CH_2 group on the five-membered ring C, both benzylic positions of this ring carrying oxygen functions.

In sibiricine[8], these functions consisted of one hydroxyl group and one ketonic carbonyl group (2.81 and 5.85 μm peaks in the i.r. spectrum of a solution in chloroform), and the structural and stereochemical problems in this case were concerned with the proper placement of these groups. The protons on ring D gave rise to n.m.r. signals that corresponded closely to those of the equivalent protons in the model ketone (6), but differed from those in fumariline (5) and indicated that the carbonyl group of sibiricine is situated adjacent to the ring protons of ring D rather than adjacent to the methylenedioxy group on that ring. Thus the carbonyl group was assigned to C-14 and the hydroxyl group to C-9, and since an appreciable NOE was observed at the proton also situated at C-9 when the *N*-methyl signal was saturated, the hydroxyl group was assigned the configuration shown in (7), in which it is on the opposite side of ring C to the *N*-methyl group.

Ochrobirine[10] was shown to have two alcohol functions (2.79 and 3.03 μm peaks in the infrared spectrum of a carbon tetrachloride solution) on ring C, and these could be acetylated to provide a diacetate which provided an n.m.r. spectrum in which critical features were better resolved than in the spectrum of the parent compound. One of the protons geminal to an acetoxyl group was shown to give rise to a signal at τ 3.45 and this was shown by a spin-decoupling experiment (removing benzylic coupling with the aromatic proton at C-13) to be at a position (C-14) adjacent to the ring protons and one hydroxyl group of ochrobirine could consequently be placed at C-14; the other proton geminal to an acetoxyl group gave rise to a signal at τ 3.64, and its location at C-9 was confirmed by demonstrating, through decoupling experiments, that its coupling with the protons at C-12 and C-13 was of the magnitude expected for *para* and *meta* benzylic coupling. From NOE studies carried out on ochrobirine itself, it was possible to show that the proton at C-14, adjacent to the ring proton at C-13, was on the same side of the molecule as the *N*-methyl group, whereas the proton at C-9 was distant from the *N*-methyl group but close to the ring A aromatic proton at C-1. From these results, the structure and stereochemistry of ochrobirine was defined to be that shown in (8) (Table 5.1).

5.2.4 Parfumine (9)

Parfumine, an alkaloid of *Fumaria parviflora* was assigned[12] structure (9) (Table 5.1) on spectroscopic grounds. The phenolic hydroxyl group and the ketone were recognised from the i.r. spectrum of the alkaloid; the conversion of the compound to an aromatic *O*-acetyl derivative and an aromatic *O*-methyl derivative confirmed that the alkaloid was phenolic and the presence of the ketone function was confirmed by the reduction of the *O*-methyl derivative of parfumine to an alcohol. The carbonyl group was assigned to C-9 rather than C-14 by correlation of the chemical shifts of the ring D protons with those of fumariline (5) and the choice of C-2 rather than C-3 for the hydroxyl group was made by comparing the chemical shift of the *O*-methyl signal present in the methyl ether and absent in the phenol with the results obtained with fumaricine (3) and fumaritine (4) from which it is apparent that the signal for the methoxyl group at C-2 appears at significantly higher field than that of the methoxyl group at C-3 [6].

5.2.5 Fumarophycine (10)

Fumarophycine was isolated[13] from *Fumaria officinalis* of Bulgarian origin and investigated by the n.m.r. and NOE techniques[14] used for other alkaloids in this series. It was shown to have the spirobenzylisoquinoline skeleton with one benzylic position on ring C unsubstituted and one substituted by an *O*-acetyl group. The NOE results showed that the unsubstituted CH_2 of ring C was at position 14 adjacent to the aromatic protons of ring D and the identity of the proton at C-1 was established by its proximity to this methylene group; the acetoxyl group was, therefore, at C-9 and its configuration with the hydrogen at C-9 *cis* to the *N*-methyl group was shown by the NOE method. Ring A carried a methoxyl group and a hydroxyl group, and the methoxyl group was placed at C-3 by the NOE method, which showed its proximity to the C-4 proton. The structure and stereochemistry shown in (10) (Table 5.1) was, therefore, assigned to fumarophycine; this structure is that of an *O*-acetylfumaritine, and, in accord with the assignment, acetylation converted both fumarophycine and fumaritine (4) to the same derivative, a diacetoxy compound. Disturbingly, however, the product of alkaline hydrolysis of fumarophycine[13] was not completely identical with fumaritine (4), nor was the *O*-methyl derivative obtained by treating the hydrolysis product with diazomethane identical with fumaricine (3), and no satisfactory explanation for the discrepancy has been advanced up to the present.

5.2.6   Fumarofine (11)

Fumarofine, a further alkaloid isolated a number of years ago from *Fumaria officinalis*[7], has also been investigated more recently by spectroscopic methods[15]. The presence of a phenolic hydroxyl group, convertible to a methyl ether by diazomethane, an acetylatable alcohol, and a carbonyl group was demonstrated by a combination of i.r. spectroscopic and synthetic methods, and the spirobenzylisoquinoline skeleton was indicated by the similarity of the mass spectral fragmentation patterns (to be discussed presently) of fumarofine and sibiricine (7) (Table 5.1). Evidence from n.m.r. spectroscopy, obtained with a solution of fumarofine in dimethyl sulphoxide-d_6 and augmented by results obtained with solutions of the methyl ether in chloroform-*d*, showed that the carbonyl group was not adjacent to the aromatic protons of ring D and was therefore placed at C-9; the presence of the hydroxyl group at C-14 was confirmed by an NOE experiment showing the proximity of the proton on the same carbon atom to the aromatic proton at C-13 of ring D. Further NOE experiments showed the proximity of the proton at C-14 to the aromatic proton at C-1 of ring A, establishing the configuration at C-14, and the proximity of the *O*-methyl group in fumarofine to the proton at C-4, placing the methoxyl at C-3 in the alkaloid. These experiments established the structure and stereochemistry shown in (11) for fumarofine and the further observation was made that the signal associated with the proton at C-1 appeared at an unusually low field and that an unusually large NOE between the proton at C-14 and one of the protons at C-5 was measured, indicating the six-membered ring B adopted a twist conformation, not found in other alkaloids of this series, allowing hydrogen bonding between the C-14 hydroxyl group and the *N*-methyl group, and bringing the C-9 carbonyl group closer to C-1 and the C-14 proton closer to C-5.

5.2.7   Mass spectrometry and other spectroscopic methods

It is clear from the account already given that n.m.r. spectroscopy has been the key instrumental technique that has provided the critical data required for the determination of the structures of the alkaloids in this series, while other spectroscopic techniques have played a subordinate diagnostic role or have been used *ex post facto* to buttress structural assignments. Reference has already been made to the use of i.r. spectroscopy to determine the presence or absence of carbonyl and hydroxyl groups. Discussions of u.v. spectra have mainly been confined, particularly in early work, to demonstrating that they are not incompatible with the presence in the alkaloid of two separate benzenoid chromophores; however, with a larger number of alkaloids and model compounds available, spectroscopic differences which, for example, distinguish C-9 ketones from C-14 ketones have been observed, and were of some value in the elucidation of the structure of fumarofine[15]. A tabulation of u.v. characteristics of members of this series has appeared as part of a more extensive correlative study[16].

Mass spectrometry served initially to provide the molecular formulae of the alkaloids and to show that their fragmentation patterns distinguished them from other benzylisoquinoline alkaloids but, as with u.v. spectroscopy, as the number of alkaloids and model compounds of established structure grew, correlations were found that were of value in the recognition of structural variants[17]. The structural feature found to have the most significant effect on the fragmentation pattern of these alkaloids was the number and character of the oxygen functions on the five-membered ring C. In the alkaloids (1, 2) which lack an oxygen function on ring C, the parent ion provides the strongest peak by far in the spectrum and there does not seem to be a clearly preferred fragmentation pathway leading to a distinctive pattern in the rest of the spectrum. On the other hand, when one hydroxyl group is present (3, 4) on ring C, cleavage of the bond between the carbon atoms carrying the OH group and the N-methyl group is the principal mode of fragmentation of the parent ion (leading to a species such as (12)), and the spectrum is dominated by ions resulting from this process; the base peak corresponds to a dihydroisoquinolinium ion (13) which has lost rings C and D but gained the hydrogen from the OH group (verified by labelling), and another significant peak, $M-31$, corresponds to loss of the CHOH unit and one hydrogen atom from ring C after the initial cleavage and may be represented by (14). This pattern is critically dependent on the availability of the hydrogen on the OH group and acetylation of the alcohol

(e.g. (10)) is enough to change the pattern drastically. Consequently, in alkaloids in which a ketone function replaces the hydroxyl group on ring C (e.g. (5)) the spectrum is characteristically different and is dominated by a peak at $M-29$, believed to correspond to the $M-31$ ion (14) of the alcohols and to have been formed by the loss of $(CO+H)$ rather than $(CHOH+H)$. When ring C carries both a hydroxyl and a carbonyl group (e.g. (7)), the parent ion provides the base peak and the rest of the rather complex spectrum can be accounted for as the superimposition of patterns resulting from fragmentation modes similar to those described for the alkaloids with either an alcohol or a ketone present. Peaks are observed corresponding to the dihydroisoquinolinium (13) and isoquinolinium ions and to $M-45$, the latter apparently being ion (14), the $M-(CO+OH)$ analogue of $M-(CO+H)$ and $M-(CHOH+H)$ in the spectra of alkaloids with a single oxygen function on ring C; a peak that is of considerably enhanced intensity when the hydroxyl and carbonyl functions are both present appears to correspond to the ion represented by (15). When both oxygen functions on ring C are hydroxyl

(15)

(e.g. (8)), a different and distinctive spectrum is observed; the parent ion and the dihydroisoquinolinium ion (13) are reduced in abundance, the $M-H_2O$ peak is significant and the base peak is at $M-47$ and appears to correspond once again to ion (14), arising this time by loss of $(CHOH+OH)$ from the molecular ion. Examination of the spectra of synthetic model compounds showed, furthermore, that the fragmentation pattern did depend in detail on the nature of the substituents on the aromatic rings.

The correlations that have resulted from this study[17] should be of diagnostic value in the future, but, with the exception of the case of fumarofine (11)[15], there has been little opportunity up to the present to apply them to new alkaloids.

5.3　TOTAL SYNTHESES

The total synthesis of spirobenzylisoquinoline alkaloids has been undertaken in a number of independent laboratories with a remarkable degree of consistency in the synthetic plan. The critical step in skeletal elaboration has concerned the formation of the spiro linkage, and two approaches, each using a Pictet–Spengler condensation with an appropriately substituted phenylethylamine, have been considered; in one approach an appropriately substituted indanone was to be constructed first and this was then to be condensed with the phenylethylamine, while in the second approach the Pictet–Spengler reaction was to be carried out with an appropriately substituted phenylpyruvic acid and the closure of the five-membered ring was to

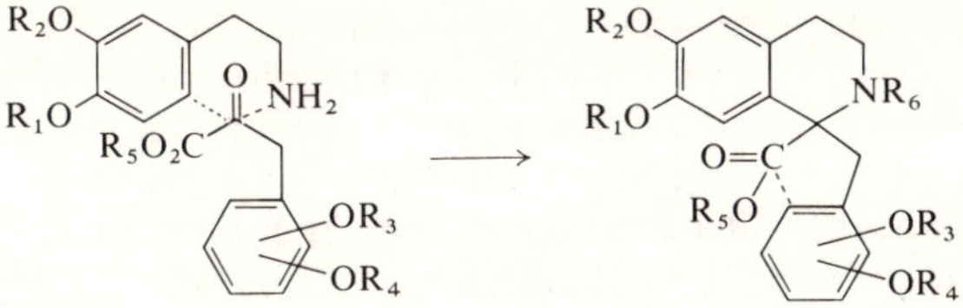

be effected in a subsequent step. The remaining questions mainly concerned the proper manipulation of oxygen functions.

5.3.1   Indanone route

5.3.1.1   *Ochotensimine and ochotensine*

The first reported synthesis of a spirobenzylisoquinoline alkaloid was that of ($\pm$)-ochotensimine[18]. Since the scope of the Pictet–Spengler reaction with ketones had received relatively little previous attention, the model compounds 2-indanone and 1,2-indanedione were allowed to react with homoveratrylamine in the presence of phosphoric acid. When it was found that the condensation was successful and that, as could be anticipated, with 1,2-indanedione the condensation product isolated resulted from reaction of the carbonyl group at C-2, the required 4,5-methylenedioxyindanedione (16) was

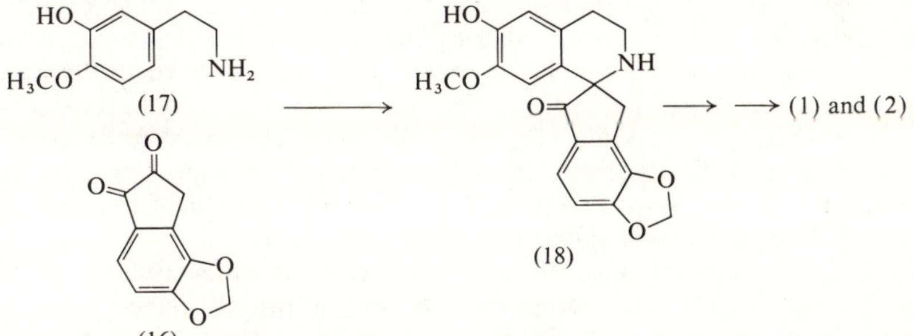

synthesised. The synthesis of this intermediate faced a problem commonly found with syntheses of these alkaloids: piperonyl derivatives are a readily available source of substituted methylenedioxybenzenes, but lead by standard synthetic methods to an isomeric methylenedioxyindanone that does not have the substitution pattern suitable for further elaboration, while the required isomer of the piperonyl precursors is not readily available. A useful route from *o*-vanillin to 2,3-methylenedioxybenzaldehyde was therefore established, and this aldehyde was then converted by standard procedures to 4,5-methylenedioxy-1-indanone, which was transformed into the dione by several routes. It was found during this study that the methylenedioxy group on the ring modified the reactivity of the substituent at C-2 in an unexpected and still unexplained manner. A ramification of this appeared in the attempted condensation of the dione with homoveratrylamine when it was found that the Pictet–Spengler cyclisation could not be effected under conditions that were used successfully in the synthesis of the model compounds. The more reactive 3-hydroxy-4-methoxyphenylethylamine (17) (which also offered a route to ochotensine) was therefore employed, and with hydrochloric acid under carefully controlled conditions this reacted with the dione (16) to afford the spiro compound (18) which, after it had been *O*-methylated with diazomethane, *N*-methylated by the Eschweiler–Clarke procedure (to form (6)), and treated with methylenetriphenylphosphorane in a Wittig reaction afforded (±)-ochotensimine (1).

Soon after the initial report of this synthesis[18], an independent synthesis following the same plan was reported[19]. In this work, exploration of the route was carried out with the indanone (19) derived from piperonal which was converted in a manner analogous to that just described to the isomers (20) of ochotensine and ochotensimine. (±)-Ochotensimine was then

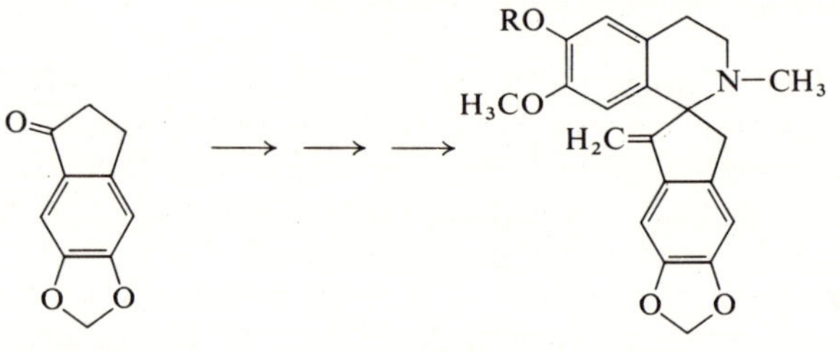

(19)   (20)

synthesised through 4,5-methylenedioxy-1-indanone by substantially the same route as that described above; the same difficulties were encountered but the manner in which they were dealt with differed in detail from those in the earlier report[18] and, furthermore, the route to ochotensine from the Pictet–Spengler product (18) was followed through. Methoxymethylation of the phenolic hydroxyl group protected it through the subsequent *N*-methylation and Wittig steps and subsequent removal of the protecting group by acid hydrolysis afforded (±)-ochotensine (2).

A further independent synthesis of ochotensine which differed in some experimental details, one being the use of a tetrahydropyranyl group to protect the phenol during the Wittig reaction, soon followed[20]. For the latter work preliminary model studies used the more easily accessible 4,5-di-

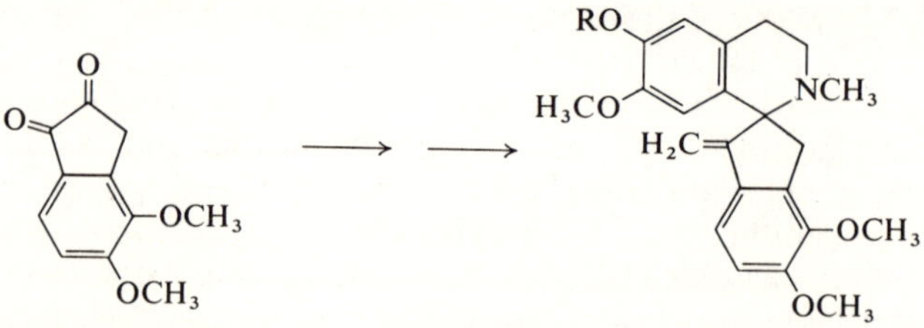

(21)   (22)

methoxy-1,2-indanedione (21), which led to the dimethoxy analogues (22) of ochotensine and ochotensimine, to explore the route.

5.3.1.2 *Fumaricine, fumaritine and fumariline*

The analogous syntheses of fumaricine (3), fumaritine (4), and fumariline (5) (Table 5.1) required the isomeric precursor to rings C and D, 6,7-methylene-dioxy-1,2-indanedione (23), which was available from piperonal through 3,4-methylenedioxydihydrocinnamic acid, which was constrained to cyclise in the desired direction by first brominating the more reactive ring position (C-6) and then closing the five-membered ring by a Friedel–Crafts reaction; debromination of the bromoindanone and oxidation of the ketone provided the required indanedione (23). After this had reacted with 3-hydroxy-4-methoxyphenylethylamine in a Pictet–Spengler cyclisation, *O*-methylation,

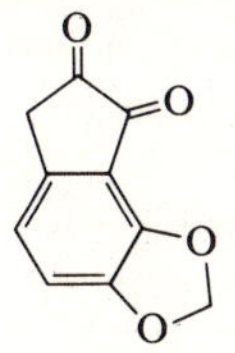

(23) 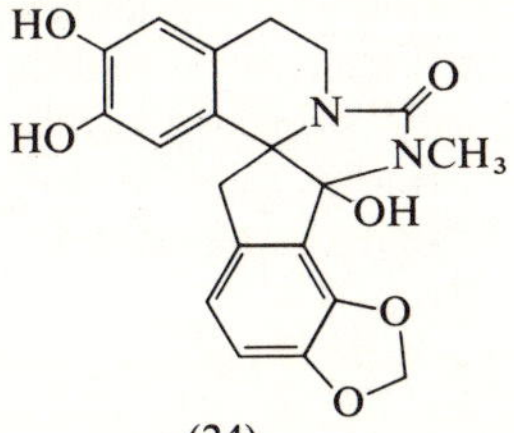   (24)

N-methylation, and reduction of the ketone with lithium aluminium hydride then afforded a pair of epimeric alcohols, from which (±)-fumaricine (3) was separated by fractional crystallisation[21]. When 3,4-dihydroxyphenyl-ethylamine was employed in the Pictet–Spengler reaction[22], the spiro product was a dihydric phenol, and treatment of this with diazomethane under con-trolled conditions led to the formation of a mixture of the dimethyl ether and the two possible monomethyl ethers; the desired monomethoxy inter-mediate was isolated by chromatography and converted to (±)-fumaritine (4) by the same route as had been used for the synthesis of fumaricine. In order to synthesise fumariline (5) the dihydric phenol obtained from the Pictet–Spengler reaction had to be converted to its methylenedioxy derivative, and because it was necessary to protect the sensitive amino ketone during this operation, the intermediate was converted to an imidazolinone (24) by treatment with methyl isocyanate. After methylenation, the protecting group was removed by lithium aluminium hydride, and *N*-methylation finally provided (±)-fumariline[22].

5.3.1.3 *Ochrobirine*

In a logical extension of this approach to the synthesis of ochrobirine, two groups have independently studied the condensation of a phenylethylamine with ninhydrin; one group[23] employed 3-hydroxy-4-methoxyphenylethyl-amine to provide the spiro dione (25) in the model study, and the other group[24] used homopiperonylamine in the same way and then *N*-methylated the spiro dione and reduced the product with sodium borohydride, obtaining a 2:1 mixture of stereoisomeric diols which, on spectroscopic grounds, was considered to have the analogue (26) of ochrobirine as the major component.

(25)

(26)

(and C-14 epimer)

Difficulties encountered in preparing the methylenedioxy-substituted ninhydrin required for the synthesis of ochrobirine itself have now been overcome[23]; 5,6-dimethoxyindanone served as a model for the approach, which was then applied to 6,7-methylenedioxyindanone (27). Conversion to the trione was accomplished by treating the indanone with *p*-nitroso-*N,N*-dimethylaniline in the presence of alkali and subjecting the product (28) to acid hydrolysis; condensation of the ninhydrin analogue (29) with homopiperonylamine afforded the spiro dione which was *N*-methylated and

(27)

(28)

(29)

(8)

reduced with sodium borohydride. The reduction in this case[23], in contrast to the result in the model series[24], was said to proceed stereospecifically to ($\pm$)-ochrobirine (8), and the difference was attributed to the presence of the methylenedioxy substituent.

5.3.1.4 *Orientation in the Pictet–Spengler condensation*

In the syntheses described so far it has been implicitly assumed that the Pictet–Spengler condensation leads to the required orientation of oxygen functions in ring A. Although a comprehensive study of this aspect of the reaction has not been reported, a study of the products obtained when 3,4-dimethoxy- and 3-hydroxy-4-methoxyphenylethylamine were allowed to react with ninhydrin under Pictet–Spengler conditions showed that the orientation of the oxygen functions was different in the two cases[17]. Although the dimethoxy compound led to the orientation of the methoxyl groups in the spiro product that was the same as in the alkaloids, ring closure was directed by 3-hydroxy-4-methoxyphenylethylamine to the position *ortho* to the hydroxyl group and produced (30); under the conditions used, this

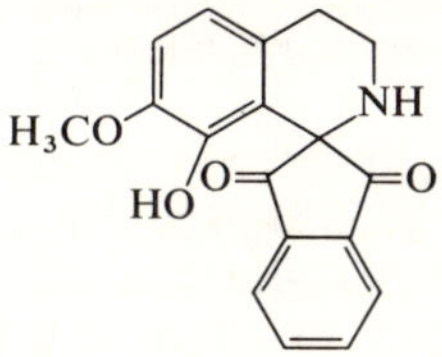

(30)

substitution pattern differed from that assigned in all other cases, including that described above for the same reactants under neutral conditions[23]. An account[25] has appeared of a study of the cyclisation of 3-hydroxy-*N*-methylphenylethylamine with a variety of aldehydes and ketones in neutral (refluxing ethanol), acidic (concentrated hydrochloric acid in ethanol) or basic (pyridine or triethylamine) medium; tetrahydroisoquinolines were isolated, and it was concluded, principally on the basis of the u.v. spectra of the products, that ring closure had occurred at a position *para* to the hydroxyl group, and consequently that cyclisation of these hydroxyphenyl-ethylamines and carbonyl compounds gave corresponding tetrahydro-isoquinolines in neutral, basic or acidic media. The isolation of (30), as described above, may be a rare and exceptional case resulting from the par-ticular combination of reactants and reaction conditions used, but it does illustrate that the factors which govern directing effects in this reaction are not completely understood.

5.3.2   Phenylpyruvate route

The alternative approach of cyclising a substituted phenylethylamine with a substituted phenylpyruvic acid followed by closure of the indanone ring has been generally less successful. The first stage, reaction with the phenyl-pyruvic acid, has been accomplished satisfactorily[19, 21], but difficulty has been encountered in effecting the subsequent Friedel–Crafts cyclisation without cleaving the aromatic oxygen functions required for the synthesis of the natural alkaloids. In one case[26], despite poor yields in this cyclisation,

(31)   (32)

which was carried out with 3-hydroxy-4-methoxyphenylethylamine and 3,4-methylenedioxyphenylpyruvic acid to form (31), the route has been followed through to an isomer, (32), of ochotensine (2).

5.3.3 Dihydro-Reissert compound route

One synthetic approach[27] which did not use a Pictet–Spengler reaction to form the spiro junction depended on the alkylation of a dihydro-Reissert compound (33) with 2,3-methylenedioxybenzyl chloride to form the intermediate (34), and the results are of intrinsic interest because they demonstrate the utility of this reaction. However, after hydrolysis of (34), the route

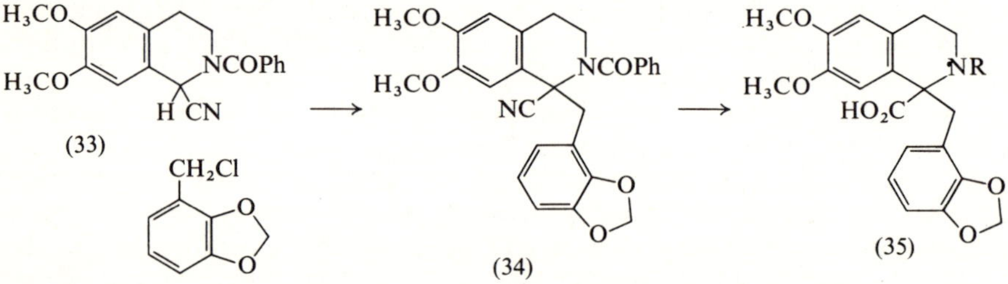

converges on the phenylpyruvic acid approach, and depends on a Friedel–Crafts cyclisation of (35) to close ring C; here again conditions to effect this step satisfactorily were not found, and the approach was abandoned.

5.4 BIOGENETIC CONSIDERATIONS

No report of a study of the biosynthesis of the spirobenzylisoquinoline alkaloids has yet appeared, but there has inevitably been biogenetic speculation based on the recognition of a possible relationship between those structural features that are unique to these alkaloids and those found in other benzylisoquinoline alkaloids of more firmly established biosynthetic origins. It is clear that all of the spirobenzylisoquinoline alkaloids can be viewed as benzylisoquinolines which have incorporated a one-carbon bridge leading to the spiro union, but there is no experimental evidence yet available bearing on the mode by which the plant carries this out or even on the biosynthetic relationship between the alkaloids ochotensimine (1) and ochotensine (2), with an exocyclic methylene group on ring C, and the alkaloids with oxygen functions on ring C. In one case, however, biogenetic speculation has led to an *in vitro* test of the feasibility of the hypothesis and an intrinsically interesting outcome of this work has been the construction of the molecular skeleton by a synthetic route substantially different from those described above[28].

The biogenetic speculation stemmed from the observation that the dihydroberberine derivative (36) bears an obvious relationship to protoberberine alkaloids such as corydaline and protopine alkaloids such as corycavine, that protoberberine and protopine alkaloids occur alongside the spirobenzylisoquinoline alkaloids, and that a route from (36) to the ochotensimine skeleton could be envisaged. The hypothetical route required a base-catalysed cleavage, using the acidic character of one phenolic hydroxy group as shown by the arrows, reclosure of the intermediate (37) to the spirane (38), utilising the electrophilicity of its enone part and the nucleophilicity of its hydroxypolyene part (which is only incidentally also a substituted enamine), and then tautomerism of (38) to provide the ochotensimine skeleton (39). Modifications of the details of this route, including a variant

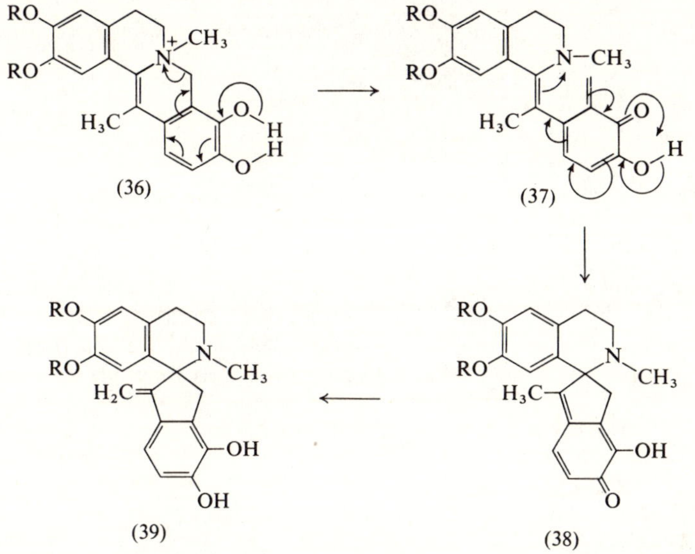

(36)

(37)

(39)

(38)

which utilises the acidity of a hydroxyl group in ring A, have been suggested, but none differs in a fundamental way from the cleavage–closure sequence just outlined, and the test of the *in vitro* feasibility of the route required the construction of a molecule corresponding to (36). However, it was recognised that the synthesis of a protoberberine with precisely the correct substitution pattern in ring C faced difficulties related to those encountered in the synthesis of similarly substituted indanones (Section 5.3.1), while a protoberberine with an alternative distribution of oxygen functions was more accessible and would still provide a test of the hypothesis regarding the rearrangement.

The protoberberine (40) was therefore synthesised by established methods, reduced by lithium aluminium hydride to a dihydroprotoberberine, which afforded (41) (R = CH$_2$Ph) on methylation. This material, with the phenolic groups protected, did not undergo rearrangement when treated with alkali, but when the benzyl groups had been removed by treatment with hydrobromic acid, the dihydric phenol (41) (R = H) produced underwent rearrangement in alkaline ethanol. However, the material isolated at this stage was not the phenol (44) (R = H), but its tautomer (43), the product expected from the cleavage–closure sequence analogous to that postulated initially, but lacking the final tautomerisation step. Tautomerisation of (43) to (44)

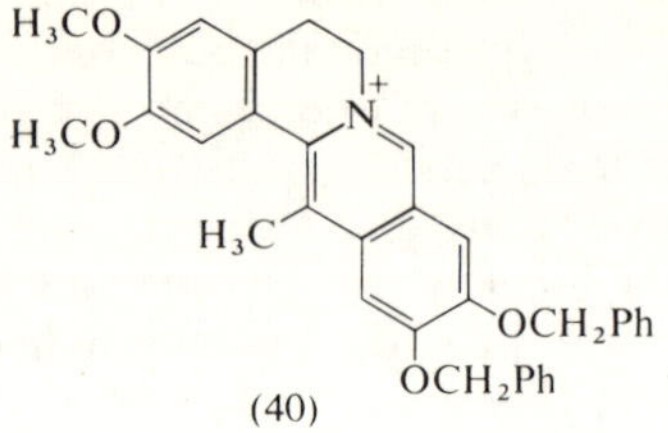

(40)

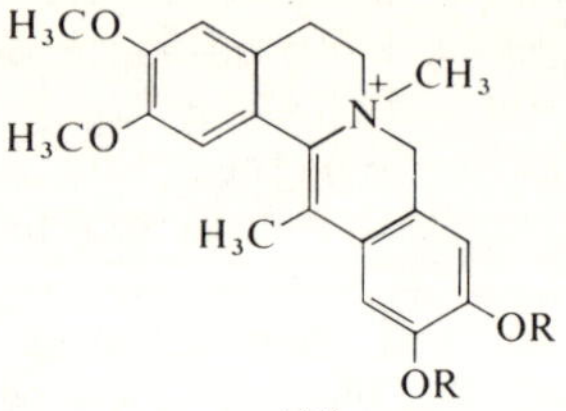

(41)

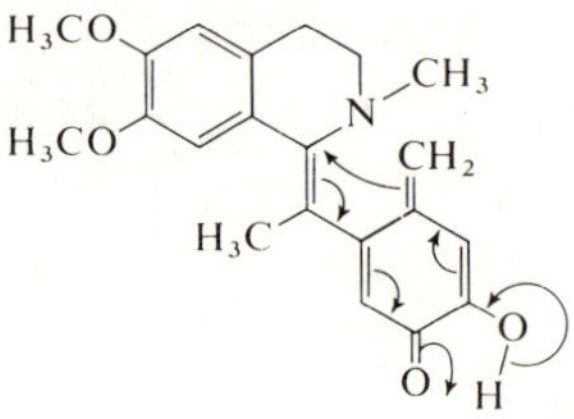

(42)

(R = H), could be effected simply by dissolving the material in a solvent such as dimethyl sulphoxide, and the product could be trapped as its di-acetate (44) (R = Ac). However, (44) (R = H) could not be obtained pure since it readily reverted to its tautomer (43), and in polar solvents which

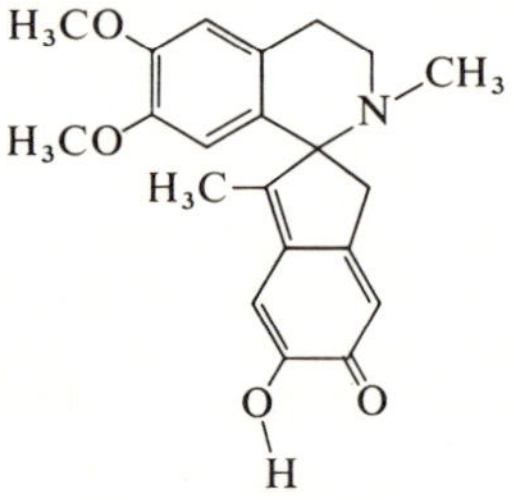

(43)

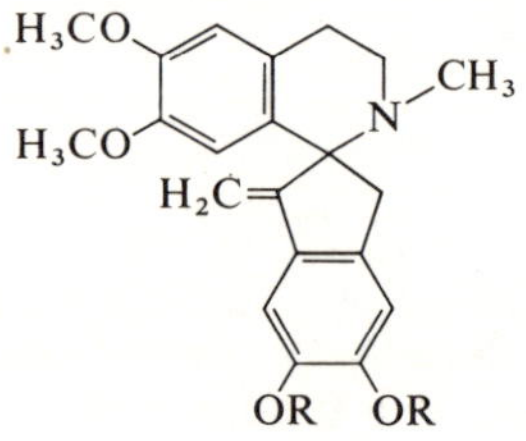

(44)

could act as proton donors, it was shown that this interesting tautomeric equilibrium lay entirely on the side of the quinone methide (43).

This synthesis has not yet been applied to the actual case represented by (36), but it is clear that the work already described illustrates that this synthetic route to ochotensimine (1) appears feasible, and that the plant could make use of a biosynthetic analogue.

5.5 ADDENDUM

A review[29] of this field became available after this manuscript had been prepared.

The absolute configuration of the alkaloids (+)-ochotensimine (1), (+)-ochotensine (2), (+)-fumariline (5), and (+)-ochrobirine (8) has recently been determined[30] by the aromatic chirality method[31] and in each of these cases it has been shown that the configuration drawn in the present review (see Table 5.1) does in fact correspond to the correct absolute configuration of the natural alkaloid.

Two alkaloids, corydaine and corpaine, have been isolated from *Corydalis paczoskii* and studied by spectroscopic methods, including the NOE technique[32]; corydaine has been assigned structure (45) and corpaine structure (46).

Further studies of the protoberberine—spirobenzylisoquinoline rearrangement (as in (36) → (39)) have been reported[33].

References

1. Manske, R. H. F. (1940). *Can. J. Res. B,* **18,** 75
2. McLean, S. and Lin, M. -S. (1964). *Tetrahedron Lett.,* 3819; McLean, S., Lin, M. -S. and Manske, R. H. F. (1966). *Can. J. Chem.,* **44,** 2449
3. McLean, S., Lin, M. -S., Macdonald, A. C. and Trotter, J. (1966). *Tetrahedron Lett.,* 185; Macdonald, A. C. and Trotter, J. (1966). *J. Chem. Soc., B,* 929
4. Anet, F. A. L. and Bourn, A. J. R. (1965). *J. Amer. Chem. Soc.,* **87,** 5250
5. Bell, R. A. and Saunders, J. K. (1968). *Can. J. Chem.,* **46,** 3421
6. Saunders, J. K., Bell, R. A., Chen, C. -Y., MacLean D. B. and Manske, R. H. F. (1968). *Can. J. Chem.,* **46,** 2873, 2876; MacLean, D. B., Bell, R. A., Saunders, J. K., Chen, C. -Y. and Manske, R. H. F. (1969). *Can. J. Chem.,* **47,** 3593
7. Manske, R. H. F. (1938). *Can. J. Res., B,* **16,** 438; Manske, R. H. F. (1969). *Can. J. Chem.,* **47,** 1103
8. Manske, R. H. F., Rodrigo, R., MacLean, D. B., Gracey, D. E. F. and Saunders, J. K. (1969). *Can. J. Chem.,* **47,** 3585
9. Manske, R. H. F. (1936). *Can. J. Res., B,* **14,** 354
10. Manske, R. H. F., Rodrigo, R. G. A., MacLean, D. B., Gracey, D. E. F. and Saunders, J. K. (1969). *Can. J. Chem.,* **47,** 3589
11. Manske, R. H. F. (1939). *Can. J. Res., B,* **17,** 89, 95
12. Israilov, A., Yunusov, M. S. and Yunusov, S. U. (1969). *Dokl. Akad. Nauk SSSR,* **189,** 1262; Eng. (Consultants Bureau) p. 999
13. Mollov, N. M., Yakinov, G. I. and Panov, P. P. (1967). *C. R. Acad. Bulg. Sci.,* **20,** 557
14. Castillo, M., Saunders, J. K., MacLean, D. B., Mollov, N. M. and Yakimov, G. I. (1971). *Can. J. Chem.,* **49,** 139
15. Yu, C. K., Saunders, J. K., MacLean, D. B. and Manske, R. H. F. (1971). *Can. J. Chem.,* **49,** 3020
16. Santavy, F., Hruban, L., Simanek, V. and Walterova, D. (1970). *Collect. Czech. Chem. Commun.,* **35,** 2418
17. Yu, C. K. and MacLean, D. B. (1971). *Can. J. Chem.,* **49,** 3025
18. McLean, S., Lin, M. -S. and Whelan, J. (1968). *Tetrahedron Lett.,* 2425; (1970). *Can. J. Chem.,* **48,** 948
19. Irie, H., Kishimoto, T. and Uyeo, S. (1968). *J. Chem. Soc., C,* 3051
20. Kelly, R. B. and Beckett, B. A. (1969). *Can. J. Chem.,* **47,** 2501
21. Kishimoto, T. and Uyeo, S. (1969). *J. Chem. Soc., C,* 2600
22. Kishimoto, T. and Uyeo, S. (1971). *J. Chem. Soc., C,* 1644
23. Kametani, T., Takano, S. and Hibino, S. (1968). *Yakugaku Zasshi,* **88,** 1123; *Chem. Abstr.,* (1969), **70,** 37955m; Kametani, T., Hibino, S. and Takano, S. (1971). *Chem. Comm.,* 925
24. Manske, R. H. F. and Ahmed, Q. A. (1969). *Can. J. Chem.,* **48,** 1280
25. Kametani, T., Kigasawa, K., Hiiragi, M. and Ishimaru, H. (1971). *J. Chem. Soc., C,* 2632
26. Kametani, T., Takano, S., Hibino, S. and Terui, T. (1969). *J. Heterocycl. Chem.,* **6,** 49
27. Shamma, M. and Jones, C. D. (1970). *J. Org. Chem.,* **35,** 3119
28. Shamma, M. and Jones, C. D. (1969). *J. Amer. Chem. Soc.,* **91,** 4009; (1970), **92,** 4943; Shamma, M. and Nugent, J. F. (1970). *Tetrahedron Lett.,* 2625

29. Shamma, M. (1971). *The Alkaloids* (R. H. F. Manske, editor), Vol. XIII, 165, (New York: Academic Press)
30. Nakanishi, K., personal communication
31. Harada, N., Nakanishi, K. and Tatsuoka, S. (1969). *J. Amer. Chem. Soc.,* **91,** 5896
32. Fesenko, D. A. and Perel'son, M. E. (1971) *Khim. Prir. Soedin,* **7,** 166; Baisheva, K. S., Fesenko, D. A., Perel'son, M. E. and Rostotskii, B. K. (1970). *Khim. Prir. Soedin,* **6,** 574; Baisheva, K. S., Fesenko, D. A., Rostotskii, B. K. and Perel'son, M. E. (1970). *Khim. Prir. Soedin,* **6,** 456
33. Shamma, M. and Nugent, J. F. (1971). *Chem. Comm.,* 1642

6
Benzylisoquinoline and Homobenzylisoquinoline Alkaloids

T. KAMETANI and K. FUKUMOTO

Tohoku University, Sendai, Japan

6.1 INTRODUCTION

The isoquinoline alkaloids[1] are classified into simple isoquinoline (1), alkylisoquinoline (2), phenylisoquinoline (3), benzylisoquinoline (4) and phenethylisoquinoline alkaloids (5) according to the basic skeleton. The benzylisoquinoline series is subdivided into 1-benzyl-(4a) and 2-benzylisoquinoline alkaloids (4b).

The 1-benzylisoquinoline alkaloids, derived biogenetically from tyrosine or phenylalanine, are transformed to pavine, isopavine, bisbenzylisoquinoline,

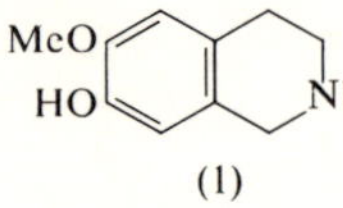

(1)
Corypalline

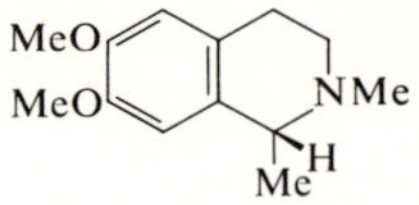

(2a)
Carnegine

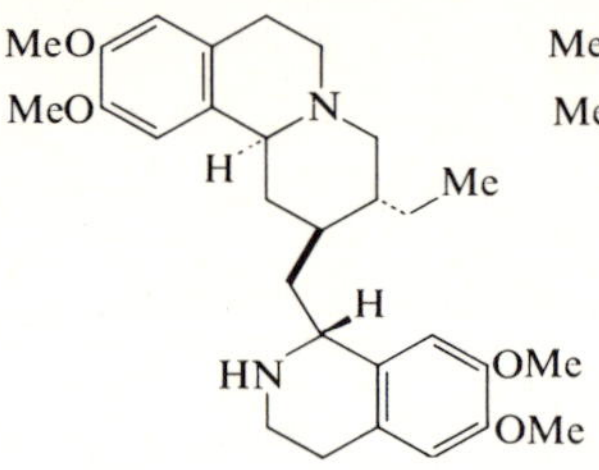

(2b)
Emetine

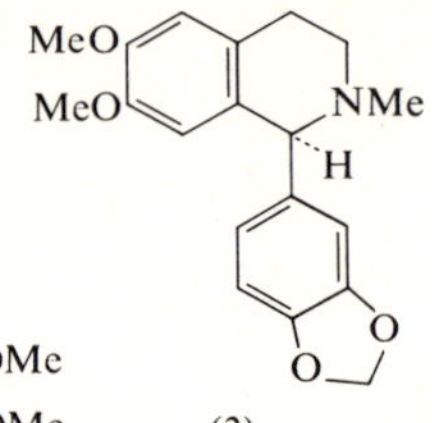

(3)
Cryptostyline I

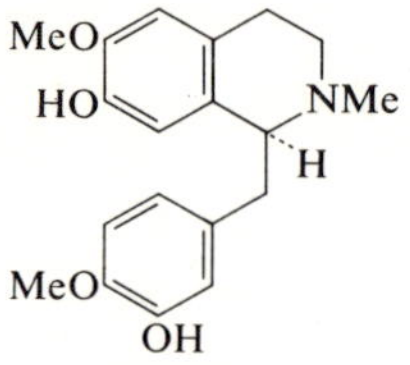

(4a)
Reticuline

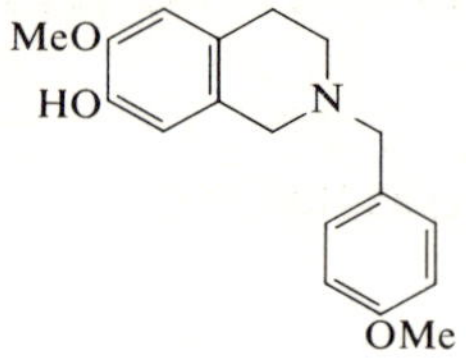

(4b)
Sendaverine

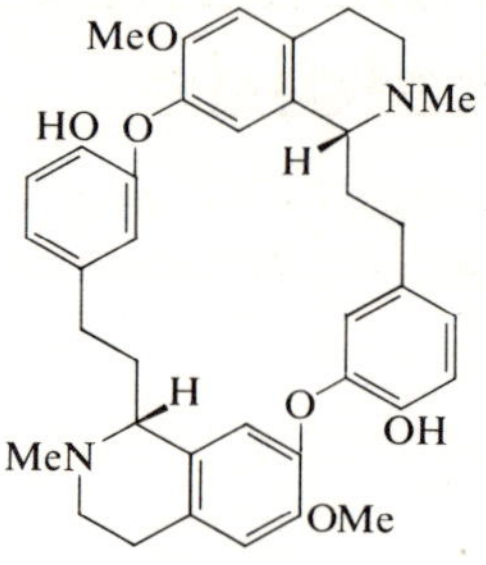

(5)
Melanthioidine

cularine, proaporphine, aporphine, protoberberine, protopine, phthalideisoquinoline, ochotensine, rhoeadine, benzophenanthridine alkaloids, morphine,
hasubanan, benzopyrrocoline, erythrina, and protostephanine by phenolic
oxidation[2] and modification with several plant enzymes[3].

Many contributions to isoquinoline research have been reported earlier[4].
In this chapter the structural elucidation, total synthesis, biogenesis, and
biosynthesis of isoquinoline alkaloids will be surveyed briefly.

6.2 STRUCTURAL ELUCIDATION

6.2.1 1 - Benzylisoquinoline alkaloids

The classical method for the structural elucidation of 1-benzylisoquinoline
alkaloids is the application of Hofmann degradation and oxidation.

If the alkaloid is phenolic, it is first *O*-methylated or *O*-ethylated. The latter
transformation usually serves the dual purpose of protecting the phenolic
group and ascertaining the position of the phenolic hydroxyl group(s). For
example, *O,O*-diethyltembetarine iodide (7), derived from (+)-tembetarine (6),
which has two methoxyl and two hydroxyl groups, underwent Hofmann
degradation and then oxidation of the product to give the known *O*-ethyl-
isovanillic acid (8), thus confirming the position of the methoxyl and the
phenolic hydroxyl groups in the benzylic moiety. Also, *O,O*-diethyl compound
(7) was converted to the tertiary base (9) with ethanolamine, and oxidation
of this gave the known isocarbostyril derivative (10)[5].

(6) (+)-Tembetarine

EtI

(9)

$HO(CH_2)_2NH_2$

(7)

I⁻

Hofmann degradation

[O]

(8)

HO_2C

[O]

(10)

The absolute configuration of this type of alkaloids has been determined by Corrodi and Hardegger, who ozonised (−)-N-norlaudanosoline to N-β-carboxyethyl-L-aspartic acid of a known absolute configuration[6]. The determination of the absolute configuration of other alkaloids has been carried out by a comparison with (−)-N-norlaudanosoline or its derivatives. Recently, spectroscopic data (mass[7], n.m.r.[8–11], i.r. and u.v. spectra[12]) have been applied to structural elucidation. An analysis of the effect of oxygen substitution on u.v. absorption in the benzylisoquinolines has been made[13], and mass spectral fragmentation patterns of this group have been studied[7].

The o.r.d.[14–18] and c.d. spectra[19] are effective in the determination of the chirality of the alkaloids. 1-Benzylisoquinolines of the L-series show positive o.r.d. Cotton effects in the region 280–240 nm.

The structure of cinnamolaurine was assigned to (−)-1,2,3,4-tetrahydro-1-(4-hydroxybenzyl)-2-methyl-6,7-methylenedioxyisoquinoline by the following spectral evidence without the use of degradation methods. The n.m.r. spectrum showed an N-methyl, a methylenedioxy group, and two isolated aromatic protons in addition to A_2B_2 type aromatic protons. The mass spectrum revealed the presence of 1,2,3,4-tetrahydro-2-methyl-6,7-methylenedioxyisoquinoline and hydroxybenzyl residue, thus leading to structure (11)[21].

The study of this class of alkaloids has led to the finding of new members of the group, which includes among others the quaternary bases[22, 27] and the glycosidic bases[28, 29], shown in Figure 6.1.

Bisbenzylisoquinoline alkaloids[30] are grouped into several types according to the mode of coupling and the number of the biphenyl ether functions as follows. Some alternations are observed in the position of the biphenyl ether bond and the patterns of oxygenation. These are illustrated in Figure 6.2.

The structural elucidation of this group of alkaloids follows the usual way; oxidation of the triethyl ether of magnoline (12), $C_{36}H_{40}N_2O_6$, which

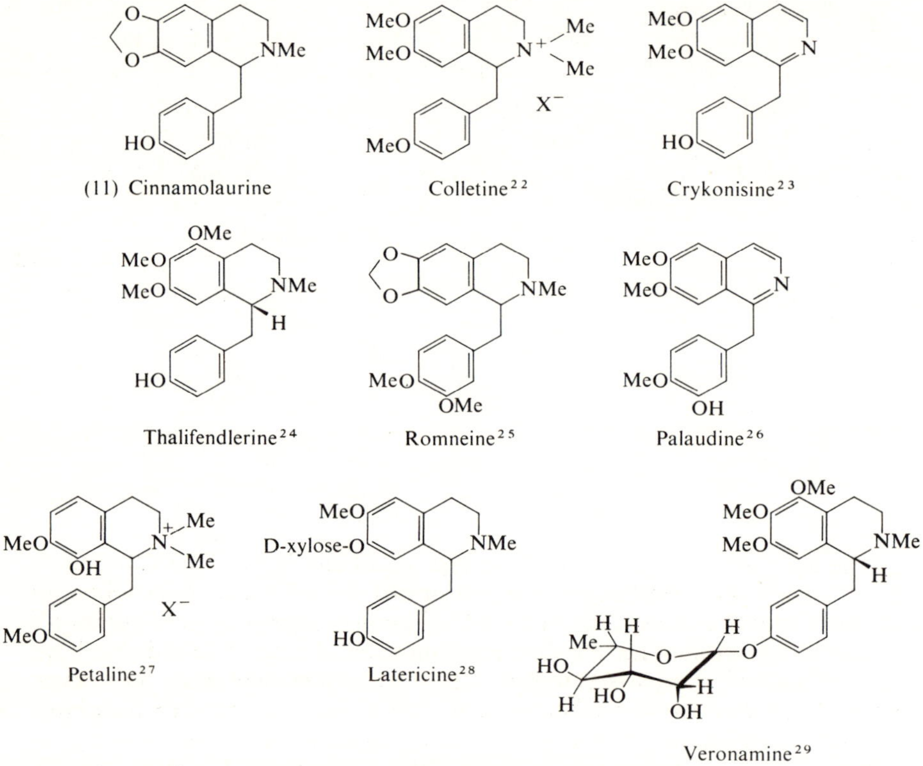

(11) Cinnamolaurine   Colletine[22]   Crykonisine[23]

Thalifendlerine[24]   Romneine[25]   Palaudine[26]

Petaline[27]   Latericine[28]   Veronamine[29]

Figure 6.1

contains two tertiary *N*-methyl, two *O*-methyl and three phenolic hydroxyl groups, gave the diphenyl ether dicarboxylic acid and the isocarbostyril. Assuming that the carbonyl and carboxyl groups represent the residues obtained by fission of the original alkaloid, the structure (12) follows logically for magnoline[31].

The method of Hofmann degradation is also applicable, and a typical example of the bisbenzylisoquinoline alkaloids is shown in Figure 6.3[32].

The absolute configuration of the asymmetric centres was easily determined by the configuration of the cleaved products by sodium in liquid ammonia[30, 33] and by the o.r.d. spectrum of the alkaloids[14, 16, 34]. Moreover, the fragmentation pattern of the mass spectrum indicates the 'type' of the bisbenzylisoquinoline alkaloids[30, 35–41]. The study of the n.m.r. spectrum may not only locate the methoxyl groups but can also reduce the stereochemical possibilities[30, 42]. The n.m.r. examinations have shown that, because of proximity effects caused by the ring systems, the chemical shifts of aromatic methoxyl groups on macrocyclic bisbenzylisoquinoline alkaloids can aid in deciding the type of oxygen bridge present[43].

Among the several new bisbenzylisoquinoline alkaloids isolated since 1967, the most interesting is repanduline (13)[44], isolated from *Daphnandra repandula*.

The cularine type of alkaloids is characterised by an oxepine ring and 7,8-dioxygenated isoquinoline systems. The structure of cularine, $C_{20}H_{23}NO_4$, which has an *N*-methyl, three *O*-methyls and a biphenyl ether group, was assigned by systematic Hofmann degradation and reductive cleavage of the

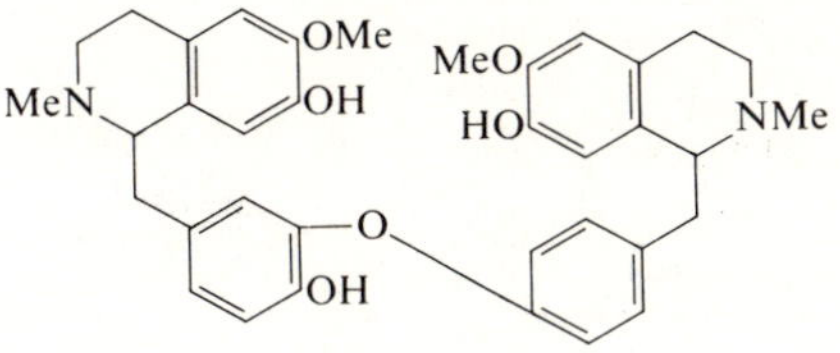

(12) Magnoline

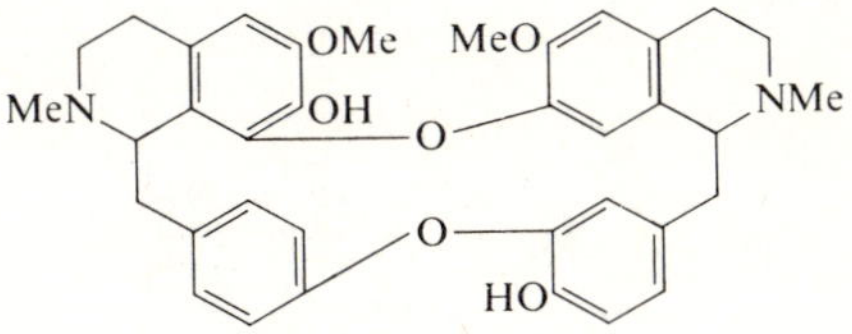

Berbamine–Oxyacanthine Type

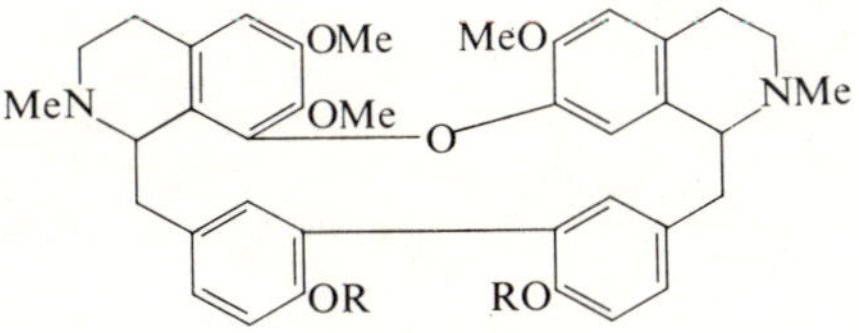

Rodiasine Type

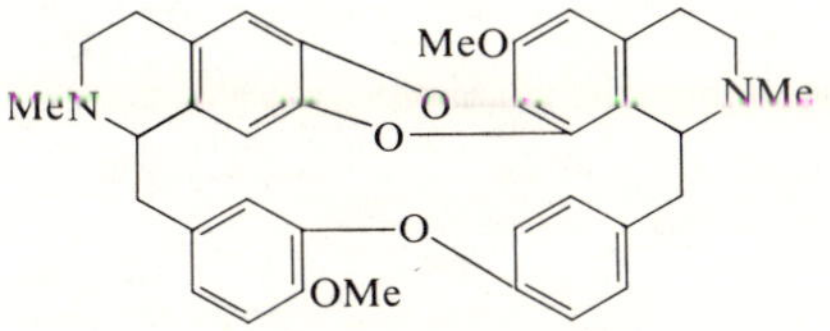

Trilobine Type

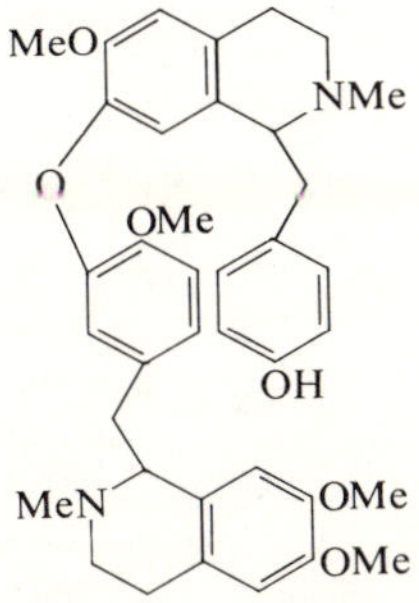

Liensinine Type

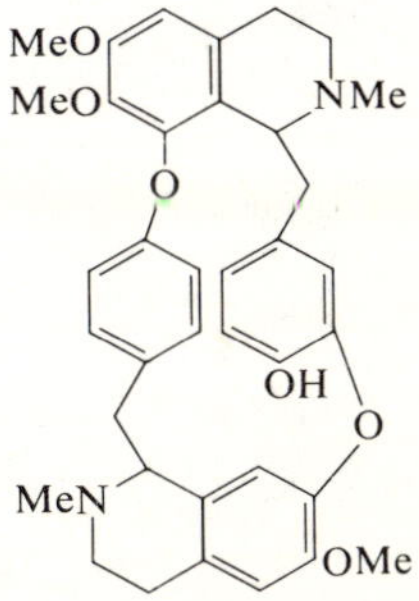

Curine Type

Figure 6.2

biphenyl ether linkage[45]. N.M.R. spectroscopy and the o.r.d. curve of the reduction product established the absolute configuration of cularine[19]. Recently, an opposite result was obtained by o.r.d. studies of the degradation product[20]. The structures of the related alkaloids were determined by their correlation to cularine and comparison of their mass spectra with those of cularine[7,46]. Recently, cancentrine was elucidated by x-ray analysis of its degradation product[47].

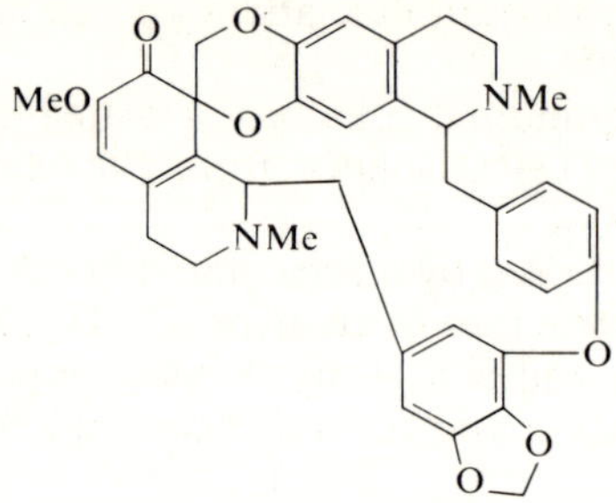

(13) Repanduline

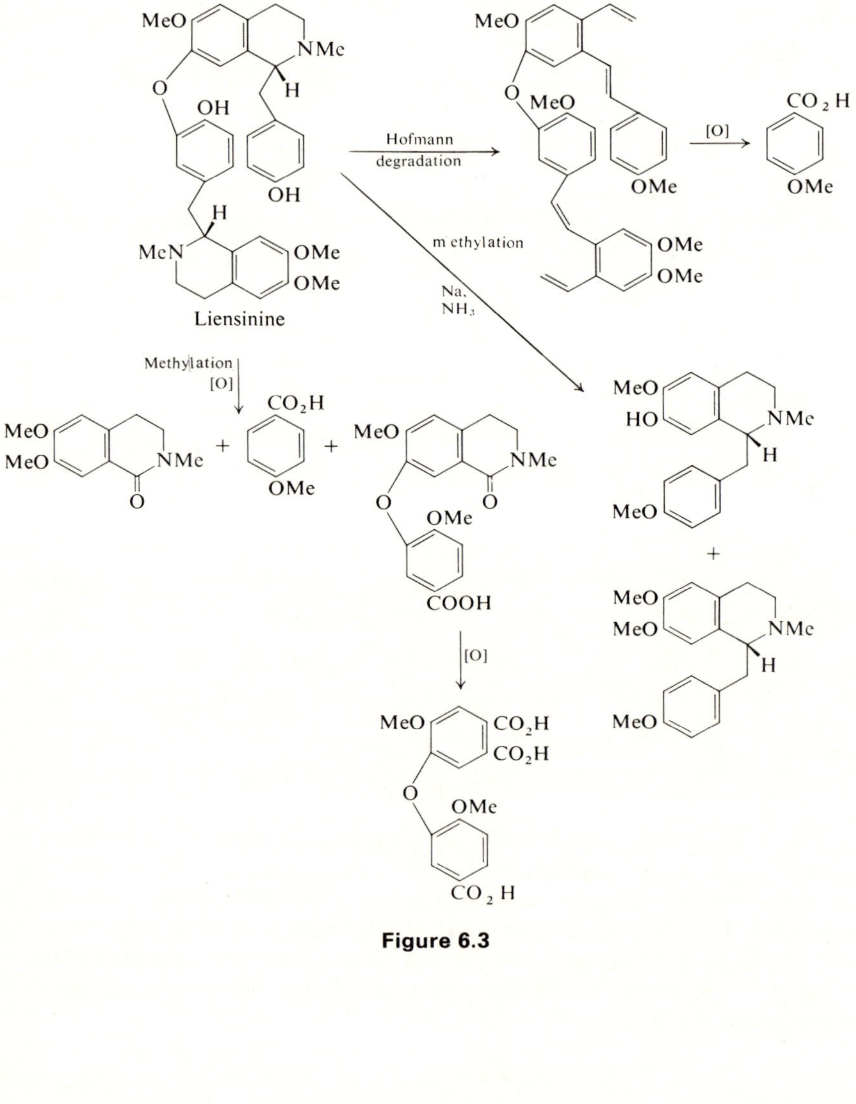

Liensinine

Hofmann degradation

methylation

Na, NH₃

Methylation
[O]

[O]

Figure 6.3

Cancentrine

	R¹	R²		R³	R⁴
Cularine	Me	Me		Me	Me
Cularicine	H		—CH₂—		Me
Cularidine	H	Me		Me	Me
Cularimine	Me	Me		Me	H

The pavine type of alkaloids, represented by argemonine, is characterised by the cyclo-octane ring system. The structure of argemonine was determined by spectral data and direct comparison of the alkaloid with *N*-methyl-pavine[48, 49] obtained from papaverine by reduction with tin and hydrochloric acid[50]. The absolute configuration of this base has been thoroughly established by chemical correlation with L-aspartic acid and by o.r.d. spectral measurements[51–54]. Amurensine and amurensinine show the isopavine structure determined by exhaustive Hofmann degradation and oxidation methods. The position of the oxygenation functions was indicated by spectral evidence[55]. Rates of methiodide formation provide a facile means for differentiating between pavine and isopavine bases, since the latter quaternises at a faster rate[56].

Argemonine

Amurensine R = H
Amurensinine R = Me

The aporphine alkaloids[57] are characterised by the presence of a 9,10-di-hydrophenanthrene system, and their structural elucidation is carried out by the usual methods. Oxidation of the alkaloid yields isocarbostyril and an alkoxylated phthalic acid. Hofmann degradation on a quaternary base gives an alkoxylated vinylphenanthrene, oxidation of which yields an alkoxy-phenanthrene carboxylic acid. Then decarboxylation of phenanthrene carboxylic acid gives phenanthrene, which may be compared with known compounds. The degradation of bulbocapnine illustrated the above principles[58–59].

Investigation of the u.v.[12, 57, 60] and n.m.r. spectra[42, 57, 62–66] of the aporphines is the most effective method in their structural elucidation. The o.r.d. spectrum has also been used to determine the absolute configuration of aporphine alkaloids[67, 68]. Jackson and Martin reported marked abnormalities, probably due to steric distortion of the biphenyl ring system[69]. Also the absolute configuration of the aporphines is determined by a conversion of (−)-laud-anosine of known absolute configuration to (−)-glaucine[70]. Shamma has pointed out that the aporphine alkaloids are not planar, but exist as either of two stereochemical arrangements (14) or (15)[60]. A correlation has been recognised between the pattern of hydroxyl group substitution and the sign of rotation (i.e. the direction of permanent twist of the biphenyl system) of the aporphines[61].

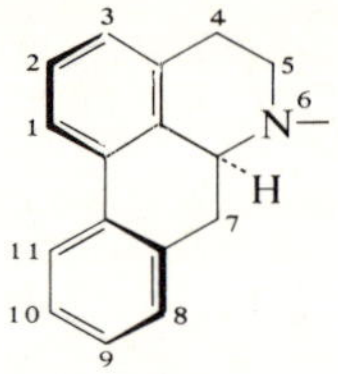

(14) 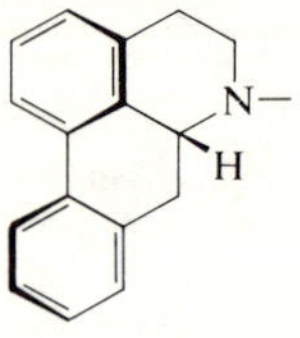   (15)

The mass spectral fragmentation pattern affords information important to structural elucidation; a molecular ion peak is very strong, and constant peaks at m/e 152 and 165 serve a useful diagnostic purpose[7]. Representative aporphines are shown in Figure 6.4.

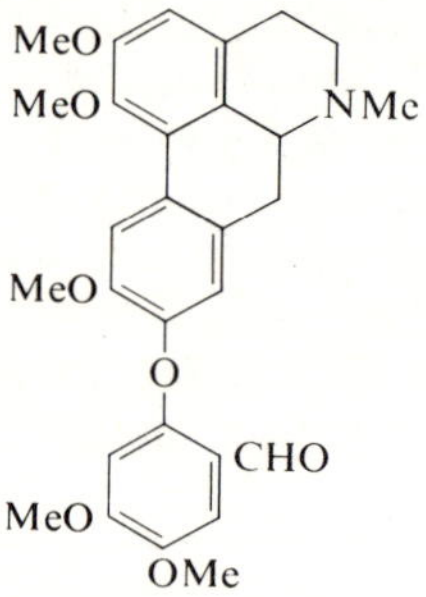

Hernandaline[69]

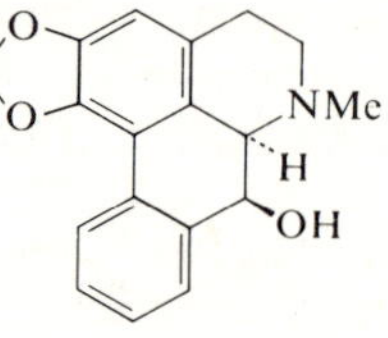

Ushinsunine[70]

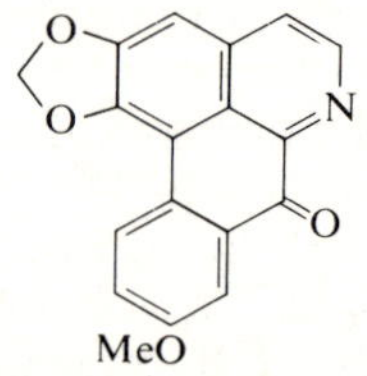

9-Methoxyliriodenine[71]

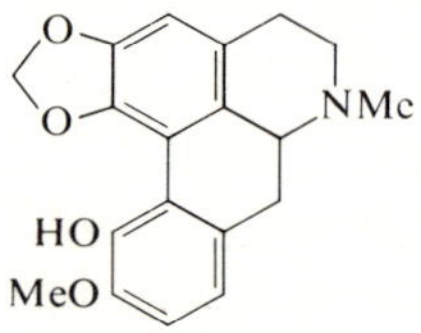

Bulbocapnine

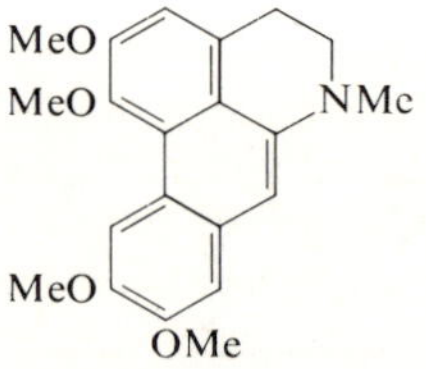

Dehydroglaucine[72]

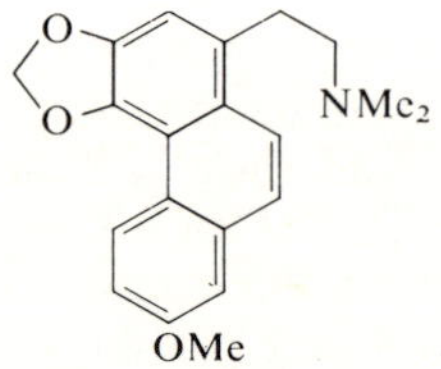

Uvariopsine[73]

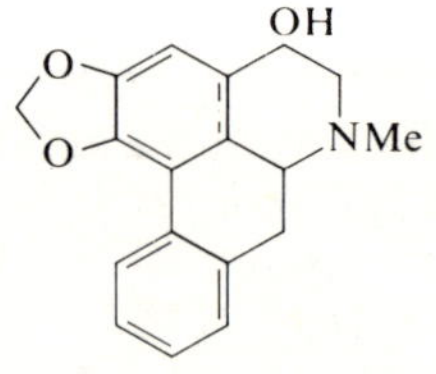

Steporphine[74]

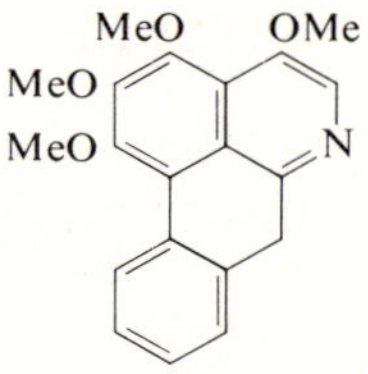

Imenine[75]

Figure 6.4

The proaporphine alkaloids[78, 79], which, before the isolation of the pro-aporphine alkaloids from nature, had been suggested by Barton to be the biogenetic precursors to some aporphine alkaloids, possess a tetracyclic nucleus incorporating a cross-conjugated cyclohexadienone or related system. Either system could be easily characterised by i.r., u.v., n.m.r. and mass spectra[78, 79].

Pronuciferine, $C_{19}H_{21}NO_3$, which has one N-methyl and two O-methyl groups, undergoes a dienone–phenol rearrangement when heated with aqueous sulphuric acid to give (−)-1,2-dimethoxy-10-hydroxyaporphine (16). A mixture of epimeric dienol (17) from pronucferine is treated with mineral acid to give the dienol–benzene rearrangement product, (−)-nuciferine (18), which has a known absolute configuration[80].

Sodium in liquid ammonia effects a reductive cleavage of pronuciferine to afford the known (−)-armepavine (19), structure (20)[81] must consequently be assigned to pronuciferine.

Crotonosine, $C_{17}H_{17}NO_3$, is also determined by usual methods[82–84].

All studies[85–87] of mass spectral fragmentation patterns of proaporphine bases show that these bases contain a fragment at $m/e = 165.069$ ($C_{13}H_9$), probably due to the perinaphthenyl cation[86]. In the o.r.d. spectrum of proaporphines, the sign of the Cotton effects depends not only upon the configuration at the 1-position of the isoquinoline ring, but also on the substitution of the spiro-dienone; more reliable determination of absolute configuration can be made from the signs and position of c.d. bands[83]. The characteristics of mass, i.r., u.v. and n.m.r. spectra are summarised by Bernauer[79] and Stuart[78].

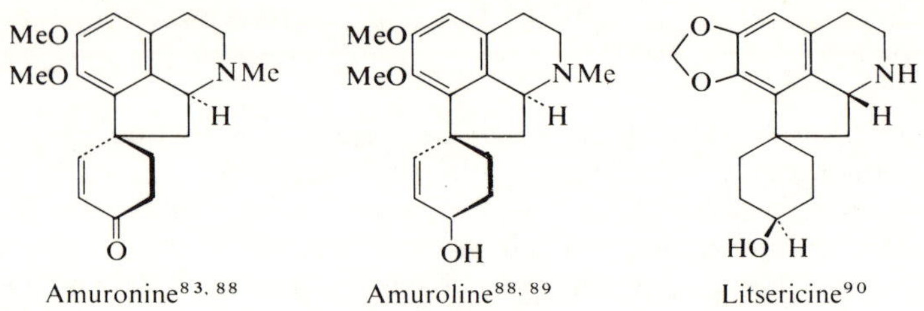

Figure 6.5

Figure 6.5 lists several types of reduced proaporphines found in nature.

The protoberberine alkaloids are subdivided into dehydro and tetrahydro bases, possessing a dibenzoquinolizidine system with oxygen functions in positions 2, 3, 9 and 10.

Their structural elucidation is usually carried out on the tetrahydro base by oxidation and Hofmann degradation. The structure of capaurine, $C_{21}H_{25}O_5N$, was determined by the oxidation method[91] and is supported by x-ray analysis[92]. The structural elucidation of thalictricavine is carried out by Hofmann degradation and by study of the oxidation product[93].

Mass spectrometry is a powerful tool for the structural investigation of the protoberberines[7, 94].

In some cases, n.m.r.[95] and u.v. spectra[13, 96, 97] are applicable to the determination of the position of substituents. The stereochemistry at C-13a has been determined by i.r. (Bohlmann bands in the region of 2800 cm^{-1})[98] and n.m.r. spectral methods[99] and by the relative rate of quaterrisation[56]. The absolute configuration of the tetrahydroprotoberberines is elucidated by correlation with $(-)$-N-norlaudanosine[6].

The tetrahydroprotoberberines, having only a single asymmetric centre at the C-13a, are assigned the S-configuration in the leavorotatory series and R-configuration in the dextrorotatory. The o.r.d. spectrum[16, 18, 100, 101] has also been used to assign the absolute configuration of these alkaloids. The data accumulated from the tetrahydroprotoberberines with no substituent at the C-13 position have been used to determine the stereochemistry at the C-13a position of corydaline and related alkaloids. Some alkaloids in this group are listed in Figures 6.6 and 6.7.

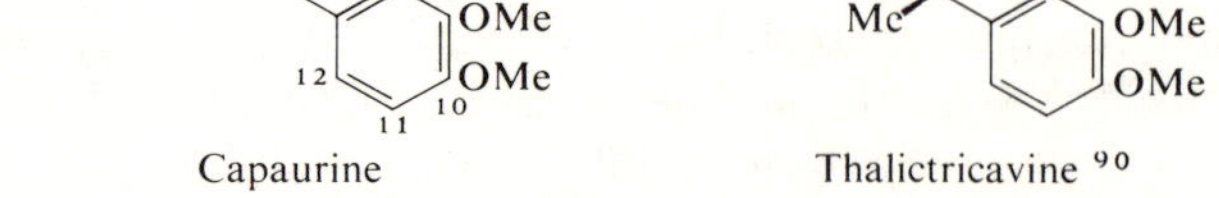

Figure 6.6

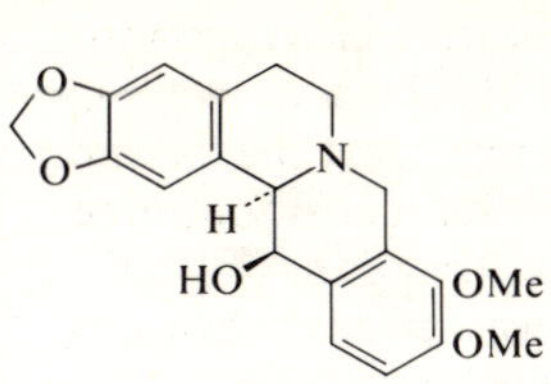

Ophiocarpine[101] 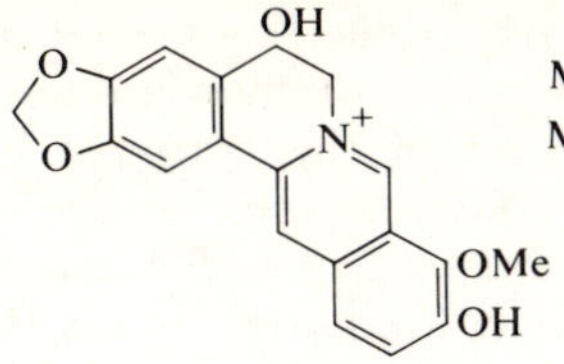   Thalidastine[102] 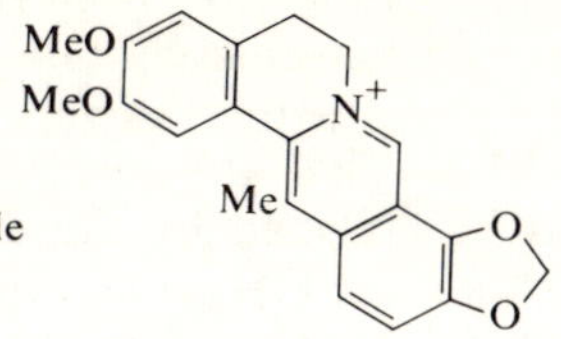   Dehydrothalictrifoline[103]

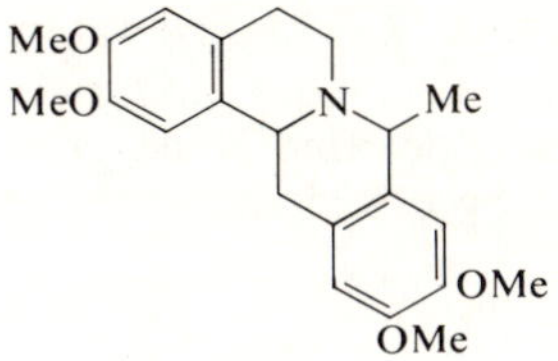

Coralydine[1] 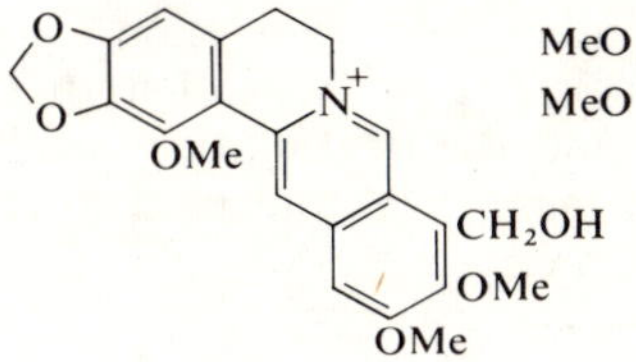   Alborine[104, 105] 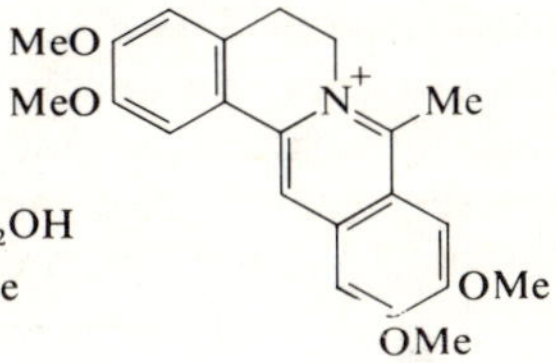   Dehydrogenated Coralydine[105]

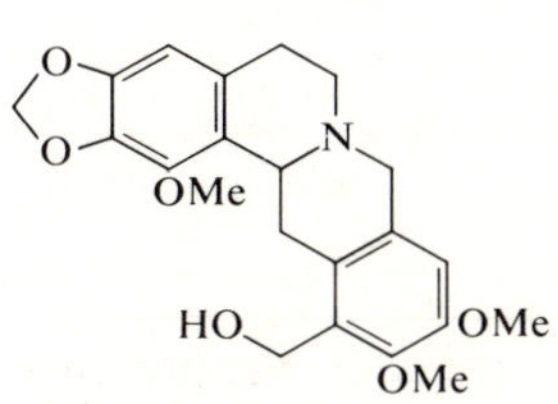

Mecambrine[106]

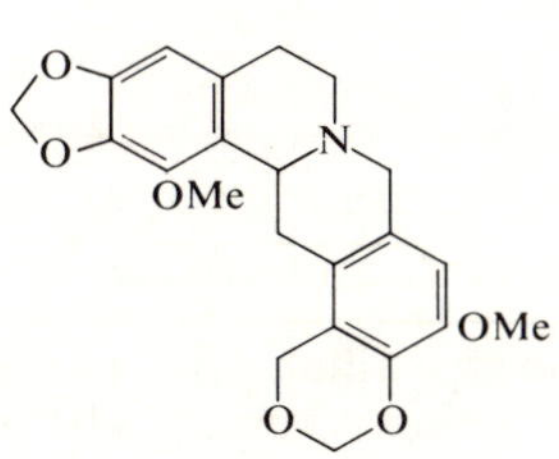

Orientalidine[106] 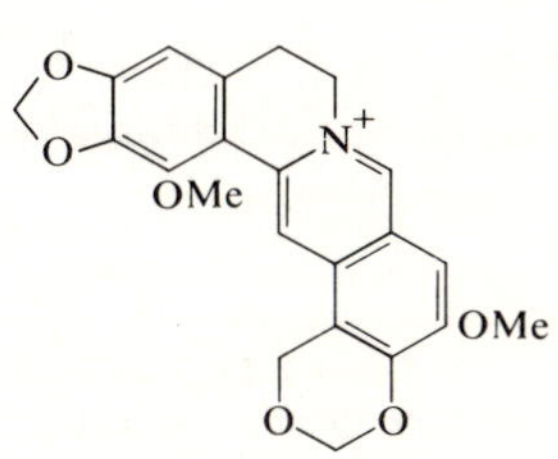   Alkaloid PO-4[106]

Figure 6.7

The protopine type alkaloids are characterised by the presence of a 10-membered ring system which shows a transannular reaction between the basic nitrogen and the carbonyl group. The structural assignment of the protopine alkaloids is achieved by degradation methods and conversion into the protoberberine alkaloids[108]. I.R., n.m.r. and mass spectra have been used to assign the structure of coulteropine[109]. Recently, fumaridine has been assigned the structure (21)[110].

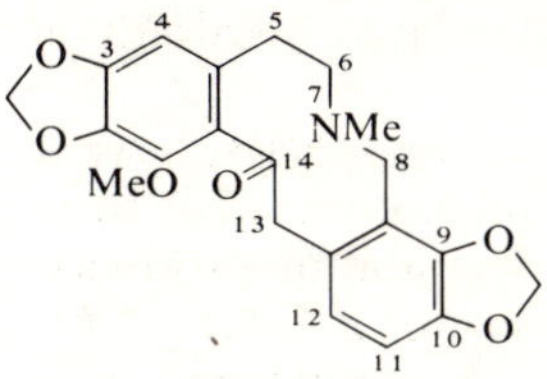

Coulteropine 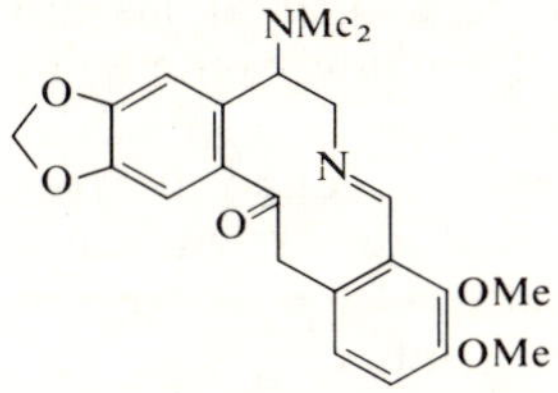   (21) Fumaridine

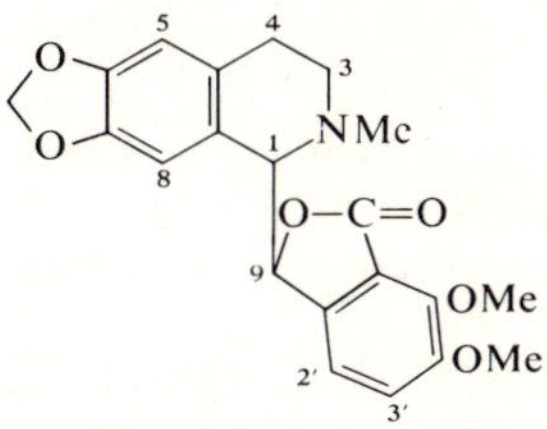

Hydarastine

The phthalideisoquinoline type alkaloids[111] are the 1-(2-methyl-1,2,3,4-tetrahydroisoquinolyl)phthalides substituted by alkoxyl and/or hydroxyl groups in positions 6, 7, 8, 4′ and 5′. The structure of the phthalideisoquinoline alkaloids has been determined by a mild oxidative cleavage into two fragments, which contain all the carbon atoms of the base as well as nitrogen and by the conversion of this into the known protoberberine alkaloids[112]. The absolute configuration of the phthalideisoquinoline alkaloids is assigned by the chemical proof[113] and spectral data[102, 113–116].

The rhoeadine type of alkaloid, found only in the *Papaveraceae* species, is characterised by the system of a seven-membered heterocyclic ring containing a nitrogen and a six-membered heterocyclic ring containing oxygen as an acetal.

The structural determination of rhoeadine, $C_{21}H_{21}NO_6$, was carried out by Hofmann degradation[117]. On the basis of studies of i.r., u.v., n.m.r. and mass spectra of rhoeadine and the degradation products, rhoeadine is assigned structure (22)[117]. The mass[118, 119] and o.r.d. spectroscopy[120] of this type of alkaloid have been studied.

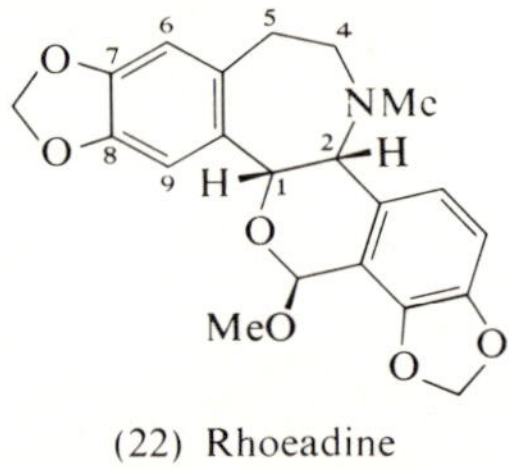

(22) Rhoeadine

The morphinandienone type alkaloids[121] are characterised by the presence of a cross-conjugated α-methoxycyclohexadienone system, which is easily ascertained by i.r., u.v., n.m.r. and mass spectra[121]. The n.m.r. spectrum of amurine, $C_{19}H_{19}O_3N$, indicated the presence of an *N*-methyl and a methyl-enedioxy group and two isolated aromatic protons and also revealed the

ABX pattern based on $\overset{\displaystyle |}{\underset{\displaystyle |}{\overset{-N}{=C}}}{\Large>}CH{-}CH_2{-}Ar$ system. Hofmann degradation

of amurine resulted in the assignment of (23) to this alkaloid[122]. The absolute configuration of amurine is determined from the study of the o.r.d. curve of tetrahydroamurine[122]. The spectroscopy[83, 125–127] of this type of alkaloid has been summarised by Stuart[121].

Morphinandienone and related alkaloids are listed in Figure 6.8. The structural elucidation of morphine, sinomenine, and related alkaloids is explained in detail in textbooks[4].

The hasubanan type alkaloids, found in *Menispermaceae* species, have a novel ring system. The structural assignment of hasubanonine (24) rests largely on spectral data obtained from the base and its degradation products. The n.m.r. and i.r. spectral studies indicate the presence of a $C\cdot CH_2\cdot CO\cdot C(OMe){:}C(OMe)$-system, two aromatic protons, one *N*-methyl and two *O*-methyl groups. Oxidation of the base yields hemipinic acid. On Hofmann degradation one methoxyl group of hasubanonine is demethylated giving an alkali-soluble methine-base, which on acetolysis loses the nitrogen-containing side chain and water to give acetylhasubanol showing a positive Gibbs' test. Acetylhasubanol may be hydrolysed to hasubanol, which on methylation

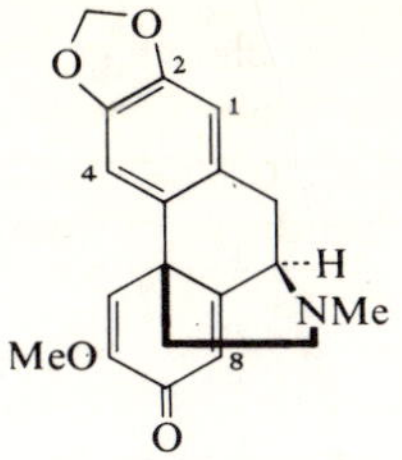

(23) Amurine 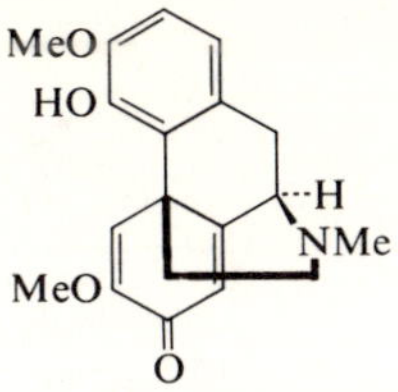   Salutaridine[121] 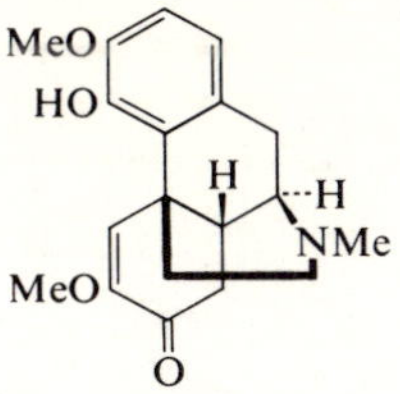   8,14-Dihydrosalutaridine[122] 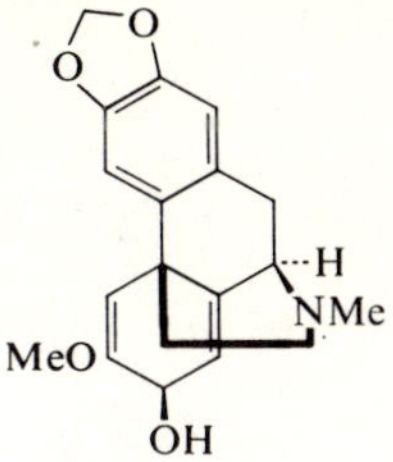   Nudaurine[120]

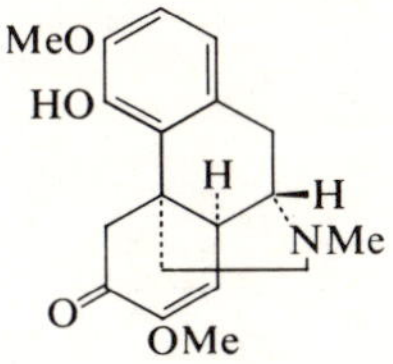

Sinomenine[4] 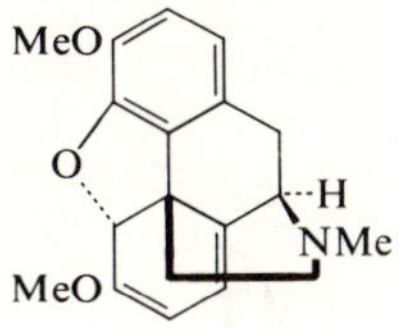   Thebaine[4] 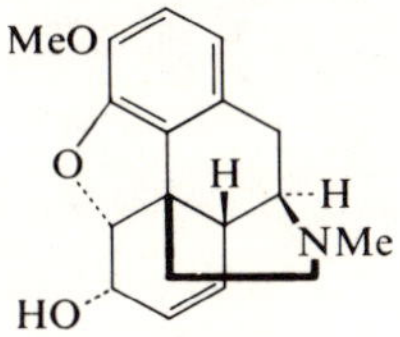   Codeine[4] 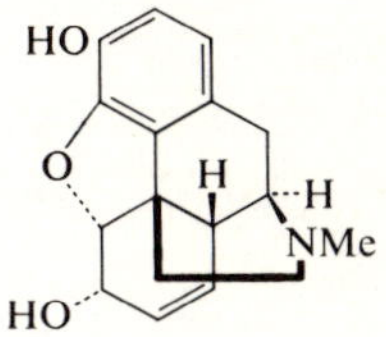   Morphine[4]

Figure 6.8

affords 3,4,6,8-tetramethoxyphenanthrene. The reduction of the base with sodium borohydride yields two epimeric dihydrohasubanonines A and B, which give the conjugated carbonyl compound (25) by demethanolisation with hydrobromic acid. Reduction of (25) with zinc-amalgam in hydrochloric acid gives the olefinic and saturated compounds (26). The saturated (26) is prepared from the olefinic compound by hydrogenation. The saturated compound is shown to be enantiomeric by an o.r.d. curve comparison with product (27) obtained by successive reduction and O-methylation of dihydro-indolinecodeine (28). The latter compound (28) was previously obtained by reduction of 14-bromocodeinone with sodium borohydride and its structure determined. In the n.m.r. spectral studies the signal attributable to the C-9H of the morphinan derivatives appears at τ 6.35–6.95, but no signal due to this proton is observed in this region for the hasubanan series. This finding is suggestive of the ethanamine bridge binding at C-13 and C-14 in the absolute configuration[128, 129].

Several hasubanan alkaloids are listed in Figure 6.9. The novel and interesting structures ((29) and (30)) of acutumine and acutumidine are proposed on the basis of a series of degradative studies, and n.m.r., i.r., u.v. and mass spectra of the bases and the degradation products[132]. Moreover, these structures are confirmed by x-ray analysis of acutumine and acetylacutumine[133].

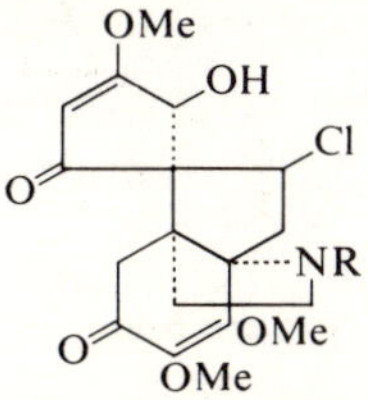

(29) Acutumine   R = Me
(30) Acutumidine R = H

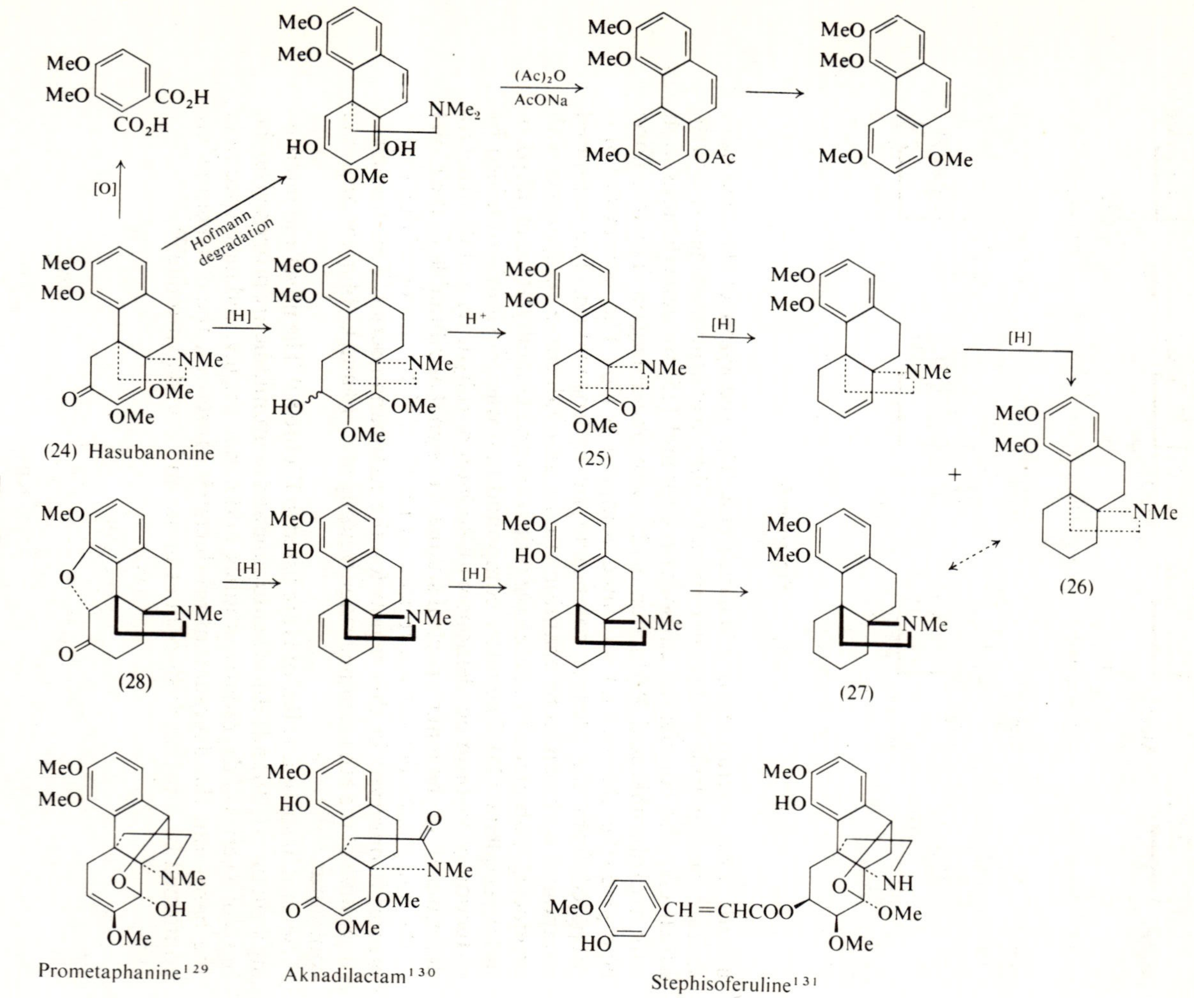

MeO
MeO
CO2H
CO2H
[O]
Hofmann degradation
MeO
MeO
HO
OH
NMe2
OMe
(Ac)2O
AcONa
MeO
MeO
MeO
OAc
MeO
MeO
MeO
OMe
MeO
MeO
O
OMe
OMe
NMe
(24) Hasubanonine
[H]
MeO
MeO
HO
OMe
OMe
NMe
H+
MeO
MeO
O
OMe
NMe
(25)
[H]
MeO
MeO
NMe
[H]
MeO
MeO
NMe
(26)
+
MeO
O
O
NMe
(28)
[H]
MeO
HO
NMe
[H]
MeO
HO
NMe
MeO
MeO
NMe
(27)
MeO
MeO
O
OH
NMe
OMe
Prometaphanine[129]
MeO
HO
O
O
OMe
OMe
NMe
Aknadilactam[130]
MeO
HO
CH=CHCOO
O
OMe
OMe
NH
Stephisoferuline[131]

Figure 6.9

The benzophenanthridine alkaloids, found in *Papaveraceae* species, are subdivided into two classes, aromatic bases and alcoholic bases exemplified by sanguinarine and chelidonine. The structural elucidation of this type of alkaloid is usually achieved by zinc-dust distillation, Hofmann degradation and oxidation[4]. This structure and its stereochemistry is proved by spectral studies[97], and the absolute configuration was determined by c.d. curves of the base[101].

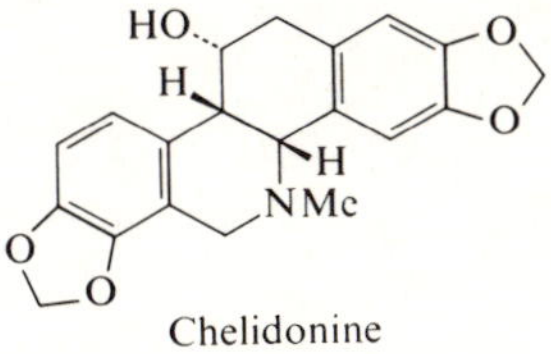

Chelidonine

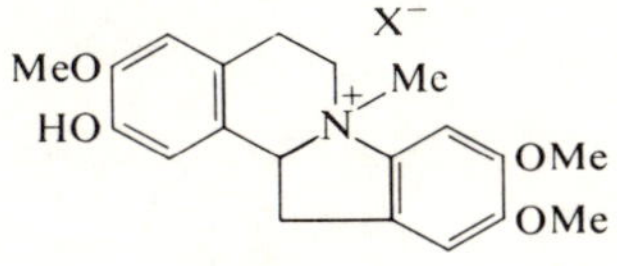

Cryptaustoline

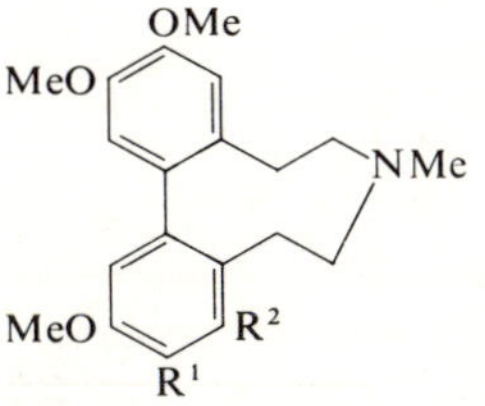

Protostephanine R[1] = H, R[2] = OMe
Erybidine   R[1] = OMe, R[2] = H

The dibenzopyrrocoline alkaloids, cryptowoline and cryptaustoline, were isolated from Australian *Lauraceae* species, and their structures were elucidated by Hofmann degradation[134]. The positions of the phenolic hydroxyl groups are determined by a similar reaction of the *O*-ethyl ether, which gives 5-ethoxy-2-ethyl-4-methoxybenzaldehyde. As aids for the elucidation of structures, mass spectrometric studies have been carried out on cryptaustoline and related compounds[135].

Protostephanine and erybidine are alkaloids having a dibenz [*d,f*]azonine system; the same ring system was obtained by the reaction of thebaine with phenylmagnesium bromide. Degradation reactions by the Hofmann method, oxidation, and zinc dust distillation established the structure of protostephanine[136]. This assignment was proved by n.m.r. analysis[137]. The structure of erybidine is also determined by Hofmann degradation and n.m.r., i.r. and u.v. spectra of base and degradation products[138].

Aromatic erythrina alkaloids[4], found in several *Erythrina* and some of *Menispermaceae* species, have a tetracyclic spiro-amine system with two conjugated bonds or one double bond. The u.v. spectra[139] of erysopine, related bases and the tetrahydro-bases show the presence of a heteroannular conjugated diene and tetrahydroisoquinoline system, respectively. Oxidation of erysotrine gives *m*-hemipinic acid and the bases yield an indole on fusion with potassium hydroxide[140]. Mild acidic treatment of the bases leads to demethoxy-bases showing a conjugated triene system with a loss of an aliphatic methoxyl group. Hydrogenation of these products yields the tetrahydro-bases, the *O*-methyl ether of which was synthesised by several groups.

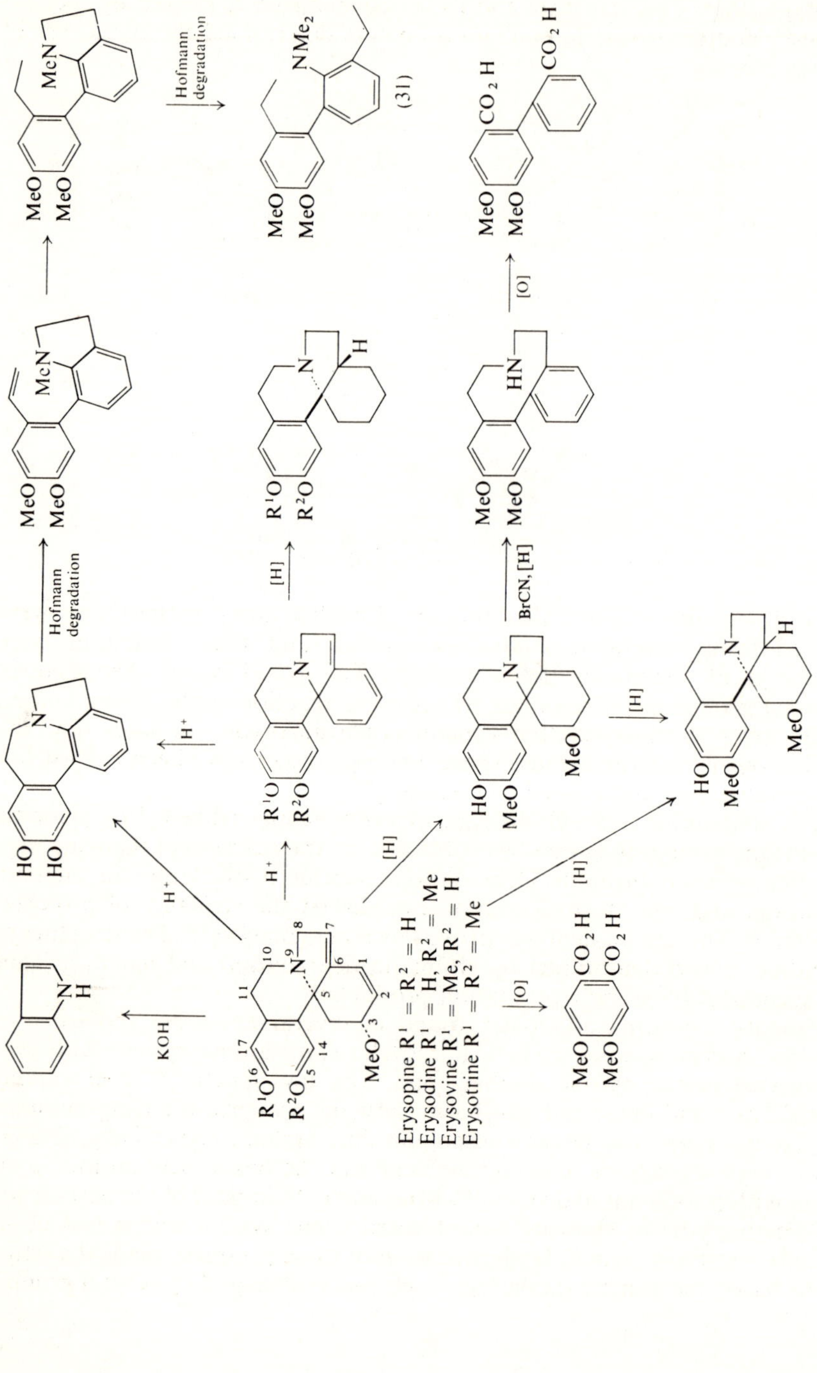

Hofmann degradation
MeN
MeO
MeO
Hofmann degradation
NMe₂
(31)
CO_2H
CO_2H
[O]
BrCN, [H]
[H]
HN
H
R^1O
R^2O
H⁺
HO
HO
N
KOH
H⁺
[H]
[H]
MeO
Erysopine $R^1 = R^2 = H$
Erysodine $R^1 = H, R^2 = Me$
Erysovine $R^1 = Me, R^2 = H$
Erysotrine $R^1 = R^2 = Me$
[O]

Drastic acidic treatment brings about demethoxylation and demethylation. Subsequent rearrangement gives apoerysopine, which is also formed from the demethoxy-bases by acid treatment. Systematic Hofmann degradation of apoerysopine dimethyl ether, followed by hydrogenation, leads to (31)[139–141]. Hydrogenation of the alkaloid yields either dihydro- or tetrahydro-bases, depending on the conditions[143]. The dihydro-base is degraded with cyanogen bromide, followed by reduction, to afford the optically inactive base, which is proved by oxidation to the dicarboxylic acid[144]. Moreover, Hofmann degradation of the tetrahydro-bases yields the methine bases containing a vinyl group conjugated with a benzene ring.

The structural assignment[4] of these bases by the chemical method is confirmed by x-ray analysis of erythraline hydrobromide, which decides the absolute configuration (3R, 5S) of these alkaloids[145]. Recently, the absolute configuration of erythratine has been established to be 2S, 3S, 5S[146]. Spectroscopic studies aided the structural elucidation[1, 147]. Several erythrina alkaloids are listed in Figure 6.10.

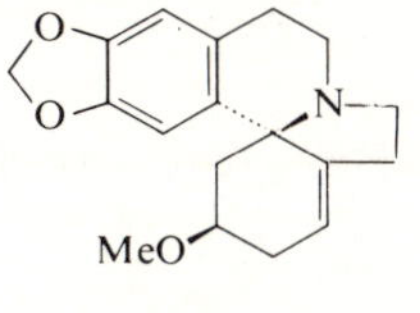

Erythramine[148]

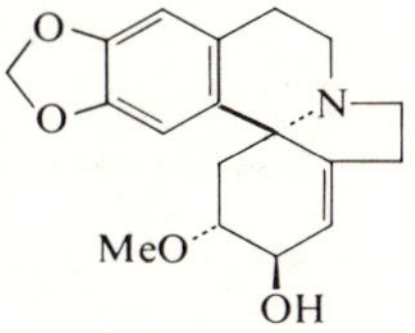

Erythratine[146]

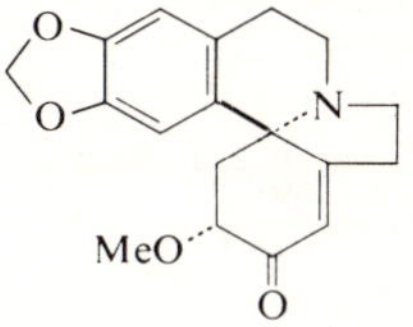

2-Ketoerythramine[146]

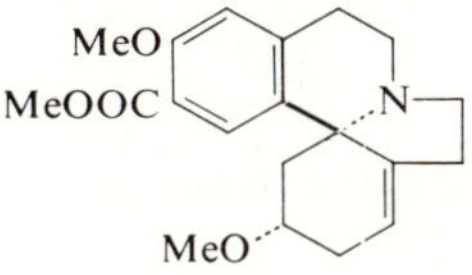

Erythroculine[149]

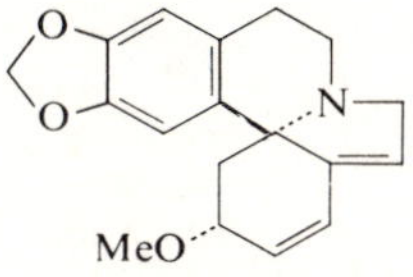

Erythrinine[150]

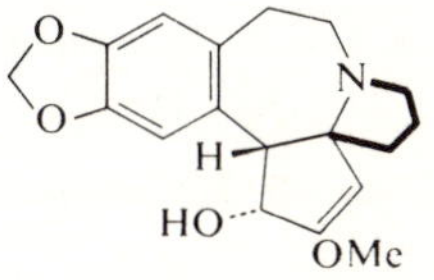

Cephalotaxine[151]

Figure 6.10

6.2.2 2-Benzylisoquinoline alkaloids

Sendaverine, $C_{18}H_{21}NO_3$, isolated from *Corydalis aurea*, is one of the bases presented in the original sample of corpaverine[152, 153]. On the basis of the oxidation product from corpaverine reported by Manske and the i.r., u.v., n.m.r. and mass spectral data, sendaverine is represented as *N*-benzyl-tetrahydroisoquinoline[154, 155]. The i.r., n.m.r. and mass spectra of sendaverine and related compounds have been studied[156], and a second example of this type, a new alkaloid, corgoine, has been isolated from *Corydalis gortshakovii*[157].

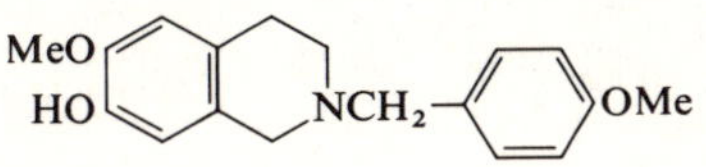

Sendaverine

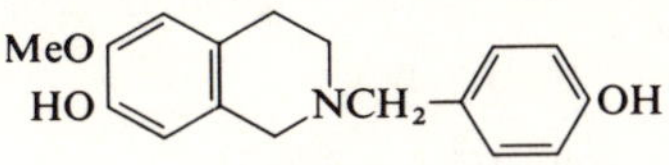

Corgoine

6.2.3   Phenethylisoquinoline alkaloids

Phenethylisoquinoline alkaloids are classified into five group bases on structural differences, homomorphinandienone (32), bisphenethylisoquinoline (33), homoproaporphine (34), homoaporphine (35) and homoerythrina alkaloids (36). These alkaloids are related to the benzylisoquinoline alkaloids such as morphinandienone, bisbenzylisoquinoline, proaporphine, aporphine and erythrina alkaloids. Although colchicine (37) and its derivatives also belong to the phenethylisoquinoline alkaloid group, these alkaloids are not included in this review as they have already been described by Manske[4].

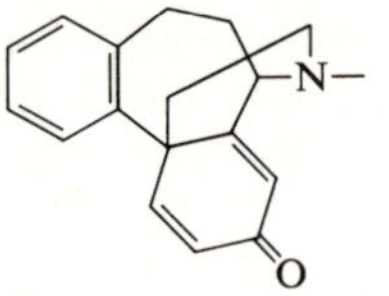

(32) Homomorphinandienone

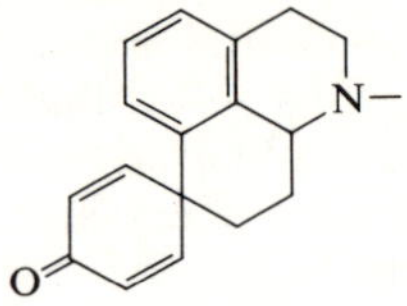

(34) Homoproaporphine

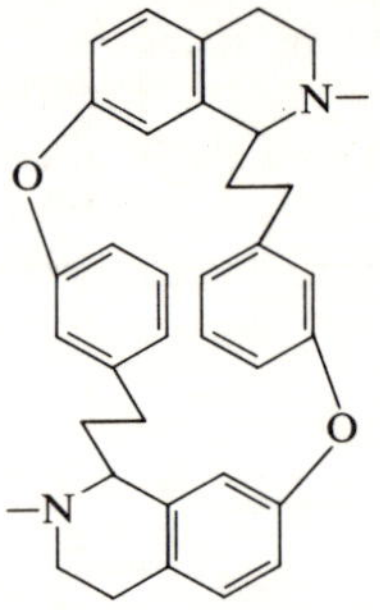

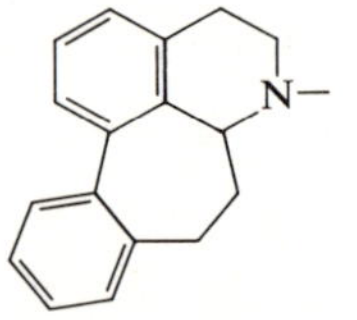

(33) Bisphenethylisoquinoline

(35) Homoaporphine

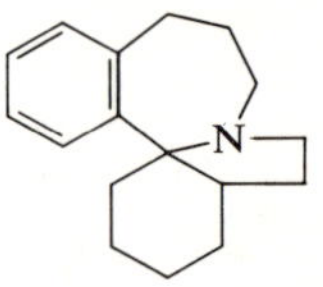

(36) Homoerythrina

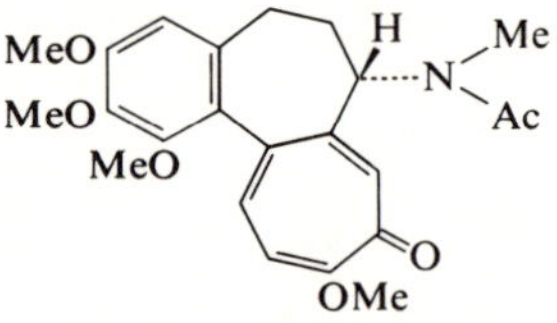

(37) Colchicine

Phenethylisoquinoline alkaloids have been isolated from six families, namely *Androcymbium, Colchicum, Kreysigia, Bulbocodium, Schelhammera,* and *Phelline.* Androcymbine (38) and *O*-methylandrocymbine (39) were isolated from the leaves of *Androcymbium melanthioides* and *Colchicum autumnale.* Oxidation of (39) gives 3,4,5-trimethoxyphthalic anhydride, while reaction with sodium in liquid ammonia yielded the phenethyltetrahydroisoquinoline derivative (40), the structure of which was confirmed by synthesis. Compound (40) showed a positive Cotton effect in the 278–265 nm region proving that it has the *S*-configuration. Moreover, androcymbine and salutaridine have a mirror-image like optical rotatory curve. The position of the phenolic hydroxyl group was assigned by analogy with 3-demethyl-

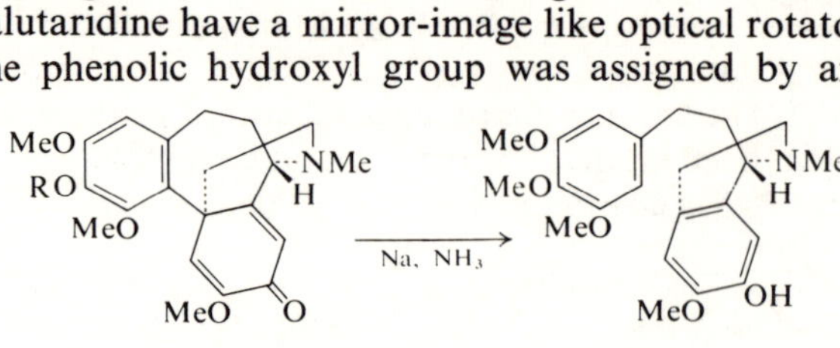

(38) Androcymbine   R = H

(39) *O*-Methylandrocymbine   R = Me

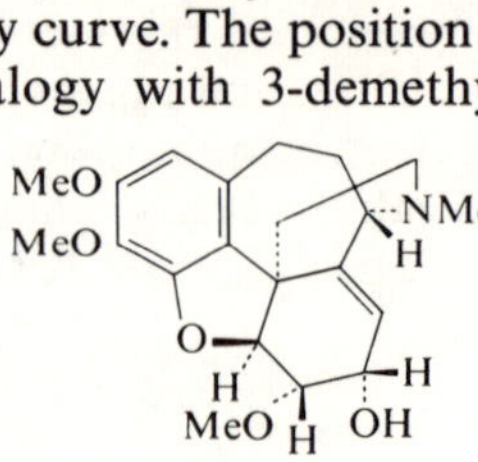

(40)

(41) Kreysiginine

colchicine. The absolute configuration of androcymbine must, therefore, be represented as structure (38)[158].

Kreysiginine (41), which is enantiomeric with alkaloid CC-21, is related as a ring A homologue to the morphine group of alkaloids such as thebaine. Mild Jones' oxidation of kreysiginine yielded an enone, which on treatment with a base gave a dienone. *O*-Methylation of this gave *O*-methylandrocymbine. The configuration between C—5H and C—6H of kreysiginine was determined to be *trans*-diaxial from the n.m.r. spectrum, and the hydroxyl group must be axial[159, 160]. Moreover, the absolute chirality of kreysiginine, defined by x-ray analysis[161], is the same as that of androcymbine[162].

The only bisphenethylisoquinoline type is melanthioidine (42) which was isolated from *Androcymbium melanthioides* together with androcymbine. The symmetry of the bisphenethylisoquinoline molecule is such that reductive cleavage of *O,O*-dimethylmelanthioidine with sodium in liquid ammonia yielded almost exclusively the phenolic isoquinoline (43)[12, 13], which showed a negative first Cotton effect. Previous o.r.d. measurements on tetrahydroisoquinoline chromophores established the illustrated *R*-configuration, and indicated that the molecule is in a head-to-tail arrangement (42)[163, 164].

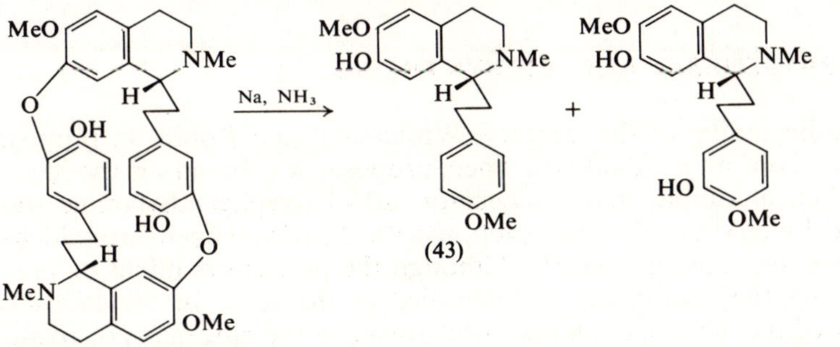

(42) Melanthioidine

(43)

The structures of homoproaporphine alkaloids, kreysiginone[165, 166], dihydrokreysiginone[165], and bulbocodine[167] were elucidated by spectroscopic methods and biogenetic arguments. The structures of the homoaporphine

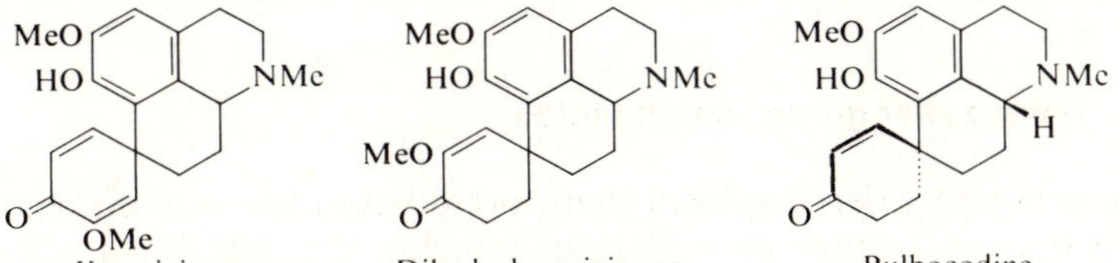

Kreysiginone

Dihydrokreysiginone

Bulbocodine

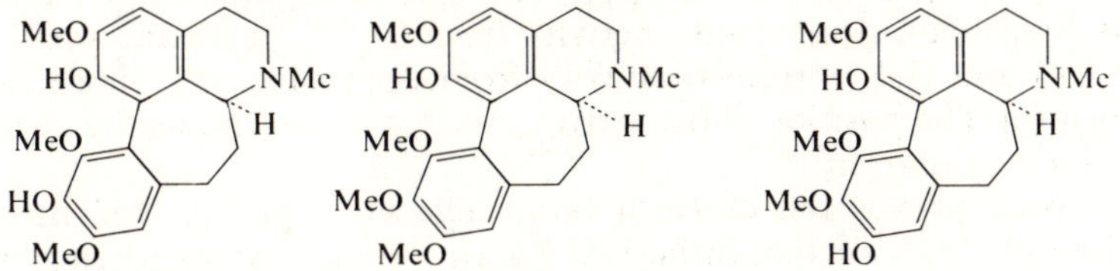

Multifloramine

Kreysigine

Floramultine

alkaloids, multifloramine, kreysigine and floramultine, were determined by total synthesis[168] of these alkaloids and by biogenetic consideration, described later. The bases were assigned the *S*-configuration by comparison with a synthetic sample[169]. Recently, (−)-kreysigine has been isolated from *Bulbocodium vernus*[167].

The homoerythrina alkaloid, schelhammerdine, was isolated from *Schelhammera pedunculata* in addition to several similar alkaloids[170–173]. The structure of these alkaloids and the relative stereochemistry at all the centres other than C-2 were determined by n.m.r. spectral assignment and chemical reaction. The complete structure and absolute configuration (2*S*, 3*S*, 5*S*) was confirmed by x-ray analysis of schelhammerine hydrobromide[170–172].

MeO

Schelhammerdine

6.3 BIOGENESIS AND BIOSYNTHESIS

At the beginning of this century, Winterstein and Robinson proposed the biogenesis of many alkaloids. Their proposal was based on the idea that a hypothetical simple, active precursor, dihydroxyphenethylamine was condensed by a Mannich type reaction with dihydroxyphenylacetaldehyde to give isoquinoline alkaloids[174]. Although the precursors might be present in the plant, they could not be identified at the time. In the 1950s Barton proposed the 'phenol oxidation mechanism' in the biogenesis of alkaloids[175]. The biphenyl ether bond in bisbenzylisoquinoline and the biphenyl bond in aporphine type alkaloids were formed by oxidative coupling of the simple phenolic bases by action of an enzyme present in the plant. The biogenesis of all isoquinoline alkaloids seems to be rationally explained by both theories, and some of these biogenetic mechanisms have been proved by tracer work.

6.3.1 1-Benzylisoquinoline alkaloids

Winterstein and Trier suggested that norlaudanosoline was biosynthesised from two C_6–C_2 units, derived from tyrosine through dihydroxyphenylalanine[176]. Tracer experiments with 2-[^{14}C]-tyrosine and 1-[^{14}-C]-dopamine established the specificity of incorporation of these substrates into papaverine and its biogenetic equivalents. Activity from 2-[^{14}C]-tyrosine was located at the C-3 and C-1 positions; activity from dopamine at a single site, the C-3 position. The position of the active carbon was determined by systematic Hofmann degradation[177].

This work proves the classical biogenetic concept that the benzylisoquinoline alkaloids are biosynthesised from a common intermediate, tyrosine. In support of a suggestion that the biosynthesis of isoquinoline alkaloids involves peptide chains (the head turning and biting its tail), a model sequence, (44)–(48), designed to simulate this process has been investigated. Compound (44) was treated with the masked phenylpyruvate (45) to give the diamide (46) which cyclised easily to a tetrahydroisoquinoline (47), the hydrolysis of which gave (48). Presumably, if nature does indeed take a course analogous to this laboratory model, 1-benzyl-1-carboxyisoquinoline derivates such as (48), may well exist in benzylisoquinoline-producing plants[178].

When oxidative coupling occurs between carbon and oxygen in the plant, there are two possibilities; one is intramolecular carbon–oxygen bond formation and the other is intermolecular carbon–oxygen bond formation. The former example occurs in the formation of cularine alkaloids[179, 180] and the latter explains the formation of bisbenzylisoquinoline alkaloids[181].

Cularine

Epistephanine is biosynthesised from tyrosine and $(-)$-*N*-methyl-coclaurine[181]. The pavine type of alkaloids, for example, argemonine, are co-existent with reticuline in the plant. This fact suggests that pavine alkaloids would be biosynthesised from reticuline through the 2,3-immonium intermediate[182, 183]. Dyke has discussed the biogenesis of pavine and isopavine alkaloids[184].

1,2,10,11- and 1,2,9,10-Tetrasubstituted aporphine alkaloids, (49) and (50) may be formed by oxidative cyclisation of benzyltetrahydroisoquinoline[174]. However, 1,2-disubstituted (51) and 1,2,10-trisubstituted aporphine (52) alkaloids could not be interpreted by the simple biogenetic mechanism above. In 1957, Barton[175] and Battersby[185] presumed a 'dienone–phenol rearrangement' and a 'dienol–benzene rearrangement' to take place in these biogeneses. This hypothesis has been proved recently by the incorporation of labelled $(+)$-orientaline into isothebaine[84, 186] and the isolation of crotono-sine having a dienone structure from *Croton linearis*[187–189]. Recently, Battersby proved that some aporphine alkaloids are biosynthesised from protosinomenine[190].

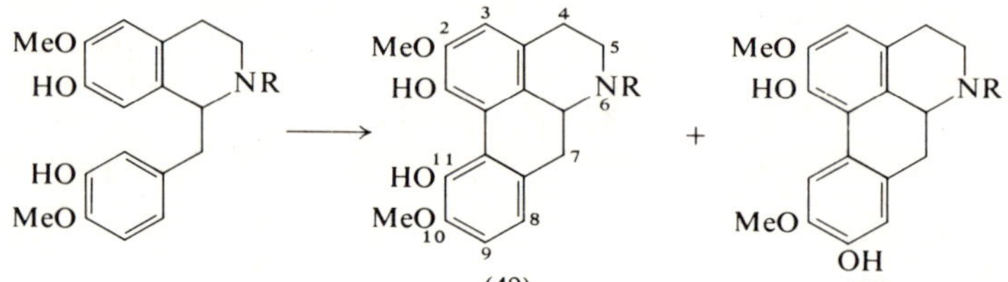

(49)

(50)

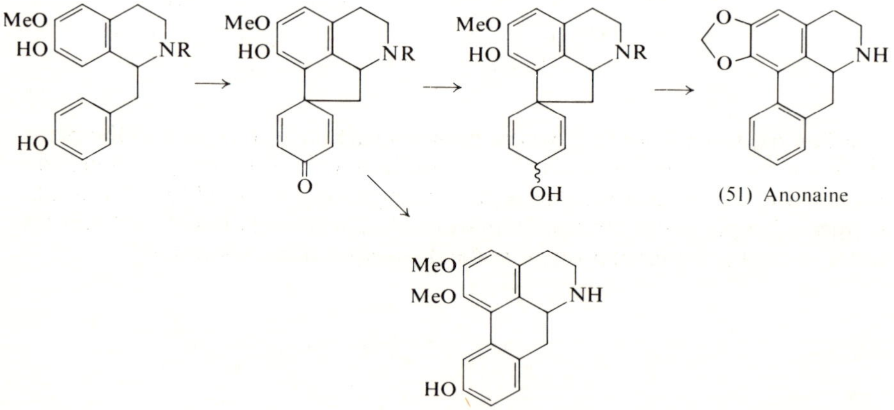

(51) Anonaine

(52) Tuduranine

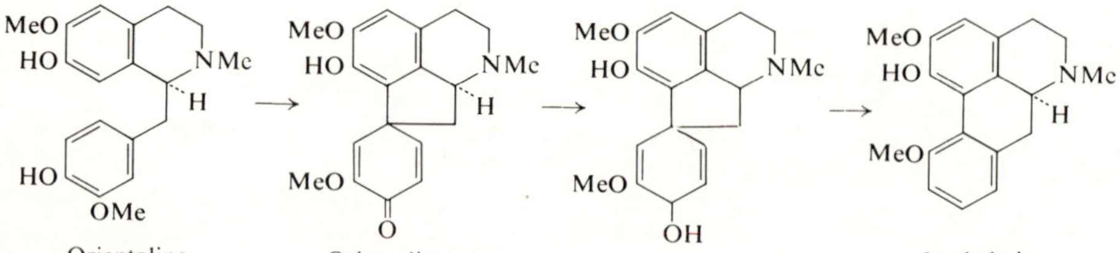

Orientaline

Orientalinone

Isothebaine

Moreover, bulbocapnine is biosynthesised from reticuline by phenolic oxidative coupling[191, 192]. As the second oxidative coupling mode of reticuline, the bond formation between the benzyl residue and C-4a position gives the morphinandienone type alkaloids, such as flavinantine[193] and salutaridine[2, 3]. The latter alkaloid is then reduced to salutaridinol-I, which loses water to form thebaine. Thebaine is converted to codeinone and then to codeine, and morphine is formed by demethylation of codeine.

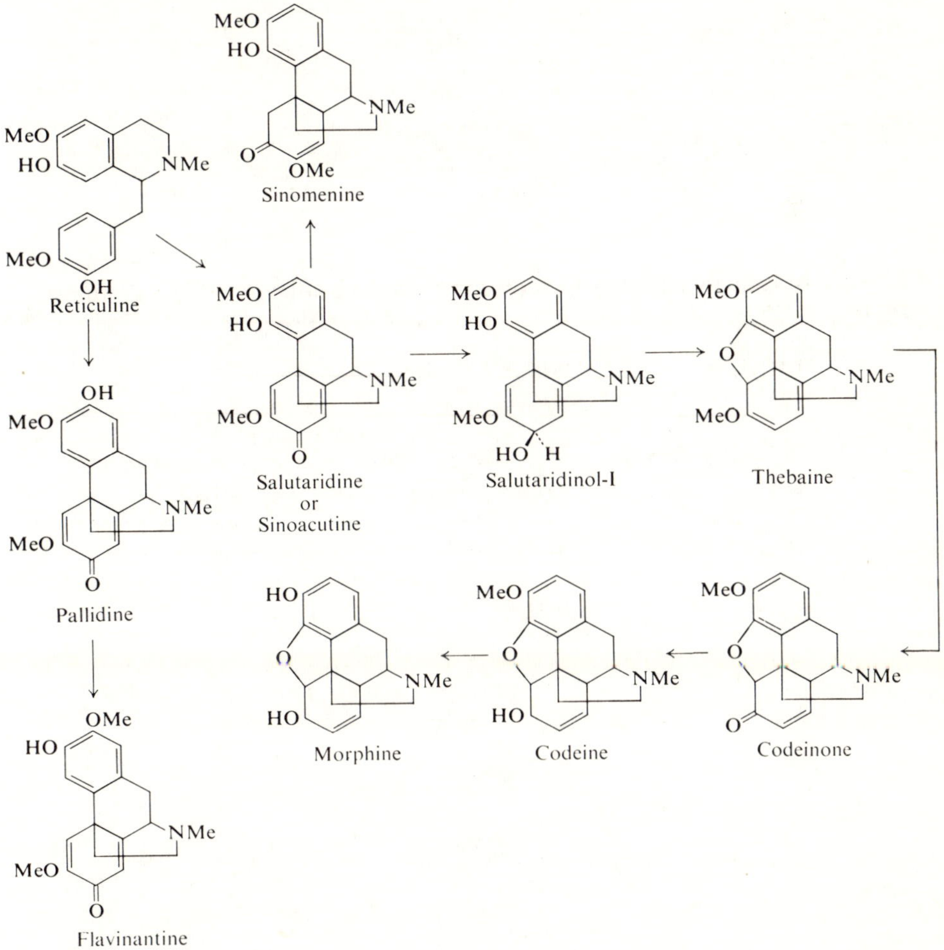

Reticuline

Sinomenine

Pallidine

Salutaridine
or
Sinoacutine

Salutaridinol-I

Thebaine

Morphine

Codeine

Codeinone

Flavinantine

Sinomenine is biosynthesised from sinoacutine, an antipode of salutaridine[194]. Protostephanine is also biosynthesised from the morphinandienone type compounds (53) as shown[195].

The hasubanan alkaloids could be biosynthesised by phenolic oxidation of the benzylisoquinoline (54) through the dienone (55) as shown[2], but the route from the morphinandienone (56) is also possible[30]. A decision between various possible routes will depend upon critical tracer experiments. A

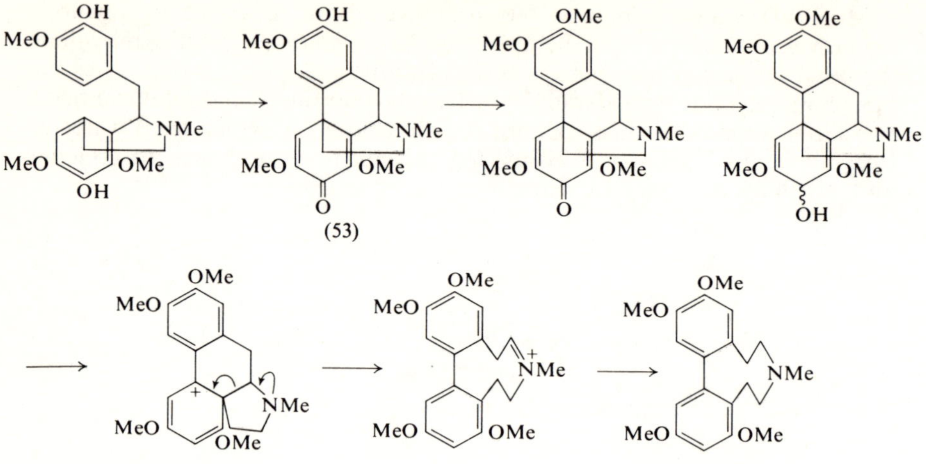

(53)

Protostephanine

possible biogenetic pathway to acutumine via the bisdienone (57) is discussed by Barton[194].

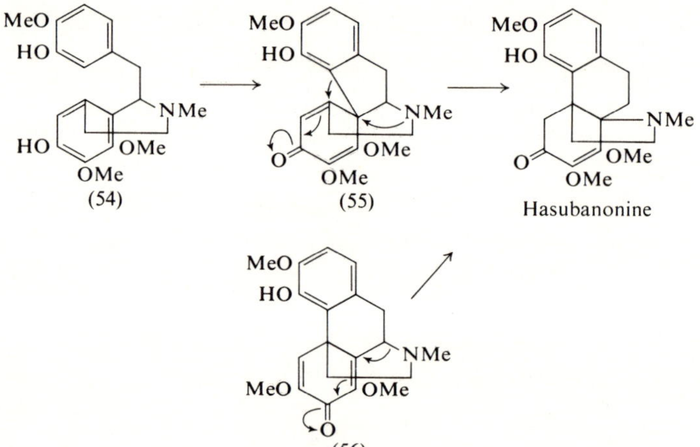

(54)   (55)   Hasubanonine

(56)

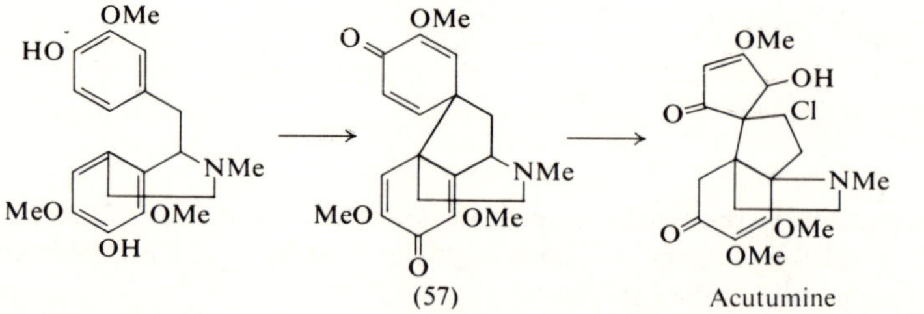

(57)   Acutumine

The erythrina alkaloids are biosynthesised from N-norprotosinomenine, a benzylisoquinoline, through the dienone (58) and biphenyl (59) derivatives[146, 196].

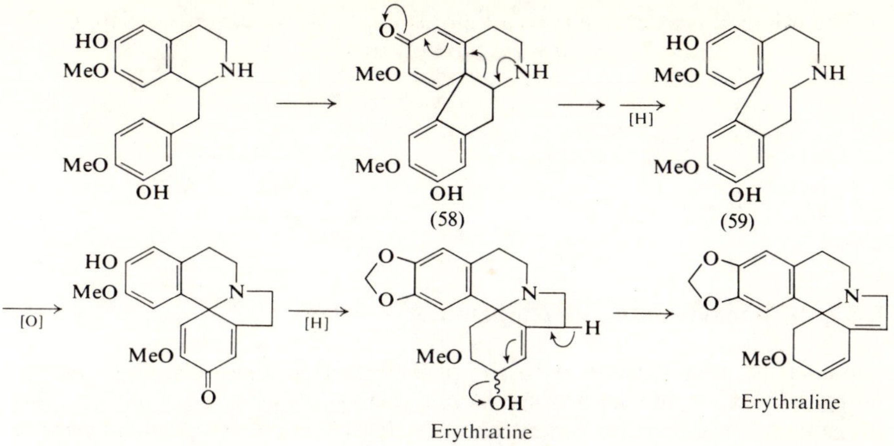

(58)

(59)

Erythratine

Erythraline

The protoberberine, protopine, phthalide and benzophenanthridine alkaloids, with an additional C_1 unit in the benzylisoquinoline system, are biosynthesised from a common precursor, (+)-reticuline or its equivalents as proved by tracer experiments[197, 200]. 2-[^{14}C]-Tyrosine[201–203] is incorporated into narcotine, berberine, hydrastine and chelidonine and scoulerine is an intermediate from reticuline to protopine, narcotine and chelidonine[204, 205]. 'The berberine bridge' is formed by oxidative coupling of a *N*-methyl group derived from methionine[185, 197, 199, 206, 207].

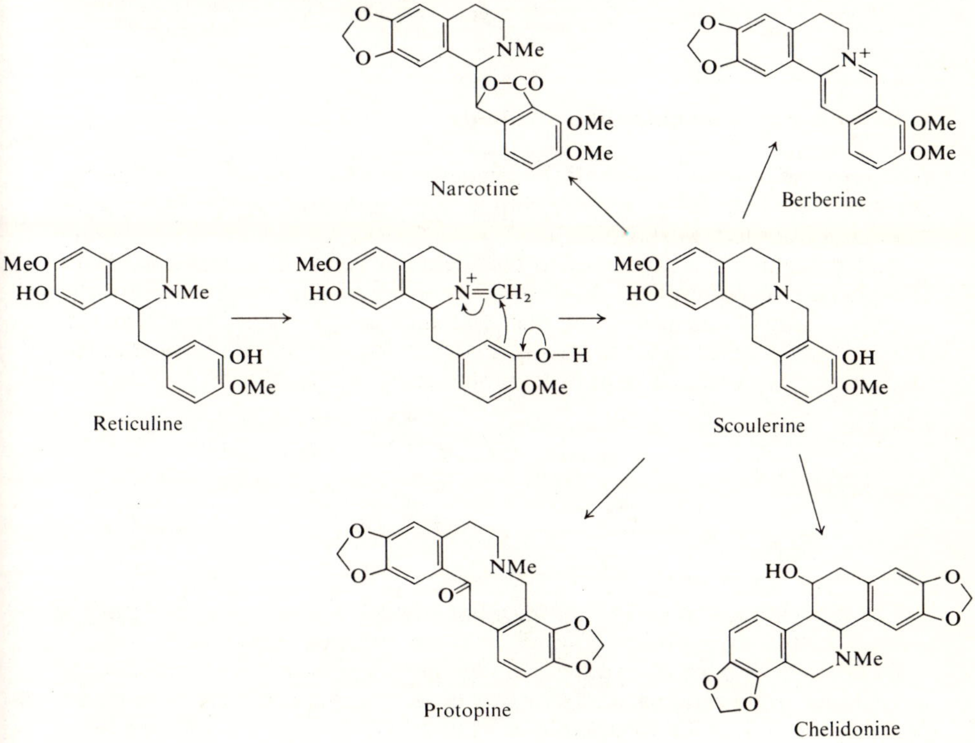

Narcotine

Berberine

Reticuline

Scoulerine

Protopine

Chelidonine

The dibenzopyrrocoline type alkaloids could be biosynthesised by C—N oxidative coupling of laudanosine type precursors[2].

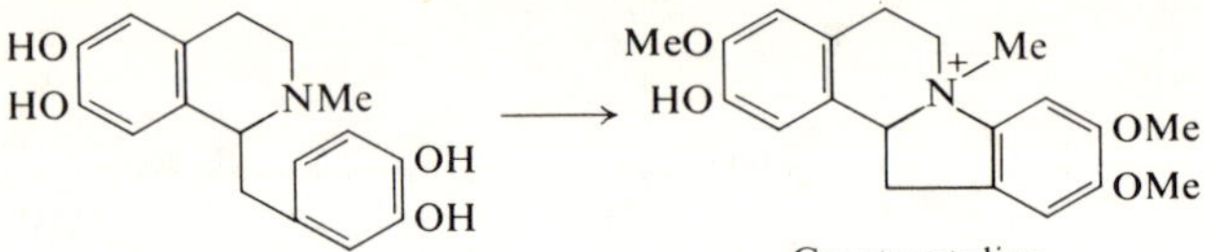

Cryptaustoline

6.3.2 2-Benzylisoquinoline alkaloids

Sendaverine is co-existent with the protoberberine alkaloids; thus, senda-verine would be biosynthesised from 1-benzylisoquinoline. One possible biogenetic route involves oxidation of *N*-methylcoclaurine. Radical pairing would give the aziridine-type compound (60), which could be cleaved reductively between the C—C bond by an enzyme.

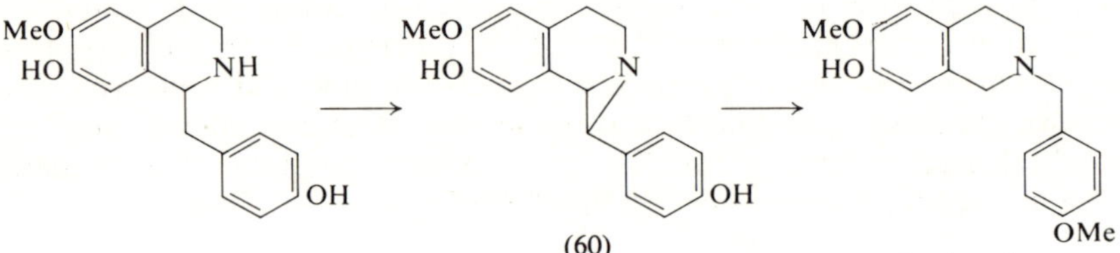

(60)

Sendaverine

6.3.3 Phenethylisoquinoline alkaloids

Although the biosynthesis of all the phenethylisoquinoline alkaloids has not yet been studied fully, androcymbine and homoaporphine have been examined by tracer works.

Androcymbine is a precursor to colchicine, which is biosynthesised from the phenethylisoquinoline. Moreover, tyrosine is incorporated into the tropolone ring of colchicine, and phenylalanine into the A ring of colchicine. These tracer studies provide strong evidence for several of the postulated steps in the biosyntheses of androcymbine and colchicine[208, 209].

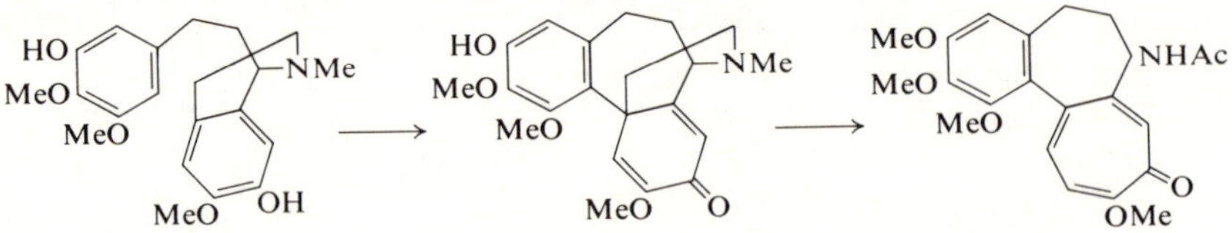

Colchicine

The biosynthesis of (−)-melanthioidine almost certainly involves phenol oxidation, and diphenolic isoquinoline is the required substrate[164].

By analogy with the biosynthesis of several aporphine alkaloids, the homo-aporphines could arise naturally by way of homoproaporphines or by direct coupling of diphenolic isoquinoline. In order to distinguish between these possibilities, the 3-[^{14}C]-diphenolic isoquinolines ((61)–(63)) were administered to *Kreysigia multiflora*. The good incorporation of (61), compared with very low efficiency of (63), is in accord with the mechanism involving direct coupling. These results imply that floramultine is the first homoaporphine alkaloid to be formed[210].

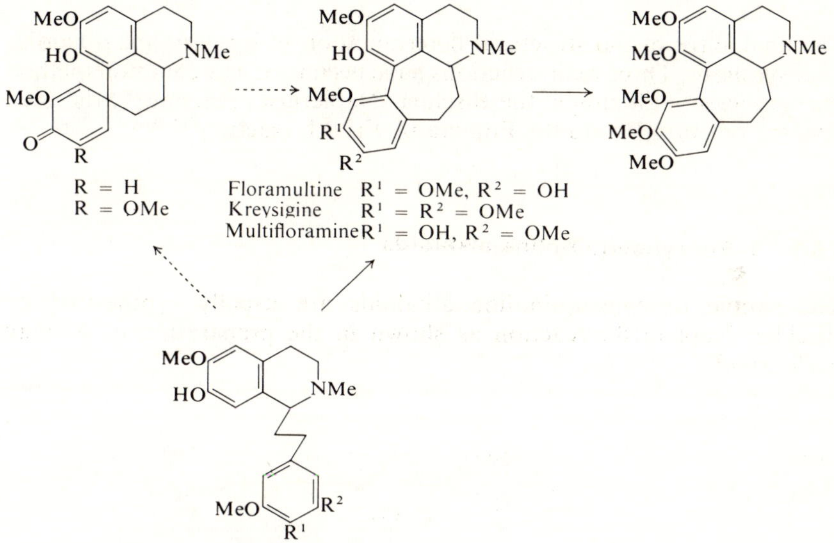

R = H
R = OMe

Floramultine   R^1 = OMe, R^2 = OH
Kreysigine   R^1 = R^2 = OMe
Multifloramine R^1 = OH, R^2 = OMe

(61) R^1 = OMe, R^2 = OH
(62) R^1 = OH, R^2 = H
(63) R^1 = OH, R^2 = OMe

It seems likely that the ring system of the homoerythrina alkaloids is derived from the diphenolic phenethylisoquinoline (64) by a route analogous to that

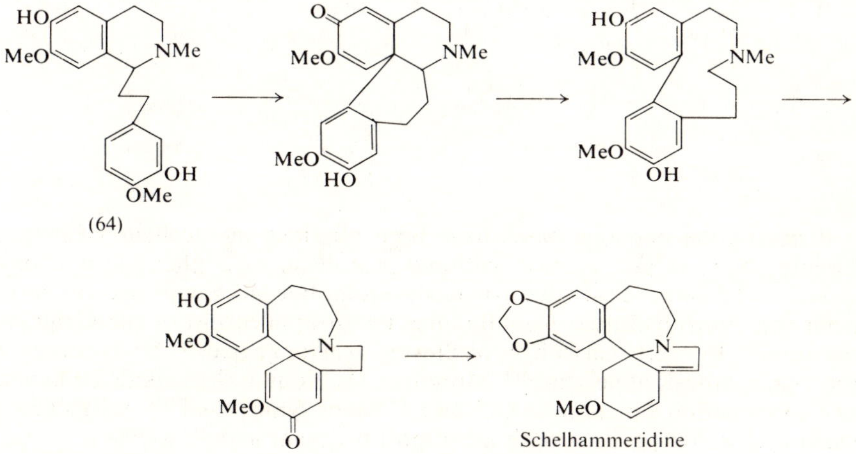

(64)

Schelhammeridine

involved in the formation of the erythrina alkaloids, for which a 1-benzyl-1,2,3,4-tetrahydroisoquinoline precursor has been established.

6.4   TOTAL SYNTHESIS

The final stage of the structural determination of isoquinoline alkaloids is total synthesis. Three main reactions have been used as a common method in the synthesis of alkaloids; the Bischler–Napieralski reaction[211], the Pictet–Spenger reaction[212], and the Pomeranz–Fritsch reaction[213, 214].

6.4.1   1-Benzylisoquinoline alkaloids

The simple benzylisoquinoline alkaloids are usually synthesised by a Bischler–Napieralski reaction as shown in the preparation of *N*-methyl-coclaurine[215].

N-Methylcoclaurine

Recently, the phenolic bases have been obtained in excellent yield by a Bischler–Napieralski reaction without protection of a phenolic hydroxyl group[216]. In some case, the 1-benzylisoquinoline alkaloids are prepared from the 1-unsubstituted isoquinolines by an application of the Grignard reaction[217], Reissert reaction[218], or Stevens rearrangement[219, 220] as shown in the total synthesis of petaline[217]. Moreover, the isoquinoline alkaloids having one oxygenation function at C-7 and C-8 are synthesised by protecting a usual cyclisation position with a bromosubstituent as follows[220].

The 1-benzylisoquinolines synthesised by the methods described above are the starting materials for the preparation of several types of isoquinoline alkaloids.

The bisbenzylisoquinoline alkaloids such as dauricine are synthesised classically by a double Bischler–Napieralski reaction of the bis-amide which has a biphenyl ether bond prepared by the Ullmann reaction[221, 222].

The second synthetic method of the bisbenzylisoquinoline alkaloids is an Ullmann reaction of the phenolic isoquinoline with the bromoisoquinoline[223], synthesised by the usual method. Magnoline is synthesised by this method[224]. Considerable effort has been expended in the synthesis of the bisbenzyl-isoquinoline alkaloids having two[221, 225–227] or three biphenyl ether linkages[228]. The synthesis of phaeanthine illustrates the type of approach which is used with this type of bisbenzylisoquinoline alkaloid[229].

Phenolic oxidative coupling of phenolic 1-benzylisoquinoline is also applicable to the synthesis of the bisbenzylisoquinoline alkaloids. This method provides a biogenetic type synthesis. Bobbitt and his associates have demonstrated that electrolytic oxidation of *N*-ethoxycarbonyl-*N*-norarme-pavine yields dauricine after reduction of the oxidation product[230].

Cularine and related alkaloids have been synthesised by the Pomeranz–Fritsch and similar reactions[231, 232]. The most interesting synthesis of cularine involves the phenolic oxidative coupling of a diphenolic benzylisoquinoline; this follows a biosynthetic pathway[179, 180]. Another synthesis of cularine by intramolecular Ullmann reaction of the phenolic bromo-isoquinoline[233] has been reported.

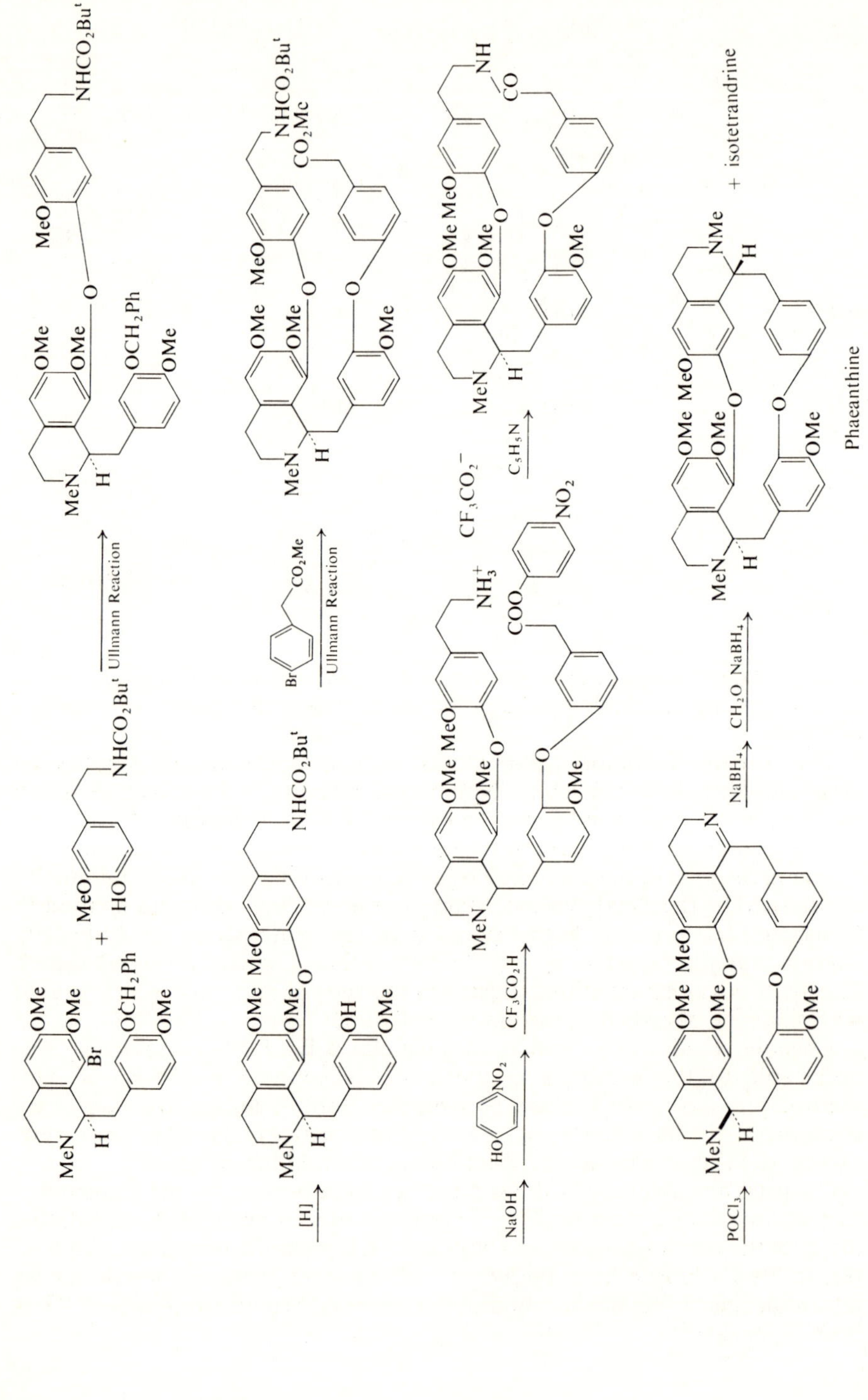

Ullmann Reaction

Ullmann Reaction

[H]

NaOH

CF_3CO_2H

C_5H_5N

$POCl_3$

$NaBH_4$

CH_2O $NaBH_4$

Phaeanthine

+ isotetrandrine

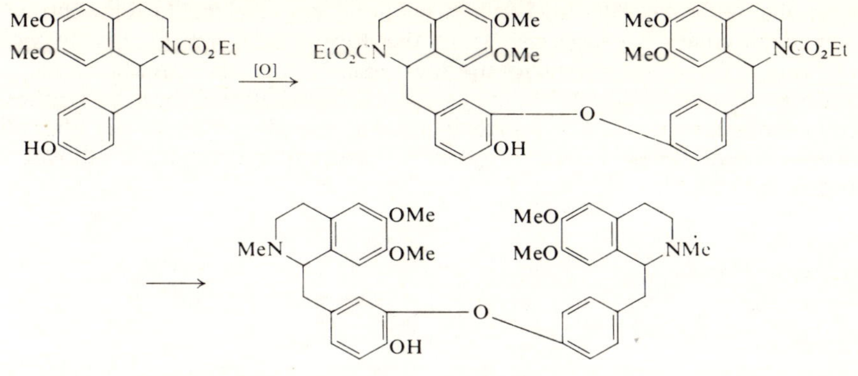

Dauricine

The pavine type alkaloid, argemonine, is synthesised from papaverine by reduction with tin and hydrochloric acid, followed by N-methylation[50].

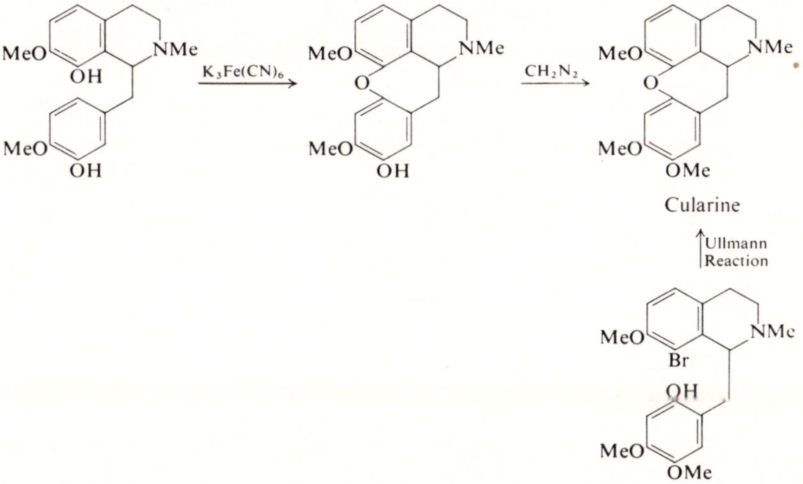

Cularine

Dyke also synthesised several isopavine alkaloids[184]. The selective reduction of methiodide from the papaverine analogue, followed by acid treatment, gives an N-methylpavine type compound[234, 235].

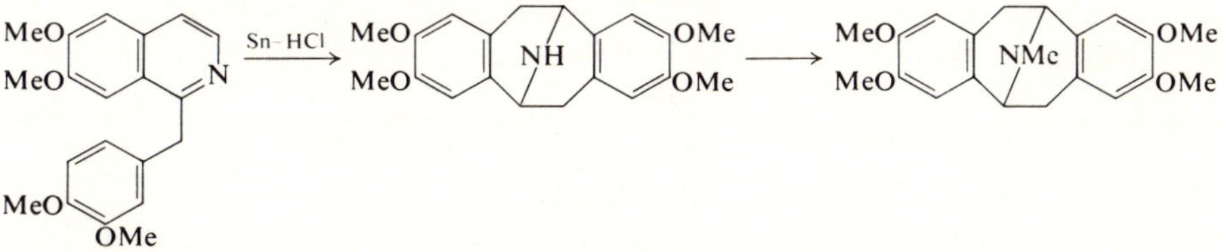

Proaporphine type alkaloids are synthesised by four methods. The first one is the classic method reported by Bernauer as shown in the total synthesis of pronuciferine[89]. The second method is a biogenetic type synthesis by phenolic oxidative coupling of the diphenolic isoquinoline. Orientalinone[236] and glaziovine[237] are obtained from the appropriate compounds. Pschorr reaction of the 8-aminoisoquinolines also yields the proaporphine in addition to other products[238]. An example of this reaction is the total synthesis of homolinearisine[239]. The fourth method is the photolysis of the phenolic bromoisoquinolines, in the presence of sodium hydroxide, as shown in Figure 6.11[240, 241].

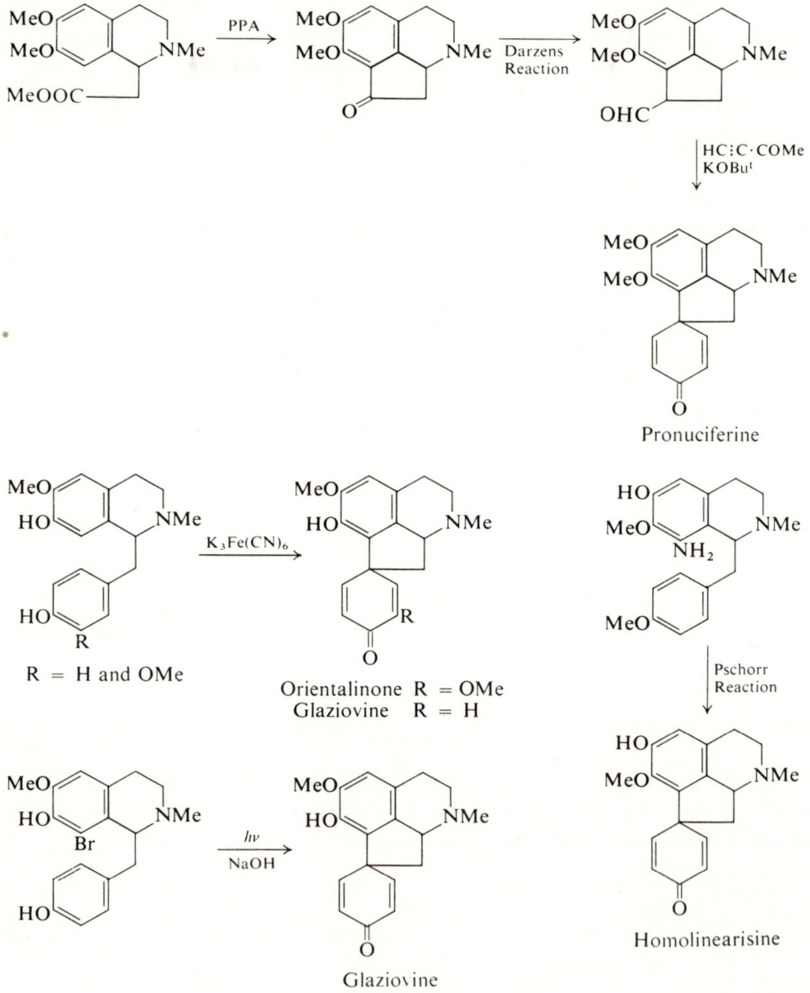

Figure 6.11

The aporphine alkaloids have been synthesised by Pschorr reaction[242] of 1-(2′-aminobenzyl)isoquinolines, which are usually prepared by four methods[243–246]. Although Pschorr reaction of the 8-aminoisoquinolines[238, 247]

gives the aporphines in moderate yield, it is difficult to make the 8-amino-
isoquinolines. The classic Pschorr reaction is the thermal decomposition of
the diazotised isoquinolines; but, recently, the synthesis of the aporphine
alkaloids by photolysis of the diazonium salts has been reported. The latter
method is superior to the former yield[248, 249]. The total synthesis of the
aporphine alkaloids by photolytic cyclisation of the 1-benzylisoquinolines
is reported as shown in Figure 6.12[250–253].

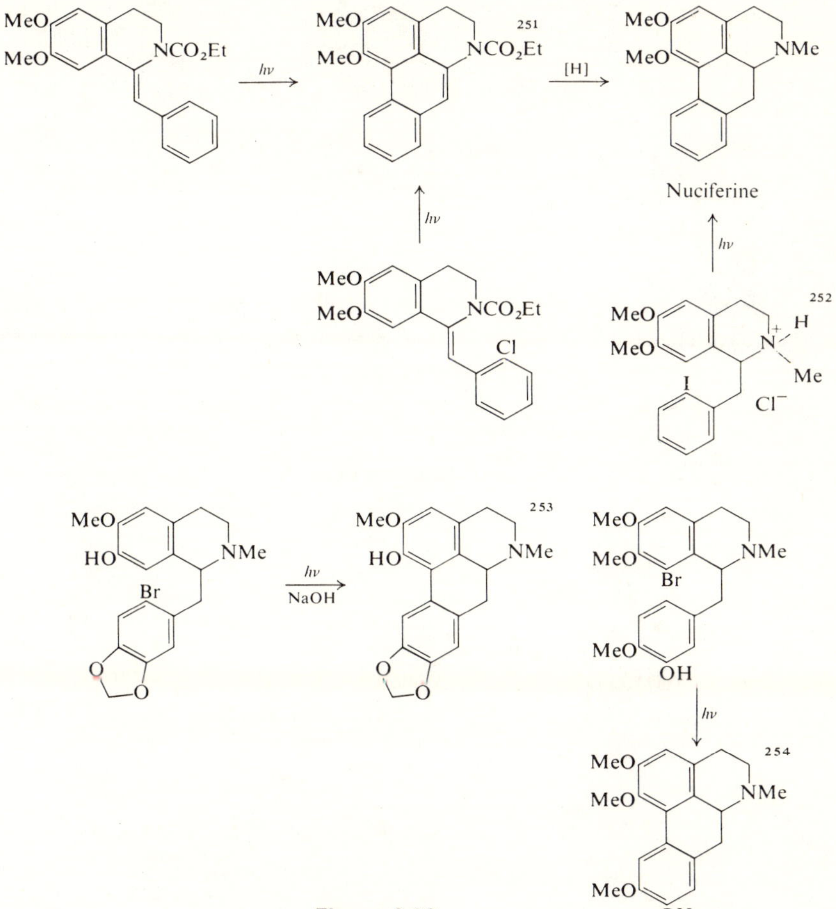

Figure 6.12

Biogenetic type synthesis of the aporphine alkaloids by phenolic oxidation
has been described by several authors[2] shown in Figure 6.13[84, 255, 256].

The most common method for the synthesis of the tetrahydroprotoberberines is the Pictet–Spengler reaction of a benzyltetrahydroisoquinoline
with formalin in the presence of an acid[212]. When the substituents on the
benzyl moiety are methoxy or benzyloxy, cyclisation occurs at the *para*
position to the substituent to give the 10,11-disubstituted tetrahydroprotoberberines. But in case of hydroxy substituents, a mixture of 9,10- and

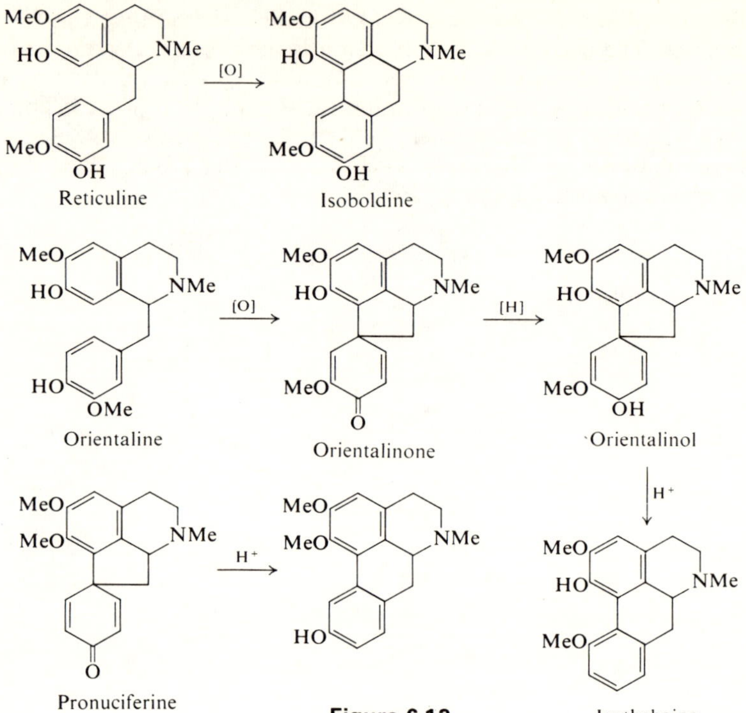

214

Reticuline [O] → Isoboldine

Orientaline [O] → Orientalinone [H] → ·Orientalinol

Pronuciferine H⁺ →

Figure 6.13

Isothebaine

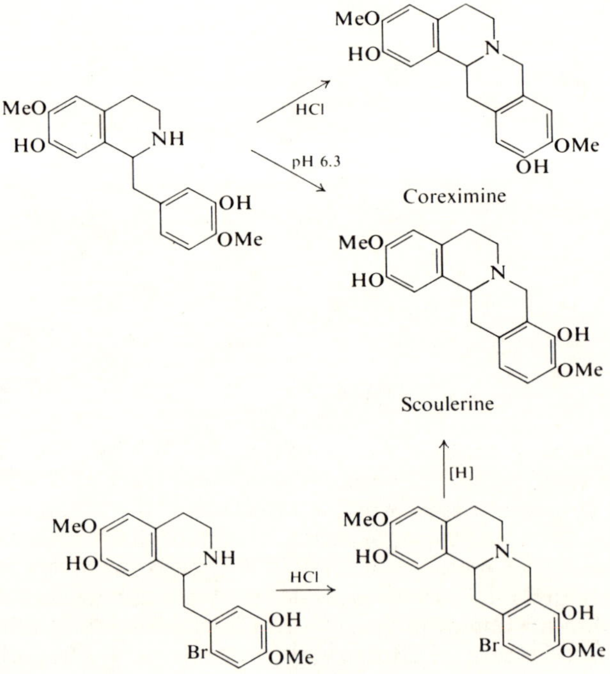

HCl → Coreximine

pH 6.3 →

Scoulerine

[H]

HCl →

10,11-disubstituted ones is obtained. Two methods by using the Pictet–Spengler reaction are reported for the 9,10-disubstituted compounds; the first involves control of pH in the cyclisation[257, 285], and the second is to protect the usual cyclisation site by a bromine atom[259].

Moreover several routes are described in order to prepare 9,10-disubstituted protoberberines. Battersby has developed a new synthesis of the tetrahydro-protoberberine from the N-phenethylisoquinolines as follows[261].

Transannular reaction of protopine and its analogue is applicable in the preparation of protoberberines, as shown in Figure 6.14[262, 263].

In special cases, the Pomeranz–Fritsch reaction[213] modified by Bobbitt[214] is used in the synthesis of this group of alkaloids[264]. 2-Acetyl-1-benzylidene-isoquinoline could be cyclised by irradiation to protoberberine, which is converted into β-coralydine[265]. Protopine type alkaloids are synthesised from the tetrahydroprotoberberines by two methods[266, 267].

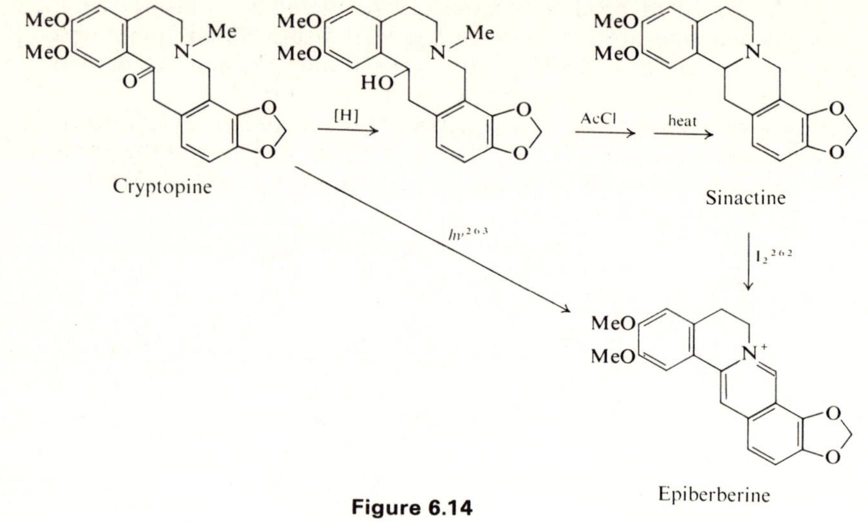

Cryptopine

Sinactine

Epiberberine

Figure 6.14

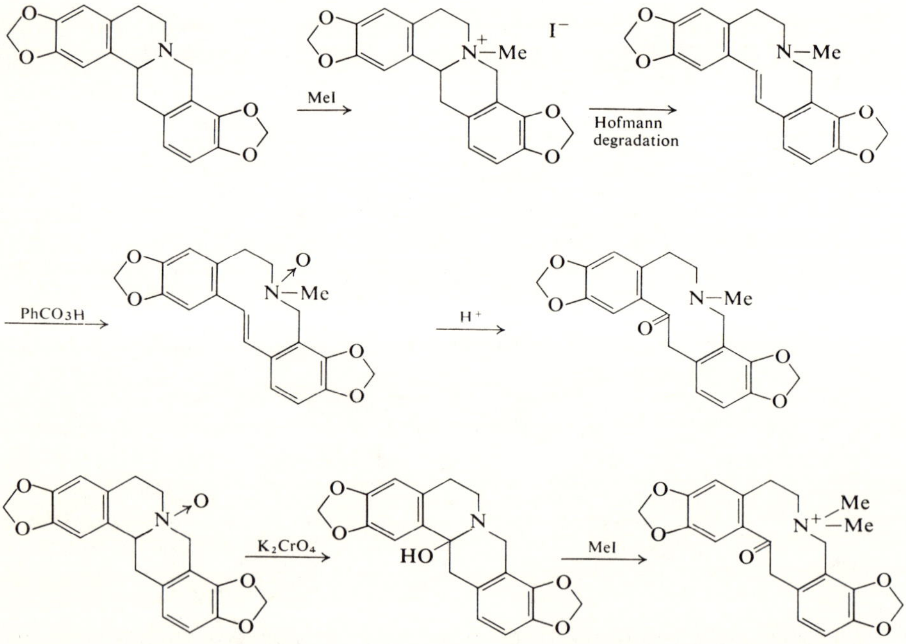

The classic synthesis[268] of the phthalideisoquinolines is through condensation of 3,4-dimethoxy-6-nitrophthalide with hydrastinine; an improved synthesis involves the condensation of meconine-α-carboxylic acid with N-methylphenethylamine[269]. Both reactions yield a mixture of two diastereomers.

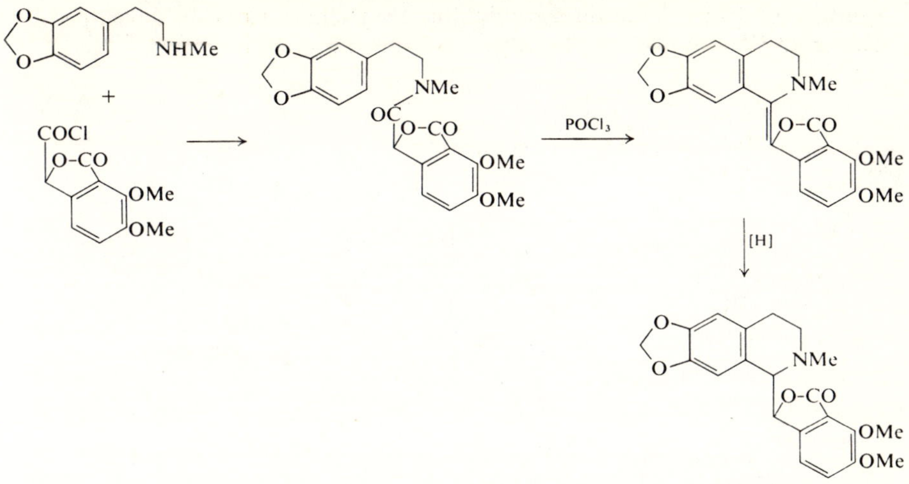

Hydrastine

The first total synthesis of the rhoeadine type alkaloids, which is through an enlargement of the isoquinoline ring in ochotensine analogue (65) by a Wagner–Meerwein rearrangement, is summarised as follows[270]. Brossi has achieved the conversion of phthalide alkaloid into rhoeadine type compound[271].

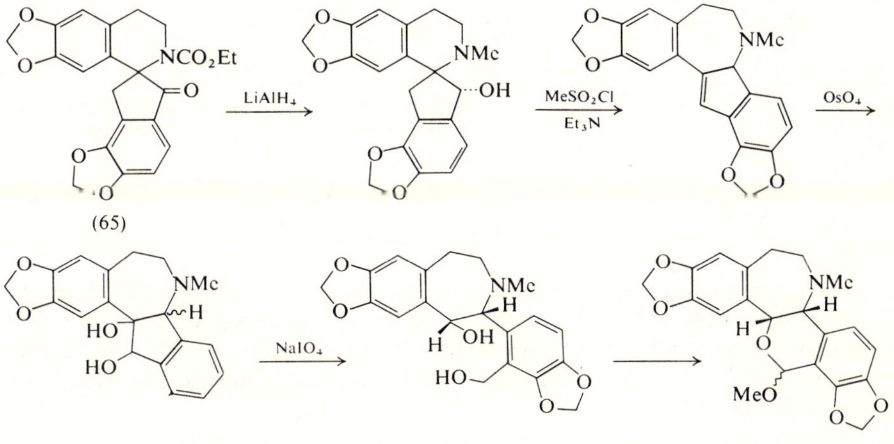

(65)

Rhoeadine

The total synthesis of morphine and codeine were first achieved by Gates[272], and the second synthesis of dihydrothebainone and, hence, of morphine has been achieved by Ginsburg[273]. Barton has also carried out a simple biogenetic-type synthesis of morphine and its related alkaloids; phenolic oxidative coupling of reticuline gives salutaridine, which is reduced to salutaridinol. Dehydration of this product with acidic catalyst affords thebaine, which is converted into codeine and morphine[274]. Salutaridine is synthesised from the aminoisoquinoline by Pschorr reaction developed by Kametani[275], and also by photolysis of 2'-bromoreticuline[276].

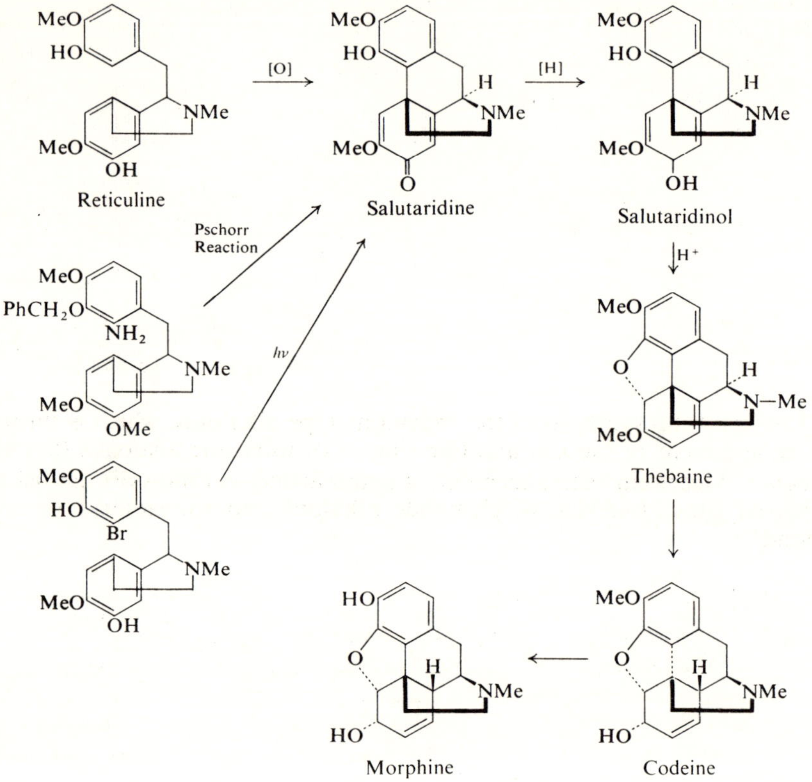

Reticuline

Pschorr Reaction

Salutaridine

Salutaridinol

Thebaine

Morphine

Codeine

Pallidine

The several morphinandienone type alkaloids are synthesised by phenolic oxidative coupling[255, 277, 279], Pschorr reaction[279], photo-Pschorr reaction[249], benzyne reaction[280] and photolytic dehydrobromination[252], as shown below. Total synthesis of this type of alkaloid is summarised by Kametani and Fukumoto[281]

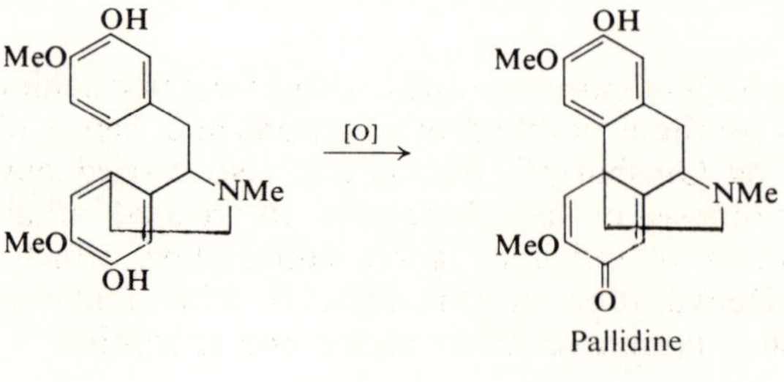

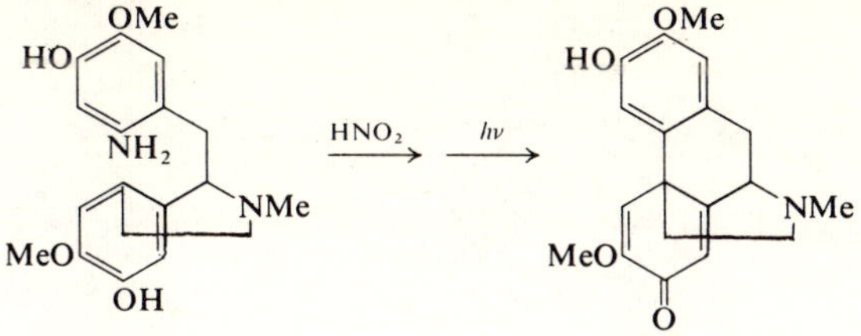

Flavinantine

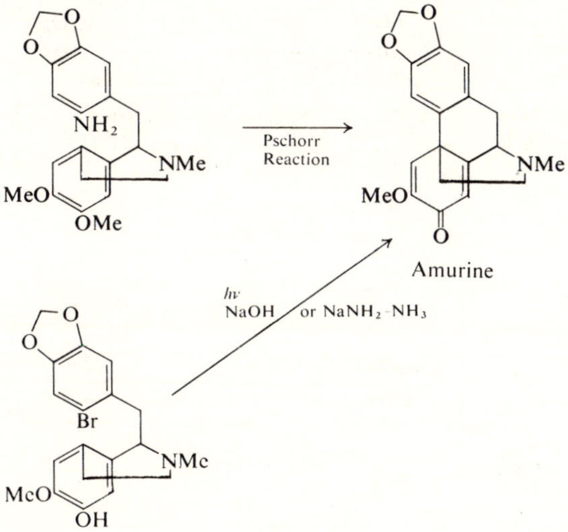

Pschorr Reaction

Amurine

hv
NaOH or $NaNH_2$–NH_3

The first complete synthesis of cepharamine has been reported by Inubushi[282] as shown in Figure 6.15.

Two interesting approaches to this alkaloid have been reported; one involving a formal total synthesis[283] and the other a model experiment[284]. Hasubanonine has also been synthesised by Inubushi[285].

Chelerythrine has been synthesised by Bailey[286] after studies on the general preparation of the dibenzophenanthridine system by Robinson[287]. Dyke[288] synthesised nitidine by the Pschorr reaction of the amino-compound which is obtained by Pomeranz–Fritsch reaction modified by Bobbitt.

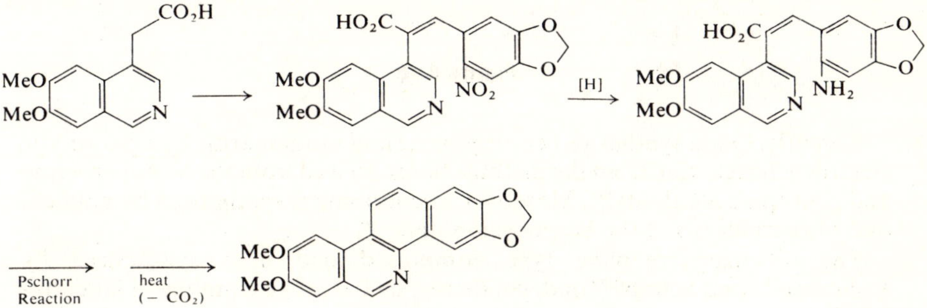

Pschorr Reaction

heat $(-CO_2)$

Nitidine

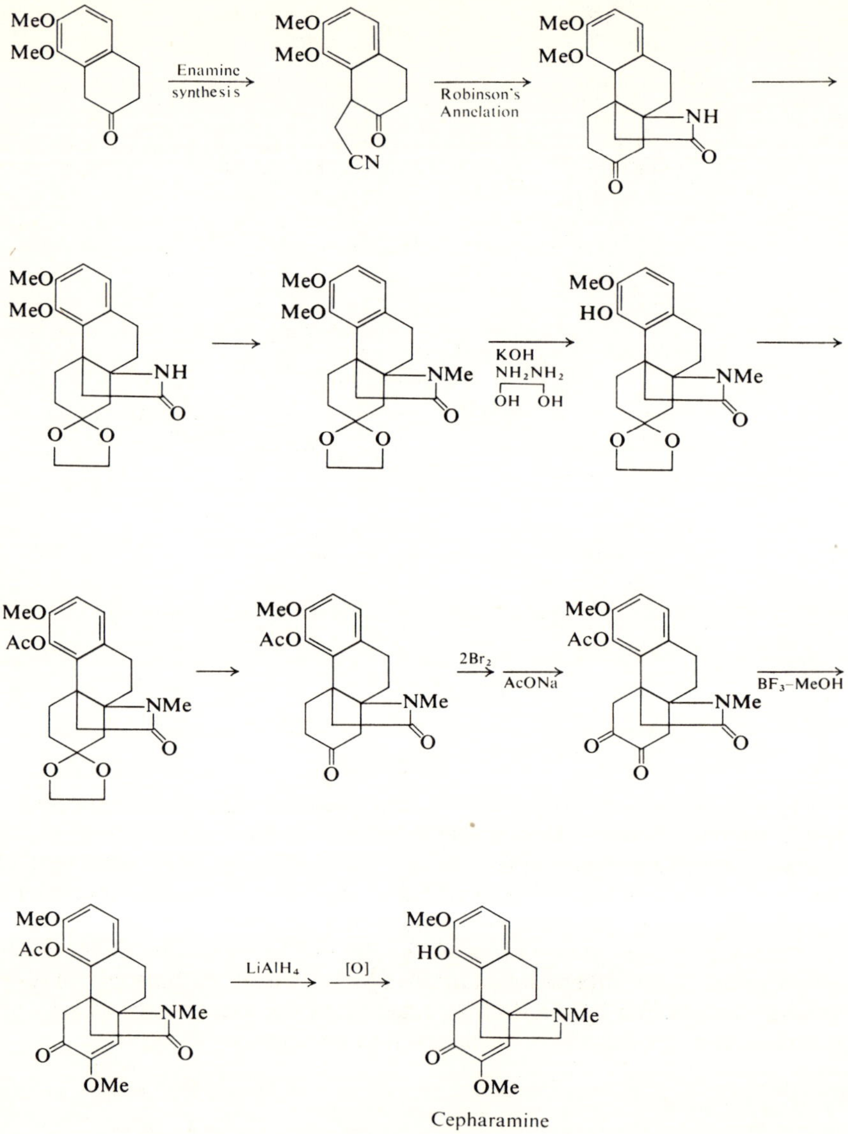

Cepharamine

Figure 6.15

Recently, Onda synthesised chelerythrine and sanguinarine by a photolytic electrocyclic reaction from the methine bases derived from the protoberberine and protopine alkaloids[289]. Moreover, chelidonine is synthesised by application of thermolysis of the benzocyclobutane[290].

The dibenzopyrrocoline type compound had been synthesised by Robinson[291] and Schöpf[292], independently, by oxidative coupling of laudanosine before cryptowoline and cryptaustoline were isolated in nature. Benington[293] and Kametani[294] synthesised cryptowoline from halogenoisoquinoline by a benzyne reaction.

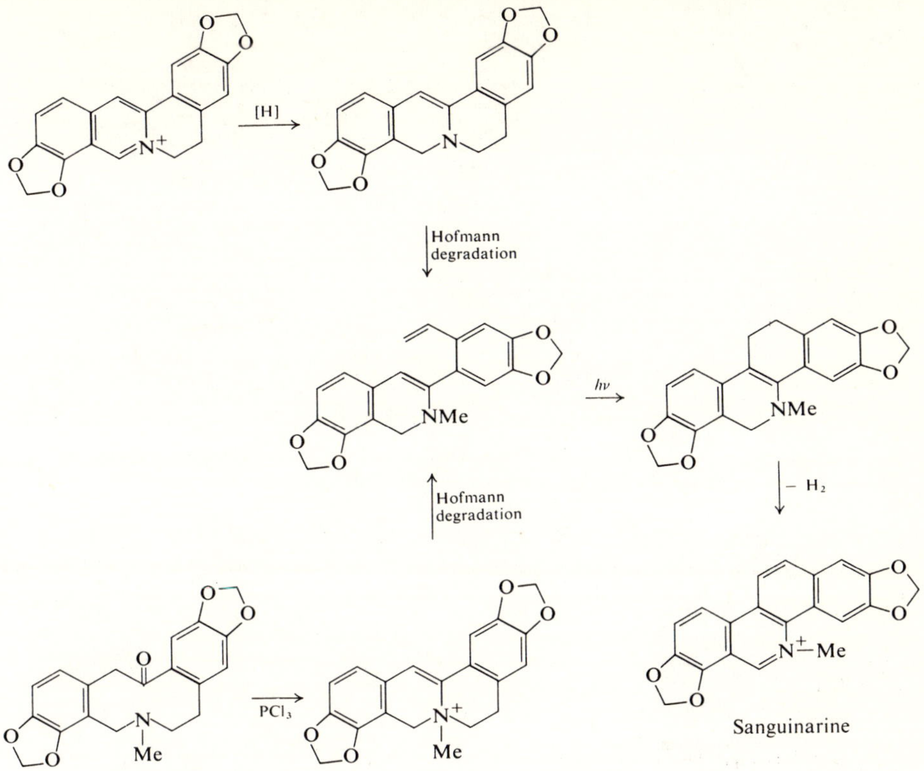

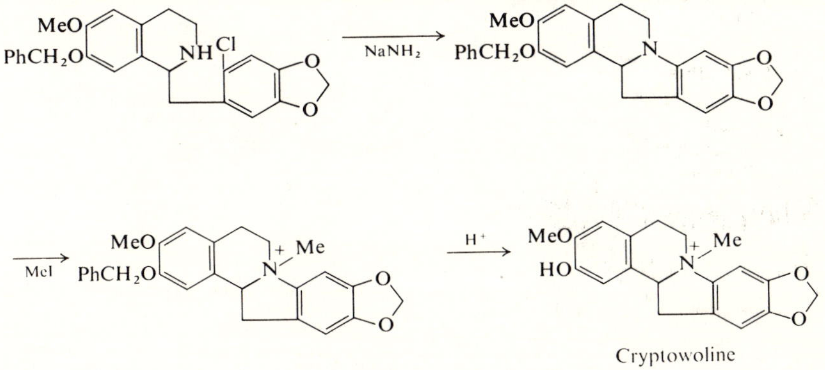

Total synthesis of protostephanine has been achieved by Takeda[137] and Brossi[295] by classic methods. The route used by the latter group is outlined as follows.

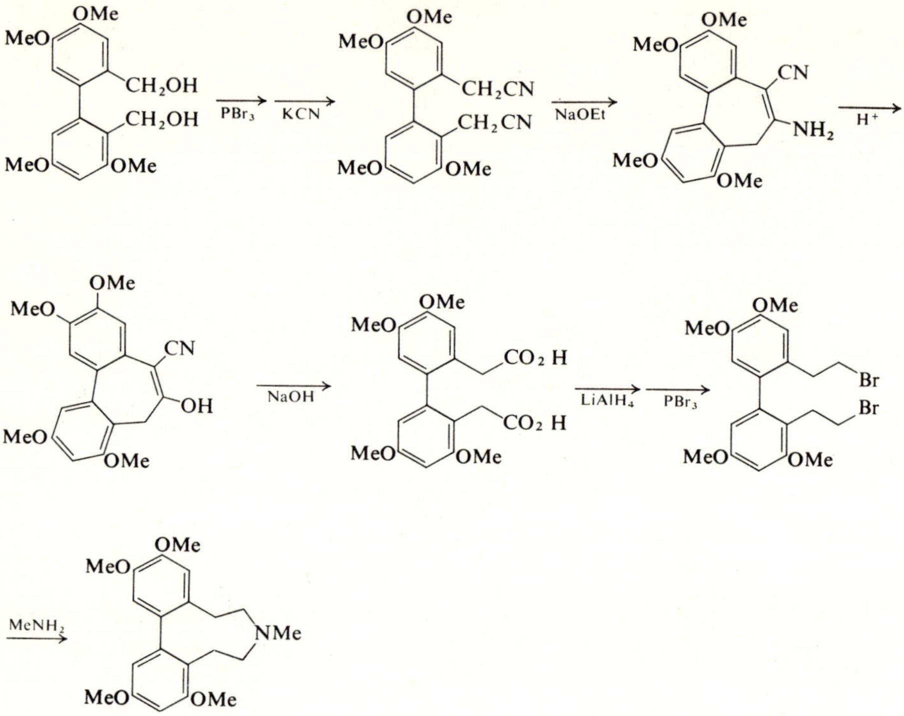

Protostephanine

An interesting and biogenetic type synthesis of protostephanine has been described by Battersby[195] as follows.

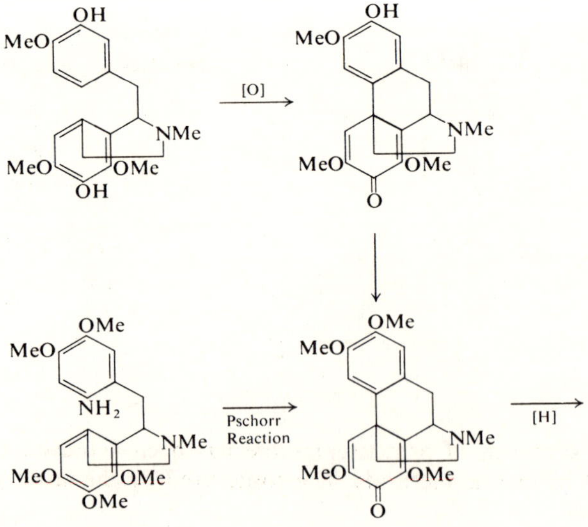

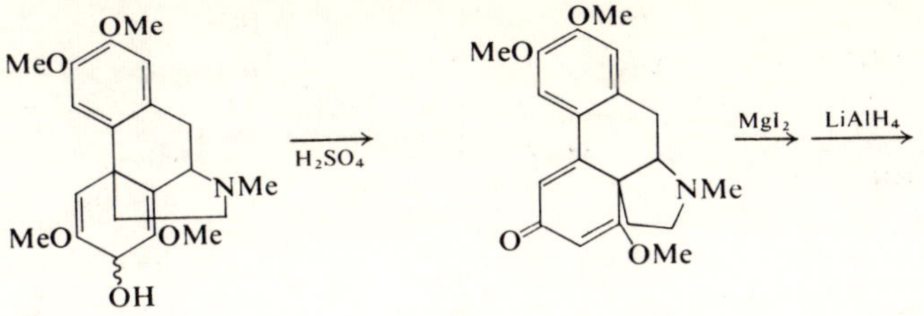

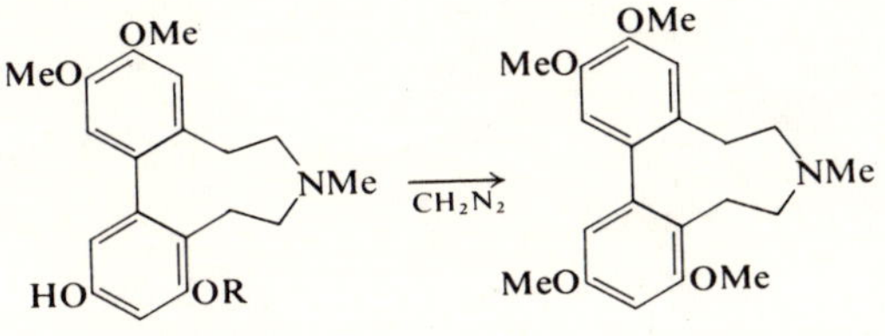

R = H and Me

Mondon[296] succeeded in carrying out a synthesis of erysotrine from Δ^1-7-acetoxylactam by applying Hansen's method

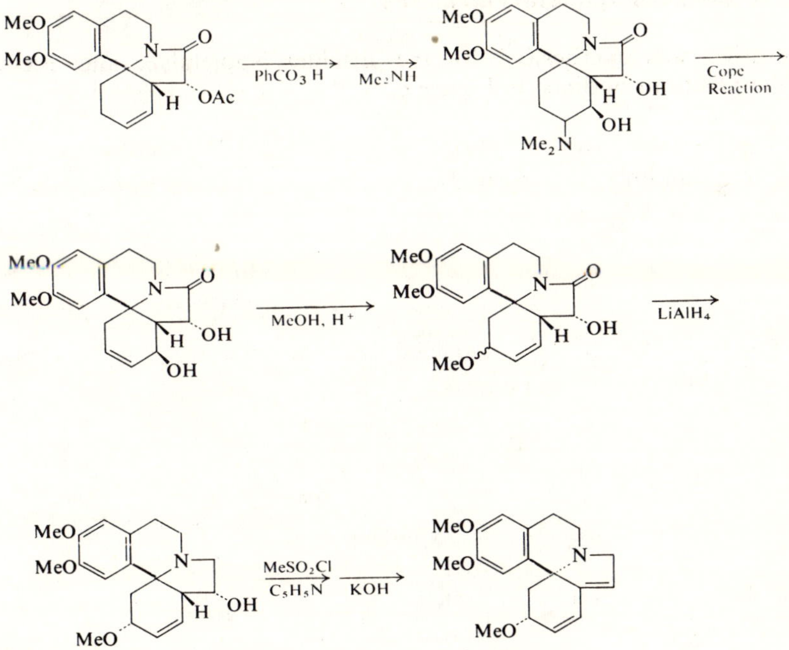

Erysotrine

Dihydroerysodine has been synthesised biogenetically by Mondon[297] through erythrinadienone[298, 299].

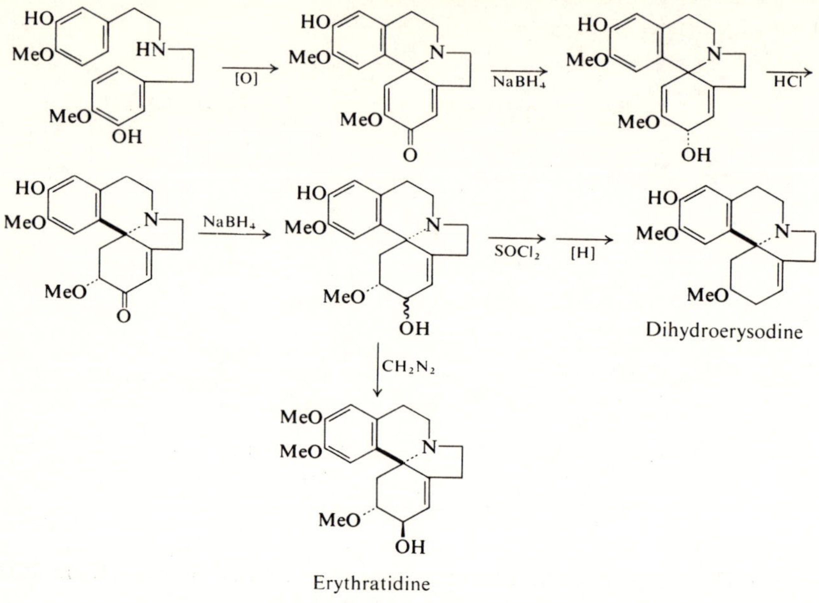

Dihydroerysodine

Erythratidine

6.4.2 2-Benzylisoquinoline alkaloids

Sendaverine was synthesised by the Bischler–Napieralski and Pictet–Spengler reactions outlined in Figure 6.16 [155].

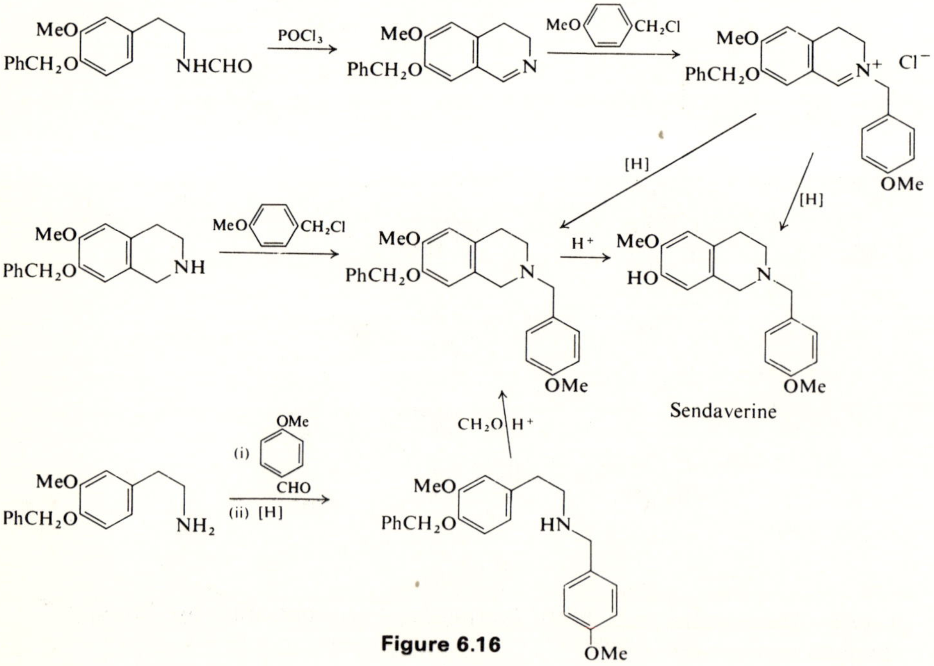

Figure 6.16

Sendaverine

6.4.3   Phenethylisoquinoline alkaloids

Melanthioidine has been synthesised by a double Ullmann reaction[300] from phenolic bromoisoquinoline by Battersby. The biogenetic type synthesis of melanthioidine was also examined[301].

Melanthioidine

A total synthesis of androcymbine and *O*-methylandrocymbine was achieved by a photo-Pschorr reaction from aminoisoquinoline[302], and by photolytic dehydrobromination of bromoisoquinoline[303, 304]. The andro-cymbine skeleton is obtained according to biogenetic theory by phenolic oxidative coupling[305].

R = H Androcymbine
R = Me *O*-Methylandrocymbine

Homoaporphine alkaloids are synthesised through phenolic oxidative coupling followed by dienone–phenol rearrangement[168, 169, 306]. Kreysiginone, a homoproaporphine alkaloid, was also synthesised by phenol oxidation[165, 306].

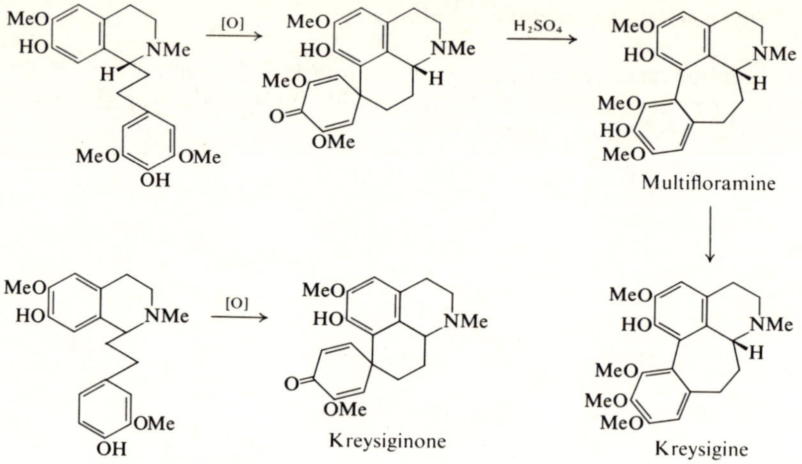

Multifloramine

Kreysiginone

Kreysigine

Photolysis is applicable to the synthesis of homoaporphine alkaloids, as shown in a preparation of kreysigine[247, 303].

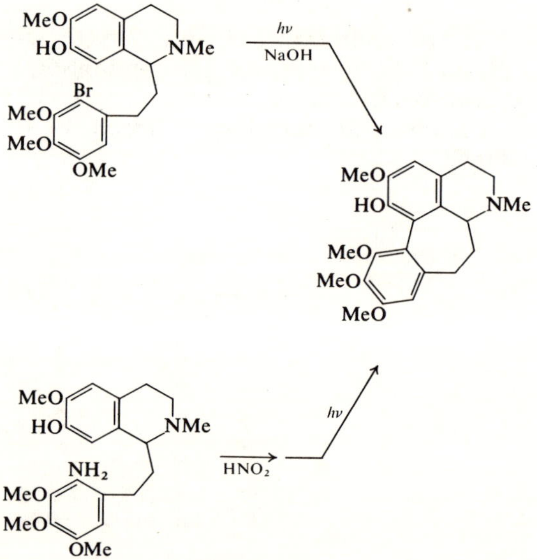

Although the homoerythrina alkaloids have not been synthesised, an interesting synthetic approach along biogenetic lines has been reported[307]

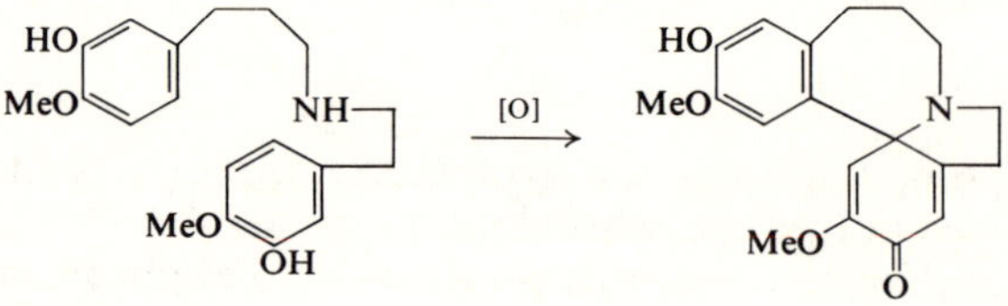

References

1. Kametani, T. (1968). *The Chemistry of the Isoquinoline Alkaloids.* (Tokyo: Hirokawa Publishing Company, Inc. and Amsterdam: Elsevier Pub. Co.)
2. Taylor, W. I. and Battersby, A. R. (1967). *Oxidative Coupling of Phenols.* (New York: Marcel Dekker)
3. Mothes, K. and Schütte, H. R. (1969). *Biosynthese der Alkaloide.* (Berlin: VEB Deutsche Verlag der Wissenschaften)
4. Earlier reviews: Manske, R. H. F., *The Alkaloids,* Vol. IV, VII, IX, X and XII
5. Albonico, S. M., Kuck, A. M. and Deulofeu, V. (1965). *Annalen,* **685,** 200
6. Corrodi, H. and Hardegger, E. (1956). *Helv. Chim. Acta,* **39,** 889
7. Budzikiewicz, H., Djerassi, C. and Williams, D. H. (1964). *Structure Elucidation of Natural Products by Mass Spectrometry, Vol. I; Alkaloids,* 173–183. (San Francisco: Holden-Day, Inc.)
8. Tomita, M., Shingu, T., Fujitani, K. and Furukawa, H. (1965). *Chem. Pharm. Bull. (Tokyo),* **13,** 921
9. Furukawa, H., Yang, T. H. and Lin, T. J. (1965). *J. Pharm. Soc. Japan,* **85,** 472
10. Sanchez, E. and Comin, J. (1967). *Tetrahedron,* **23,** 1139
11. Dalton, D. R., Cava, M. P. and Buck, K. T. (1965). *Tetrahedron Letters,* 2687
12. Sangster, A. W. and Stuart, K. L. (1965). *Chem. Rev.,* **65,** 69
13. Hruban, L. and Šantavý, F. (1967). *Collect. Czech. Chem. Commun.,* **32,** 3414
14. Battersby, A. R., Bick, I. R. C., Klyne, W., Jennings, J. P., Scopes, P. M. and Vernengo, M. J. (1965). *J. Chem. Soc.,* 2239
15. Albonico, S. M., Comin, J., Kuck, A. M., Sanchez, E., Scopes, P. M., Swan, R. J. and Vernengo, M. J. (1966). *J. Chem. Soc. C,* 1340
16. Craig, J. C. and Roy, S. K. (1965). *Tetrahedron,* **21,** 401
17. Craig, J. C., Martin-Smith, M., Roy, S. K. and Stenlake, J. B. (1966). *Tetrahedron,* **22,** 1335
18. Kametani, T. and Ihara, M. (1968). *J. Chem. Soc. C,* 1305
19. Bhacca, N. S., Craig, J. C., Manske, R. H. F., Roy, S. K., Shamma, M. and Slusarchyk, W. A. (1966). *Tetrahedron,* **22,** 1467
20. Kunitomo, J., Morimoto, K., Yamamoto, K., Yoshikawa, Y., Azuma, K. and Fujitani, K. (1971). *Chem. Pharm. Bull. (Tokyo),* **19,** 2197
21. Gellert, E. and Summons, R. E. (1969). *Tetrahedron Letters,* 5055
22. Theumann, D. and Comin, J. (1966). *An. Asoc. Quim. Argent.,* **54,** 217
23. Lu. S.-T. (1967). *J. Pharm. Soc. Japan,* **87,** 1278
24. Shamma, M. and Dudock, B. S. (1968). *J. Pharm. Sci.,* **57,** 262
25. Preininger, V., Cross, A. D., Murphy, J. W., Šantavý, F. and Toube, T. (1969). *Collect. Czech. Chem. Commun.,* **34,** 875
26. Brochmann-Hanssen, E. and Hirai, K. (1968). *J. Pharm. Sci.,* **57,** 904
27. McCorkindale, N. J., McCulloch, A. W., Magrill, D. S., Caddy, B., Martin-Smith, M., Smith, S. J. and Stenlake, J. B. (1969). *Tetrahedron,* **25,** 5475
28. Preininger, V., Cross, A. D. and Šantavý, F. (1966). *Collect. Czech. Chem. Commun.,* **31,** 3345
29. Shamma, M., Kelly, M. G. and Podczasy, M. A. (1969). *Tetrahedron Letters,* 4951
30. Tomita, M. (1967). *Celebration Publication for the Retirement of Professor Masao Tomita, 1926–1967.* (Tokyo: Hirokawa Publishing Company, Inc.)
31. Proskurnina, N. F. and Orekhov, A. P. (1940). *J. Gen. Chem. USSR,* **10,** 707
32. P'an, P,-C., Chou, Y.-L., Sun, T.–C. and Kao, Y.-S. (1963). *Sci. Sinica,* **11,** 321. [*Chem. Abstr.* (1964). **58,** 3467; Hsieh, Y.-Y., Chen, W.-C. and Kao, Y.-S. (1964). *Sci. Sinica,* **12,** 2018 [*Chem. Abstr.* (1965). **62,** 9183]
33. Tomita, M. and Kunitomo, J. (1962). *J. Pharm. Soc. Japan,* **82,** 734, 741
34. Moiseeva, G. P., Ismailov, Z. F. and Yunusov, S. Yu. (1970). *Khim. Prir. Soedin.,* **6,** 705
35. DeJongh, D. C., Shrader, S. R. and Cava, M. P. (1966). *J. Amer. Chem. Soc.,* **88,** 1052
36. Tomita, F., Furukawa, H., Kikuchi, T., Kato, A. and Ibuka, T. (1966). *Chem. Pharm. Bull. (Tokyo),* **14,** 232; Tomita, M., Kikuchi, T., Fujitani, K., Kato, A., Furukawa, H., Aoyagi, Y., Kitano, M. and Ibuka, T. (1966). *Tetrahedron Letters,* 857
37. Chan, K. C., Evans, M. T. A., Hassall, C. H. and Sangster, A. M. W. (1967). *J. Chem. Soc. C,* 2479
38. Milne, G. W. A. and Plimmer, J. R. (1966). *J. Chem. Soc. C,* 1966

39. Grundon, M. F. and McGarvey, J. E. B. (1966). *J. Chem. Soc. C,* 1082
40. Baldas, J., Porter, Q. N., Bick, I. R. C. and Vernengo, M. J. (1966). *Tetrahedron Letters,* 2059
41. Shamma, M., Dudock, B. S., Cava, M. P., Rao, K. V., Dalton, D. R., DeJongh, D. C. and Shrader, S. R. (1966). *Chem. Commun.,* 7
42. Bick, I. R. C., Harley-Mason, J., Sheppard, N. and Vernengo, M. J. (1961). *J. Chem. Soc.,* 1896
43. Baldas, J., Porter, Q. N., Bick, I. R. C., Douglas, G. K., Falco, M. R., deVries, J. X. and Yunusov, S. Yu. (1968). *Tetrahedron Letters,* 6315
44. Harley-Mason, J., Howard, A. S., Taylor, W. I., Vernengo, M. J., Bick, I. R. C. and Clezy, P. S. (1967). *J. Chem. Soc. C,* 1948; Bick, I. R. C., Bowie, J. H., Harley-Mason, J. and Williams, D. H. (1967). *J. Chem. Soc. C,* 1951; Aoki, K. and Harley-Mason, J. (1967). *J. Chem. Soc. C,* 1957
45. Manske, R. H. F. (1950). *J. Amer. Chem. Soc.,* **72,** 55
46. Kametani, T., Shibuya, S., Kibayashi, C. and Sasaki, S. (1966). *Tetrahedron Letters,* 3215
47. Clark, G. R., Manske, R. H., Palenik, G. J., Rodrigo, R., MacLean, D. B., Baczynskyj, L., Gracey, D. E. F. and Saunders, J. K. (1970). *J. Amer. Chem. Soc.,* **92,** 4998
48. Martell, Jr., M. J., Soine, T. O. and Kier, L. B. (1963). *J. Amer. Chem. Soc.,* **85,** 1022
49. Stermitz, F. R., Lwo, S.-Y. and Kallos, G. (1963). *J. Amer. Chem. Soc.,* **85,** 1551
50. Battersby, A. R. and Binks, R. (1955). *J. Chem. Soc.,* 2888
51. Barker, A. C. and Battersby, A. R. (1967). *J. Chem. Soc. C,* 1317
52. Chan, R. P. K., Craig, J. C., Manske, R. H. F. and Soine, T. O. (1967). *Tetrahedron,* **23,** 4209
53. Cervinka, O., Favrova, A. and Novak, V. (1966). *Tetrahedron Letters,* 5375
54. Martell, Jr., M. J., Soine, T. O. and Kier, L. B. (1967). *J. Pharm. Sci.,* **56,** 973
55. Šantavý, F., Hruban, L. and Maturová, M. (1966). *Collect. Czech. Chem. Commun.,* **31,** 4286
56. Shamma, M., Jones, C. D. and Weiss, J. A. (1969). *Tetrahedron,* **25,** 4347
57. Shamma, M. and Slusarchyk, W. A. (1964). *Chem. Rev.,* **64,** 59
58. Gadamer, J. and Kuntze, F. (1911). *Arch. Pharm.,* **249,** 598
59. Späth, E., Holter, H. and Posega, R. (1928). *Ber.,* **61,** 322
60. Shamma, M. (1960). *Experientia,* **16,** 484
61. Shamma, M. and Hillman, M. J. (1969). *Experientia,* **25,** 544
62. Goodwin, S., Shoolery, J. N. and Johnson, L. F. (1958). *Proc. Chem. Soc. (London),* 306
63. Vernengo, M. J. (1963). *Experientia,* **19,** 294
64. Highet, R. J. and Highet, P. F. (1965). *J. Org. Chem.,* **30,** 902
65. Pachler, K. G. R., Arndt, R. R. and Baarschers, W. H. (1965). *Tetrahedron,* **21,** 2159
66. Baarschers, W. H. and Pachler, K. G. R. (1965). *Tetrahedron Letters,* 3451
67. Djerassi, C., Mislow, K. and Shamma, M. (1962). *Experientia,* **18,** 53
68. Craig, J. C. and Roy, S. K. (1965). *Tetrahedron,* **21,** 395
69. Jackson, A. H. and Martin, J. A. (1966). *J. Chem. Soc. C,* 2181
70. Faltis, F. and Alder, E. (1951). *Arch. Pharm.,* **284,** 281
71. Cava, M. P., Bessho, K., Douglas, B., Markey, S. and Weisbach, J. A. (1966). *Tetrahedron Letters,* 4279
72. Yang, T.-H. (1962). *J. Pharm. Soc. Japan,* **82,** 798
73. Casagrande, C. and Merotti, G. (1970). *Farmaco, Ed. Sci.,* **25,** 799 [*Chem. Abstr.,* (1971). **74,** 23047n]
74. Kiryakov, H. G. (1968). *Chem. and Ind.,* 1807
75. Bouquet, A., Cava, A. and Paris, R. R. (1970). *Compt. Rend. Acad. Sci., Ser. C,* **271,** 1100 [*Chem. Abstr.* (1971) **74,** 23043h]
76. Kunitomo, J., Okamoto, Y., Yuge, E. and Nagai, Y. (1969). *Tetrahedron Letters,* 3287
77. Glick, M. D., Cook, R. E., Cava, M. P., Srinivasan, M., Kunitomo, J. and DaRocha, A. I. (1969). *Chem. Commun.,* 1217
78. Stuart, K. L. and Cava, M. P. (1968). *Chem. Rev.,* **68,** 321
79. Bernauer, K. and Hofheinz, W. (1968). *Progress in the Chemistry of Organic Natural Products,* **26,** 246
80. Bernauer, K. (1963). *Helv. Chim. Acta,* **46,** 1783; Bernauer, K. (1964). *Helv. Chim. Acta,* **47,** 2122

81. Cava, M. P., Nomura, K., Schlessinger, R. H., Buck, K. T., Douglas, B., Raffauf, R. F. and Weisbach, J. A. (1964). *Chem. and Ind.*, 282; Cava, M. P., Nomura, K., Talapatra, S. K., Mitchell, M. J., Schlessinger, R. H., Buck, K. T., Beal, J. L., Douglas, B., Rauffauf, R. F. and Weisbach, J. A. (1968). *J. Org. Chem.*, **33**, 2785

82. Haynes, L. J., Stuart, K. L., Barton, D. H. R. and Kirby, G. W. (1966). *J. Chem. Soc. C,* 1676

83. Snatzke, G. and Wollenberg, G. (1966). *J. Chem. Soc. C,* 1681

84. Battersby, A. R., Brocksom, T. J. and Ramage, R. (1969). *Chem. Commun.,* 464

85. Tomita, M., Kato, A., Ibuka, T., Furukawa, H. and Kozuka, M. (1965). *Tetrahedron Letters,* 2825

86. Baldwin, M., Loudon, A. G., Maccoll, A., Haynes, L. J. and Stuart, K. L. (1967). *J. Chem. Soc. C,* 154

87. Tomita, M., Kato, A., Ibuka, T., Furukawa, H., Asada, S. and Kozuka, M. (1967). *Shitsuryo Bunseki,* **15,** 104 [*Chem. Abstr.* (1968). **68,** 6685]

88. Flentje, H., Döpke, W. and Jeffs, P. W. (1966). *Pharmazie,* **21,** 379

89. Bernauer, K. (1968). *Helv. Chim. Acta,* **51,** 1119

90. Nakasato, T. and Asada, S. (1966). *J. Pharm. Soc. Japan,* **86,** 134

91. Manske, R. H. F. (1947). *J. Amer. Chem. Soc.,* **69,** 1800

92. Shimanouchi, H., Sasada, Y., Ihara, M. and Kametani, T. (1969). *Acta Crystallogr.,* **B25,** 1310

93. Tani, C., Takao, N., Takao, S. and Tagahara, K. (1962). *J. Pharm. Soc. Japan,* **82,** 751

94. Chen, C. Y. and MacLean, D. B. (1968). *Can. J. Chem.,* **46,** 2501

95. Kaneko, H. and Naruto, S. (1971). *J. Pharm. Soc. Japan,* **91,** 101

96. Shamma, M., Hillmann, M. J. and Jones, C. D. (1969). *Chem. Rev.,* **69,** 779

97. Hruban, L., Šantavý, F. and Hegerová, S. (1970). *Collect. Czech. Chem. Commun.,* **35,** 3420

98. Bohlmann, F. (1958). *Ber.,* **91,** 2157

99. Uskoković, M., Bruderer, H., Von Planta, C., Williams, T. and Brossi, A. (1964). *J. Amer. Chem. Soc.,* **86,** 3364

100. Lyle, G. G. (1960). *J. Org. Chem.,* **25,** 1779

101. Snatzke, G., Hrbek, Jr., J., Hruban, L., Horeau, A. and Šantavý, F. (1970). *Tetrahedron,* **26,** 5013

102. Ohta, M., Tani, H. and Morozumi, S. (1964). *Chem. Pharm. Bull. (Tokyo),* **12,** 1072; Ohta, M., Tani, H., Morozuni, S. and Kodaira, S. (1964). *Chem. Pharm. Bull. (Tokyo),* **12,** 1080

103. Shamma, M. and Dudock, B. S. (1965). *Tetrahedron Letters,* 3825

104. Taguchi, H. and Imazeki, I. (1964). *J. Pharm. Soc. Japan,* **84,** 955

105. Pfeifer, S. and Thomas, D. (1966). *Pharmazie,* **21,** 701

106. Preininger, V., Hruban, L., Šimánek, V. and Šantavý, F. (1970). *Collect. Czech. Chem. Commun.,* **35,** 124

107. Šinánek, V., Preininger, V., Sedmerá, P. and Šantavý, F. (1970). *Collect. Czech. Chem. Commun.,* **35,** 1440

108. Manske, R. H. F. (1954). *The Alkaloids,* **4,** 147

109. Stermitz, F. R., Chen, L. and White, J. I. (1966). *Tetrahedron,* **22,** 1095

110. Israilov, I. A., Yunusov, M. S. and Yunusov S. Yu. (1970). *Khim. Prir. Soedin.,* **6,** 588 [*Chem. Abstr.* (1971). **74,** 42528m]

111. Pinder, A. R. (1964). *Chem. Rev.,* **64,** 551

112. Manske, R. H. F. (1954). *The Alkaloids,* **4,** 168

113. Battersby, A. R. and Spencer, H. (1965). *J. Chem. Soc.* 1087

114. Safe, S. and Moir, R. Y. (1964). *Can. J. Chem.,* **42,** 160

115. Bláha, K., Hrbek, Jr., J., Kovář, J., Pijewska, L. and Šantavý, F. (1964). *Collect. Czech. Chem. Commun.,* **29,** 2328

116. Huang, W.-K., Chang, C.-C. and Lin, G.-S. (1965). *Hua. Hsueh Hsueh Pao,* **31,** 470 [*Chem. Abstr.* (1966). **64,** 15936 d]

117. Šantavý, F., Kaul, J. L., Hruban, L., Dolejš, L., Hanuš, V., Bláha, K. and Cross, A. D. (1965). *Collect. Czech. Chem. Commun.,* **30,** 3479

118. Pfeifer, S., Bhanerjee, S. K., Doleyš, L. and Hanuš, V. (1965). *Pharmazie,* **20,** 45

119. Dolejš, L. and Hanuš, V. (1967). *Tetrahedron,* **23,** 2997

120. Šantavý, F., Hrbek, Jr., J. and Bláha, K. (1967). *Collect. Czech. Chem. Commun.,* **32,** 4452

121. Stuart, K. L. (1971). *Chem. Rev.,* **71,** 47
122. Döpke, W., Flentje, H. and Jeffs, P. W. (1968). *Tetrahedron,* **24,** 4459
123. Barton, D. H. R., Kirby, G. W., Steglich, W., Thomas, G. M., Battersby, A. R., Dobson, T. A. and Ramuz, H. (1965). *J. Chem. Soc.,* 2423
124. Haynes, L. J., Husbands, G. E. M. and Stuart, K. L. (1968). *J. Chem. Soc. C,* 951
125. Audier, H., Fetizon, M., Ginsburg, D., Mandelbaum, A. and Rüll, Th. (1965). *Tetrahedron Letters,* 13
126. Wheeler, D. M. S., Kinstle, T. H. and Rinehart, Jr., K. L. (1967). *J. Amer. Chem. Soc.,* **89,** 4494
127. Stuart, K. L. and Chambers, C. (1967). *Tetrahedron Letters,* 2879
128. Tomita, M., Ibuka, T., Inubushi, Y., Watanabe, Y. and Matsui, M. (1964). *Tetrahedron Letters,* 2937; (1965). *Chem. Pharm. Bull. (Tokyo),* **13,** 538
129. Kupchan, S. M., Suffness, M. I., White, D. N. J., McPhail, A. T. and Sim, G. A. (1968). *J. Org. Chem.,* **33,** 4529
130. Tomita, M., Inubushi, Y. and Ibuka, T. (1967). *J. Pharm. Soc. Japan,* **87,** 381
131. Kupchan, S. M. and Suffness, M. I. (1970). *Tetrahedron Letters,* 4975
132. Tomita, M., Okamoto, Y., Kikuchi, T., Osaki, K., Nishikawa, M., Kamiya, K., Sasaki, Y., Matoba, K. and Goto, K. (1971). *Chem. Pharm. Bull. (Tokyo),* **19,** 770
133. Tomita, M., Okamoto, Y., Kikuchi, T., Osaki, K., Nishikawa, M., Kamiya, K., Sasaki, Y., Matoba, K. and Goto, K. (1967). *Tetrahedron Letters,* 2421
134. Ewing, J., Hughes, G. K., Ritchie, E. and Taylor, W. C. (1953). *Aust. J. Chem.,* **6,** 78
135. Kametani, T. and Ogasawara, K. (1968). *Chem. Pharm. Bull. (Tokyo),* **16,** 1498
136. Takeda, K. (1958). *Ann. Rep. ITSUU Lab.,* **9,** 24, and references cited herein
137. Takeda, K. (1963). *Ann. Rep. ITSUU Lab.,* **13,** 45
138. Ito, K., Furukawa, H. and Tanaka, H. (1971). *Chem. Pharm. Bull. (Tokyo),* **19,** 1509
139. Carmack, M., McKusick, B. C. and Prelog, V. (1951). *Helv. Chim. Acta,* **34,** 1601
140. Folkers, K., Koniuszy, F. and Shavel, Jr., J. (1951). *J. Amer. Chem. Soc.,* **73,** 589
141. Wiesner, K., Valenta, Z., Manson, A. J. and Stonner, F. W. (1955). *J. Amer. Chem. Soc.,* **77,** 675
142. Mondon, A. and Menz, H.-U. (1964). *Tetrahedron,* **20,** 1729
143. Prelog, V., Wiesner, K., Khorana, H. G. and Kenner, G. W. (1949). *Helv. Chim. Acta,* **32,** 453
144. Prelog, V., McKusick, B. C., Merchant, J. R., Julia, S. and Wilhelm, M. (1956). *Helv. Chim. Acta,* **39,** 498
145. Nowacki, W. and Bonsma, G. F. (1958). *Z. Krist.,* **110,** 89
146. Barton, D. H. R., James, R., Kirby, G. W., Turner, D. W. and Widdowson, D. A. (1968). *J. Chem. Soc. C,* 1529
147. Boar, R. B. and Widdowson, D. A. (1970). *J. Chem. Soc. B,* 1591
148. Folkers, K. and Koniuszy, F. (1939). *J. Amer. Chem. Soc.,* **61,** 1232
149. Inubushi, Y., Furukawa, H. and Ju-ichi, M. (1970). *Chem. Pharm. Bull. (Tokyo),* **18,** 1951
150. Ito, K., Furukawa, H. and Tanaka, H. (1970). *Chem. Commun.,* 1076
151. Powell, R. G., Weisleder, D., Smith, Jr., C. R. and Wolff, I. A. (1969). *Tetrahedron Letters,* 4081
152. Kametani, T., Ohkubo, K., Noguchi, I. and Manske, R. H. F. (1965). *Tetrahedron Letters,* 3345; Kametani, T., Ohkubo, K. and Noguchi, I. (1966). *J. Chem. Soc. C,* 715
153. Kametani, T., Ohkubo, K. and Manske, R. H. F. (1966). *Tetrahedron Letters,* 985
154. Kametani, T. and Ohkubo, K. (1965). *Tetrahedron Letters,* 4317
155. Kametani, T. and Ohkubo, K. (1967). *Chem. Pharm. Bull. (Tokyo),* **15,** 608
156. Kametani, T., Ohkubo, K. and Takano, S. (1967). *J. Pharm. Soc. Japan,* **87,** 563
157. Ibragimova, M. U., Yunusov, M. S. and Yunusov, S. Yu. (1970). *Khim. Prir. Soedin.,* **6,** 638 [*Chem. Abstr.* (1971). **74,** 54046r]
158. Battersby, A. R., Herbert, R. B., Pijewska, K. and Šantavý, F. (1965). *Chem. Commun.,* 228
159. Battersby, A. R., Munro, M. H. G., Bradbury, R. B. and Šantavý, F. (1968). *Chem. Commun.,* 695
160. Fridrichsons, J., Mackay, M. F. and Mathieson, A. McL. (1968). *Tetrahedron Letters,* 2887
161. Hart, N. K., Johns, S. R., Lamberton, J. A. and Saunders, J. K. (1968). *Tetrahedron Letters,* 2891

162. Beecham, A. F., Hart, H. K., Johns, S. R. and Lamberton, J. A. (1968). *Aust. J. Chem.*, **21**, 2829
163. Battersby, A. R., Herbert, R. B. and Šantavý, F. (1965). *Chem. Commun.*, 415
164. Battersby, A. R., Herbert, R. B., Mo, L. and Šantavý, F. (1967). *J. Chem. Soc. C*, 1739
165. Battersby, A. R., McDonald, E., Munro, M. H. G. and Ramage, R. (1967). *Chem. Commun.*, 934
166. Kametani, T., Satoh, F., Yagi, H. and Fukumoto, K. (1970). *J. Chem. Soc. C*, 382
167. Šantavý, F., Sedmera, P., Snatzke, G. and Reichstein, T. (1971). *Helv. Chim. Acta*, **54**, 1084
168. Battersby, A. R., Bradbury, R. B., Herbert, R. B., Munro, M. H. G. and Ramage, R. (1967). *Chem. Commun.*, 450
169. Brossi, A., O'Brien, J. and Teitel, S. (1969). *Helv. Chim. Acta*, **52**, 678
170. Johns, S. R., Kowala, C., Lamberton, J. A., Sioumis, A. A. and Wunderlich, J. A. (1968). *Chem. Commun.*, 1102
171. Fitzgerald, J. S., Johns, S. R., Lamberton, J. A. and Sioumis, A. A. (1969). *Aust. J. Chem.*, **22**, 2187
172. Johns, S. R., Lamberton, J. A. and Sioumis, A. A. (1969). *Aust. J. Chem.*, **22**, 2219
173. Langlois, N., Das, B. C. and Portier, P. (1969). *Compt. Rend. Acad. Sci., Ser., C*, **269**, 639
174. Robinson, R. (1955). *The Structural Relations of Natural Products*. (Oxford: Clarendon Press)
175. Barton, D. H. R. and Cohen, T. (1957). *Festschrift Authur Stoll*, 117. (Basel: Birkhauser)
176. Winterstein, E. and Trier, G. (1910). *Die Alkaloide*. (Berlin: Gebr. Bornträger)
177. Battersby, A. R. and Harper, B. J. T. (1962). *J. Chem. Soc.*, 3526
178. Krejcarek, G. E., Dominy, B. W. and Lawton, R. G. (1968). *Chem. Commun.*, 1450
179. Jackson, A. H. and Stewart, G. W. (1971). *Chem. Commun.*, 149
180. Kametani, T., Fukumoto, K. and Fujihara, M. (1971). *Chem. Commun.*, 352
181. Barton, D. H. R., Kirby, G. W. and Wiechers, A. (1966). *J. Chem. Soc. C*, 2313
182. Stermitz, F. R. and Seiber, J. N. (1966). *J. Org. Chem.*, **31**, 2925
183. Stermitz, F. R. and McMurtrey, K. D. (1969). *J. Org. Chem.*, **34**, 555
184. Dyke, S. F. and Ellis, A. C. (1971). *Tetrahedron*, **27**, 3803
185. Battersby, A. R. (1963). *Proc. Chem. Soc. (London)*, 189
186. Battersby, A. R., Brown, R. T., Clements, J. H. and Iverach, G. G. (1965). *Chem. Commun.*, 230; Battersby, A. R. and Brown, T. H. (1966). *Chem. Commun.*, 170
187. Barton, D. H. R., Bhakuni, D. S., Chapman, G. M., Kirby, G. W., Haynes, L. J. and Stuart, K. L. (1967). *J. Chem. Soc. C*, 1295
188. Stuart, K. L. and Graham, L. (1971). *Chem. Commun.*, 392
189. Barton, D. H. R., Bhakuni, D. S., Chapman, G. M. and Kirby, G. W. (1967). *J. Chem. Soc. C*, 2134
190. Battersby, A. R., McHugh, J. L., Staunton, J. and Todd, M. (1971). *Chem. Commun.*, 985
191. Blaschke, G. (1968). *Arch. Pharm.*, **301**, 432
192. Blaschke, G. (1970). *Arch. Pharm.*, **303**, 358
193. Stuart, K. L., Teetz, V. and Franck, B. (1969). *Chem. Commun.*, 333
194. Barton, D. H. R., Kirby, A. J. and Kirby, G. W. (1968). *J. Chem. Soc. C*, 929
195. Battersby, A. R., Bhatnagar, A. K., Hackett, P., Thornber, C. W. and Staunton, J. (1968). *Chem. Commun.*, 1214
196. Barton, D. H. R., Boar, R. B. and Widdowson, D. A. (1970). *J. Chem. Soc. C*, 1213
197. Barton, D. H. R., Hesse, R. H. and Kirby, G. W. (1965). *J. Chem. Soc.*, 6379
198. Battersby, A. R. and Hirst, M. (1965). *Tetrahedron Letters*, 669
199. Battersby, A. R., Francis, R. J., Ruveda, E. A. and Staunton, J. (1965). *Chem. Commun.*, 89
200. Battersby, A. R., Francis, R. J., Hirst, M. and Staunton, J. (1963). *Proc. Chem. Soc. (London)*, 268
201. Battersby, A. R. and McCaldin, D. J. (1962). *Proc. Chem. Soc. (London)*, 365
202. Gear, J. R. and Spenser, I. D. (1963). *Can. J. Chem.*, **41**, 783
203. Leete, E. (1963). *J. Amer. Chem. Soc.*, **85**, 473
204. Battersby, A. R., Francis, R. J., Hirst, M., Southgate, R. and Staunton, J. (1967). *Chem. Commun.*, 602
205. Battersby, A. R., Hirst, M., McCaldin, D. J., Southgate, R. and Staunton, J. (1968). *J. Chem. Soc. C*, 2163
206. Barton, D. H. R. (1963). *Proc. Chem. Soc. (London)*, 293

207. Gupta, R. N. and Spenser, I. D. (1965). *Can. J. Chem.*, **43**, 133
208. Battersby, A. R., Herbert, R. B., McDonald, E., Ramage, R. and Clements, J. H. (1966). *Chem. Commun.*, 603, and references cited herein
209. Barker, A. C., Battersby, A. R., McDonald, E., Ramage, R. and Clements, J. H. (1967). *Chem. Commun.*, 390
210. Battersby, A. R., Böhler, P., Munro, M. H. G. and Ramage, R. (1969). *Chem. Commun.*, 1066
211. Whaley, W. M. and Govindachari, T. R. (1951). *Organic Reactions*, **6**, 74
212. Whaley, W. M. and Govindachari, T. R. (1951). *Organic Reactions*, **6**, 151
213. Gensler, W. J. (1951). *Organic Reactions*, **6**, 191
214. Bobbitt, J. M., Roy, D. N., Marchand, A. and Allen, C. W. (1967). *J. Org. Chem.*, **32**, 2225
215. Hellmann, H. and Elser, W. (1961). *Annalen*, **639**, 77
216. Teitel, S. and Brossi, A. (1968). *J. Heterocyclic Chem.*, **5**, 825
217. Grethe, G., Uskoković, M. R. and Brossi, A. (1968). *J. Org. Chem.*, **33**, 2500
218. Gibson, W., Popp, F. D. and Noble, A. C. (1966). *J. Heterocyclic Chem.*, **3**, 99
219. Grethe, G., Lee, H. L., Uskoković, M. R. and Brossi, A. (1970). *Helv. Chim. Acta*, **53**, 873
220. Kametani, T., Kobari, T., Fukumoto, K. and Fujihara, M. (1971). *J. Chem. Soc. C*, 1796
221. Kametani, T., Kusama, O. and Fukumoto, K. (1968). *J. Chem. Soc. C*, 1798
222. Kametani, T. and Fukumoto, K. (1964). *J. Chem. Soc.*, 6141
223. Tomita, M., Furukawa, H., Lu, S.-T. and Kupchan, S. M. (1967). *Chem. Pharm. Bull. (Tokyo)*, **15**, 959
224. Kametani, T., Iida, H. and Sakurai, K. (1969). *J. Chem. Soc. C*, 500, and references cited herein
225. Tomita, M., Fujitani, K., Aoyagi, Y. and Kajita, Y. (1968). *Chem. Pharm. Bull. (Tokyo)*, **16**, 217
226. Tomita, M., Fujitani, K. and Aoyagi, Y. (1968). *Chem. Pharm. Bull. (Tokyo)*, **16**, 62
227. Fujita, E. and Sumi, A. (1970). *Chem. Pharm. Bull. (Tokyo)*, **18**, 2591
228. Tomita, M., Ueda, S. and Teraoka, A. (1963). *J. Pharm. Soc. Japan*, **83**, 87
229. Inubushi, Y., Masaki, Y., Matsumoto, S. and Takami, F. (1969). *J. Chem. Soc. C*, 1547
230. Bobbitt, J. M. and Hallcher, R. C. (1971). *Chem. Commun.*, 543
231. Kametani, T. and Fukumoto, K. (1963). *J. Chem. Soc.*, 4289
232. Kametani, T., Shibuya, S., Seino, S. and Fukumoto, K. (1964). *J. Chem. Soc.*, 4146
233. Ishiwata, S., Fujii, T., Miyagi, N., Satoh, Y. and Itakura, K. (1970). *Chem. Pharm. Bull. (Tokyo)*, **18**, 1850
234. Stermitz, F. R. and Seiber, J. N. (1966). *Tetrahedron Letters*, 1177
235. Barker, A. C. and Battersby, A. R. (1967). *J. Chem. Soc. C*, 1317; Chen, C.-H., Soine, T. O. and Lee, K.-H. (1970). *J. Pharm. Sci.*, **59**, 1529
236. Battersby, A. R., Brown, T. H. and Clements, J. H. (1965). *J. Chem. Soc.*, 4550
237. Kametani, T. and Yagi, H. (1967). *J. Chem. Soc. C*, 2182
238. Ishiwata, S., Itakura, K. and Misawa, K. (1970). *Chem. Pharm. Bull. (Tokyo)*, **18**, 1219
239. Ishiwata, S. and Itakura, K. (1970). *Chem. Pharm. Bull. (Tokyo)*, **18**, 1841
240. Horii, Z., Nakashita, Y. and Iwata, C. (1971). *Tetrahedron Letters*, 1167
241. Kametani, T., Sugahara, T., Sugi, H., Shibuya, S. and Fukumoto, K. (1971). *Chem. Commun.*, 724; Kametani, T., Sugi, H., Shibuya, S. and Fukumoto, K. (1971). *Chem. and Ind.*, 818
242. DeTar, D. F. (1957). *Organic Reactions*, **9**, 409
243. Kametani, T. and Noguchi, I. (1967). *J. Chem. Soc. C*, 1440
244. Shamma, M. and Slusarchyk, W. A. (1965). *Tetrahedron Letters*, 1509
245. Narayanaswami, S., Prabhaker, S. and Pai, B. R. (1969). *Indian J. Chem.*, **7**, 945
246. Gibson, M. S., Prenton, G. W. and Walthew, J. M. (1970). *J. Chem. Soc. C*, 2234
247. Ishiwata, S. and Itakura, K. (1969). *Chem. Pharm. Bull. (Tokyo)*, **17**, 2261
248. Kametani, T., Koizumi, M., Shishido, K. and Fukumoto, K. (1971). *J. Chem. Soc. C*, 1923
249. Kametani, T., Sugi, H., Shibuya, S. and Fukumoto, K. (1971). *Chem. Pharm. Bull. (Tokyo)*, **19**, 1513
250. Cava, M. P., Mitchell, M. J., Havlicek, S. C., Lindert, A. and Spangler, R. J. (1970). *J. Org. Chem.*, **35**, 175
251. Kupchan, S. M. and Kanojia, R. M. (1966). *Tetrahedron Letters*, 5353

252. Kametani, T., Shibuya, S., Sugi, H., Kusama, O. and Fukumoto, K. (1971). *J. Chem. Soc. C,* 2446
253. Kametani, T., Sugahara, T., Sugi, H., Shibuya, S. and Fukumoto, K. (1971). *Tetrahedron* **27,** 5993
254. Kupchan, S. M., Moniot, J. L., Kanojia, R. M. and O'Brien, J. B. (1971). *J. Org. Chem.,* **36,** 2413
255. Kametani, T., Kozuka, A. and Fukumoto, K. (1971). *J. Chem. Soc. C,* 1021 and references cited herein
256. Tomita, M. and Kozuka, M. (1969). *J. Pharm. Soc. Japan,* **86,** 871
257. Battersby, A. R., Southgate, R., Staunton, J. and Hirst, M. (1966). *J. Chem. Soc. C,* 1052
258. Kametani, T., Fukumoto, K., Terui, T., Yamaki, K. and Taguchi, E. (1971). *J. Chem. Soc. C,* 2709
259. Kametani, T. and Ihara, M. (1967). *J. Chem. Soc. C,* 520; *J. Chem. Soc. C, 1968,* 1305
260. Bradsher, C. K. and Dutta, N. L. (1961). *J. Org. Chem.,* **26,** 2231
261. Battersby, A. R., LeCount, D. J., Garratt, S. and Thrift, R. I. (1961). *Tetrahedron,* **14,** 46
262. Perkin, Jr., W. H. (1918). *J. Chem. Soc.,* **113,** 492
263. Dominguez, X. A., Delgado, J. G., Reeves, W. P. and Gardner, P. D. (1967). *Tetrahedron Letters,* 2493
264. Brown, D. W., Dyke, S. F., Hardy, G. and Sainsbury, M. (1968). *Tetrahedron Letters,* 5177
265. Lenz, G. R. and Yang, N. C. (1967). *Chem. Commun.,* 1136
266. Haworth, R. D. and Perkin, Jr., W. H. (1926). *J. Chem. Soc.,* 1769
267. Bentley, K. W. and Murray, A. W. (1963). *J. Chem. Soc.,* 2497
268. Hope, E. and Robinson, R. (1914). *J. Chem. Soc.,* **105,** 1456, 2085
269. Haworth, R. D., Pinder, A. R. and Robinson, R. (1950). *Nature (London),* **165,** 529
270. Irie, H., Tani, S. and Yamane, H. (1970). *Chem. Commun.,* 1713
271. Klötzer, M., Teitel, S., Blount, J. F. and Brossi, A. (1971). *J. Amer. Chem. Soc.,* **93,** 4321
272. Gates, M. and Tschudi, G. (1956). *J. Amer. Chem. Soc.,* **78,** 1380
273. Elad, D. and Ginsburg, D. (1954). *J. Chem. Soc.,* 3052
274. Barton, D. H. R., Bhakuni, D. S., James, R. and Kirby, G. W. (1967). *J. Chem. Soc. C,* 128
275. Kametani, T., Ihara, M., Fukumoto, K. and Yagi, H. (1969). *J. Chem. Soc. C,* 2030
276. Kametani, T., Nemoto, H., Nakano, T., Shibuya, S. and Fukumoto, K. (1971). *Chem. and Ind.,* 788
277. Franck, B., Lubs, J. and Dunkelmann, G. (1967). *Angew. Chem.,* **79,** 989, 1066
278. Kametani, T., Fukumoto, K., Kozuka, A., Yagi, H. and Koizumi, M. (1969). *J. Chem. Soc. C,* 2034
279. Kametani, T., Fukumoto, K. and Sugahara, T. (1969). *J. Chem. Soc. C,* 801
280. Kametani, T., Shibuya, S., Kigasawa, K., Hiiragi, M. and Kusama, O. (1971). *J. Chem. Soc. C,* 2712
281. Kametani, T. and Fukumoto, K. (1971). *J. Heterocyclic Chem.,* **8,** 341; Kametani, T. and Fukumoto, K. (1972). *Acc. Chem. Res.,* **5,** 212
282. Inubushi, Y., Kitano, M. and Ibuka, T. (1971). *Chem. Pharm. Bull. (Tokyo),* **19,** 1820
283. Keely, Jr., S. L., Martinez, A. J. and Tahk, F. C. (1970). *Tetrahedron,* **26,** 4729
284. Evans, D. A., Bryan, C. A. and Wahl, G. M. (1970). *J. Org. Chem.,* **35,** 4122
285. Ibuka, T., Tanaka, K. and Inubushi, Y. (1970). *Tetrahedron Letters,* 4811
286. Bailey, A. S. and Worthing, C. R. (1956). *J. Chem. Soc.,* 4535
287. Bailey, A. S. and Robinson, R. (1950). *J. Chem. Soc.,* 1375
288. Dyke, S. F., Sainsbury, M. and Moon, B. J. (1968). *Tetrahedron,* **24,** 1467
289. Onda, M., Yonezawa, K. and Abe, K. (1971). *Chem. Pharm. Bull. (Tokyo),* **19,** 31
290. Oppolzer, W. and Keller, K. (1971). *J. Amer. Chem. Soc.,* **93,** 3836
291. Robinson, R. and Sugasawa, S. (1932). *J. Chem. Soc.,* 789
292. Schöpf, C. and Thierfelder, K. (1932). *Annalen,* **497,** 22
293. Benington, F. and Morin, R. D. (1967). *J. Org. Chem.,* **32,** 1050
294. Kametani, T. and Ogasawara, K. (1967). *J. Chem. Soc. C,* 2208
295. Pecherer, B. and Brossi, A. (1967). *J. Org. Chem.,* **32,** 1053
296. Mondon, A. and Nestler, H. J. (1964). *Angew. Chem.,* **76,** 651
297. Mondon, A. and Ehrhardt, M. (1966). *Tetrahedron Letters,* 2557
298. Gervay, J. E., McCapra, F., Money, T., Sharma, G. M. and Scott, A. I. (1966). *Chem. Commun.,* 142

299. Barton, D. H. R., Boar, R. B. and Widdowson, D. A. (1970). *J. Chem. Soc. C*, 1208 and references cited herein
300. Kametani, T., Takano, S. and Haga, K. (1968). *Chem. Pharm. Bull. (Tokyo)*, **16,** 663
301. Kametani, T., Takano, S. and Kobari, T. (1969). *J. Chem. Soc. C*, 9
302. Kametani, T., Koizumi, M. and Fukumoto, K. (1971). *J. Chem. Soc. C*, 1792; *J. Org. Chem.*, **36,** 3729
303. Kametani, T., Satoh, Y., Shibuya, S., Koizumi, M. and Fukumoto, K. (1971). *J. Org. Chem.*, **36,** 3733
304. Kametani, T. and Koizumi, M. (1971). *J. Chem. Soc. C*, 3796
305. Kametani, T., Fukumoto, K., Koizumi, M. and Kozuka, A. (1969). *J. Chem. Soc. C*, 1295
306. Kametani, T., Satoh, F., Yagi, H. and Fukumoto, K. (1968). *J. Org. Chem.*, **33,** 690
307. Kametani, T. and Fukumoto, K. (1968). *J. Chem. Soc. C*, 2156

7
Steroid Alkaloids

G. G. HABERMEHL
Darmstadt Institute of Technology, German Federal Republic

7.1 STEROIDAL ALKALOIDS FROM ANIMALS

One of the most striking facts in alkaloid chemistry during the last decade was that alkaloids can be obtained not only from plants but also from animals. These compounds occur in the skin gland secretions of amphibia, and they obviously are part of a protecting system of the skin against microorganisms[1-4]. These alkaloids are not biogenic amines or amino acid derivatives and all of them represent types of compounds which have not yet been found elsewhere in Nature. Among these alkaloids two groups belong to steroids.

7.1.1 Salamandra alkaloids

7.1.1.1 Chemistry

To obtain the salamandra alkaloids the animals are anaesthetised and then the skin glands are sucked out by means of a glass tube connected to a water pump[5]. The crude secretion promply stiffens to a gum-like mass, which is ground with sand and then extracted with 80% ethanol[6]. On evaporation of this extract a mixture of impure alkaloids is obtained. The alkaloids can be separated by crystallisation of salts or derivatives and also by chromatography on silica gel with cyclohexane–diethylamine–methanol (95:5:15) as solvent. Chemical investigation (for a review see Schöpf[7] or Habermehl[8]) did not lead to an unambiguous structure of the main alkaloid, samandarin. This structure was established by an x-ray structure analysis[9]. The structures of all the other salamandra alkaloids have been determined in the same way, or

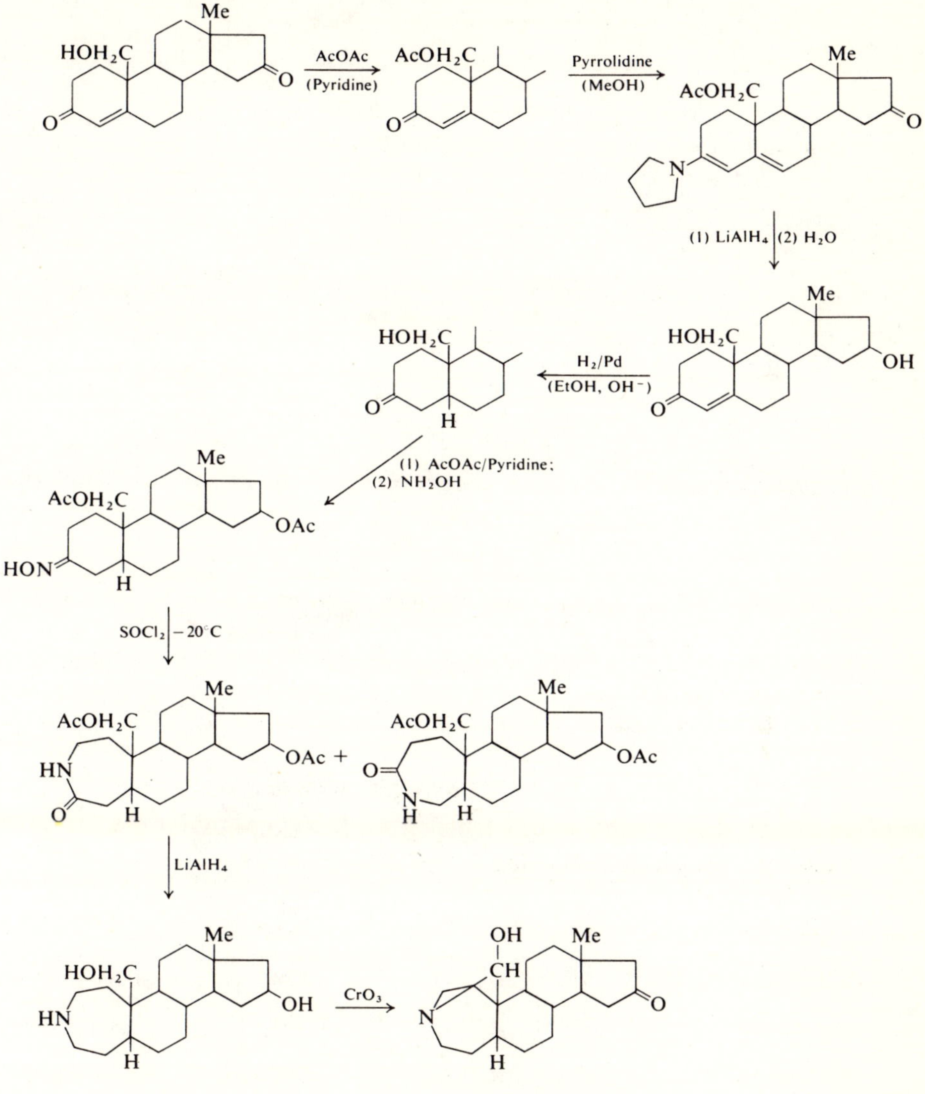

Figure 7.1 Synthesis of cycloneosamandion

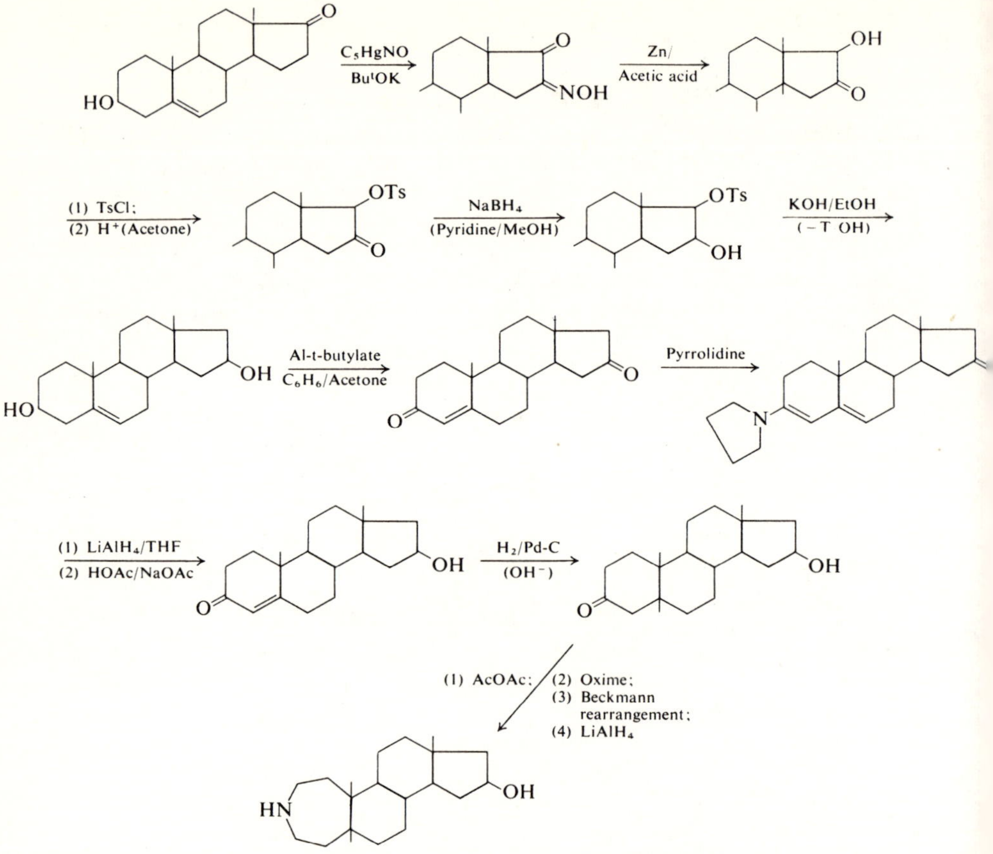

Figure 7.2 Synthesis of samanin

Table 7.1 Salamandra alkaloids

Name	Formula	m.p. °C	References
Samandarin	$C_{19}H_{31}NO_2$	188	Schöpf *et al.*[5]; Wölfel *et al.*[9]
Samandaron	$C_{19}H_{29}NO_2$	190	Schöpf *et al.*[5, 10]; Wölfel *et al.*[9]
Samandaridin	$C_{21}H_{31}NO_3$	290	Schöpf *et al.*[10]; Habermehl[11, 12]
O-Acetyl-samandarin	$C_{21}H_{33}NO_3$	159	Habermehl[13]
Samandenon	$C_{22}H_{31}NO_2$	191	Habermehl[14]
Samandinin	$C_{24}H_{39}NO_3$	170	Habermehl *et al.*[15]
Samanin	$C_{19}H_{33}NO$	197	Habermehl *et al.*[16]
Cycloneosamandion	$C_{19}H_{29}NO_2$	119	Habermehl *et al.*[17]; Göttlicher *et al.*[18]
Cycloneosamandaridin	$C_{21}H_{31}NO_3$	282	Habermehl *et al.*[19]

Four of these alkaloids have been synthesised: samandaron[20], samandaridin[12], samanin[16] and cycloneosamandion[17]. The pathway of the syntheses of samanin and cycloneosamandion is given in Figures 7.1 and 7.2.

by spectroscopic data. An outline of these compounds is given in Table 7.1
and in Figures 7.1 and 7.2.

7.1.1.2 Biogenesis

As could be found from experiments with radioactive labelled compounds,
salamandra alkaloids are formed like other steroids from the acetate via
cholesterol[21]. The enlargement of ring A may proceed after a fission between
carbon atoms 2 and 3; the inserted nitrogen comes from glutamine. The
different types of side chains of the alkaloids can be formed by degradation
of the side chain at C-17 of cholesterol.

7.1.2 Phyllobates alkaloids

7.1.2.1 Batrachotoxin

Four toxic steroidal alkaloids, batrachotoxin, homobatrachotoxin, pseudo-
batrachotoxin and batrachotoxinin A, have been isolated from skin extracts

Table 7.2 Alkaloids from *Phyllobates aurotaenia*

Name	Formula	Toxicity LD_{50}/g kg^{-1} mouse
Batrachotoxin	$C_{31}H_{42}N_2O_6$	2
Batrachotoxinin A	$C_{24}H_{35}NO_5$	1000
Homobatrachotoxin	$C_{32}H_{44}N_2O_6$	3
Pseudobatrachotoxin	$C_{24}H_{33}NO_4$	—

of the Colombian arrow poison frog, *Phyllobates aurotaenia*[22-25] as well as
from other frogs of the genus Phyllobates, namely *P. vittatus* and *P. lugubris*.
The relative toxicities of these steroids are shown in Table 7.2. The structure
of batrachotoxinin A has been determined by x-ray diffraction analysis[25,26].
The result is given in structure (1).

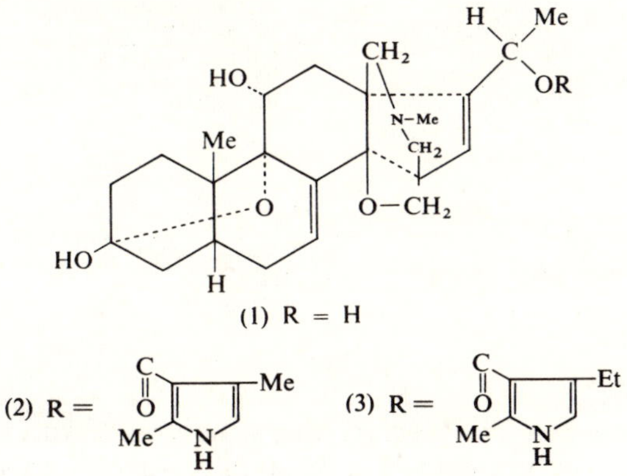

(1) R = H

(2) R =

(3) R =

On hydrolysis batrachotoxin yields batrachotoxinin A, and vice versa a partial synthesis of batrachotoxin could be performed[22, 27]. The anhydride of 2,4-dimethylpyrrole-3-carboxylic acid with ethyl chloroformate reacted with batrachotoxinin A (1) to yield batrachotoxin (2). Homobatrachotoxinin (3) is the ester of methylethylpyrrolecarboxylic acid and batrachotoxinin A.

7.2 PLANT ALKALOIDS

Steroidal alkaloids are found in several families, in detail: Solanaceae, Liliaceae, Buxaceae and Apocynaceae. In some genera of Liliaceae, such as *Veratrum*, *Fritillaria* and *Zygadenus* alkaloids with a modified steroid skeleton, namely C–nor-D-homosteroids are found. On the whole, more than a hundred alkaloids are known. This article deals with the recent progress of the last 10 years. For earlier work in this field, see References 28 and 29.

7.2.1 Liliaceae

7.2.1.1 Genus: Veratrum

Veratrum album subsp. *lobelianum* contains numerous alkaloids of jerveratrum, ceveratrum and solanidan type. Some of the minor alkaloids have been investigated during recent years.

(a) *Verazine* – This compound was first isolated by Adam *et al.*[30] and later

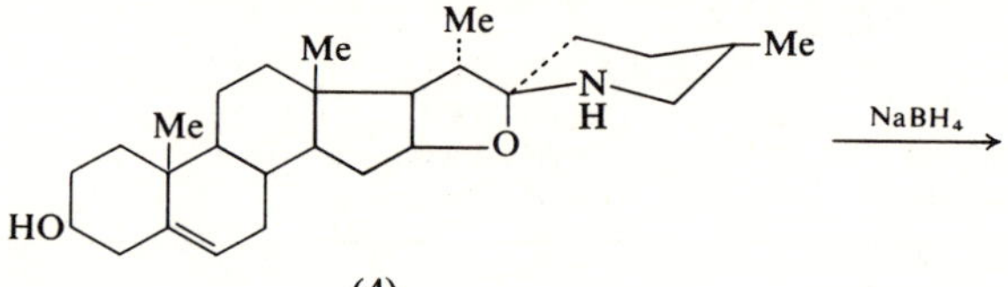

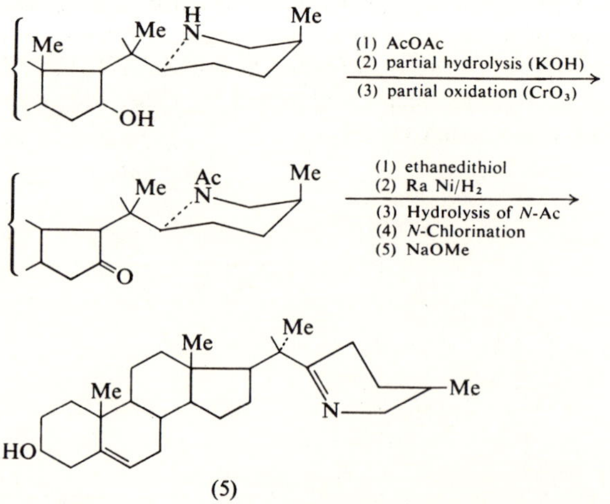

elucidated by Adam *et al.*[31]. Most of the structural features come from spectroscopic data. The presence of a 22,26-imino-cholestan skeleton was proved by catalytic hydrogenation, which yielded a mixture of two stereo-isomeric tetrahydro-verazines. Both of these were identical in every respect with (22S:25S), and (22R:25S) respectively 22,26-imino-5α-choleston-3β-ol, had been synthesised earlier[32]. Verazine thus is (25S)-22,26-imino-cholesta-5,22(N)-dien-3β-ol, a *Veratrum* alkaloid of a novel structure type related to the solanum alkaloids. The only difference between verazine and tomatillidine consists of a different configuration at C-25 as well as the absence of the 24-keto group. Unsubstituted cholestan alkaloids such as verazine possess biogenetic interest. Compounds of this type should be key substances in the biogenetic pathway of cholestan alkaloids as well as of ceveratrum and jerveratrum alkaloids[29, 35]. Verazine (5) has been synthesised from tomatid-5-en-3β-ol (4)[36]. The reaction scheme is shown.

(b) *Veralkamine* — The structure of this alkaloid has been established by chemical and physical investigations[37, 38]. From selenium dehydrogenation — yielding 2-ethyl-5-methylpyridine and γ-methylcyclopentenophenantrene — the steroidal nature of this alkaloid as well as the absence of a C-nor-D-homo skeleton characteristic for most of the other *Veratrum* bases was demonstrated. In addition to the chemical results an x-ray crystal structure analysis of the hydroiodide was performed[39]. From these investigations veralkamine (6) was found to be a steroidal alkaloid of a novel type, as a 17β-methyl-18-nor-17-isocholestan derivative. Its structure [(22S:25S)-22,26-epimino-17β-methyl-18-norcholesta-5,12-dien-3β,16β-diol] is represented by the following formula:

(6)

(c) *Veralinine* — Veralinine (7) belongs to the same structural type of alka-loid as veralkamine; it is 16-desoxy-veralkamine. The structure elucidation again was performed with the aid of i.r., n.m.r. and mass spectral data.

(7)

(d) *Veramine* — A third alkaloid of the group with a 17β-methyl-18-nor-17-isocholestan skeleton, is veramine (8), (25S)-17β-methyl-18-nor-17-iso-16α-0,22β-N-spirosola-5,12-dien-3β-ol. From the selenium dehydrogenation it appeared that this alkaloid belongs to the steroidal alkaloids (see Section 7.1). More information came from spectral data. The difference between veramine and veralkamine is a spiroaminoketal group in the former instead of a piperidine ring in the latter. This spiroaminoketal group shows the same characteristic reactions as known from spirosolan alkaloids.

Thus, on acetylation by heating with acetic anhydride/pyridine the ring F was opened[42] to yield an acetyl compound. On reduction with LiAlH$_4$ however, ring E was opened to yield the 16α-ol. In this case the piperidine ring remains unchanged[43-45]. By treatment with *N*-chlorosuccinimide and following treatment with alkali[46, 47], veramine was formed again.

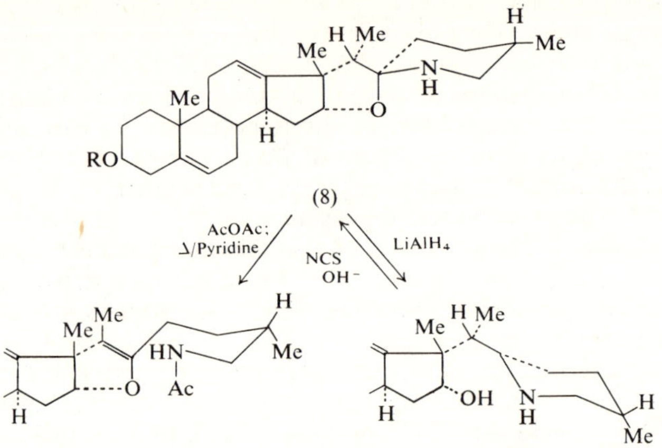

(8)

(e) *Veracintine* — Veracintine (9)[48] again is a representative of a novel structural type as it contains a pyrroline heterocycle in the side chain at C-17. The structure of this alkaloid has been established from spectral data of the base, the di-, and tetrahydro-derivative, and the *N,O*-diacetyl compound. As follows from these results, veracintine possesses the structure of 20-(2-methyl-1-pyrrolin-5-yl)-pregn-5-en-3β-ol:

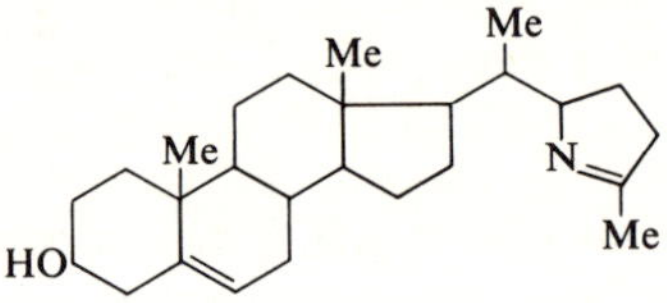

(9)

(f) *3-Acetyl-15-veratroyl-germine* — Tomko and Vassova[49] were able to isolate two novel esters of the well known alkaloid germin from *Veratrum album*. These esters, acetylveratroylgermine (10) and veratroylgermine were saponified by methanolic sodium hydroxide to yield veratric acid, germine and, in the former case, acetic acid. The structure elucidation came from i.r., n.m.r., and mass spectral data as well as from the optical rotation.

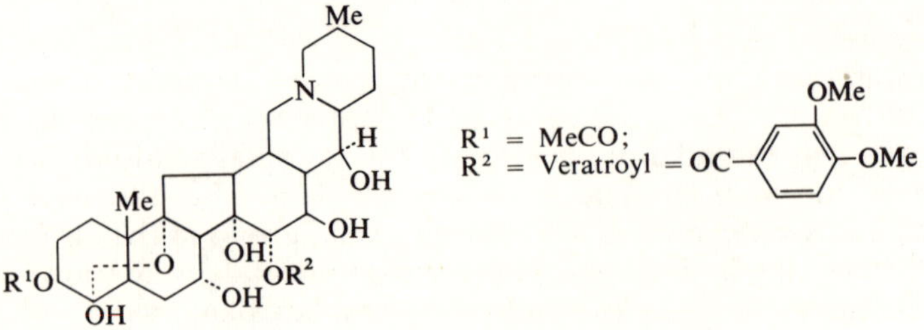

R^1 = MeCO;
R^2 = Veratroyl = OC—⟨ ⟩—OMe, OMe

(10)

Table 7.3 Outline of alkaloids from Liliaceae, genus *Veratum*

Alkaloid	Formula	m.p. °C	D (degrees)	Salts, derivatives, °C	Occurrence	Remarks
Verazine	$C_{27}H_{43}NO$	176–178	−89.7		*Veratrum album* subsp. *lobelianum*	
Veralkamine	$C_{27}H_{43}NO_2$			HJ = 260		Base: amorphous N,O,O-triacetate, m.p. 152–154°C
Veramine	$C_{27}H_{41}NO_2$		−93.9			amorphous
Veralinine	$C_{27}H_{43}NO$	124–126	−80.0			
Veracintine	$C_{26}H_{41}NO$	196–201	+ 7.5			
3-Acetyl-15-veratroylgermine	$C_{38}H_{53}NO_{12}$		+ 6.6			amorphous

7.2.2 Buxaceae

The Buxaceae are a group of plants very rich in alkaloids. According to microchemical investigations by Hegnauer[50] all of the species of the six genera (*Buxus, Notobuxus, Pachysandra, Sarcococca, Styloceras* and *Simmondsia*) investigated so far, contain alkaloids. These alkaloids, too, are closely related to each other, and they possess certain specific structural properties, such as an amino function at C-3, and a C_2N-side chain at C-17. A good part of them, moreover, possesses an oxygen function at C-16 and one or two methyl groups at C-4. They are links between lanosterol type and cholesterol type compounds, and it seems very likely that the methyl groups are lost during the biogenetic conversion from cycloartenol to Buxus alkaloids. Regarding the skeleton, Buxus alkaloids differ from other steroidal alkaloids by a C-10, C-9, C-19-cyclopropane ring (like cycloartenol) or, instead of this, a seven-membered ring B.

7.2.2.1 *Buxus*

(a) *Buxus balearica*—Two alkaloids have been isolated from *Buxus balearica*: Baleabuxine[51]; and Baleabuxidine[52]. The structures of both of

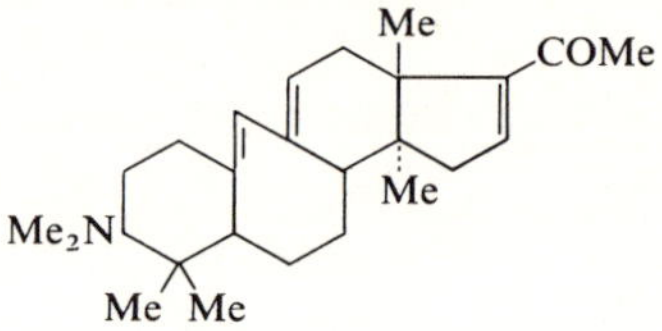

Buxpsiine

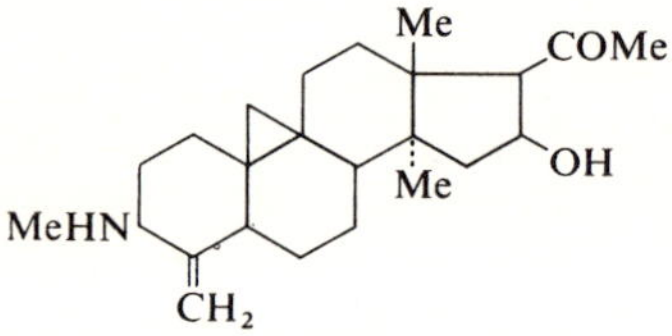

Buxtauine

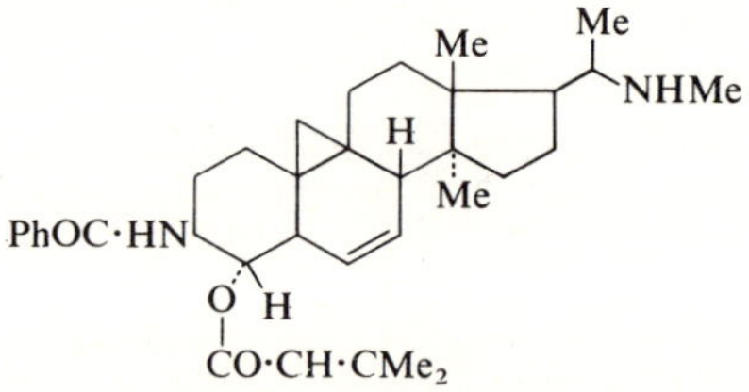

Buxaltine

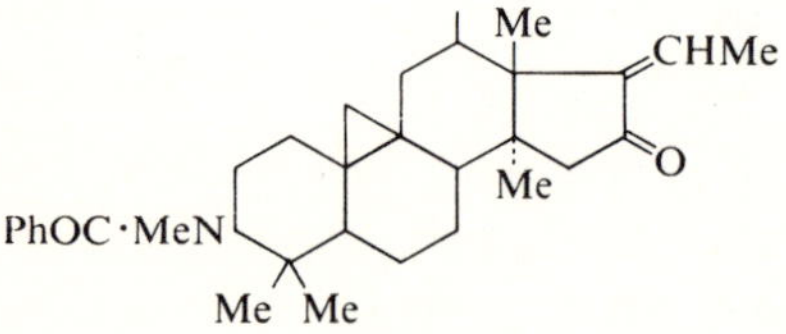

Buxanine

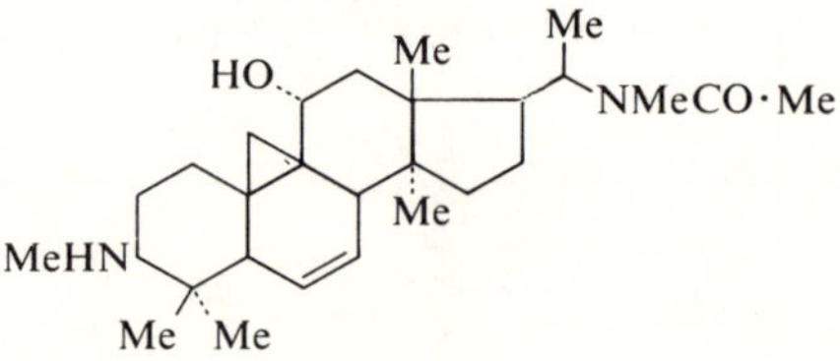

Buxiramine

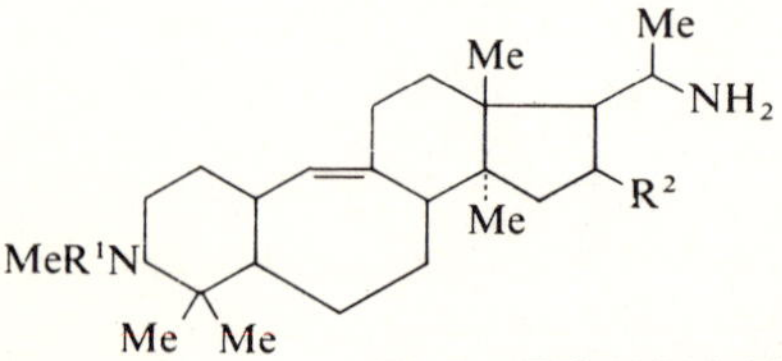

Buxamine R^1 = Me; R^2 = H
Buxaminol R^1 = Me; R^2 = OH
Norbuxamine R^1 = H; R^2 = H

Figure 7.3 Alkaloids isolated from *Buxus sempervirens*

these alkaloids were determined by chemical investigations and by aid of i.r., n.m.r. and mass spectrometric data. They are closely related and differ only in so far as one of the C-4 methyl groups of baleabuxine is converted into a CH_2OH group in baleabuxidine.

(b) *Buxus sempervirens*—A large number of alkaloids has been isolated from this plant. Most of the structures have been determined by the groups of Döpke, Votický and Tomko, Stauffacher and Kupchan. Separation and purification was performed by chromatography, followed by recrystallisation. Structure determination came from spectroscopic data, such as i.r., n.m.r., mass spectrometry and optical rotation. The alkaloids are listed in Table 7.4. They belong to different structure types, which are biogenetically related to each other (Figure 7.3). Most of them contain the normal steroid skeleton, others as for example buxpsiine belong to the 19-nor-B-homo series, which may easily arise from the cycloartenal type of structure. Alkaloids are found with two, one, or no methyl groups at C-4, also methylene groups occur. All of them contain an amino function at C-3, which may be substituted by methyl groups; amides of benzoic acid are also found at this position. Differences in the side chain at C-17 were also observed.

(c) *Buxus microphylla*—Also from *Buxus microphylla* var. *suffruticosa* numerous alkaloids have been isolated by Nakano *et al.*[59, 60] (Table 7.4). Isolation was performed by extracting leaves and twigs with ethanol and by

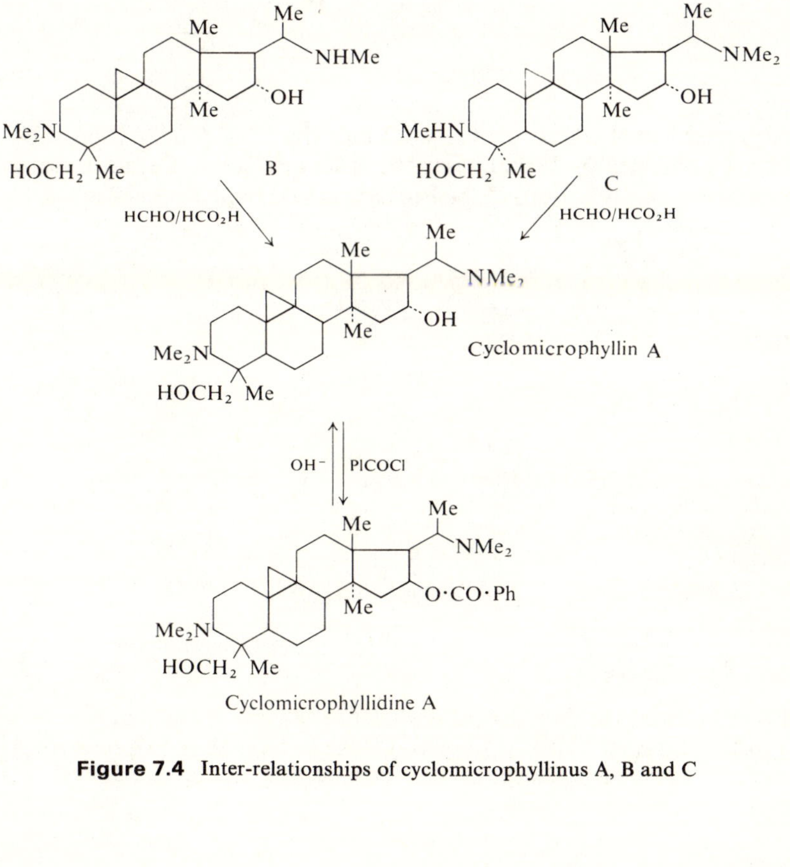

Figure 7.4 Inter-relationships of cyclomicrophyllinus A, B and C

chromatography of the extract. Structures were established by spectroscopic data as well as by chemical experiments.

Thus, the inter-relationship of cyclomicrophylline A, B and C, and cyclomicrophyllidine A was established[59] as follows from the scheme given in Figure 7.4.

Synthetic work in the field of Buxus alkaloids is only limited. An inter-conversion of cyclobuxine D (11) and cyclobuxosuffrine (12) has been reported by Nakano and Votický[61].

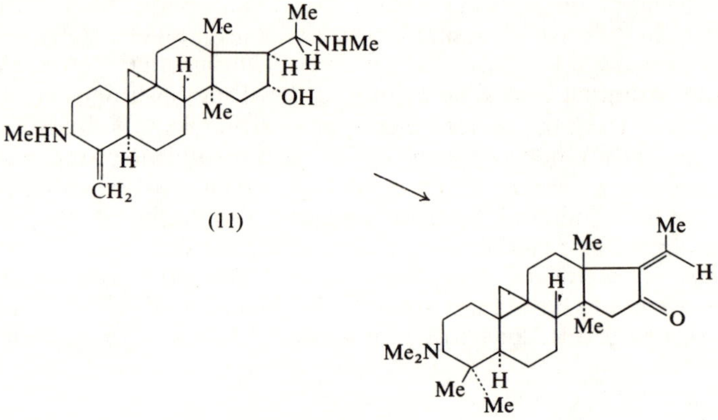

An approach to the total synthesis of baleabuxines and cycloprotobuxines has been published by Nakano et al.[62]. Baleabuxines and cycloprotobuxines represent two major groups of the Buxus alkaloids with cycloartenal skeleton:

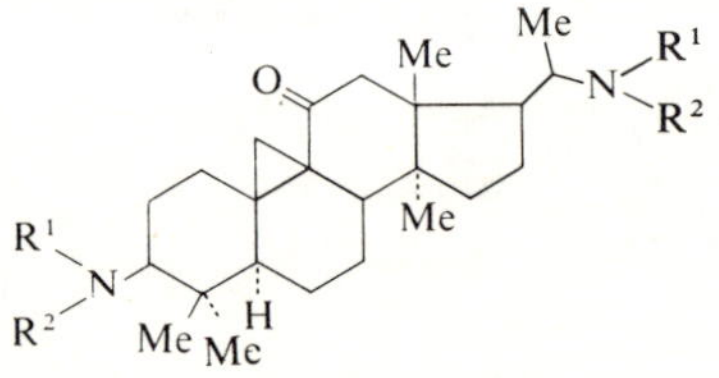

Baleabuxine-type

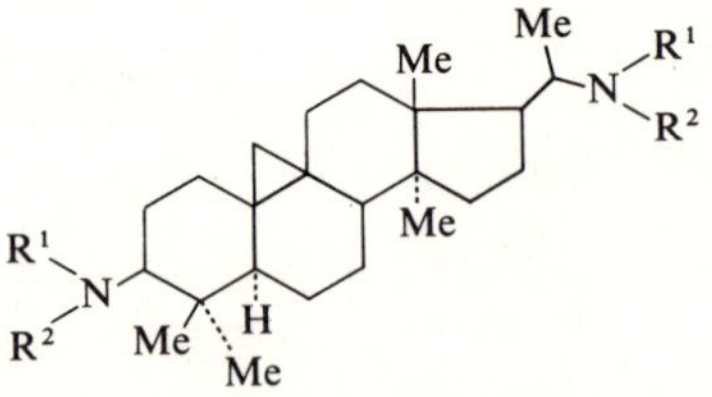

Cycloprotobuxine-type

The synthesis of the fundamental skeleton started from lanosterol, which may be degraded into 3β-acetoxy-4,4,14α-trimethyl-5α-pregnane-11,20-dione[63]. In a six-step synthesis this compound can be converted into an

11-oxo-9β-19-cyclo-compound which, after two more steps, afforded a 3,12-diketo-compound according to the following scheme:

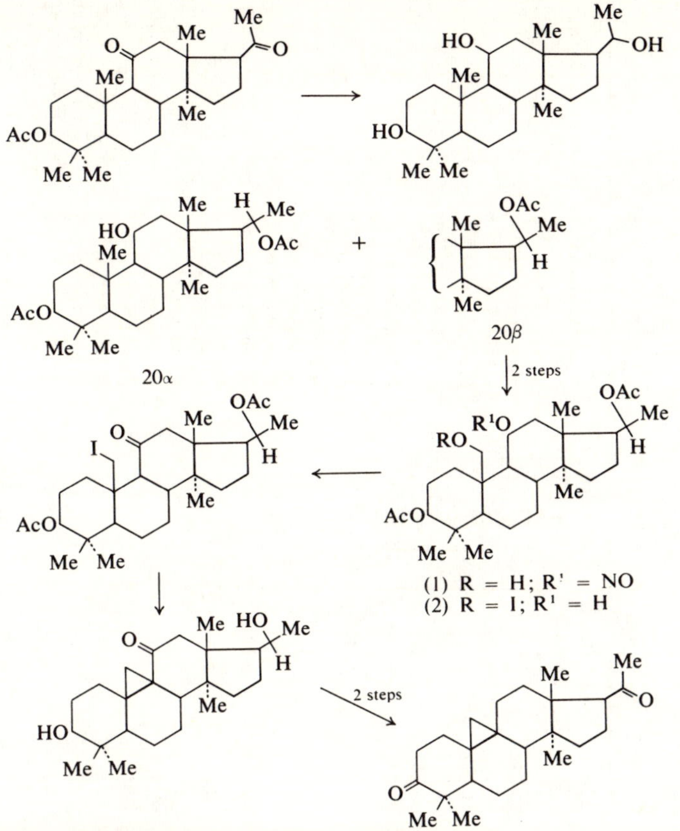

7.2.2.2 Sarcococca

(a) *Sarcococca saligna* — Two alkaloids have been isolated[64]: salignine, $C_{29}H_{52}N_2O$, and another one without name, $C_{25}H_{44}NO$. The structure of both of them is unknown.

(b) *Sarcococca pruniformis* — The alkaloids of *Sarcococca pruniformis* possess a pregnane skeleton. The substituents are not complicated, so that it is easily possible to determine the structure by aid of spectroscopic methods together with chemical degradation, e.g. Hoffmann and Emde degradation, which may yield 5α-pregnane. Kohli *et al.*[65] report on two alkaloids, A and B, the structures of which were elucidated as follows:

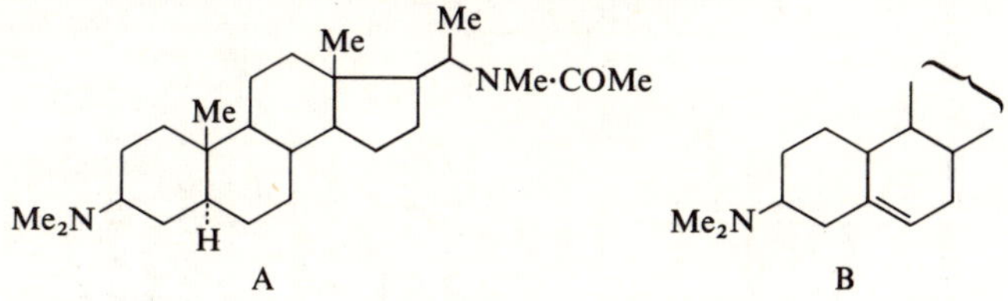

Table 7.4 Outline of Buxus alkaloids

Alkaloid	Formula	m.p. °C	D (degrees)	Salts, derivatives, (D)	Occurrence	Remarks
Baleabuxine	$C_{30}H_{50}N_2O_2$	258–259	+ 115		*Buxus balearica*[51]	
Baleabuxidine	$C_{30}H_{50}N_2O_4$	257	+ 71		*B. balearica*[52]	
Buxaltine	$C_{35}H_{48}N_2O_3$	188–191			*B. sempervirens*[53]	
Buxiramine	$C_{27}H_{44}N_2O_2$	213–215			*B. sempervirens*[53]	
Buxatine	$C_{33}H_{51}N_2O$	214–217	+ 112		*B. sempervirens*[54]	
Buxanine	$C_{32}H_{43}NO_2$	196–199	− 38		*B. sempervirens*[55]	
Buxidine					*B. sempervirens*[55]	
Buxenone	$C_{25}H_{39}NO$				*B. sempervirens*[55]	
Buxalphine	$C_{33}H_{46}N_2O_3$				*B. sempervirens*[56]	
Buxdeltine	$C_{33}H_{46}N_2O_3$				*B. sempervirens*[56]	
Buxetine	$C_{32}H_{50}N_2O_3$				*B. sempervirens*[56]	
Buxpsiine	$C_{26}H_{39}NO$	176–178	+ 105		*B. sempervirens*[56, 57]	
Buxpiine	$C_{25}H_{39}NO_2$				*B. sempervirens*[56]	
Buxtauine	$C_{24}H_{37}NO_2$				*B. sempervirens*[56]	
Buxomegine	$C_{25}H_{39}NO$				*B. sempervirens*[56]	
Bebuxine	$C_{26}H_{46}N_2O$				*B. sempervirens*[56]	= *cyclovirobuxin*
Buxene	$C_{27}H_{41}NO_3$	204			*B. sempervirens*[69]	
N-Methylbuxene	$C_{28}H_{43}NO_3$	182	− 104		*B. sempervirens*[69]	
Buxamine	$C_{26}H_{44}N_2$		− 32	(a) 210 (b) 187	*B. sempervirens*[58]	(a) Dihydrogentartrate (b) N-Isopropylidene-derivative
Norbuxamine	$C_{25}H_{42}N_2$		+ 20	(a) 300 (b) 198	*B. sempervirens*[58]	
Buxaminol	$C_{26}H_{44}N_2O$	200	+ 38	(a) 210 (b) 209	*B. sempervirens*	

Name	Formula	m.p.	$[\alpha]_D$		Source	
Cyclomicrophylline A	$C_{28}H_{48}N_2O_2$	233	− 92		*B. microphylla*[59]	
Cyclomicrophylline B	$C_{27}H_{46}N_2O_2$	252	− 65		*B. microphylla*	
Cyclomicrophylline C	$C_{27}H_{46}N_2O_2$	283	− 40		*B. microphylla*	
Dihydrocyclomicrophylline A	$C_{28}H_{50}N_2O_2$	272	+ 37		*B. microphylla*	
Dihydrocyclomicrophylline F	$C_{26}H_{46}N_2O_2$	260	+ 46		*B. microphylla*	
Cyclomicrophyllidine A	$C_{35}H_{52}N_2O_3$		−160		*B. microphylla*	amorphous base
Dihydrocyclomicrophyllidine A	$C_{35}H_{54}N_2O_3$		− 33		*B. microphylla*	amorphous base
Cyclomicrobuxine	$C_{25}H_{39}NO_2$	180	+172		*B. microphylla*	
Cyclobuxoxazine	$C_{27}H_{46}N_2O_2$	246	+ 29		*B. microphylla*[60]	
Cyclobuxine D	$C_{25}H_{42}N_2O$				*B. sempervirens*[56]	
					B. microphylla[60]	
Cyclomicrobuxeine	$C_{25}H_{37}NO$	142	+126		*B. microphylla*[60]	
Cyclobuxomicreine	$C_{25}H_{39}NO$	197	+ 37		*B. microphylla*	
Cyclomicrosine	$C_{34}H_{50}N_2O_3$	284	− 33		*B. microphylla*	
Cyclosuffrobuxinine	$C_{24}H_{35}NO$	149	− 67		*B. microphylla*	
Cyclosuffrobuxine	$C_{25}H_{37}NO$	172	− 92		*B. microphylla*	
Cyclobuxosuffrine	$C_{25}H_{39}NO$	204	− 62		*B. microphylla*	
Cyclobuxoviridine	$C_{26}H_{41}NO$	183	+ 16		*B. microphylla*	
Cyclobuxophylline	$C_{26}H_{41}NO$	196	− 72		*B. microphylla*	
Cyclomikuranine	$C_{26}H_{43}NO_2$	211	− 3		*B. microphylla*[60]	
Cycloprotobuxine A	$C_{28}H_{50}N_2$	207	+ 76		*B. microphylla*	
Salignine	$C_{29}H_{52}N_2O$	229–230	+ 18.5	·HCl:261	*Sarcococca saligna*[64]	Structure unknown
Without name	$C_{25}H_{44}NO$	136	+ 64		*S. saligna*	Structure unknown
A	$C_{26}H_{46}N_2O$	245	− 14		*S. pruniformis*[65]	
B	$C_{26}H_{44}N_2O$	233	− 56.4		*S. pruniformis*	
Saracodinine	$C_{23}H_{38}N_2$	136			*S. pruniformis*[66]	
Saracodine	$C_{26}H_{46}N_2O$	192			*S. pruniformis*[67]	
A	$C_{28}H_{48}N_2O$	150	+ 46		*S. pruniformis*[68]	
B	$C_{28}H_{47}NO_2$	274	+ 10		*S. pruniformis*[68]	
C					*S. pruniformis*[68]	Structure unknown
D					*S. pruniformis*	Structure unknown

Further alkaloids have been isolated by Chatterjee and co-workers. Saracocinine[66] was found to be 3-amino-20α-dimethylamino-3,5-pregnadiene, and saracodine is a 5α-pregnane derivative of the following structure:

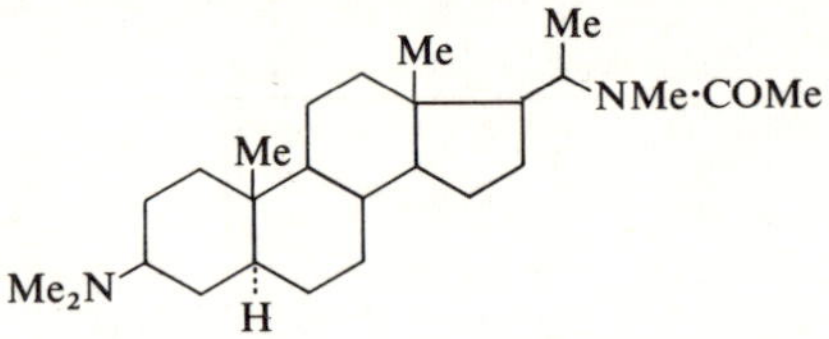

$$Me \qquad Me \qquad Me$$
$$NMe \cdot COMe$$
$$Me$$
$$Me_2N \qquad H$$

This structure is identical with that of Kohli's alkaloid A (see above) but the compounds have different melting points (cf. Table 7.4). It might be that the two alkaloids are diastereoisomers with an epimeric configuration of the C-20 asymmetric centre.

Four more alkaloids were assigned as A, B, C and D[68]. The alkaloids A and B must not be confused with alkaloids A and B of Kohli *et al.*[65]. The structures of alkaloids A and B are given as follows:

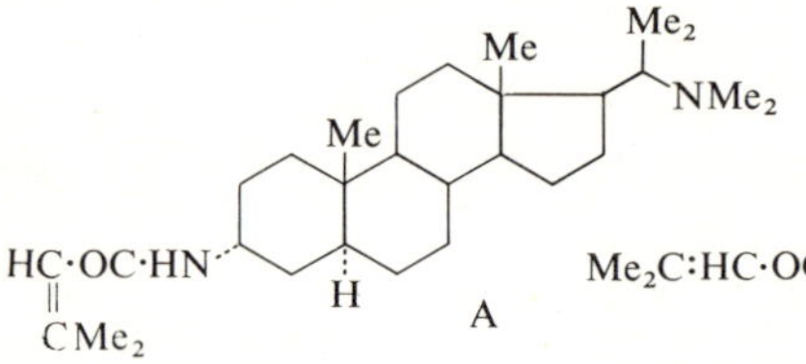

$$Me \qquad Me_2 \qquad\qquad Me \qquad Me$$
$$NMe_2 \qquad\qquad NMe_2$$
$$Me \qquad\qquad\qquad Me$$
$$HC \cdot OC \cdot HN \qquad H \qquad\qquad Me_2C{:}HC \cdot OC \cdot O \qquad H$$
$$\underset{CMe_2}{\overset{\parallel}{}} \qquad A \qquad\qquad\qquad\qquad B$$

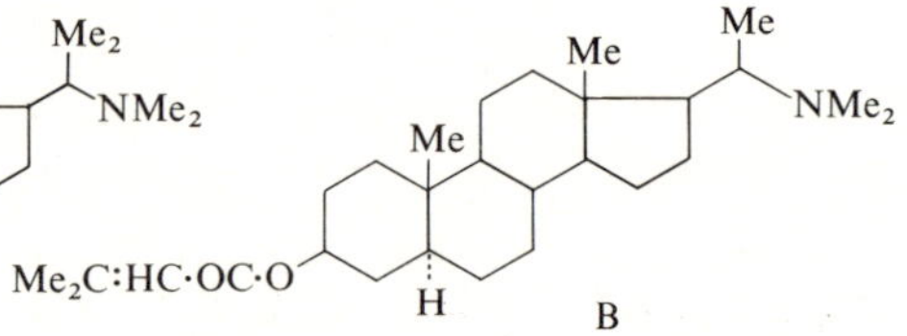

It is remarkable that both of these alkaloids contain an isoprene residue as dimethylacrylic acid, and that the nitrogen function at C-3, common for all of the other Buxus alkaloids known so far, has changed into an oxygen function.

The structures of alkaloids C and D are completely unknown.

7.2.3   Apocynaceae

Alkaloids from Apocynaceae are closely related to each other, no matter if from *Holarrhena*, *Funtumia*, *Malouetia*, *Paravallaris*, or *Kibatalia* spp. A common feature of these compounds is a heterocyclic five-membered ring system adjacent to ring D of the steroid skeleton, and containing carbon atoms C-17, C-13, and C-18. This heterocyclic ring system may be either a lactone ring with the carbonyl group at C-18 or it may be a pyrrolidine ring, or a Δ^1-pyrroline ring*. At C-3 of the skeleton, oxygen as well as amine functions are found. Other sites in the molecule for oxygen functions are C-11 and C-16.

---

*As it is well known, that Δ^1-pyrroline systems are very reactive one should observe these findings with caution and perhaps think of them as experimental artefacts. This problem, however, has not been resolved.

7.2.3.1 *Holarrhena*

Of all of the Apocynaceae the genus *Holarrhena* has been investigated most thoroughly. A very good and comprehensive review was written by Černý and Šorm[70].

The rapid growth of investigations in this field has been aided by the use of chromatography. Thus within a short time, numerous alkaloids were isolated in pure form. Holarrhena alkaloids are usually chromatographed on alumina columns by elution with benzene, diethyl ether, and diethyl ether-ethanol mixture as successive solvents. Thin-layer chromatography on silica gel plates with NH_4OH containing solvents has been studied very intensely by Lábler and Černý[71]. They found influences of substituents in the 3, 18 and 20 positions together with configurational relationship on the R_f-values.

Spectroscopic methods are very valuable for structure determination. Mass spectral investigations of Holarrhena alkaloids have been reported by L. Dolejš *et al.*[72]; according to these authors it is possible to deduce the essential features of the molecule from the mass spectral fragmentation patterns. In particular, a differentiation can be made between types I and II:

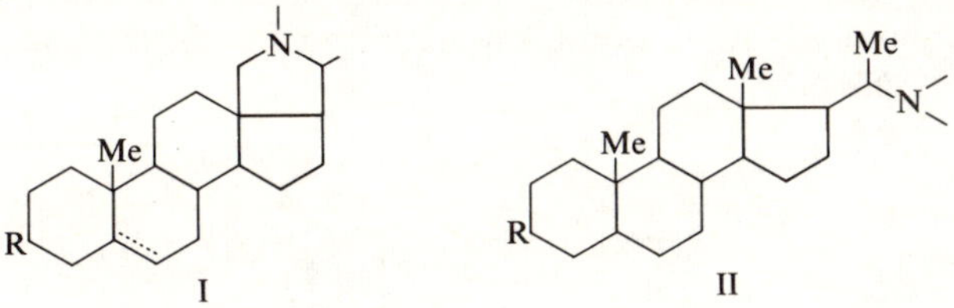

The number of methyl groups attached to nitrogen and the Δ^5-double bond can also be determined.

(a) *Holarrhena antidysenterica* — Tscheche and Ockenfeld[73] isolated two alkaloids, 7α-hydroxy-conessine and holonamine and elucidated their structures by spectroscopic data as well as by chemical reactions. For example, 7α-hydroxy-conessine (13) and conessine (14) yield the same ketone on oxidation and allylic oxidation respectively:

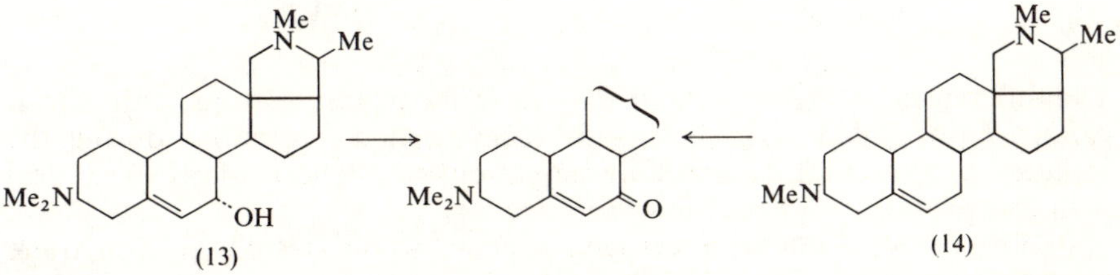

A partial synthesis starting from conessine is given in the same reference. Alkaloids without the heterocyclic ring E, but with an open side chain at C-17 are reported in a series of papers by Goutarel *et al.*, in detail: kurchiline, kurchiphylline, kurchiphyllamine, kurchialine, holadysine, holadysamine. Kurchiline[74] possesses an acetyl group at C-17.

In kurchiphylline (15), kurchiphyllamine (16) and holadysine (17) an ethyl group occurs, while kurchaline (18) and holadysamine (19) are allylic alcohols[75]:

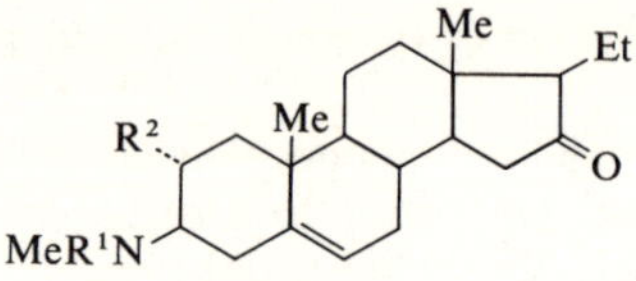

(15) R¹ = Me
      R² = OH
(16) R¹ = H
      R² = OH
(17) R¹ = H
      R² = H

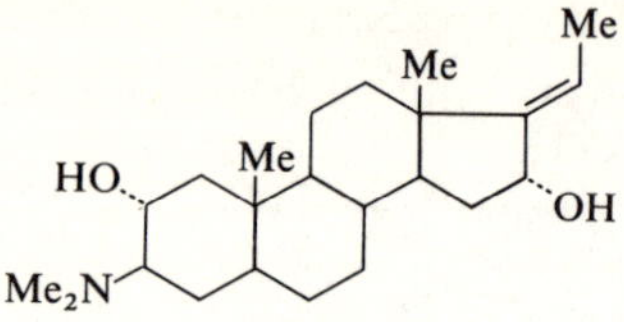

(18)

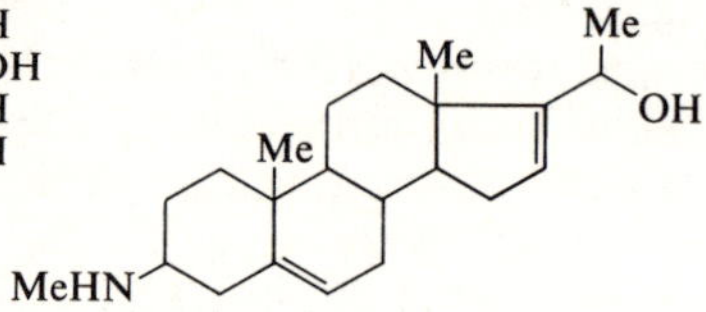

(19)

Structures were determined besides spectroscopic methods by chemical reactions and degradation to known compounds. The inter-relationship between kurchiphyllamine, kurchiline (20) and kurchiphylline is shown in the following scheme:

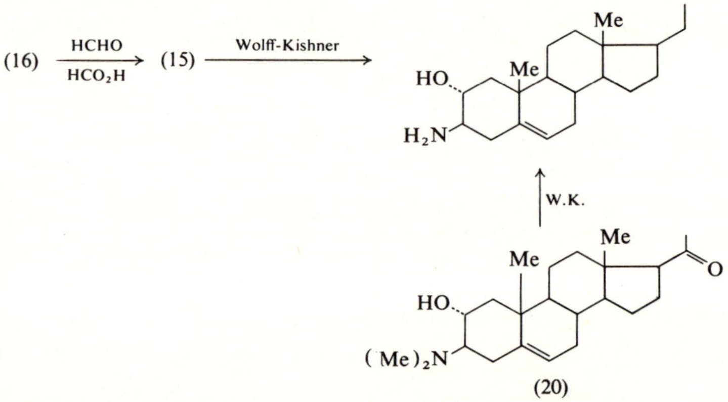

$$(16) \xrightarrow[\text{HCO}_2\text{H}]{\text{HCHO}} (15) \xrightarrow{\text{Wolff-Kishner}}$$

W.K.

(20)

The differences in the side chain may have biogenetic reasons, although it cannot be excluded, that they come from chemical reactions during the isolation process. In this respect, investigations on allylic alcohols which deal with this problem[76, 77] are of interest.

Besides the alkaloids with nitrogen attached to the steroid skeleton, quite recently some basic steroidal glycosides have been found. They are four aminoglycosteroids, holantosine A, B, C, D, and one aminoglycocardenolide, holarosine A. The holantosines A and B are the D-glycosides of 4-deoxy-4-amino-D-cymarose with two genines, holantogenine and anhydro-holanto-genine[78]. The same genines form the holantosines C and D, but as L-glycosides of 4-desoxy-4-amino-L-oleandrose[79]. Structures have been confirmed by synthesis[80]. 3β,18-Diacetoxy-14β-hydroxy-20-keto-5α-17α-pregnane (21)

yields on treatment with 3% KOH in methanol holantogenine C (22) together with holantogenine D (23).

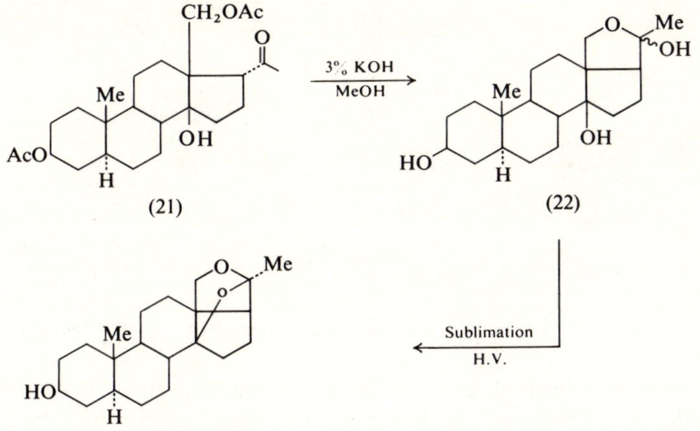

$$3\% \text{ KOH} / \text{MeOH}$$

(21)   (22)

Sublimation
H.V.

(23)

The conversion of holantosine C (24) into the anhydro product D is also
observed on recrystallisation from acetone. Methanolysis of holantosine C
with 0.05 N HCl leads to the genines of C and D, and in addition, a methyl
acetal is formed:

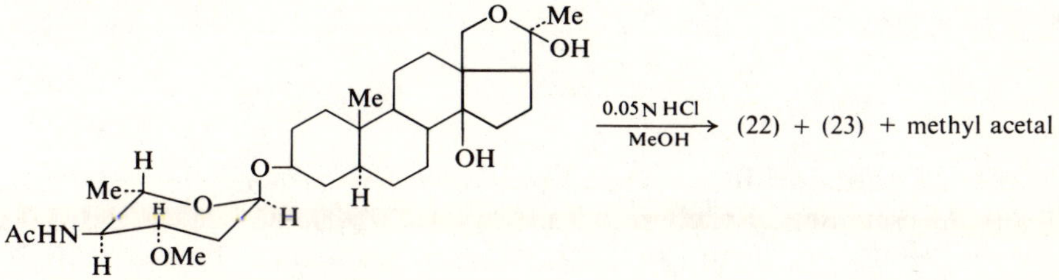

$$\xrightarrow[\text{MeOH}]{0.05\,\text{N HCl}} (22) + (23) + \text{methyl acetal}$$

(24)

Holarosine (25) is the first representative of a glycoside of an amino sugar
and a cardenolide[79]. It It is the L-glycoside of 4-deoxy-4-amino-L-oleandrose
and allo-uzarigenine:

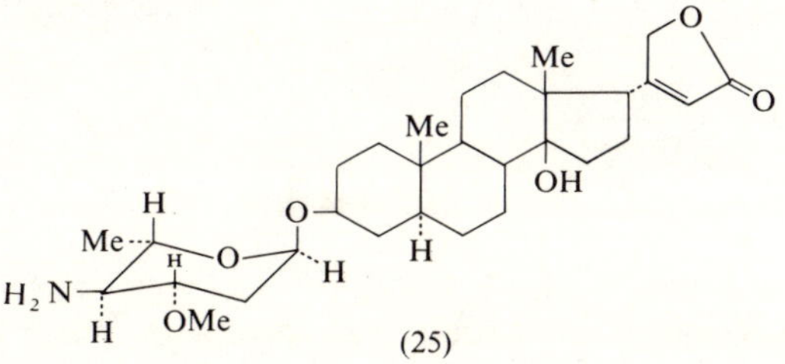

(25)

Structures of the sugar and of the genine part of the molecule were elucidated
by intensive n.m.r. studies.

(b) *Holarrhena curtisii* — From this plant basic substances were isolated which are not real alkaloids but glycosides of an aminosugar and a steroid[84]. Holacurtine (26) is 3β(4-deoxy-4-methylamino-β-D-cymaropyranosido)5α-pregnan-14β-ol-20-one, and holacurtenine the same compound but with 5α-pregn-14en-20-one as the aglycone.

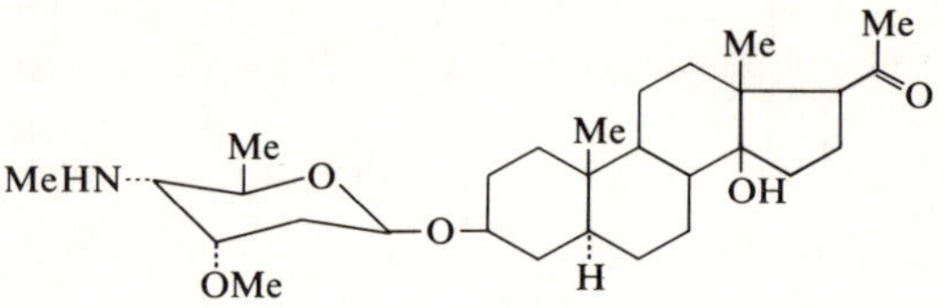

(26)

As well as the two glycosides mentioned above free aglycones are found in the plant.

(c) *Holarrhena floribunda* — Alkaloids of the conessine type have been isolated from *H. floribunda*. Holaline (27), for example, is a 5α-hydroxy-conessine[81a]. By dehydration conessine (28) is formed:

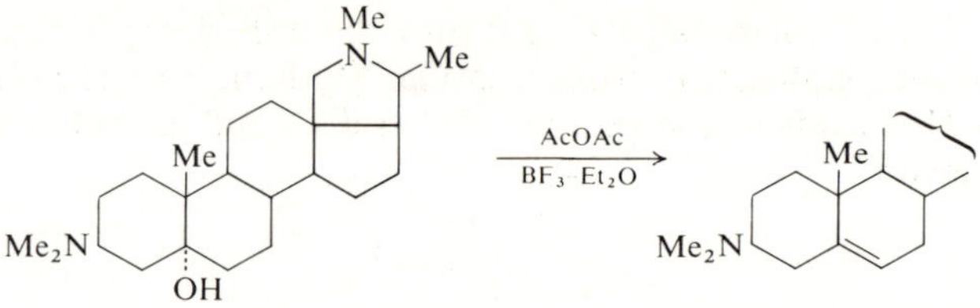

(27) (28)

Vice versa a partial synthesis of holaline starts from 3β-dimethylamino-5α,6β-dihydroxy-conanine. Reaction with $MeSO_2Cl$/pyridine yields the 5,6α-epoxide which on treatment with lithium aluminium hydride affords holaline[81b].

By a simple reaction, which is common in steroid chemistry[82] holaromine[81c] may be obtained (30), from holadienine (29):

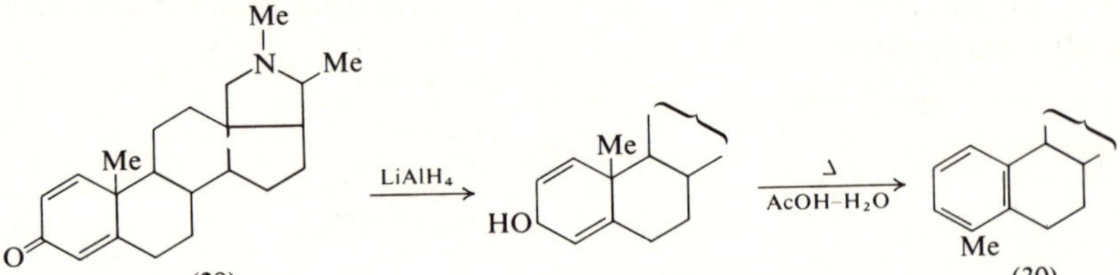

(29) (30)

The dehydration reaction causes aromatisation of ring A and migration of the C-19 methyl group from C-10 to C-4.

Besides alkaloids of the conessine type, compounds with an amino-pregnanolone structure are also found[83]:

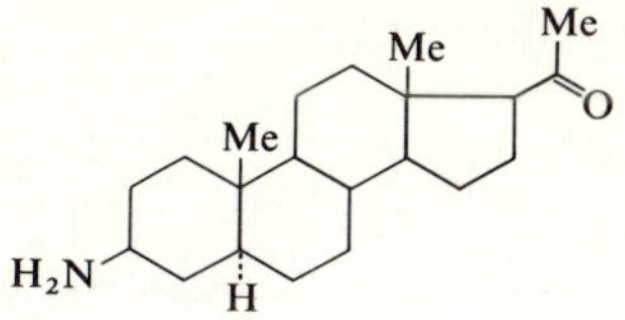

Dihydroholaphyllamine

Holaphyllamine = Δ^5-double bond

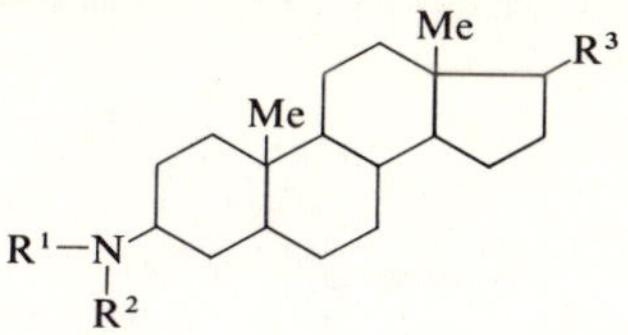

R^1 = H, R^2 = Me, R^3 = C(OH)$_2$Me
= holaphylline
R^1 = H, R^2 = Me, R^3 = CHOHMe
= holaphyllidine
R^1 = Me, R^2 = Me, R^3 = COMe
= methylholaphylline

In addition, progesterone is found in this plant. This fact leads to biogenetic considerations. On the basis of radioactive studies it was concluded that pregnenolone is a precursor of these alkaloids.

7.2.3.2 *Funtumia*

Funtumia alkaloids are of the same structural type as Holarrhena alkaloids. Studies during recent years were performed by Goutarel *et al.* Besides other alkaloids which are listed in Table 7.5 and which are of less general interest, three papers should be mentioned.

Funtudienine (31)[86] was isolated from the bark of *Funtumia latifolia*:

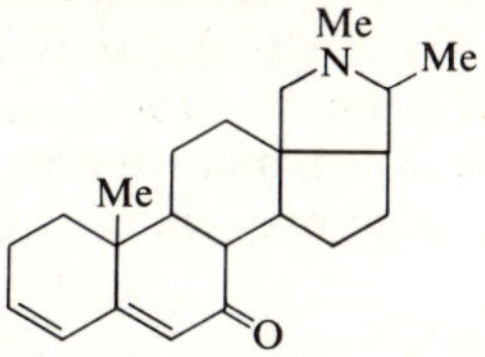

(31)

The structure is remarkable, as no oxygen or nitrogen function is found at C-3, but quite unusually at C-7. Evidence for the structure came from spectroscopic data and from conversion into known compounds.

Also from the bark, Funtessine (32), 3β-amino-5α-conan-12β-ol, was isolated[87]. The structure was confirmed by partial synthesis from dihydro-holarrhenine as follows:

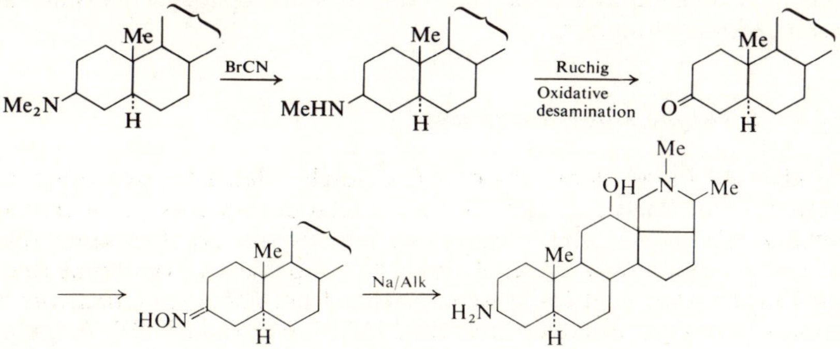

BrCN

Ruchig
Oxidative
desamination

Na/Alk

(33)   (32)

Irehline (33) is again an alkaloid with a pyrroline ring[88]

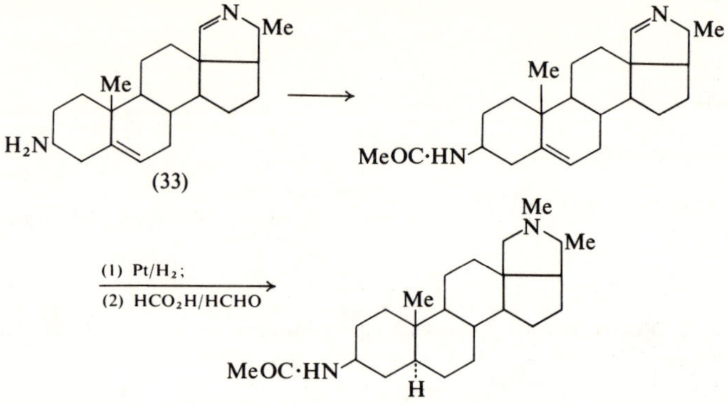

On hydrogenation of *N*-acetylirehline and subsequent methylation malou-
phyllamine (34) is formed, an alkaloid of *Maluetia baequertiana*[89].

7.2.3.3 *Maluetia*

Early work on Maluetia alkaloids is reported in a review of Goutarel and
co-workers[94]. This review mainly deals with *Maluetia bequaertiana* alkaloids.
One more paper of recent origin is on *M. arborea* and *M. tamaquariana*[95].
From *Maluetia arboreana* two alkaloids, malarborine (35) and malarboreine
(36) have been isolated and their structures elucidated.

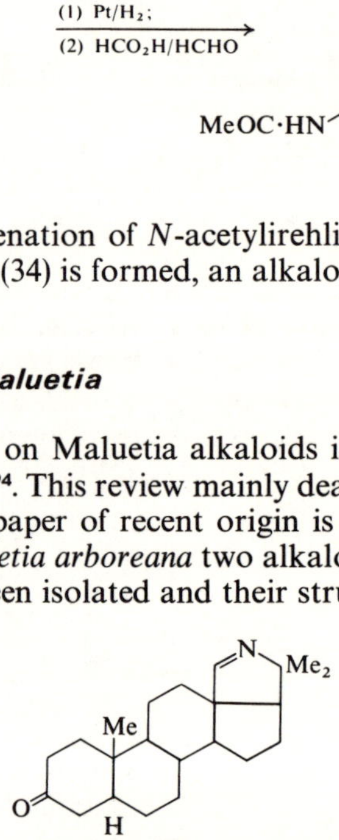

From *Maluetia tamaquariana* four alkaloids, well known from other Apocynaceae have been found: kurchessine[96], dihydrokurchessine[97], conessine[98] and
tetramethylholarrhimine[99].

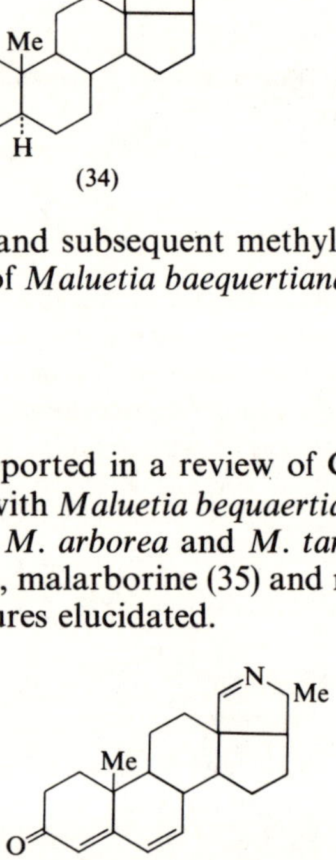

7.2.3.4 *Kibatalia and Paravallaris*

Kibatalia and Paravallaris alkaloids are closely related to each other, and
the same is true for the species. This close relationship goes so far that, e.g.
Kibatalia microphylla and *Paravallaris microphylla* are the same plant.
With one exception only, these alkaloids do not possess a pyrrolidine ring E,
but a 18-carboxylic acid 18-20 lactone. Thus from *Kibatalia gitingensis* five
alkaloids have been isolated: kibataline (37)[100], paravallarine[101], *N*-methyl-
paravallarine[101], 20-epiparavallarine[101], and 20-epi-*N*-methyl-paravallarine[102].

A sixth alkaloid, gitingensine is reported but without structure[103]. From *Paravallaris maingayi*, maingayine (38)[106] was isolated; it possesses a pyrroline ring E, and is identical with 11-desoxy-holonamine[107]:

(37)   (38)

7.2.3.5   *Biosynthesis of Holarrhena alkaloids*

Special interest was drawn towards the biosynthesis of holarrhena alkaloids. The problem of possible precursors was studied by Bennett and Heftmann[85]. They found that radioactive pregnenolone, but not progesterone was converted to holaphylline and holaphyllamine by leaves of *H. floribunda*. The amination involved direct replacement of OH by NH_2 rather than biosynthetic transamination of the ketone. This replacement reaction is analogous to the amination of sugars and hydroxypurines. Radioactive labelled progesterone was converted to alkaloids of unknown structure.

7.2.4   Alkaloids of Solanaceae

7.2.4.1   *Solanum alkaloids*

Among steroidal alkaloids the Solanum group is the best known and most investigated. This is due not only to the large number of Solanaceae plants but also to economic reasons, as Solanum alkaloids may still become competitive with diosgenine as sources of steroids.

Isolation and separation cause no difficulties as chromatography is an easy and valuable tool. Analytical work on a thin layer scale has been published recently[108].

Earlier work on Solanum alkaloids is published in several reviews[28, 109–112]. Two structure types are mainly observed, alkaloids of the solanidine type, and those of the solasodine or spirosolane type[113].

Solanidine-type   Spirosolane-type

(a) *Solanum dulcamara* — Recent work on the constituents of this plant was published by Schreiber and his group. They found a number of spirosolane alkaloids as glycosides. Hydrolysis yielded the alkaloids themselves.

Table 7.5 Outline of alkaloids from Apocynaceae

Alkaloids	Formula	m.p. °C	D (degrees)	Salts	Occurrence	Remarks
7-Hydroxyconessine	$C_{24}H_{40}N_2O$	178	− 61		*Holarrhena antidysenterica*[73]	
Holonamine	$C_{21}H_{27}NO_2$	259	− 14.8		*H. antidysenterica*[73]	
Kurchiphyllamine	$C_{22}H_{35}NO_2$	161	−211		*H. antidysenterica*[74, 75]	
Kurchiphylline	$C_{23}H_{37}NO_2$	184	+173		*H. antidysenterica*[74, 75]	
Kurchiline	$C_{23}H_{37}NO_2$	219	+ 46		*H. antidysenterica*[74]	
Kurchaline	$C_{23}H_{37}NO_2$	185	− 37		*H. antidysenterica*[75]	
Holadysine	$C_{22}H_{35}NO$	120	−199		*H. antidysenterica*[75]	
Holadysamine	$C_{22}H_{35}NO$	173	− 78		*H. antidysenterica*[75]	
Holantosin A	$C_{30}H_{49}NO_7$	261	− 28		*H. antidysenterica*[78]	Those data refer to the *N*-acetyl-compounds
Holantosin B	$C_{30}H_{47}NO_6$	290	− 29		*H. antidysenterica*[78]	
Holantosin C	$C_{30}H_{49}NO_7$	245	− 73		*H. antidysenterica*[79]	
Holantosin D	$C_{30}H_{47}NO_6$	267	− 74		*H. antidysenterica*[79]	
Holarosin A	$C_{33}H_{49}NO_7$	270	− 39		*H. antidysenterica*[79]	
Funtudienine	$C_{22}H_{31}NO$	amorph.	−247	oxime:190	*Funtumia latifolia*[86]	
Latifolinine	$C_{22}H_{33}NO$	amorph.	+190	oxime:235	*F. latifolia*[90]	
Norlatifoline	$C_{21}H_{33}NO$	192	− 26		*F. latifolia*[90]	
Funtessine	$C_{22}H_{38}N_2O$	195	+ 49		*F. latifolia*[87]	
Irehline	$C_{21}H_{32}N_2$	118	− 48		*F. latifolia*[88]	
Irehine	$C_{23}H_{39}NO$	174	− 46		*F. latifolia*[91]	
Irehamine	$C_{22}H_{37}NO$	230	− 33		*F. latifolia*[91]	
Hydroxyconessine	$C_{24}H_{40}N_2O$	157	+ 20		*F. latifolia*[92]	

Irehdiamine A	$C_{21}H_{36}N_2$	148	− 47	*Funtumia elastica*[93]
Irehdiamine B	$C_{22}H_{38}N_2$	117	− 56	*F. elastica*[93]
Malouphyllamine	$C_{24}H_{40}N_2O$	219	+ 38	*Malouetia bequaertiana*[89]
Malarborine	$C_{21}H_{31}NO$	165		*M. arborea*[95]
Malarborcine	$C_{21}H_{27}NO$	176	+ 57	*M. arborea*[95]
Kurchessine	$C_{25}H_{44}N_2$	139	− 40	*M. tamaquariana*[96]
Dihydrokurchessine	$C_{25}H_{46}N_2$			*M. tamaquariana*[97]
Conessine	$C_{24}H_{40}N_2$	124	− 2	*M. tamaquariana*[98]
Tetramethylholarrhimine	$C_{25}H_{44}N_2O$	233	− 35	*M. tamaquariana*[99]
Kibataline	$C_{23}H_{35}NO_2$	171	− 42	*Kibatalia gitingensis*[100]
Paravallarine	$C_{22}H_{33}NO_2$			*K. gitingensis, K. microphylla* *P. microphylla*[104]
N-Methyl-paravallarine	$C_{23}H_{35}NO_2$			*K. gitingensis*[101]
20-epi-*N*-methyl-paravallarine	$C_{23}H_{35}NO_2$	191	− 33	*K. gitingensis*[102]
20-epi-paravallarine	$C_{22}H_{33}NO_2$			*K. gitingensis*[101]
Paravallaridine	$C_{22}H_{33}NO_3$			*K. microphylla = P. microphylla*[105]
Gitingensine		161	− 65	*K. gitingensis*[103]
Maingayin	$C_{21}H_{27}NO$	152	+ 2.1	*Paravallaris maingavi*[106]
Holalin	$C_{24}H_{42}N_2O$	267	+ 36	*Holarrhena floribunda*[81a]
Holadienine	$C_{22}H_{33}NO$	109	80	*H. floribunda*[82]
Holaromine	$C_{22}H_{31}N$	198	90	*H. floribunda*[82]
Dihydroholaphyllamine	$C_{21}H_{35}NO$	300	79	*H. floribunda*[84]
Holaphyllidine	$C_{22}H_{37}NO$	—	—	*H. floribunda*[84]
Methylholaphylline	$C_{23}H_{35}NO$	121	29	*H. floribunda*[84]
Holacurtine	$C_{29}H_{49}NO_5$	162	42	*H. floribunda*[84]
Holacurtenine	$C_{29}H_{47}NO_4$	137	57	*H. floribunda*[84]

Tomatid-5-en-3β-ol[114] yields tomatidine (from *Lycopersicon pimpinelli-folium*)[110] on hydrogenation. Under acidic conditions the *O*, *N*-diacetyl derivative can be converted to 3β-acetoxy-pregna-5,16-dien-20-one[114].

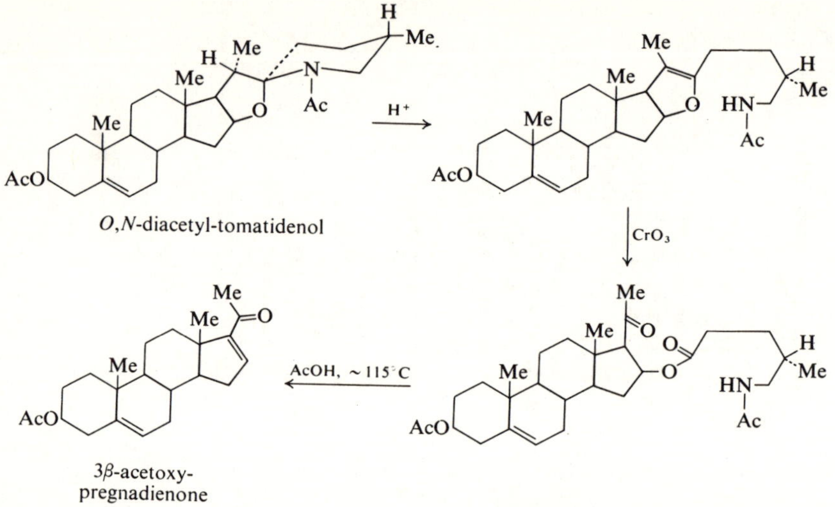

O,N-diacetyl-tomatidenol

CrO₃

3β-acetoxy-
pregnadienone

AcOH, ~115°C

A synthesis was performed[115] starting from (22S:25S)-22,26-imino-5α-cholestan-3β,16β-diol. Tomatidenol possesses interest as the aglycone of numerous glycosides which differ only by their sugar part. These glycosides are named α, β, γ_1, γ_2 and δ-solamarine[116].

Other alkaloids of the same type, isolated from *Solanum dulcamara*, are 15-hydroxy-soladulcidine, and 15α-hydroxy-tomatidenol[117].

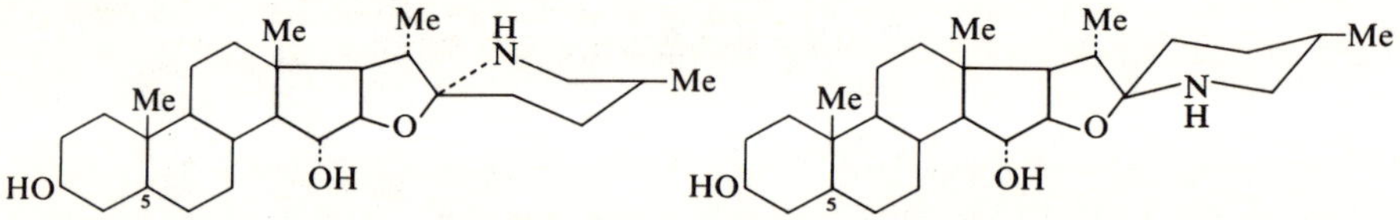

5α = 15α-Hydroxy-soladulcidine
Δ^5 = 15α-Hydroxy-solasodine

5α = 15α-Hydroxy-tomatidine
Δ^5 = 15α-Hydroxy-tomatidenol

A hydroxy group at C-15 is quite unusual in naturally occurring steroids and these four alkaloids are the first compounds with a 15α-hydroxy group. This group causes quite an unexpected hydrogenation in so far as from 15α-hydroxy-solasodine upon opening of ring E the 22,26-imino-5α-cholestane-3β,15α-16β-triole is formed, and the same is true for 15α-hydroxy-tomatidenol (39).

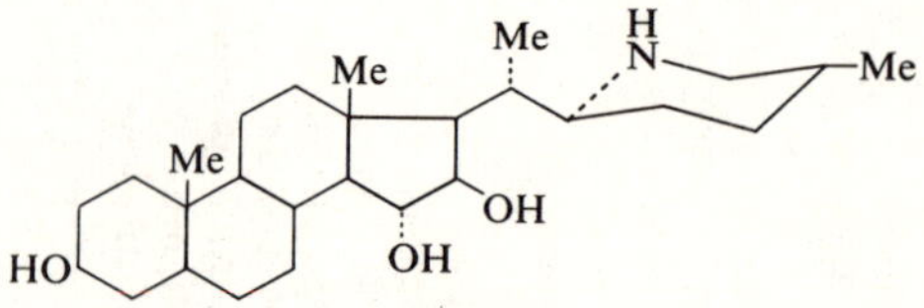

(39)

15α-Hydroxy-tomatidine and 15-hydroxy-soladulcidine can be obtained via
the *N*-chloro compound.

   (b) *Solanum congestiflorum* — From *S. congestiflorum* solanocapsine (41)
was isolated. The structure was elucidated by spectroscopic data[118] and
x-ray crystal-structure analysis[119]. Two independent syntheses, both starting
from solafloridine (40) have been published[121, 122]. The synthesis of the
Schreiber group is given below:

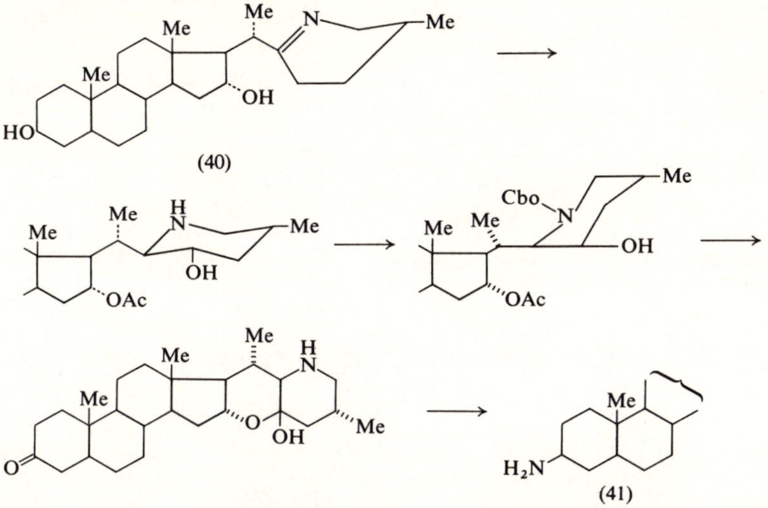

Two more alkaloids, solafloridine and solacongestidine (42), were isolated
by Sato *et al.*[120]. A conversion of solasodine to both alkaloids has been
published by the same group[123]. Further alkaloids of this plant are 23-oxo-
and 24-oxo-solacongestidine[120]. The structures of these four alkaloids were
arrived at from spectral data combined with chemical investigations:

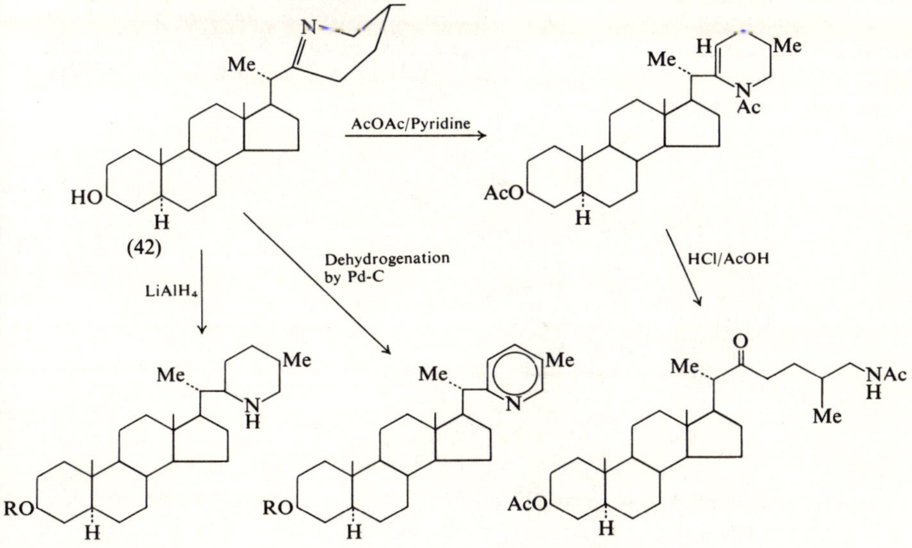

(c) *Solanum chacoense* — A number of glyco-alkaloids have been isolated from this plant and this explains the resistance of *S. chacoense* against the potato beetle, *Leptinotarsa decemlineata*. The aglycone of these substances is leptinidine (43). By chemical investigations[124, 125] as well as by i.r. and n.m.r. studies[126] its constitution and configuration has been established as follows:

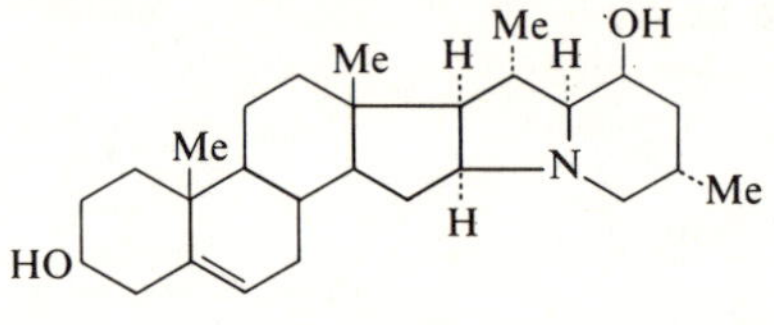

(43)

(d) *Solanum hypomalacophyllum* — This plant of the Venezuelan Andes contains solaphyllidine (44)[127] Its structure was established by x-ray crystal-structure analysis to be:

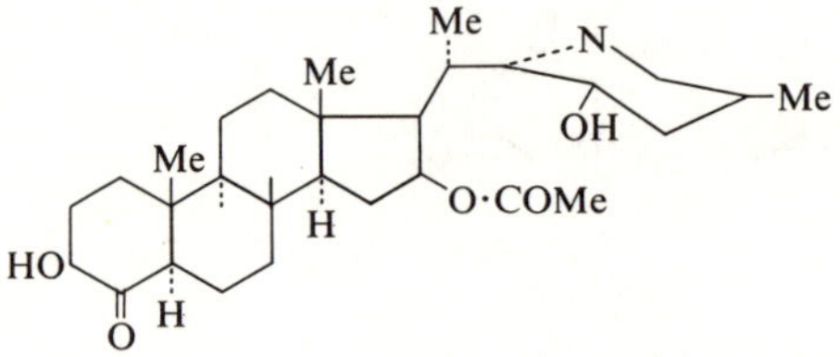

(44)

This structure represents a separate type of Solanum alkaloid, which might be a biogenetic link between the cholesterol type precursor and the commonly occurring Solanum alkaloids of the solanidine or spirosolane type.

(e) *Solanum paniculatum* — One alkaloid, jurubidine (45) could be obtained from the roots but not from the leaves of this plant[128]. Its structure was established by spectroscopic data and by desamination with HNO_2 which yielded neotigogenine (46).

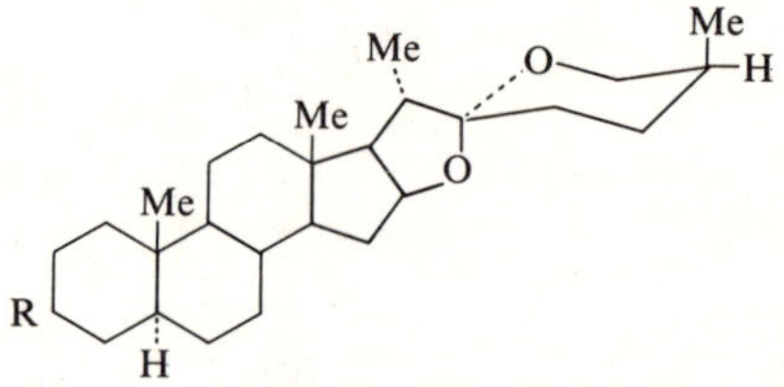

(45) R = NH₂
(46) R = OH

Thus it must be represented as (22*R*:25*S*)-3β-amino-5β-spirostane.

It is interesting that the glucose molecule in the glycoside jurubine[129] is attached to C-26, and that ring F is open. Cyclisation takes place after hydrolysis:

$$\xrightarrow[\text{or emulsine}]{H^+}$$

$$\left(\right) \xrightarrow[-H_2O]{} \quad (45)$$

Besides this alkaloid a number of closely-related steroidal saponines like neotigogenine are present in the same plant.

(f) *Solanum torvum* — There is no difference in the constituents of the plant between *S. torvum* and *S. paniculatum*[130].

(g) *Solanum armatum* — Two alkaloids, the well known solasodine[132] and solamargine (47)[133] were isolated from the fruit of the Australian Solanaceae *S. armatum*[131].

L-Rhamnose-L-Rh-D-GluO

(47)

(h) *Other Solanum spp.* — During a screening programme of Colombian Solanaceae 60 spp. were investigated chromatographically[134]. Alkaloids were found in 21 of these species; the highest amounts in *S. trachycarpum*, *S. marginatum*, *S. jubatum*, *S. mammosum*, *S. inopinium*, *S. atropurpurum*, *S. umbellatum*. No details have, however, been given.

7.2.4.2 *Lycopersicon pimpinellifolium*

Besides tomatidine, soladulcidine, tomatidadienine, a steroidal alkaloid $C_{27}H_{45}NO_3$ of unknown structure, and what is most surprising, 3β-hydroxy-5α-pregn-16-en-20-one were isolated. This is the first time a pregnane derivative has been found in a plant[135].

7.2.4.3 *Biogenesis of Solanum alkaloids*

From experiments with radioactive labelled compounds it was found[136] that acetate and mevalonate are incorporated. Thus in Solanaceae the steroidal skeleton seems to be biosynthesised by the same pathway as in other plants

Table 7.6   Alkaloids from *Solanaceae*

Alkaloid	Formula	m.p. °C	D (degrees)	Occurrence	Remarks
Tomatid-5-en-3β-ol	$C_{27}H_{43}NO_2$	238	$-$ 39	*Solanum dulcamara*[114]	
15-Hydroxy-soladulcidine	$C_{27}H_{45}NO_3$	212	$-$ 38	*S. dulcamara*[117]	
15-Hydroxy-solasodine	$C_{27}H_{43}NO_3$	216	$-$ 84	*S. dulcamara*[117]	
15-Hydroxy-tomatidine	$C_{27}H_{45}NO_3$	193	$+$ 20	*S. dulcamara*[117]	
15-Hydroxy-tomatidenol	$C_{27}H_{43}NO_3$	240	$-$ 24	*S. dulcamara*[117]	
Solanocapsine	$C_{27}H_{46}N_2O_2$	211	$+$ 28	*S. congestiflorum*[118]	
Solafloridine	$C_{27}H_{45}NO_2$	175	$+$123	*S. congestiflorum*[120]	
Solacongestidine	$C_{27}H_{45}NO$	174	$+$ 36	*S. congestiflorum*[120]	
23-Oxo-solacongestidine	$C_{27}H_{43}NO_2$	213–223	$+$ 33	*S. congestiflorum*[120]	
24-Oxo-solacongestidine	$C_{27}H_{43}NO_2$	162	$+$ 41	*S. congestiflorum*[120]	
Solaphyllidine	$C_{29}H_{47}NO_5$	170	$-$ 25	*S. hypomalacophyllum*[127]	
(22*R*:25*S*)-3β-Amino-5 spironstane = Jurubidine	$C_{27}H_{45}NO_2$	186	$-$ 79	*S. paniculatum*[128]	
Solamargine	$C_{45}H_{33}NO_{15}$	310	$-$112	*S. armatum*[131]	
Tomatidine	$C_{27}H_{45}NO_2$	207	$+$ 6	*Lycopersicon pimpinellifolium*[135]	
Soladulcidine	$C_{27}H_{45}NO_2$	206	$-$ 53	*L. pimpinellifolium*[135]	
Tomatidadiene	$C_{27}H_{41}NO$	139	$-$ 97	*L. pimpinellifolium*[135]	
Alkaloid	$C_{27}H_{45}NO_3$	173	$+$ 14	*L. pimpinellifolium*[135]	

and in animals. From other investigations it is known[137] that the alkaloid composition in *S. dulcamara* varies during the growth period. The amounts of tomatidenol, soladulcidine and solasodine were observed and determined in young, green, nearly mature, and in mature fruits. During fruit growth the relative amount of solasodine increases, but during maturation it is metabolised at a higher rate than the other alkaloids, which finally predominate in the mature fruits in which the total alkaloid content is very low.

7.2.4.4 *Pharmacologic activity*

The action of solamargine and solasonine was tested against some strains of dermatophytes[138]. Fungistatic activity was found in the following pathogenes: *Trichophyton mentagrophytes*; *Mierosporum gypseum*; *Mierosporum canis*; *Epidermophytum floccosum*; *Candida albicans*.

Guinea-pigs and rabbits were infected with the above fungi and then treated with 0.5% of glyco-alkaloid in a Vaseline–lanoline mixture (10 : 1).

In all of these cases positive results were found, so that Solanum alkaloids might have some medical value.

7.2.4.5 *Relations between structure and antibiotic activity of steroidal alkaloids*

The inhibiting action of the steroidal alkaloids soladulcidine, soladulcine, solasodine, solasonine, solamargine, Δ^5-tomatidenol, tomatine and tomatidenine against fungi was tested. All of them show an antibiotic activity against *Claviceps*, *Sclerotinia*, *Piricularia*, *Rhizoctonia* and *Polyporus*. Unaffected are *Botrytis*, *Coniophora*, *Fomes* and *Fusarium conglutinans* and *F. oxysprorum*, while *F. bulbigenum*, *Aspergillus clavatus* and *Alternaria solani* were affected by the Δ^5-unsaturated spirosolanoles only, and *Trichothecium* by the saturated ones only.

Alkaloids of the tomatidine series (spirosolanes) displayed a stronger growth inhibition than those of solanidine type. The activity of the aglyca is the same as that of the glycosides, so that the sugar component is without any importance.

7.2.5 Spectroscopic investigations in steroidal alkaloids

7.2.5.1 *N.M.R. studies*

Intensive n.m.r. studies were performed by several groups[140-142]; these studies led to information about the configuration at C-22. From these investigations the configurations at C-22 of the following compounds have been established: tomatidine, tomatiden-3β-ol, solasodine, 5α-solasodan-3β-ol together with diosgenine, sarsasapogenine, cyclopseudodiosgenine and cyclopseudosarsasapogenine. Moreover, a comparison of the signals at

C-3, C-6, C-16, C-18, C-19, C-21, C-26 and C-27 is given in a table[140]. By this means a quick reference between spectra and compounds is possible.

7.2.5.2   *Mass spectra studies*

Mass spectra of several classes of steroidal alkaloids have been studied by Budzikiewicz[143]. Due to the specific fragmentation patterns according to the site of the nitrogen atoms rather clear fragmentations are observed, and with this spectra and compounds can be correlated. Thus 3-dimethyl-amino-compounds show peaks at $m/e = 84$ and $m/e = 110$

$$m/e = 84 \qquad m/e = 110$$

20-amino-pregnanes yield a characteristic fragment of $m/e = 72$, conanine derivatives one of $m/e = 71$.

$$\text{CH}=\text{NMe}_2 \quad m/e = 72$$

$$+ \ \text{CH}_2{-}\text{N}{=}\text{CH}{-}\text{Me} \quad m/e = 71$$

Solanidine-type alkaloids undergo a multi-step cleavage resulting in two fragments:

$$m/e = 150 \qquad m/e = 204$$

Solanocapsine gives rise to three characteristic fragments:

the fragment 114 possessing the following structure:

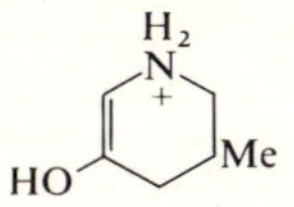

Spirosolane alkaloids also show a fragment of *m/e* 114, but in addition another one of *m/e* = 138:

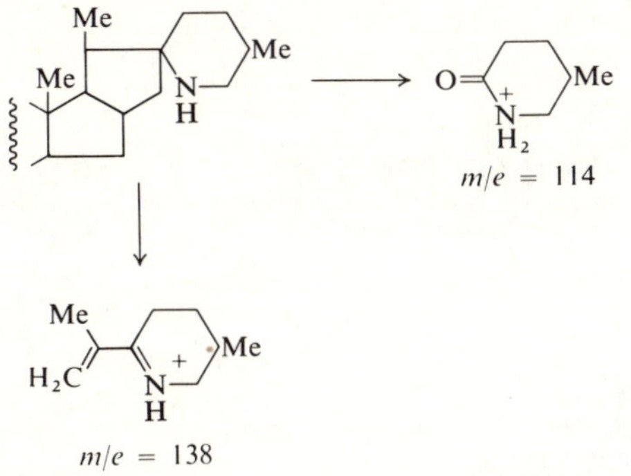

$$m/e = 114$$

$$m/e = 138$$

7.2.5.3 *Absolute configuration of solanidanes*

The absolute configuration of all solanidane alkaloids, i.e. demissidine, solanidine, leptinidine, rubijervine, isorubijervine, and veralobine was determined by an x-ray crystal-structure analysis of demissidine hydroiodide. They all possess (22R:NS) configuration, whereas the so-called 22-iso-solanidanes have the (22S:NS) configuration[144].

7.2.5.4 *Synthetic work in steroidal alkaloids*

Much work has been invested into degradation[145] and synthesis of these alkaloids. Reviews on the total synthesis of steroidal alkaloids have been published[146, 147]. In these syntheses 3β, 16β-diacetoxy-5α-pregnan-20-one underwent a Grignard type reaction with 2-lithium-5-methyl-pyridine. Thus the C_{27}-skeleton of spirosolane alkaloids could be built in one step. The further steps can be seen from the formula scheme given in Figure 7.5. Cyclisation of ring E in these syntheses can be achieved by Hofmann–Löffler–Freytag-reaction of the N-chloro-compound. Another cyclisation reaction is the photolysis of N-nitroso-22,26-iminocholestones[148]:

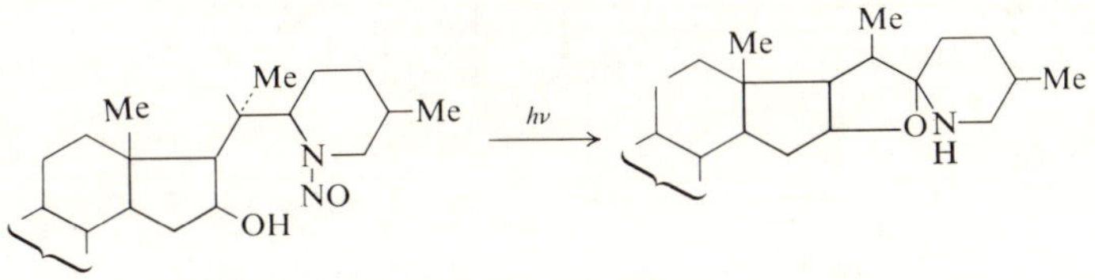

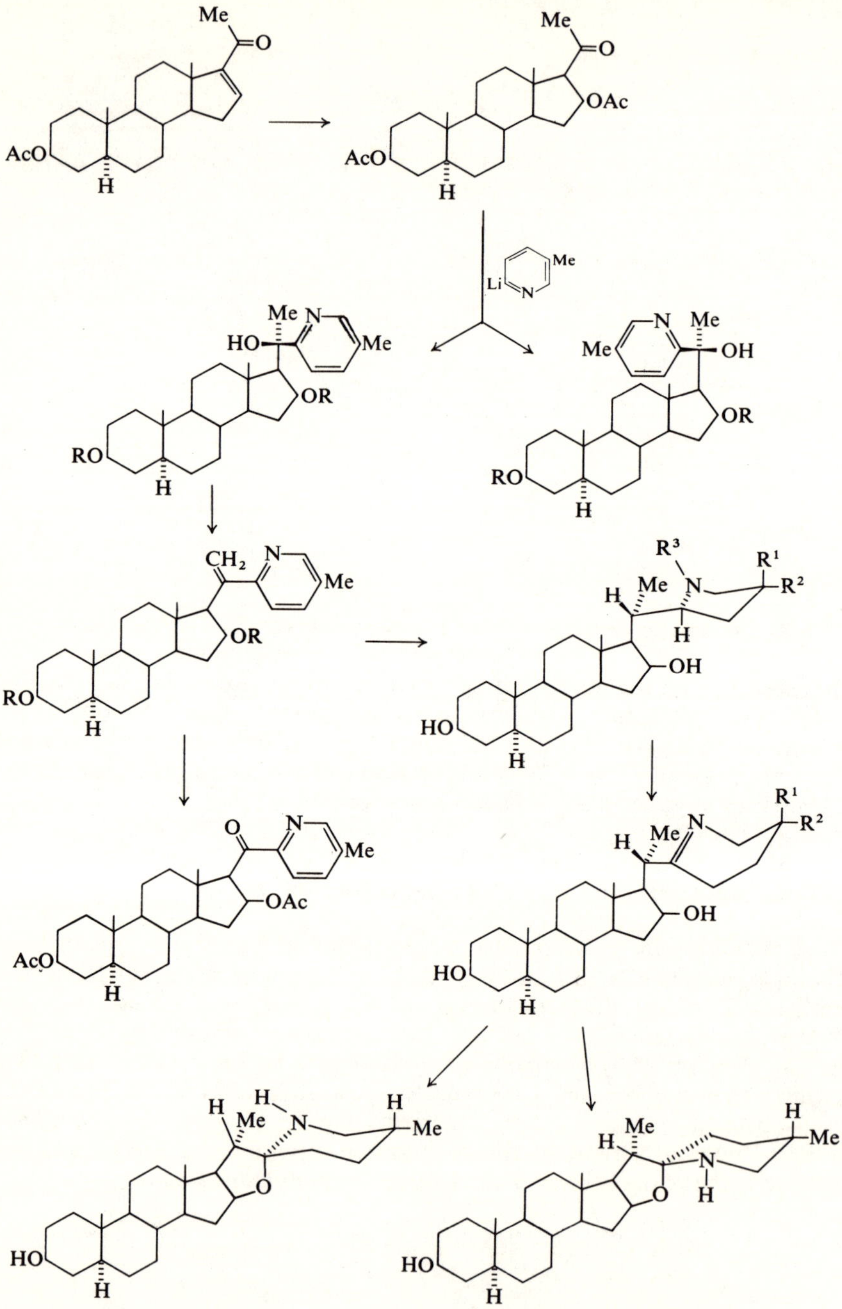

Figure 7.5 Synthesis of spirosolane alkaloids

In this way, solasodine and tomatidine were prepared. Another type of synthesis for solanidine has been published quite recently by Kessar *et al.*[149]. The ring closure of rings E and F in one step is performed by a reduction with Zn/AcOH.

The resulting lactame can be reduced to the amine by lithium aluminium hydride. Tomatid-5-ene-3β-ol and solasodine were prepared[150] by similar selective transformations.

References

1. Habermehl, G. and Preusser, H. J. (1969). *Z. Naturforsch.*, **24b**, 1599
2. Csordás, A. and Michl, H. (1969). *Toxicon*, **7**, 103
3. Habermehl, G. and Preusser, H. J. (1970). *Z. Naturforsch.*, **25b**, 1451
4. Habermehl, G., Preusser, H. J. and Haury, D. (1973). *Toxicon* (in press)
5. Schöpf, C. and Braun, W. (1934). *Liebigs Ann. Chem.*, **514**, 69
6. Habermehl, G. (1964). *Liebigs Ann. Chem.*, **680**, 104
7. Schöpf, C. (1961). *Experientia*, **17**, 285
8. Habermehl, G. (1971). In *'Venomous Animals and Their Venoms'*. (New York: Academic Press)
9. Wölfel, E., Schöpf, C., Weitz, G. and Habermehl, G. (1961). *Chem. Ber.*, **94**, 2361
10. Schöpf, C. and Koch, K. (1942). *Liebigs Ann. Chem.*, **552**, 37
11. Habermehl, G. (1963). *Chem. Ber.*, **96**, 143
12. Habermehl, G. (1963). *Chem. Ber.*, **96**, 840
13. Habermehl, G. (1963). *Chem. Ber.*, **96**, 2029
14. Habermehl, G. (1966). *Chem. Ber.*, **99**, 1439
15. Habermehl, G. and Vogel, G. (1969). *Toxicon*, **7**, 163
16. Habermehl, G. and Haaf, A. (1969). *Liebigs Ann. Chem.*, **772**, 155
17. Habermehl, G. and Haaf, A. (1968). *Z. Naturforsch.*, **23b**, 1551
18. Göttlicher, G. and Habermehl, G. (1973). *Acta Crystallogr.* (in press)
19. Habermehl, G. and Haaf, G. (1965). *Chem. Ber.*, **98**, 3001
20. Hara, S. and Oka, K. (1966). *J. Amer. Chem. Soc.*, **89**, 1041
21. Habermehl, G. and Haaf, A. (1968). *Chem. Ber.*, **101**, 198
22. Tokuyama, T., Daly, J. and Witkop, B. (1969). *J. Amer. Chem. Soc.*, **91**, 3931
23. Märki, F. and Witkop, B. (1963). *Experientia*, **19**, 329
24. Daly, J. W., Witkop, B., Bommer, P. and Biemann, K. (1965). *J. Amer. Chem. Soc.*, **87**, 124
25. Tokuyama, T., Daly, J., Witkop, B., Karle, J. L. and Karle, J. (1968). *J. Amer. Chem. Soc.*, **90**, 1917
26. Karle, J. L. and Karle, J. (1969). *Acta Crystallogr.*, **B25**, 428
27. Albuquerque, E. X., Daly, J. and Witkop, B. (1971). *Science*, **172**, 995
28. Boit, H. G. (1961). *Ergebnisse der Alkaloid-Chemie bis 1960*. (Berlin: Akademie-Verlag)
29. Schreiber, K. (1966). Über die Biochemie der Steroidalkaloide, p. 65, *Abh. dtsch. Akad. Wiss. Berlin*. (Chemie, Geologie, Biologie)
30. Adam, G., Schreiber, K., Tomko, J. and Vassová, A. (1967). *Tetrahedron*, **23**, 167
31. Adam, G., Schreiber, K. and Tomko, J. (1967). *Liebigs Ann. Chem.*, **707**, 203
32. Schreiber, K. and Adam, G. (1964). *Tetrahedron*, **20**, 1707
33. Schreiber, K. and Adam, G. (1964). *Chem. Ber.*, **97**, 2358
34. Bianchi, E., Djerassi, C., Budzikiewicz, H. and Sato, Y. (1965). *J. Org. Chem.*, **30**, 754
35. Schreiber, K. and Ripperger, H. (1962). *Liebigs Ann. Chem.*, **655**, 114
36. Adam, G., Schreiber, K. and Tomko, J. (1967). *Liebigs Ann. Chem.*, **707**, 203

37. Tomko, J., Vassová, A., Adam, G., Schreiber, K. and Höhne, E. (1967). *Tetrahedron Lett.*, 3907

38. Tomko, J., Vassová, A., Adam, G. and Schreiber, K. (1968). *Tetrahedron*, **24**, 4865

39. Höhne, E., Adam, G., Schreiber, K. and Tomko, J. (1968). *Tetrahedron*, **24**, 4875

40. Tomko, J., Brázdová, V. and Votický, Z. (1971). *Tetrahedron Lett.*, 3041

41. Tomko, J., Votický, Z., Vassová, A., Adam, G. and Schreiber, K. (1968). *Collect. Czech. Chem. Commun.*, **33**, 4054

42. Schreiber, K. (1968). 'The Alkaloids—Chemistry and Physiology', Vol. X, p.31. (New York: Academic Press)

43. Sato, Y. and Latham, H. G., Jnr. (1956). *J. Amer. Chem. Soc.*, **78**, 3146

44. Briggs, L. H. and Locker, R. H. (1950). *J. Chem. Soc.*, 3020

45. Sato, Y. and Ikekawa, N. (1961). *J. Org. Chem.*, **26**, 1945

46. Schreiber, K. and Adam, G. (1961). *Experientia*, **17**, 13

47. Schreiber, K. and Adam, G. (1963). *Liebigs Ann. Chem.*, **666**, 155

48. Tomko, J., Brázdová, V. and Votický, Z. (1971). *Tetrahedron Lett.*, 3041

49. Tomko, J. and Vassová, A. (1970). *Chemické Zvesti*, **25**, 69

50. Hegnauer, R. (1964). Chemotaxonomie der Pflanzen, Vol. 3. (Basel: Birkhäuser Verlag)

51. Herlem-Gaulier, D., Khoung-Hun-Laine, F., Stanislas, E., Gontard, R. and Magdeleine, M. L. (1965). *Bull. Soc. Chim. France*, 657

52. Herlem-Gaulier, D. and Goutarel, R. (1965). *Compt. Rend.*, **261**, 4139

53. Döpke, W. and Müller, B. (1969). *Pharmazie*, **24**, 649

54. Döpke, W., Müller, B. and Jeffs, P. W. (1967). *Naturwissenschaften*, **54**, 249

55. Döpke, W. and Müller, B. (1966). *Pharmazie*, **21**, 769

56. Votický, Z. and Tomko, J. (1966). *Abh. Dtsch. Akad. Wiss. Berlin, Kl. Chem. Geol. Biol.*, 93

57. Tomko, J., Bauerova, O., Votický, Z., Goutarel, R. and Longevialle, P. (1966). *Tetrahedron Lett.*, 915

58. Stauffacher, D. (1964). *Helv. Chim. Acta*, **47**, 968

59. Nakano, T. and Terao, S. (1965). *J. Chem. Soc.*, 4512

60. Nakano, T., Terao, S. and Saeki, Y. (1966). *J. Chem. Soc.*, 1412

61. Nakano, T. and Votický, Z. (1970). *J. Chem. Soc.*, 590

62. Nakano, T., Alonso, M. and Martin, A. (1970). *Tetrahedron Lett.*, 4929

63. Voser, W., Jeger, O. and Ruzicka, L. (1952). *Helv. Chim. Acta*, **35**, 503

64. Kiamuddin, M. and Haque, M. E. (1966). *Pakistan J. Sci. Ind. Res.*, **9**, 103

65. Kohli, J. M., Zaman, A. and Kidwai, A. R. (1967). *Tetrahedron*, **23**, 3829

66. Chatterjee, A., Mukherjee, K. S. and Dutta, C. P. (1966). *J. Indian Chem. Soc.*, **43**, 285

67. Chatterjee, A., Das, B., Dutta, C. P. and Mukherjee, K. S. (1965). *Tetrahedron Lett.* 67

68. Chatterjee, A. and Mukherjee, K. S. (1966). *Chem. Ind.*, 769

69. Döpke, W., Härtel, R. and Fehlhaber, H. W. (1969). *Tetrahedron Lett.*, 4423

70. Černý, V. and Šorm, F. (1967). *Alkaloids*, 305. (New York: Academic Press)

71. Lábler, L. and Černý, V. (1963). *Collect. Czech. Chem. Commun.*, **28**, 2932

72. Dolejš, L., Hanuš, V., Černý, V. and Šorm, F. (1963). *Collect Czech. Chem. Commun.*, **28**, 1548

73. Tscheche, R. and Ockenfeld, H. (1964). *Chem. Ber.*, **97**, 2316

74. Janot, M. M., Longevialle, P. and Goutarel, R. (1964). *Bull. Soc. Chim. France*, 2158

75. Janot, M. M., Longevialle, P. and Goutarel, R. (1966). *Bull. Soc. Chim. France*, 1212

76. Brande, E. A. (1950). *Quart. Rev. Chem. Soc.*, 404

77. Green, M. B. and Hickinbottom, W. J. (1957). *J. Chem. Soc.*, 3262, 3270

78. Janot, M. M., Qui-Khuong-Huu, Monneret, C., Kaboré, J., Hildesheim, J., Gero, S. D. and Goutarel, R. (1970). *Tetrahedron*, **26**, 1695

79. Qui-Khuong-Hun, Monneret, C., Kaboré, J., Choay, P., Tekam, J. M. and Goutarel, R. (1971). *Bull. Soc. Chim. France*, 864

80. Schaffner, B. P., Berner-Fenz, L. and Wehrli, H. (1970). *Helv. Chim. Acta*, **53**, 2266

81a. Janot, M. M., Devissagouet, P., Qui-Khuong-Huu and Goutarel, R. (1967). *Bull. Soc. Chim. France*, 4315

81b. Janot, M. M., Devissagouet, P., Qui-Khuong-Huu and Goutarel, R. (1965). *Compt. Rend. Acad. Sci. Paris*, **260**, *(8)*, 6638

81c. Janot, M. M., Devissagouet, P., Qui-Khuong-Huu and Goutarel, R. (1967). *Arch. Pharm. France*, **25**, 733

82. Inhoffen, H. H. (1964. *Molecular Rearrangements*, (P. de Mayo, editor). (New York: John Wiley Interscience)
83. Leboeuf, M., Cave, A. and Goutarel, R. (1969). *Ann. Pharm. France*, **27**, 217
84. Janot, M. M., Devissagouet, P., Qui-Khuong-Huu, Paello, J., Bisset, N. G. and Goutarel, R. (1968). *Compt. Rend. Acad. Sci. Paris, Ser. C*, **266**, 388
85. Bennett, R. D. and Heftmann, E. (1965). *Phytochemistry*, **4**, 873
86. Qui-Khuong-Huu, Monneret, C., Yassi, J. and Goutarel, R. (1964). *Bull. Soc. Chim. France*, 2169
87. Qui-Khuong-Huu, Yassi, J., Monneret, C. and Goutarel, R. (1965). *Bull. Soc. Chim. France*, 1831
88. Janot, M. M., Ming Truong-Ho, Qui-Khuong-Huu and Goutarel, R. (1963). *Bull. Soc. Chim. France*, 1977
89. Janot, M. M., Khuong-Huu-Laine, F., Magdeleine, M. J. and Goutarel, R. (1963). *Bull. Soc. Chim. France*, 641
90. Qui-Khuong-Huu, Yassi, J. and Goutarel, R. (1963). *Bull. Soc. Chim. France*, p. 2486
91. Truong-Ho, M., Monseur, X., Qui-Khuong-Huu and Goutarel, R. (1963). *Bull. Soc. Chim. France*, 2332
92. Goutarel, R., Conreur, C. and Parello, J. (1963). *Bull. Soc. Chim. France*, 2401
93. Truong-Ho, M., Qui-Khuong-Huu and Goutarel, R. (1963). *Bull. Soc. Chim. France*, 549
94. Khuong-Huu-Laine, F., Bisset, N. G. and Goutarel, R. (1965). *Ann. Pharm. France*, **23**, 395
95. Soti, F., Černý, V. and Šorm, F. (1967). *Tetrahedron Lett.*, 1437
96. Lábler, L. and Šorm, F. (1963). *Collect. Czech. Chem. Commun.*, **28**, 2345
97. Černý, V., Lábler, L. and Šorm, F. (1957). *Collect. Czech. Chem. Commun.*, **22**, 76
98. Qui-Khuong-Huu, Truong-Ho, M., Lábler, L., Goutarel, R. and Šorm, F. (1965). *Collect. Czech. Chem. Commun.*, **30**, 1016
99. Tscheche, R. and Wiensz, K. (1958). *Chem. Ber.*, **91**, 1504
100. Cave, A., Potier, P., Cave, A. and Le Men, J. (1964). *Bull. Soc. Chim. France*, 2415
101. Cave, A., Potier, P. and Le Men, J. (1967). *Ann. Pharm. France*, **25**, 107
102. Cave, A., Potier, P. and Le Men, J. (1965). *Bull. Soc. Chim. France*, 2502
103. Aquilar-Santos, G. (   ). *Philippine J. Sci.*, **94**, 217
104. Le Men, J. (1960). *Bull. Soc. Chim. France*, 860
105. Le Men, J., Kan, C. and Bengelmans, R. (1963). *Bull. Soc. Chim. France*, 597, 1306
106. Davis, J. B., Jewers, K., Manchandra, A. H. and Wood, A. B. (1970). *Chem. and Ind.*, 627
107. Tscheche, R. and Ockenfels, H. (1964). *Chem. Ber.*, **97**, 2316, 2326
108. Rönsch, H. and Schreiber, K. (1967). *J. Chromatogr.*, **30**, 149
109. Schreiber, K. (1968). *The Alkaloids, Vol. 10, p. 1*, (R. H. F. Manske, editor). (New York: Academic Press)
110. Prelog, V. and Jeger, O. (1960). *The Alkaloids*, Vol. VII, p. 343, (R. H. F. Manske and H. L. Holmes, editors). (New York: Academic Press)
111. Adam, G. and Schreiber, K. (1965). *Beiträge zur Biochemie und Physiologie von Naturstoffen.* (Jena: VEB Gustav Fischer Verlag)
112. Bognár, R. and Makleit, S. (1965). *Pharmazie*, **20**, 40
113. *Nomenclature of Solanum alkaloids*, cf. Schreiber, K. (1963). *Z. Chem.*, **3**, 346
114. Schreiber, K. and Rönsch, H. (1965). *Liebigs Ann. Chem.*, **681**, 187
115. Schreiber, K. and Rönsch, H. (1965). *Liebigs Ann. Chem.*, **681**, 196
116. Rönsch, H. and Schreiber, K. (1966). *Phytochemistry*, **5**, 1227
117. Rönsch, H. and Schreiber, K. (1965). *Liebigs Ann. Chem.*, **694**, 169
118. Ripperger, H. and Schreiber, K. (1969). *Liebigs Ann. Chem.*, **723**, 159
119. Höhne, E., Ripperger, H. and Schreiber, K. (1970). *Tetrahedron*, **26**, 3569
120. Sato, Y., Kaneko, H., Bianchi, E. and Kataoka, H. (1969). *J. Org. Chem.*, **34**, 1577
121. Nagai, M. and Sato, Y. (1970). *Tetrahedron Lett.*, 2911
122. Ripperger, H., Sych, F. J. and Schreiber, K. (1970). *Tetrahedron Lett.*, 5251
123. Kusano, G., Aimi, N. and Sato, Y. (1970). *J. Org. Chem.*, **35**, 2624
124. Kuhn, R. and Löw, I. (1961). *Chem. Ber.*, **94**, 1088, 1096
125. Kuhn, R. and Löw, I. (1962). *Chem. Ber.*, **95**, 1748
126. Schreiber, K. and Ripperger, H. (1967). *Chem. Ber.*, **100**, 1381
127. Usubillaga, A., Seelkopf, C., Karle, J. L., Daly, J. W. and Witkop, B. (1970). *J. Amer. Chem. Soc.*, **92**, 700

128. Schreiber, K., Ripperger, H. and Budzikiewicz, H. (1965). *Tetrahedron Lett.*, 3999
129. Ripperger, H., Budzikiewicz, H. and Schreiber, K. (1967). *Chem. Ber.*, **100**, 1725
130. Schreiber, K. and Ripperger, H. (1967). *Kulturpflanze*, **XV**, 199
131. Schreiber, K. and Ripperger, H. (1963). *Arch. Pharm.*, **296**, 792
132. Briggs, L. H., Brooke, E. G., Harvey, W. E. and Odell, A. L. (1952). *J. Chem. Soc.*, 3587; ibid., 1953, 2833
133. Rochelmeyer, H. (1939). *Arch. Pharm.*, **277**, 329
134. Perez-Medina, L. A., Travecedo, E. and Devia, J. E. (1964). *Plant. Med.*, **12**, 478
135. Schreiber, K. and Aurich, O. (1966). *Phytochemistry*, **5**, 707
136. Guseva, A. R. and Paseshnitchenko, V. A., *5th Proc. Int. Congr. Biochem. Moscow 1961*, **7**, 287
137. Willuhn, G. (1966). *Abh. Deut. Akad. Wiss. Berlin, Kl. Chem. Geol. Biol.*, 97
138. Alkiewicz, J., Gertig, H., Klamyszek, F., Kowalewski, Z. and Moderski, F. (1966). *Diss. Pharm. Pharmacol.*, **18**, 553
139. Wolters, B. (1964). *Arch. Pharm.*, **297**, 748
140. Boll, P. M. and Philipsborn, W. v. (1965). *Acta Chem. Scand.*, **19**, 1365
141. Rosen, W. E., Ziegler, J. B., Skabica, A. C. and Shoolery, J. N. (1959). *J. Amer. Chem. Soc.*, **81**, 1687
142. Kutney, J. P. (1963). *Steroids*, **2**, 225
143. Budzikiewicz, H. (1964). *Tetrahedron*, **20**, 2267
144. Höhne, E., Schreiber, K., Ripperger, H. and Worch, H. W. (1965). *Tetrahedron*, **22**, 673
145. Beisler, J. A. and Sato, Y. (1971). *J. Chem. Soc. C*, 149
146. Adam, G. (1963). *Z. Chem.*, **3**, 379, 403
147. Bose, S. (1965). 70th birthday Commemoration Volume Vol. II, p. 393, Calcutta
148. Adam, G. and Schreiber, K. (1966). *Tetrahedron*, **22**, 3595
149. Kessar, S. V., Rampal, A. L., Gandhi, S. S. and Mahajan, R. K. (1971). *Tetrahedron*, **27**, 2153
150. Kessar, S. V., Gupta, Y. P., Singh, M. and Mahajan, R. K. (1971). *Tetrahedron*, **27**, 2869

8
The *Amaryllidaceae* Alkaloids

P. W. JEFFS
Duke University, Durham, North Carolina

8.1 INTRODUCTION

The subject matter covered in this review is hopefully broadly representative
of research in the field but no attempt has been made to provide an exhaustive
coverage of all of the literature; instead emphasis has been given to developments which in the author's opinion are of special interest and significance
and which have taken place since the last comprehensive review of the
Amaryllidaceae alkaloids[1].

8.2 LYCORINE-TYPE ALKALOIDS

The lycorine-type alkaloids, which are named after the parent base lycorine (1),
form one of the major sub-groups of the family. The absolute configuration

of lycorine was determined by an x-ray crystallographic analysis of dihydroly-corine hydrobromide[2] and all members of the series presently known are based upon this single modification of the basic pyrrolo[*d,e*]phenanthridine nucleus. With the exception of the few examples recorded below, no major structural revisions or new bases of interest in this sub-group have been reported since the last comprehensive review of the *Amaryllidaceae* alkaloids appeared.

8.2.1 Narcissidine and related bases

Narcissidine, originally[3] assigned structure (2) and later revised[4] to (3), has been shown to be correctly represented by structure (4) from the results of single-crystal x-ray analyses of narcissidine hydrobromide[5] and methiodide[6]. This structure had been considered in an earlier study but was rejected because of the presumed presence of a 1,2-glycol from the consumption of periodate by dihydronarcissidine. The behaviour of the lycorine-type alkaloids towards periodate cleavage has since been found to be anomalous; even ring-C monohydroxy derivatives react with this reagent.

(1)

(2) R^1 = H, R^2 = Me (4) R^1 = R^2 = R^3 = Me
(3) R^1 = Me, R^2 = H (5) R^1 = R^2 = Me, R^3 = H
                          (6) R^1 = R^2 = —CH$_2$—, R^3 = H

The revision of the structure of narcissidine has required that ungiminorine (5) and parkacine (6), two alkaloids whose previous structural assignments relied upon their interrelation with narcissidine, be revised as indicated.

8.2.2 Approaches to the synthesis of lycorine-type alkaloids

A significant effort has been devoted to the synthesis of the lycorine alkaloids, unfortunately so far with only moderate success.

A variety of approaches have been tried by Hendrickson and his co-workers[7] in an unsuccessful effort to synthesise the lycorine system. The first attempt using an AC ⟶ ABC ⟶ ABCD ring construction sequence failed when the necessary steps required to construct ring D on the perhydro-phenanthridine (7) could not be effected, in an acceptable yield.

An alternative approach which envisaged cyclisation of the tricyclic amine (8), via the diazo-derivative, was frustrated by the formation of the quinazoline (9) on reduction of the nitro-compound (10). Although the authors foresaw the possibility of internal conjugate addition of the amino-group to the enone system in (8), the subsequent irreversible retro-aldol cleavage involving the rupture of ring C was not expected and its occurrence resulted in the abandonment of this route.

(7)

(8) X = NH$_2$
(10) X = NO$_2$

(9)

(11) X = N$_2^+$ BF$_4^-$

(12)

(13)

A third approach involved an attempted Pschorr cyclisation of the amide
(1) (R = H), which could in principle generate either the crinine or lycorine
system depending on the site of attack. Unfortunately neither of the desired
products were obtained, instead the product from this reaction proved to be
the benztriazinone (12) (R = H) resulting from the efficient internal capture
of the diazo-group by the secondary amide function. Surprisingly, a similar
benztriazinone (12) (R = Me) has been obtained by Kametani and colleagues[8]
from an attempted Pschorr cyclisation of the related amine (11) (R = Me,
CH$_2$ instead of CO).

The latter authors were also unsuccessful in obtaining the lycorine or
crinine systems from an attempted phenol-oxidative coupling of the oxindole
(13)[9].

8.2.2.1   α-Dihydrocaranone

The ketone (14), which is a known transformation product of the lycorine
alkaloid, caranine, has been synthesised in an attractively simple manner as
outlined schematically in Figure 8.1[10]. An attempt to cyclise the octahydro-
indolanone (15) by generating the benzyne in liquid ammonia with potassium
or sodium amide was unsuccessful and afforded the amination product (16).
The problem of amination of the intermediate benzyne was circumvented by
using lithium piperidide in tetrahydrofuran and the cyclisation product
α-dihydrocaranone (14) was obtained in 35–40% yield. The stereochemistry
at C-11b of this compound, which was confirmed by conversion to the known
γ-lycorane (17), unfortunately does not correspond to that found in the
natural series. Since it is known that base equilibration of the 1-ketones in
the C-3a α-dihydro-lycorine series leads exclusively to products containing a

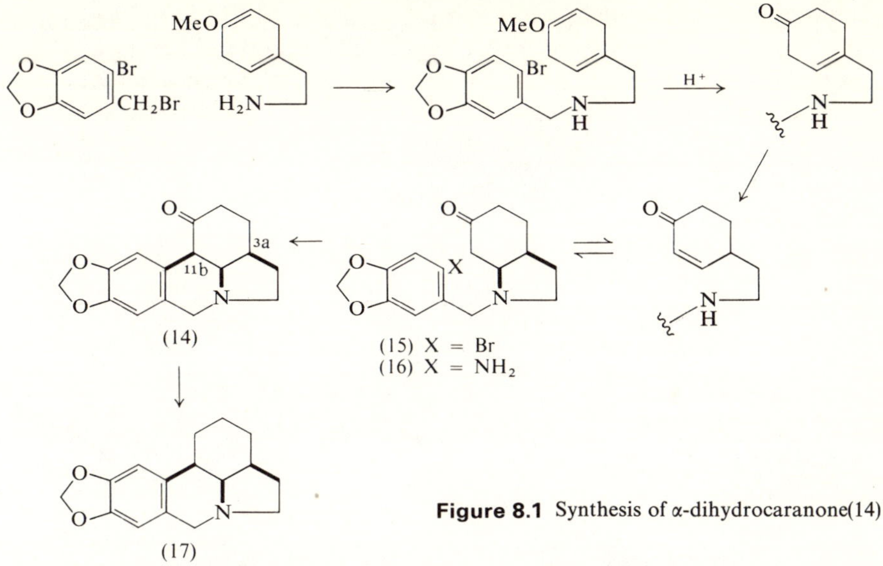

Figure 8.1 Synthesis of α-dihydrocaranone(14)

(14)

(15) X = Br
(16) X = NH$_2$

(17)

cis-B, C ring fusion, adaptation of this route holds little promise for effecting the synthesis of a compound containing the required stereochemistry.

A recent attempt[11] to synthesise the lycorine system by a well-conceived route involving a base-catalysed cyclisation of the substituted pyrrolidinone (18) unfortunately also resulted in formation of the product (19) containing the undesired *cis*-B, C stereochemistry.

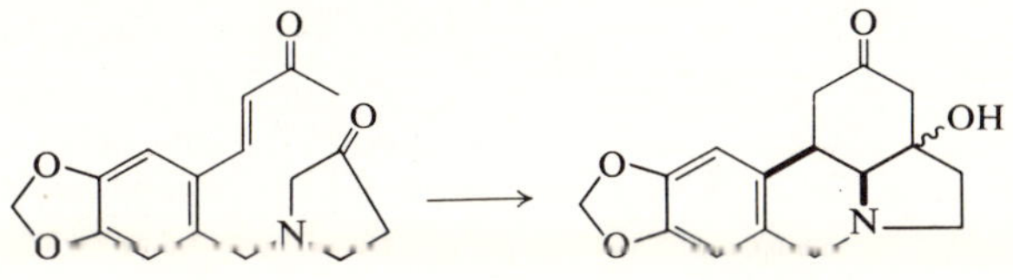

(18)

(19)

8.2.2.2 *Dihydrolycorine*

The recent synthesis of (±)-dihydrolycorine accomplished by Uyeo and his collaborators[12] is portrayed in Figure 8.2. Like previous syntheses of other *Amaryllidaceae* alkaloids by this group (Sections 8.3.4 and 8.6.4.5), the approach is hinged upon the utilisation of a Schmidt reaction in the synthetic sequence.

The tetralone ester (20), obtained as one of the products from the Friedel–Craft cyclisation of a mixture of the two half esters (21), was converted to the corresponding alcohol (22) and subjected to the Schmidt reaction to give the desired isocarbostyril (23) (R = OH). Homologation of the side-chain in a standard manner afforded the carboxylic acid (23) which was transformed by iodolactonisation to (24). When the iodolactone was treated with acetic acid–acetic anhydride it afforded the iodoacetate (25). The retention of configuration at C-2 observed in this reaction is presumably the result of an intramolecular displacement by the amido-nitrogen on the lactone carbonyl, a process which is made particularly favourable by the proximity of these

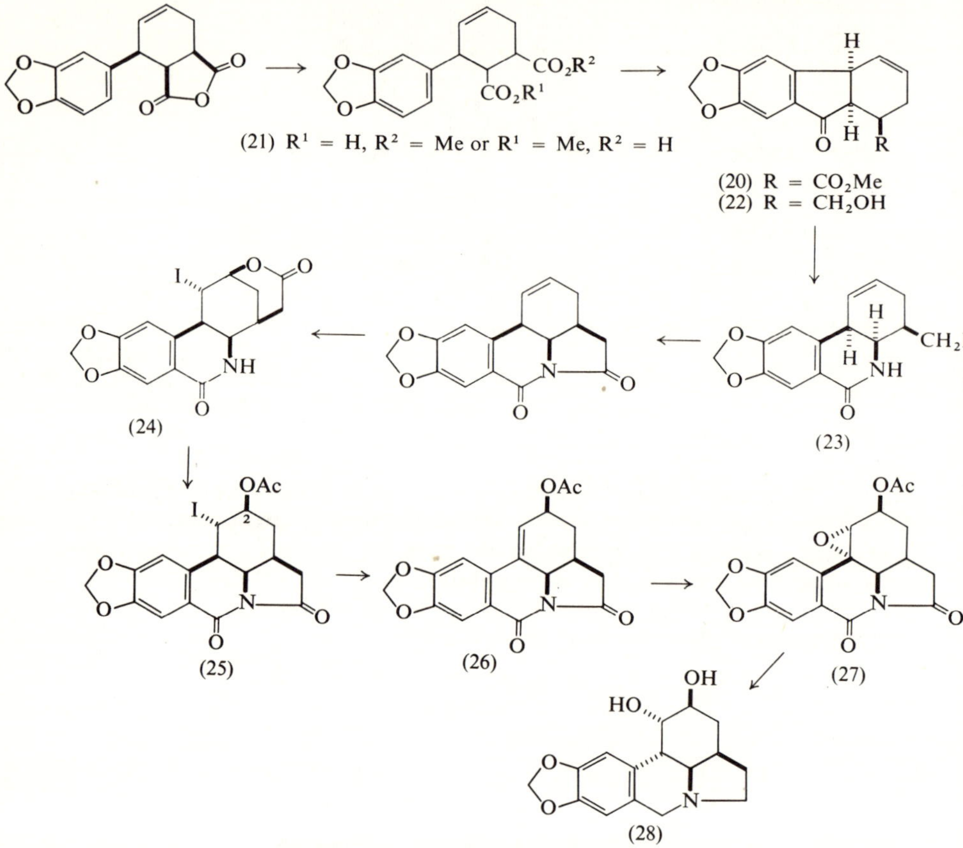

(21) R^1 = H, R^2 = Me or R^1 = Me, R^2 = H

(20) R = CO_2Me
(22) R = CH_2OH

(24)

(23)

(25)

(26)

(27)

(28)

Figure 8.2   Synthesis of dihydrolycorine(28)

groups in this molecule. Despite the *cis* arrangement of the C-11b hydrogen and the C-1 iodine in (25), the desired olefin (26) was obtained on treatment of (25) with lithium chloride in dimethylformamide. The C-2 acetoxyl was retained to assist in directing the subsequent epoxidation of (26) from the α-face and in the event, this proved to be successful since the product obtained from treatment of the olefin with *m*-chloroperbenzoic acid showed n.m.r. properties in accord with the α-epoxide (27). Although hydride reduction of (27) was expected to afford (±)-dihydrolycorine directly, the authors experienced considerable difficulty in effecting this transformation, and the best conditions that could be devised using lithium aluminium hydride and zinc chloride gave only a 5% yield of (±)-dihydrolycorine (28). A possible drawback to this last step which may account for the poor yield of dihydroly-corine is the requirement to effect epoxide ring opening by hydride attack at the quaternary C-11b carbon. This synthesis constitutes the first total synthesis of a natural lycorine alkaloid since *O,O*-diacetyldihydrolycorine, to which the synthetic product was converted on acetylation, occurs naturally and is known as nartazine[13].

8.3 GALANTHAMINE-TYPE ALKALOIDS

8.3.1 Chlidanthine

Previous evidence had established the functional groups of this phenolic alkaloid and indicated that chlidanthine (29) was related to galanthamine (30) from its conversion to apogalanthamine (31) on treatment with acid. The assignment of its structure as (29) has been confirmed[14] through the conversion of chlidanthine and galanthamine to their methyl ether methiodides, which proved to be identical. The formation of chlidanthine methyl ether methiodide was accomplished in the usual way employing dimethyl sulphate, followed by potassium iodide. However, the use of this procedure with galanthamine failed and the following indirect methylation route was devised. Galanthamine was converted to its C-3 epimer (32) in hot mineral acid and the latter with thionyl chloride gave the 3α-chloro-compound (33) in a reaction which proceeded stereospecifically with retention of configuration. Displacement of the C-3 chloro-substituent with methoxide ion afforded the desired galanthamine *O*-methyl ether (34) which was converted to its methiodide to complete the sequence.

(29)

(30) R^1 = OH, R^2 = H
(32) R^1 = H, R^2 = OH
(33) R^1 = H, R^2 = Cl
(34) R^1 = OMe, R^2 = H
(35) R^1 = R^2 = O

(31)

A recent report indicates the *Chlidanthus fragrans*, which produces galanthamine and chlidanthine, also contains *O*-methylgalanthamine[15].

8.3.2 Oxidation of galanthamine

A re-investigation[16] has been undertaken of the previously unidentified by-product, $C_{17}H_{19}NO_4$, obtained from the chromic acid oxidation of galanthamine to narwedine (35).

The product, which is also obtained from oxidation of narwedine with manganese dioxide, has been assigned structure (36) from its spectral properties. Although (36) was optically active, its partially racemic nature was indicated from an increase in both its melting point and optical rotation on repeated recrystallisation. Its formation has been rationalised by a sequence which involves benzylic oxidation of narwedine to the carbinolamine as the initial step. This carbinolamine, perhaps as a consequence of relief of strain is then required to undergo rapid ring opening to the aminoaldehyde intermediate (37) to avoid the expected further oxidation at the

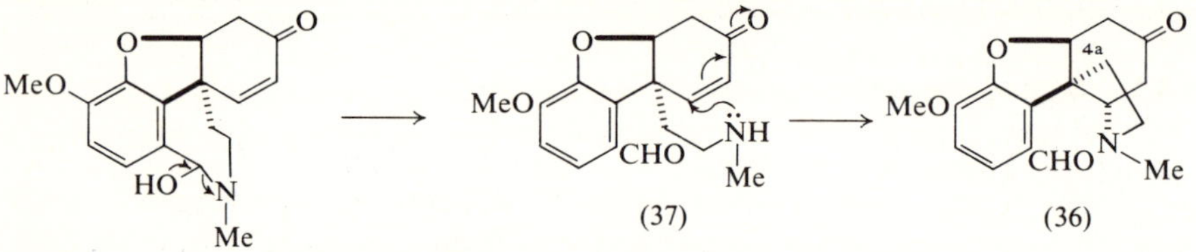

benzylic position. The possibility of the latter compound reverting to the carbinolamine is prevented by the intervention of the more favourable Michael reaction which completes the reaction to give the observed aldehyde (36).

8.3.3 Synthesis of galanthamine

The first synthesis of galanthamine accomplished by Barton and Kirby[17] from phenol-oxidative coupling of *N*-methylnorbelladine occurred in poor yield (1.4%). Subsequent studies on model systems have demonstrated that the yields of oxidative coupling products are greatly enhanced by protection of the basic nitrogen through the formation of amides[18, 19]. In terms of synthesising galanthamine this advantage is partly offset by the tendency of the amides to give products arising from *para-para* coupling. Kametani and his co-workers have employed this knowledge to good purpose in effecting several syntheses of galanthamine via the phenolic-oxidative coupling route. In the first example[20], oxidative coupling of the bromobenzamide (38) using potassium ferricyanide and sodium hydrogen carbonate in a two-phase system of chloroform and water, proved remarkably successful in giving the crystalline bromoenone (39) in 40% yield. Reduction of (39) with lithium aluminium hydride afforded ($\pm$)-galanthamine (50%) and ($\pm$)-3-*epi*galanthamine (40%).

Despite this initial success, subsequent attempts to employ this approach to the synthesis of the galanthamine ring system through oxidative coupling of bromo-amides closely related to (38) has proved less successful; the desired coupled products are obtained in poor yield[21]. Some indication of the

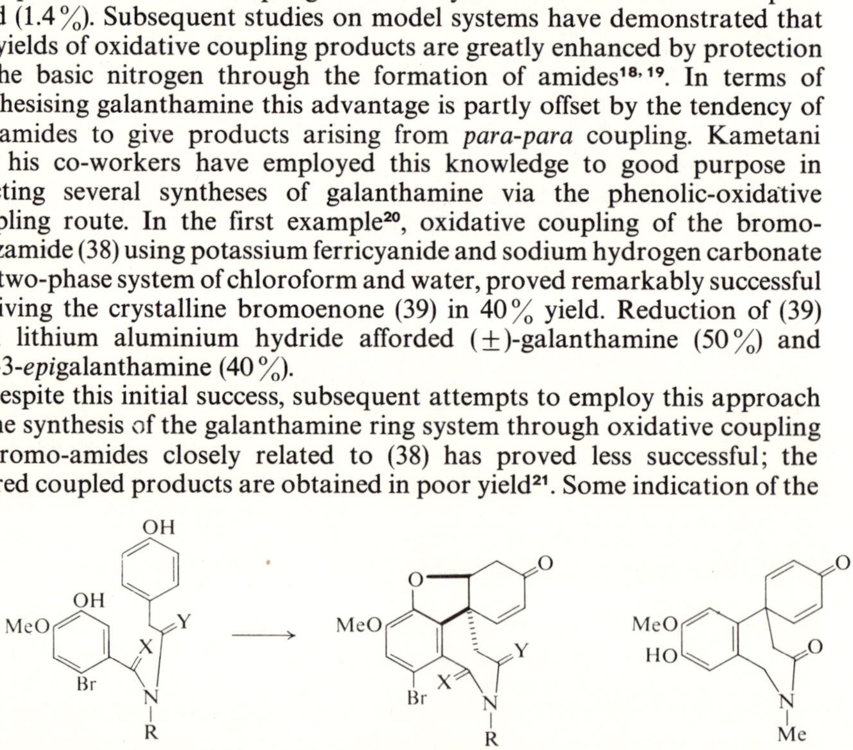

(38) R = Me, X = O, Y = H$_2$ (39) R = Me, X = O, Y = H$_2$ (44)
(40) R = Me, X = H$_2$, Y = O   (41) R = Me, X = H$_2$, Y = O
(42) R = Me, X = H$_2$, Y = O, Br = H   (42) R = Me, X = H$_2$, Y = O, Br = H

unpredicability of the success of this method is illustrated by the phenol oxidative coupling of the bromo-amide (40) which although closely related to (39) proceeds to give only a 2.0% yield of the enone (41).

Interestingly, oxidation of the diphenolic amide (42), which lacks a suitably disposed bromine to block the favoured *para–para* coupling affords an acceptable 5% yield of the desired narwedine-type enone (43) together with 10% of the expected dienone (44)[22].

The attention devoted to synthesis in the galanthamine series has received its impetus from the finding that galanthamine has analgesic activity comparable to morphine[21].

8.3.4 Synthesis of lycoramine

Two syntheses of (±)-lycoramine have been reported by Uyeo and his co-workers[23, 24]. Both syntheses are very lengthy, involving more than 20 steps, and limitations on space do not permit an adequate discussion of these to be presented. The more satisfactory route[24], of which the late stages are illustrated in Figure 8.3 starts from 2,3-dimethoxybenzaldehyde and affords (±)-lycoramine in 0.67% yield.

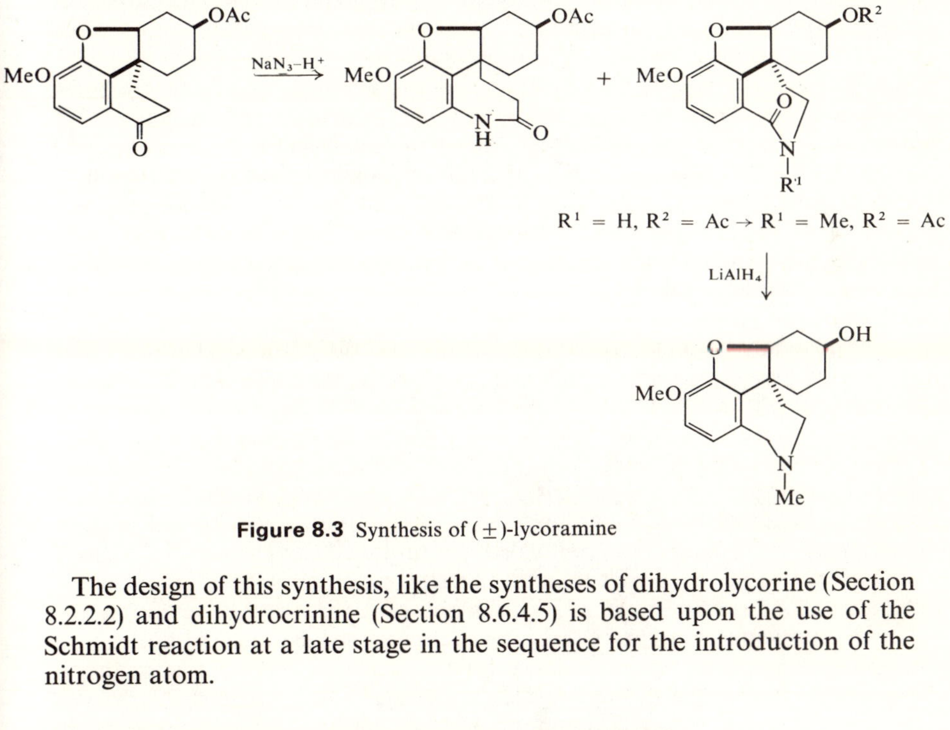

$$R^1 = H, R^2 = Ac \rightarrow R^1 = Me, R^2 = Ac$$

Figure 8.3 Synthesis of (±)-lycoramine

The design of this synthesis, like the syntheses of dihydrolycorine (Section 8.2.2.2) and dihydrocrinine (Section 8.6.4.5) is based upon the use of the Schmidt reaction at a late stage in the sequence for the introduction of the nitrogen atom.

8.4 PRETAZETTINE AND RELATED ALKALOIDS

8.4.1 Pretazettine

A re-investigation[25] of the alkaloids of *Sprekelia formosissima* and *Ismene calithina* has demonstrated that tazettine (45), previously reported as one of the major alkaloids of these plants, is an artefact. Using an isolation procedure which avoided the usual strongly basic conditions employed in a typical extraction procedure, the major alkaloid found was a new non-crystalline base, pretazettine (46). Pretazettine, although stable in dilute acids is found to rearrange to tazettine under a variety of conditions. The observation that chromatography over alumina or treatment with dilute NaOH afforded rapid and complete conversion to tazettine suggested that pretazettine, and not tazettine, is the natural alkaloid in these plants. In support of this contention the absence of tazettine could be concluded from an examination of the thin-layer chromatogram of the alkaloid extract.

A comparison of the chemical and physical properties of pretazettine with a labile base, 'isotazettine', which had been isolated from *Leucojum aestivum* some 10 years earlier by Proskurnina[26], suggested the identity of these compounds. This was corroborated by the results of an independent[27] spectral study of 'isotazettine' which led to an assignment of its structure corresponding to that proposed for pretazettine from the results of the more complete study by Wildman and Bailey[25].

The elucidation of the structure of pretazettine* was greatly aided by a comparison of its ^{1}H n.m.r. spectra with that of tazettine. An AB pattern attributable to the benzylic methylene hydrogens, which is centred at δ 4.78 in the spectrum of tazettine, is absent in the corresponding region of the pretazettine spectrum which instead possesses a one-proton singlet at δ 6.02. The presence of a benzylic hydroxyl in pretazettine suggested by these results was confirmed by its oxidation with MnO_2 to the expected lactone (47). However, the major product from this reaction was the *N*-demethylated lactone (48). The absence of the benzylic hydrogen signal in the region of δ 6.0 in the n.m.r. spectra of these compounds coupled with their i.r. and u.v. spectral properties provided support for the proposed structures. The formation of (48) by oxidative cleavage of the *N*-methyl group was established by the formation of this compound on treatment of the intermediate lactone (47) with MnO_2.

Independent chemical evidence for the structures of the lactones (47) and (48) was provided by relating them with haemanthidine (49). The bridgehead lactam (50), obtained by MnO_2 oxidation of haemanthidine, was converted with sodium acetate–acetic acid buffer to the amino acid (51) which spontaneously recyclised to the lactone (48). Since a considerable body of evidence existed to support the assignment of the stereochemistry of the 11-hydroxyl group in haemanthidine, the conversion of the amino acid to the lactone under such mild conditions permitted the important assignment of a *trans* B:D ring fusion in the latter and thereby established the same stereochemistry for pretazettine at this ring juncture. Evidence for the configuration of the benzylic hydroxyl of pretazzetine is less certain. Although the n.m.r. spectrum showed no evidence for the existence of both C-8 epimeric forms as is observed with some crinine alkaloids having a benzylic hydroxyl, the formation of two C-8 epimeric methyl acetals from pretazettine in

---

*It has been suggested that this name is preferred to isotazettine since the prefix *iso* in the tazettine series has been commonly used to designate compounds possessing a 3α-orientated oxygen substituent.

refluxing acidic methanol has been cited in favour of the mobility of the benzylic hydroxyl in this alkaloid. In the reviewer's opinion this does not constitute proof that both C-8 epimeric forms of pretazettine exist under neutral conditions in solution.

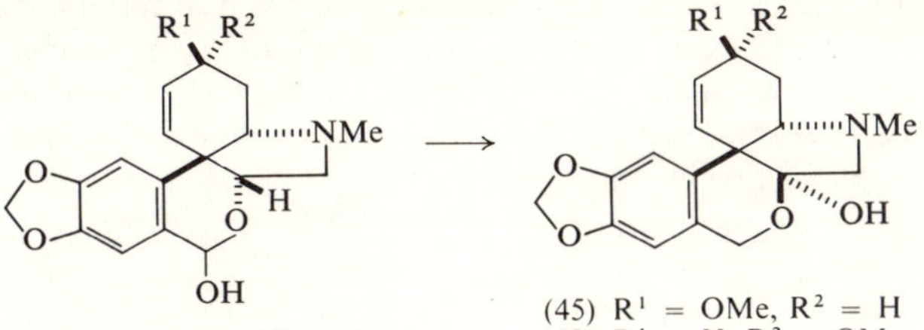

(46) R^1 = OMe, R^2 = H
(54) R^1 = H, R^2 = OMe

(45) R^1 = OMe, R^2 = H
(53) R^1 = H, R^2 = OMe

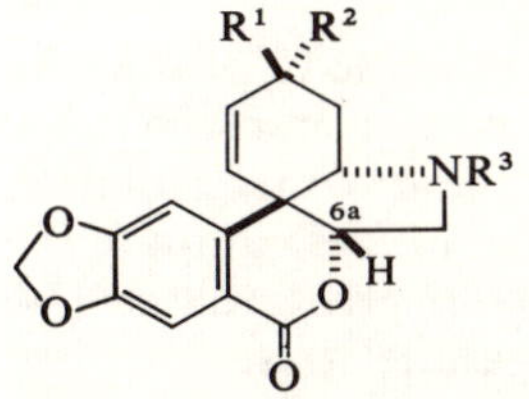

(47) R^1 = OMe, R^2 = H, R^3 = Me
(48) R^1 = OMe, R^2 = R^3 = H
(55) R^1 = H, R^2 = OMe, R^3 = Me
(56) R^1 = R^3 = H, R^2 = OMe
(57) R^1 = H, R^2 = OMe, R^3 = CO$_2$Et

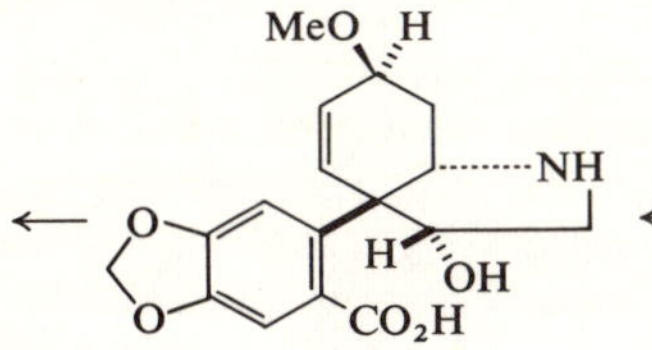

(51)

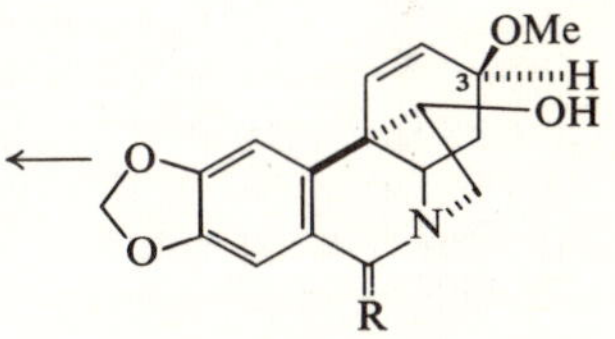

(49) R = H, OH
(50) R = O

In a subsequent study[28], a further set of remarkably facile reactions which are responsible for the interconversion of several compounds representing the crinine and [2]benzo[3,4*g*]indole skeletons was uncovered. Characteristic

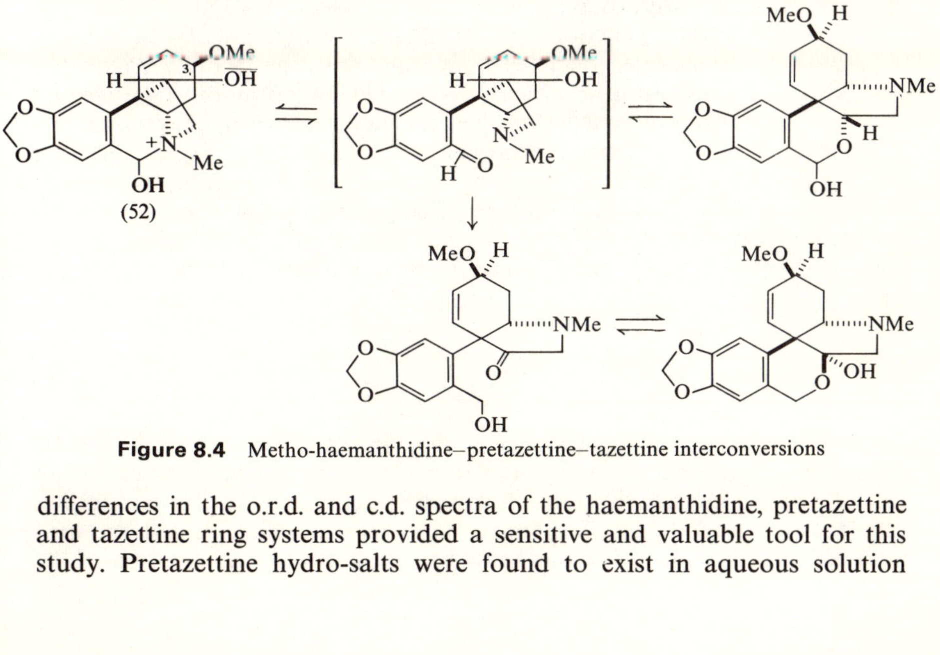

(52)

Figure 8.4 Metho-haemanthidine–pretazettine–tazettine interconversions

differences in the o.r.d. and c.d. spectra of the haemanthidine, pretazettine and tazettine ring systems provided a sensitive and valuable tool for this study. Pretazettine hydro-salts were found to exist in aqueous solution exclusively as haemanthidine metho-salts (52). The reversibility of this process was demonstrated by the isolation of pretazettine from this solution on basification and immediate extraction into chloroform. If extraction with chloroform was omitted, the competing, and irreversible, process by which haemanthidine metho-salts and pretazettine are converted to tazettine via the transient carbinolamine tautomer occurred to completion within 1 h. Therefore, of the various rearrangements which the carbinolamine form must undergo as the central figure of this kaleidoscopic picture, that of hydride transfer has to be the slowest.

A summary of this remarkable series of interconversions is presented schematically in Figure 8.4.

8.4.2 Precriwelline

The discovery of pretazettine cast doubt upon the authenticity of the rare base criwelline (53) which had been previously reported to occur in *Crinum powellii*. A re-investigation[25] of this plant employing the modified isolation procedure used in the isolation of pretazettine provided no indication of the presence of criwelline, instead a new amorphous base, precriwelline (54) was isolated. The chemical properties of precriwelline parallel those described for pretazettine.

8.4.3 Macronine and related bases

Although the structure of the lactone base, macronine (55) was elucidated in 1964, the stereochemistry at the 6a-position was left undesignated[29]. Since oxidation of precriwelline affords macronine the stereochemistry at this position has been assigned as shown[25].

3-*Epi*macronine (47), encountered during the investigation of the chemistry of pretazettine, has been detected as a trace constituent of *Sprekelia formosissima*[25]. Examination of *Crinum erubescens* by the modified procedure used in the isolation of pretazettine, has afforded *N*-demethylmacronine (56) and a trace alkaloid possessing the rather unusual structure (57). The structure of the latter was confirmed from its preparation from *N*-demethylmacronine by treatment with ethyl chloroformate. The possibility that this substance is an artefact resulting from *N*-demethylmacronine and the solvents, chloroform and ethanol, used in the isolation procedure was pointed out by the authors[32].

8.5 TAZETTINE

Despite the recent work which has shown that tazettine can no longer be regarded as a natural alkaloid, the study of the chemistry and structure of this compound has provided the chemist with an interesting and continuing challenge for over 40 years. The last remaining feature concerning the relative stereochemistry of tazettine was settled in 1961 with the elucidation of the stereochemistry of the hemi-ketal ring[30] and this assignment, despite being disputed[31], has proved to be correct (see below). The absolute configuration thus remained as the only structural detail not firmly established. An early proposal[32] for the absolute configuration of tazettine had been made which was based upon an application of the empirical Mills' rule. However, since a similar proposal based on this rule in the crinine series had been challenged, Highet and Highet[33] have carried out a further study of the problem.

8.5.1 Hofmann degradation of dihydrotazettine and the absolute configuration of tazettine

Previous studies had led to the assignment of the double bond in the Hofmann product from dihydrotazettine (58) to the non-conjugated 2′,3′-position on the basis of u.v. spectral evidence. This assignment was placed in doubt by examination of the n.m.r. spectrum of the Hofmann product which showed only a single olefinic hydrogen signal. Reconciliation[34] of this seemingly conflicting evidence was arrived at by the reassignment of the double bond to the 1′,2′-position, as in structure (59). It was suggested that non-bonded interactions between the cyclohexenyl ring and the *ortho* side-chain in this compound would prevent the attainment of co-planarity of the 1′,2′-double bond and the aromatic ring. Support for this proposal was provided by the observation that the model compound 1-*o*-tolylcyclohexene also failed to exhibit a styrene chromophore in its u.v. spectrum.

(58)

(59) R = CO·CH₂·NMe₂
(61) R = H

(60)

(62)

A most important consequence of the revision of the structure of dihydrotazettine methine was that it provided through its conversion to $R(+)$-β-methoxyadipic acid (60) by ozonolysis of the alcohol (61), the first unequivocal answer to the absolute configuration of tazettine.

In making the assignment of the absolute stereochemistry it was necessary to establish that the 1′,2′-double bond in dihydrotazettine methine originated from the C-4a and C-12b carbons of the alkaloid rather than from the C-1 and C-12b positions. The latter possibility was excluded by the results obtained from the Hofmann degradation of 1,2-d_2-dihydrotazettine methiodide (62) which was shown to afford the product containing two deuterium atoms located at C-5' and C-6' position from n.m.r. and mass spectral studies. The course of the Hofmann reaction was interpreted from these results as involving a four-centre elimination (see arrows).

In view of the previously well defined interrelationship[1] of tazettine with pretazettine and representative members of the (+)-crinine and montanine ring systems, the absolute configurations of these alkaloid series are also uniquely defined as a result of this work.

The absolute configuration of tazettine has been confirmed subsequently by a single-crystal x-ray analysis of tazettine methiodide[35].

8.6   CRININE-TYPE ALKALOIDS

8.6.1   Absolute configuration and o.r.d. and c.d. spectra of *Amaryllidaceae* alkaloids

Alkaloids based on both enantiomeric modifications of the parent 5,10b-ethananophenanthridine ring system of the crinine alkaloids are found and some attention has been given to determining absolute configurations in this series, using chiroptical, chemical, and most recently, x-ray methods. The absolute configurations of the parent members (−)-crinine (63) and its enantiomer (+)-vittatine have been placed on a firm basis by the elucidation of the absolute configuration of tazettine (Section 8.5). These assignments have subsequently been confirmed[36] by recent x-ray structure determinations of haemanthamine *p*-bromobenzoate (64; R^1 = *p*-BrC$_6$H$_4$·CO—, R^2 = H) and 6-hydroxybuphanidrine methiodide (Section 8.6.2). Interestingly, these assignments are in agreement with the original proposals[37] made on the basis of Mills' rule.

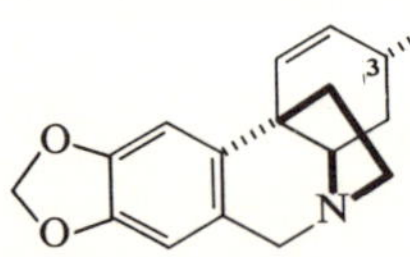

(63)

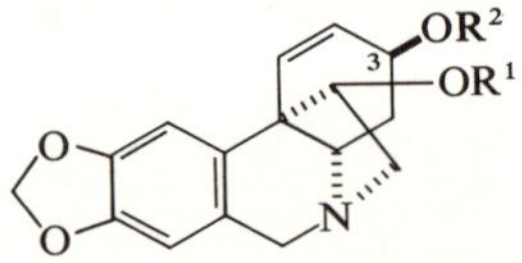

(64) R^1 = H, R^2 = Me
(70) R^1 = R^2 = H

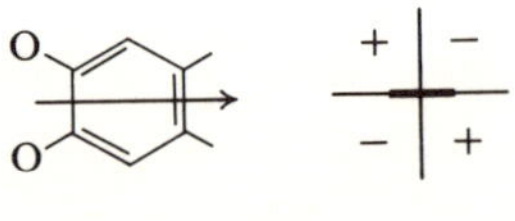

(65)

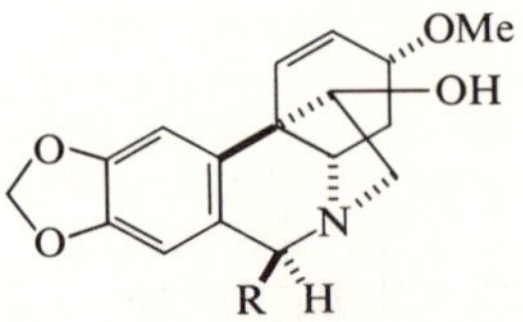

(66)

In contrast, the application of o.r.d. and c.d. for assignment of absolute configuration from Cotton effects produced by the aromatic chromophores has met with mixed success in the *Amaryllidaceae* field. A quadrant rule (see 65) proposed by Kuriyama *et al.*[38] for the ^1L_b band (which appears in the region of 290 nm in the c.d. spectra of alkaloids possessing a methylene-dioxyphenyl chromophore) gives results in the lycorine, crinine, homoly-corine and tazettine series which are in agreement with the absolute configurations derived from more recent work. However, the more general application of this rule has been recently questioned[39]. A second quadrant rule suggested by DeAngelis and Wildman[40] for the ^1L_a c.d. band (240 nm) is only valid in the *Amaryllidaceae* series for the lycorine and galanthamine alkaloids; the absolute stereochemistries indicated for the crinine, tazettine and montanine groups are opposite to those established by chemical and subsequent x-ray studies. A more recent approach[39] in which the hypothesis that the sign and even the magnitude of the Cotton effect of the ^1L_a and ^1L_b bands (which are normally opposite in sign) is determined by the chiral sphere nearest the aromatic ring has been examined. Unfortunately an adequate discussion of this rather sophisticated method is beyond the scope of this review. An example of its success may be illustrated by the fact that it is capable of predicting the sign and the relative magnitudes of the Cotton effects in the c.d. spectra of the C-3 epimers, haemanthamine (64) [289(+3.42), 244(−3.96)] and crinamine (66; R = H) [290(+3.51), 245(−2.84)]. A key feature of the method is that separate sector rules are needed for the ^1L_a and ^1L_b bands and careful attention is required in designating the local symmetry of the chromophore, which in aromatic systems changes with substitution pattern.

Despite the difficulties in providing a theoretical interpretation of the c.d. and o.r.d. spectra of *Amaryllidaceae* alkaloids, the shape of the curves for each of the various ring systems is quite distinctive. Used in this empirical fashion the method provides a sensitive and powerful tool for assigning alkaloids of unknown structure to the appropriate sub-group and may even yield information on relative stereochemistry at certain centres[41].

8.6.2   6-Hydroxybuphanidrine and 6-hydroxypowelline

The plant *Nerine bowdenii* has proved to be a rich source of crinine alkaloids[1]. A reinvestigation[42] of the alkaloid fraction from this species has led to the isolation of two new bases which have been characterised as 6-hydroxybuph-anidrine (67; R^1 = OH, R^2 = Me) and 6-hydroxypowelline (67; R^1 = OH, R^2 = H). Conversion of the former to the known alkaloid buphanidrine (67; R^1 = H, R^2 = Me) on successive treatment with thionyl chloride and lithium aluminium hydride served to establish all of the structural and stereochemical features of the new alkaloid except the position and stereochemistry of the hydroxyl group. The location of the hydroxyl at the C-6 position was established by oxidation of the alkaloid to the bridgehead lactam (67; C=O at C-6 instead of R,H) with manganese dioxide. The spectral properties [v, 1690 cm^{-1}; λ, 266 nm (ε 20 400), 238(7200), 287(1500), 321(1350)] of this compound as an aryl conjugated ketone were in full accord with those expected for a compound of this structure and compared well with the bridge-head lactams obtained from oxidation of the 6-hydroxy derivatives of crinamine and haemanthamine.

One of the unique characteristics of 6-hydroxycrinamine (66; R = OH) and haemanthidine (49) is their behaviour in solution, in which they are both known to exist as a mixture of equilibrating C-6 epimers[43]. It was surprising therefore to find that the n.m.r. spectrum of 6-hydroxybuphanidrine showed only a single resonance at δ 5.31 for the benzylic proton and no evidence was

obtained from spectral or chemical properties to suggest that it existed other than as a single epimer in solution.

Some evidence for the configuration of the C-6 hydroxyl group was obtained by the finding that the *O*-acetate (67) (R^1 = OAc, R^2 = Me) underwent solvolysis in methanol to give the methyl ether (67; R^1 = OMe, R^2 = Me). The stereochemistry of this compound was assigned on the grounds that both steric and electronic considerations suggest that the capture of methanol by the benzylic carbonium ion would lead to the 6α-*quasi*-axial isomer. Since methylation of the potassium salt of 6-hydroxybuphanidrine with methyl toluene-*p*-sulphonate afforded the same methyl ether under conditions which the C-6 oxygen bond is expected to remain intact, the 6α-configuration was ascribed to the benzylic hydroxyl group and the alkaloid was depicted as (67; R^1 = OH, R^2 = Me). Unlike 6-hydroxycrinamine and haemanthidine, 6-hydroxybuphanidrine exhibits the normal behaviour of a carbinolamine in giving an aldehyde (68) when treated with methyl iodide and base.

The structure of 6-hydroxybuphanidrine has been confirmed by a single-crystal x-ray analysis of its methiodide salt[36].

Elucidation of the structure of the second new base, 6-hydroxypowelline (67; R^1 = OH, R^2 = H) followed from the conversion of the mono-acetyl derivative (67; R^1 = OH, R^2 = Ac) to powelline and from a comparison the n.m.r. spectral properties of the alkaloid and its derivatives with the spectra of analogous compounds derived from 6-hydroxybuphanidrine. 6-Hydroxy-powelline also showed no evidence of existing as mixture of equilibrating C-6 epimers in solution.

8.6.3   11-Hydroxyvvittatine and maritidine

Although full details of the elucidation of the structure of these alkaloids have not yet appeared, evidence presented in a review[1] has given ample support

for the structure of maritidine (69) and a recent synthesis of ($\pm$)-maritidine
has since provided confirmation of its structure (Section 8.6.4.3). In the same
review, 11-hydroxyvittatine has been represented by structure (70) and,
while its reported easy conversion to apohaemanthamine (71) in acid leaves
little doubt as to the correctness of the gross details of its structure and the
configuration of the C-11 hydroxyl, no firm evidence is presented for the
assigned stereochemistry at the C-3 position. The reported rearrangement
of 11-hydroxyvittatine to montanine (Section 8.7.1) would seem to support
the proposed stereochemical assignments but certain features of this reaction
are in need of further elaboration.

8.6.4 Synthesis

The initial successes of Muxfeldt[45] and Whitlock[46] and their co-workers in
synthesising ($\pm$)-crinine have been followed by a number of other syntheses
of representatives of this sub-group. Both Hendrickson[47] and Uyeo[48] have
utilised intermediates developed from their previous forays into the synthesis
of *Amaryllidaceae* alkaloid systems in accomplishing elegant, if somewhat
lengthy, syntheses of ($\pm$)-haemanthidine and ($-$)-crinine, respectively. Two
other syntheses, which bear some similarity in the overall plan of approach in
assembling the crinine ring system but differ in the manner of execution,
have led to expeditious routes to ($\pm$)-dihydro*epi*crinine [3-*epi*elwesine][49]
and ($\pm$)-haemanthamine[50]. A biogenetically-patterned synthesis[44] employing
phenol-oxidative coupling, and an analogous photochemical aryl coupling
reaction[51] have led to simple and efficient routes to two crinine type alkaloids.

8.6.4.1 ($+$)-3-Epi*elwesine* [($+$)-*dihydro*epi*crinine*]

An approach designed upon the utilisation of the generally useful concept of
annelation of an endocyclic enamine, which was independently conceived
by Curphey[52], Stevens[53], and Tahk[54], and first employed in a successful

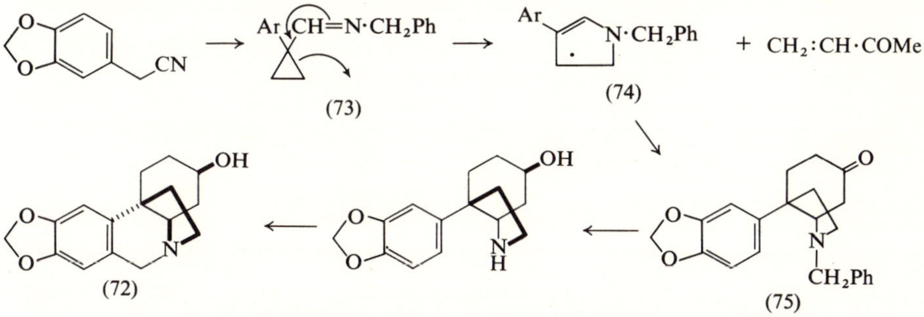

Figure 8.5 Synthesis of ($\pm$)-3-*epi*elwesine(($\pm$)-dihydroepicrinine)(72)

synthesis of the *Aizoacea* alkaloid, mesembrine, has been extended (Figure
8.5) by Stevens and DuPree[49] to the synthesis of ($\pm$)-3-*epi*elwesine (72), a
minor alkaloid of *Galanthus elwesii*. 6-Phenylpiperonyl cyanide was converted

in three steps to the cyclopropyl imine (73) which underwent an acid-catalysed thermal rearrangement on heating with ammonium chloride to the endo-cyclic enamine (74). Annelation of the enamine with methyl vinyl ketone occurred in the expected manner to give the *cis*-octahydroindole (75). Debenzylation of the major epimer obtained from the sodium borohydride reduction of (75) gave the secondary amine (Figure 8.5) which was subjected to a Pictet–Spengler cyclisation with formaldehyde to complete the synthesis of (±)-3-*epi*elwesine.

8.6.4.2   Haemanthamine

An attractively simple route to this alkaloid has been recorded[50]. The synthesis, which is portrayed schematically in Figure 8.6, is founded upon the construction of the *cis*-hydroindole intermediate (76) by a 2+4 cyclo-addition reaction

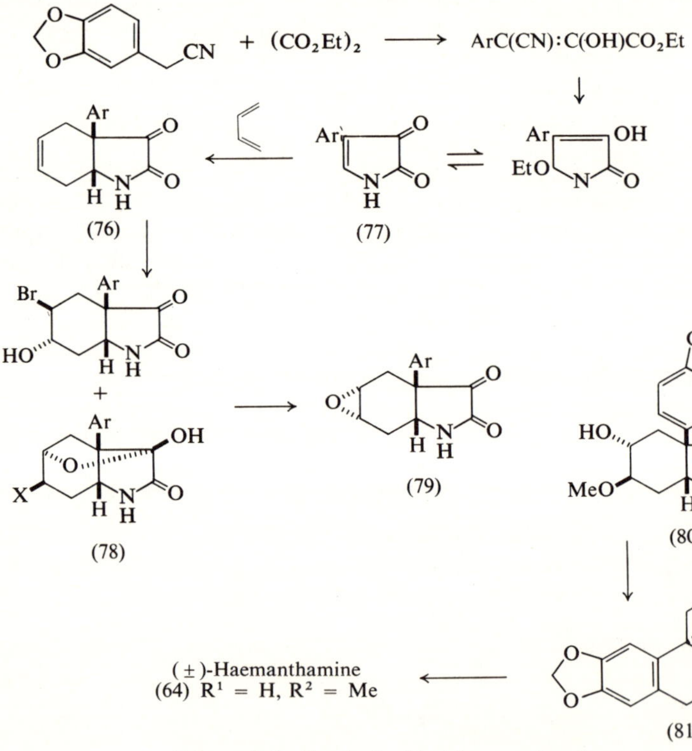

Figure 8.6   Synthesis of (±)-haemanthamine

with butadiene in which the aryl pyrroline-3,4-dione is used as a novel dienophile. Treatment of (76) with *N*-bromoacetamide gave a mixture of a bromo-acetal (78); (X = Br) and a bromohydrin, each of which afforded the same epoxide (79) when treated with methoxide. Opening of the epoxide with borontrifluoride in methanol created the hemiacetal methyl ether (78; X = OMe) which was reduced with lithium aluminium hydride to a single product (80). The stereochemistry of the C-3 hydroxyl generated in the reduction of the hemiacetal was ascribed as shown on the basis of its subsequent conversion to haemanthamine. A Pictet–Spengler reaction of (80) with formaldehyde provided the 11-hydroxy crinine derivative (81; R = H) which was converted to (±)-haemanthamine by a base induced elimination of the monotosylate (81; R = tosyl).

8.6.4.3   *Maritidine*

A synthesis[44] employing phenol-oxidative coupling has provided a short and efficient route to this alkaloid. *O*-Methylnorbelladine (82; R = H) was converted to its *N*-trifluoroacetyl derivative (82; R = COCF$_3$) to protect the nitrogen and presumably to enhance the formation of the *para–para* coupling product (Section 8.3.3). Phenolic oxidation of *N*-trifluoroacetyl-*O*-methyl-norbelladine using vanadium oxytrichloride as the oxidant gave the dienone (83) in 28% yield. Hydrolysis of the trifluoroacetyl group liberated the secondary amine which underwent the expected spontaneous cyclisation to (84). Reduction of the latter with sodium borohydride, followed by treatment of the product with diazomethane gave (±)-3-*epi*maritidine (85). The necessary inversion of configuration at C-3 required to complete the synthesis of (±)-maritidine was accomplished by refluxing 3-*epi*maritidine in hydrochloric

(82)

(83)

(84) R^1 = R^2 = O
(85) R^1 = OH, R^2 = H

(±)-maritidine (69)

acid. This last step parallels the epimerisation reaction used in the synthesis of crinine[46] and results from the known tendency of cyclohexenyl cations to combine with nucleophiles in an axial manner; in the crinine system attack at C-3 is preferred to that of C-1 for steric reasons.

The *quasi*-axial configuration of the C-3 hydroxyl in maritidine which was established by the synthesis is in agreement with the tentative assignment made previously from consideration of its c.d. spectrum[41]. A comparison of the *J*(2,3) values in the n.m.r. spectra of maritidine and its C-3 epimer was in agreement with the assigned structures and consistent with previous values obtained for C-3 epimers in this series[55].

8.6.4.4 (±)-**Epi***crinine*

Photocyclisation of aryl bromides has formed a useful alternative to phenol-oxidative coupling in recent laboratory syntheses of a number of alkaloid systems (Chapter 6). In applying this approach, Kametani and his co-workers[59] have accomplished an extremely simple synthesis of (±)-*epi*crinine (86). Irradiation of an ethanolic sodium hydroxide solution of 2-bromo-*N*-(4-hydroxy-phenethyl)-4,5-methylenedioxybenzylamine (87) gave (±)-oxocrinine (88) in 5% yield. Reduction of oxocrinine over lithium aluminium

hydride gave (±)-*epi*crinine. In view of the experience with similar photo-cyclisations, it is apparent that oxocrinine is formed via the intermediate dienone. What is not clear, however, are the mechanistic details of the photo-processes which lead to the formation of the dienone, either in this example or others which are related to it. The maritidine intermediate (84) employed in the Schwartz synthesis has also been prepared by the photochemical route from photolysis of the bromobenzylaminophenol (87; —O·CH$_2$·O— = MeO, MeO)[56].

8.6.4.5   (—)-Dihydrocrinine

The starting point for the synthesis[48] of this alkaloid utilised the lactam (89). This compound has been obtained previously[57] in an 11-step reaction sequence in which a Schmidt reaction of a spirotetralone had been employed

to introduce the nitrogen atom. A bromination–dehydrobromination sequence served to provide the unsaturated keto-lactam (90) which was cyclised in an interesting manner by subjecting it to treatment with ethylene glycol and toluene-*p*-sulphonic acid to give the acetal (91) as the major product. Conversion of the latter to (±)-dihydrooxocrinine (92; R^1 = R^2 = O) was accomplished employing standard reactions. Resolution of the racemate

followed by a Meerwin–Pondorf reaction of each of the enantiomers gave
$(-)$-dihydrocrinine (92; $R^1 = H$, $R^2 = OH$) and $(+)$-dihydrocrinine.

8.6.4.6 $(\pm)$- *Haemanthidine*

Hendrickson and his co-workers[47] have extended a previous approach[58] in
completing the synthesis of this alkaloid. The lactam acid (93) available from
the previous synthesis, provided the new point of embarkation for the
synthesis of haemanthidine. The plan used allowed for the initial functionalisation of the cyclohexene ring by taking advantage of the angular carboxyl
group to exert the required stereochemical control. This was accomplished
starting with the conversion of (93) into the iodo-lactone (94), which was then
transformed to the epoxy acid salt (95) in base. The latter was then converted

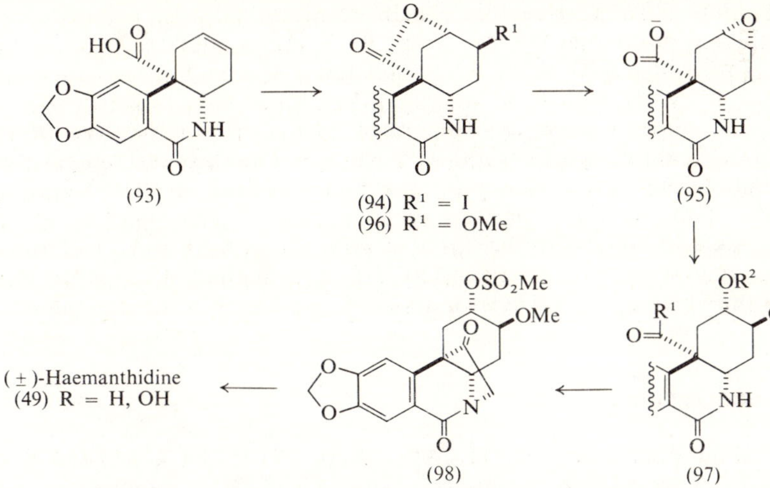

Figure 8.7 Synthesis of $(\pm)$-haemanthidine

to the methoxy-lactone (96) by treatment with boron trifluoride in methanol.
Saponification of the lactone and conversion to the phenacyl ester (97);
$R^1 = O \cdot CH_2 \cdot COPh$, $R^2 = H$) sufficed to liberate the C-2 hydroxyl to allow
mesylation to (97; $R^1 = O \cdot CH_2 \cdot COPh$, $R^2 = SO_2Me$). At this point the
oxygenated two-carbon bridge was next constructed via an acid-catalysed
cyclisation of the diazoketone (97) ($R^1 = CHN_2$, $R^2 = SO_2Me$) to afford
the 6,11-diketone (98). The decision to effect the formation of ring D before
elimination of the methanesulphonate group was based upon the rationale
that the axially disposed C-2 methanesulphonate group would assist in
reversing the trend of hydride reduction found in the 11-ketones of the
natural series which leads to a predominance of the wrong epimer.

The propensity for 6,11-dihydroxycrinines to undergo internal Cannizaro
rearrangement to the tazetting ring system under basic conditions was
considered in the decision which led to the use of disiamylborane for the
reduction of the 6,11-diketone. The product from this reaction was not
isolated but acetylated directly before elimination of the methanesulphonate

group with 1,5-diazabicyclo[3.4.0]non-5-ene to give, after de-acetylation of the products, ($\pm$)-haemanthidine (20%) and its 11-epimer (5%). The success in retaining the C-2 methanesulphonate group to influence the course of the hydride reduction was evident from the small amount of the C-11 epimer produced. Since haemanthidine has been converted to pretazettine and to tazettine, the synthesis described may also be claimed to represent a formal synthesis of these compounds. A direct synthesis of nortazettine (45; NMe = NH) was in fact achieved from the successive treatment of the 6,11-diketone (98) with sodium borohydride and hot alkali.

8.7 ALKALOIDS BASED ON THE 5,11-METHANOMORPHANTHRIDINE SERIES

The alkaloids of known structure which belong to this small sub-group are represented by montanine (99; R^1 = Me, R^2 = H), coccinine (100; R^1 = Me, R^2 = H), manthine (100; R^1 = R^2 = Me) and a recent addition, pancracine (99; R^1 = R^2 = H). Structural studies in this series have relied in part on correlations derived from conversion of 11-hydroxy derivatives of the (+)-crinine alkaloids to the montanine skeleton[59]. The absolute configuration of the montanine series, therefore, is directly related to the (+)-crinine alkaloids. Consequently, the · recent unequivocal determination of the absolute stereochemistry of the latter series (Section 8.5.1 and 8.6.1) allows the single enantiomorphic modification of the 5,11-methanomorphanthridine nucleus on which all of the known alkaloids are based, to be designated as shown.

8.7.1 Pancracine

The alkaloid pancracine, $C_{16}H_{17}NO_4$ (m.p. 272–273 °C), isolated from certain *Narcissus* species is identical with 'Alkaloid 6' obtained from *P. maritinum*[60]. More recently its isolation from *Rhodophiala bifida* has been reported[61] together with montanine, the major alkaloid in this species. Structural studies[62] have led to its characterisation as the 5,11-methanomorphanthridine base (99; R^1 = R^2 = H). Assignment of pancracine to the montanine class was suggested by the finding that hydrolysis of montanine with hydrobromic acid afforded pancracine. The n.m.r. spectrum of O,O-diacetylpancracine (99; R^1 = R^2 = Ac) showed a close resemblance to the spectrum of montanine and when subjected to a detailed study employing double resonance methods it was possible to make a complete assignment of all of the proton signals on the basis of the proposed structure. However, the n.m.r. spectral results did not provide a firm indication of the stereochemical details and further evidence for the structure of pancracine was sought from chemical studies as follows. Controlled hydrolysis of the O,O-diacetate (99; R^1 = R^2 = Ac) provided both mono-acetates, (99; R^2 = H, R^2 = Ac) and (99; R^1 = Ac, R^2 = H), which were differentiated on the basis of the oxidation of the former to an $\alpha\beta$-unsaturated ketone (101).

The same unsaturated ketone was also derived from oxidation of a compound assigned structure (100; R^1 = H, R^2 = Ac), which was reportedly obtained from the rearrangement of the methanesulphonate of 3-*O*-acetyl-11-hydroxyvittatine (102). Unfortunately the details of this latter reaction have not yet been provided. Since the two mono-acetates affording the $\alpha\beta$-unsaturated ketone were not identical, they could only differ in configuration at the C-2 position. On this basis the mono-acetate derived from pancracine was assigned structure (99; R^1 = H, R^2 = Ac) and the representation of pancracine as the $2\alpha,3\beta$-dihydroxy system (99; R^1 = R^2 = H) followed.

While these data taken in conjunction with some additional spectral results would seem to provide good evidence for the proposed structure of pancracine there are several features concerned with the rearrangement of the methanesulphonate (102) which are in need of further clarification. In an earlier review[1], the methanesulphonate was reported to rearrange in the presence of methoxide to montanine, a reaction which, incidentally, constitutes the evidence on which the stereochemistry of the C-3 hydroxyl group in 11-hydroxyvittatine is presently based. When aqueous bicarbonate was used instead of methoxide for the solvolysis of the methanesulphonate a rearranged mono-acetate was obtained. Hydrolysis of this mono-acetate gave a diol which in view of the course of previous rearrangement of (102) with methoxide, was reasonably attributed the $2\alpha,2\beta$-*trans*-dihydroxy structure (99; R^1 = R^2 = H). However, the melting point for this product is over 100 °C lower than that of pancracine to which an identical structure was subsequently attributed[62]; thus the identity of these compounds seems improbable. The implication of the results of this more recent work would suggest that the diol obtained by rearrangement of the methanesulphonate in bicarbonate solution has the hydroxyl groups in a *cis*-relationship as in structure (100; R^1 = R^2 = H). If this is so, then the different stereochemical course of the rearrangement of (102) observed with methoxide when compared to aqueous base is most surprising and in need of further study.

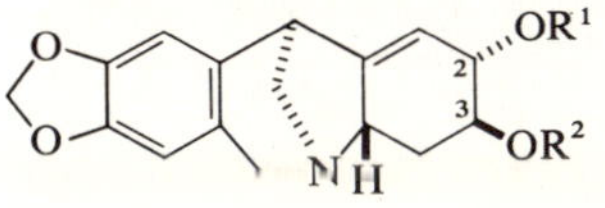

(99) 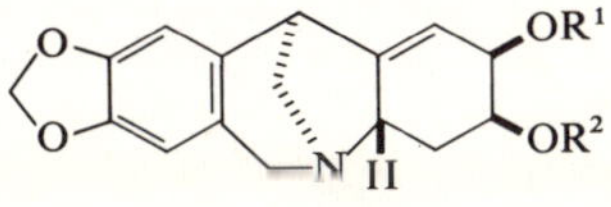 (100) 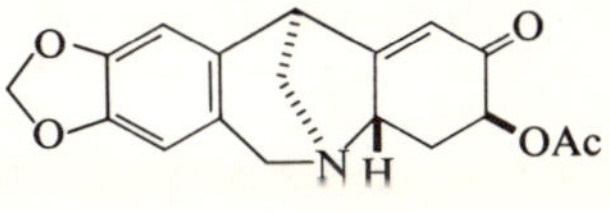 (101)

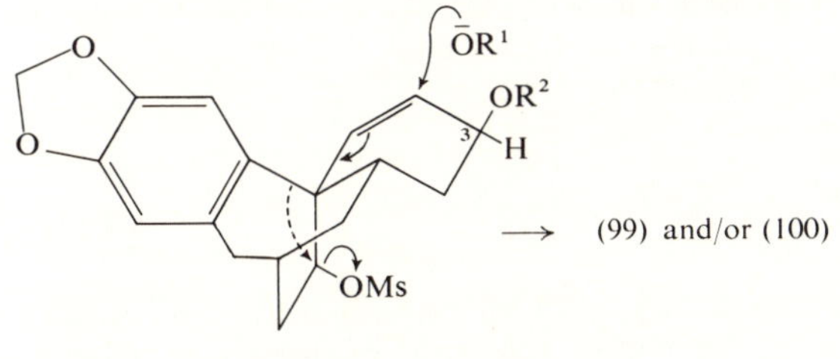

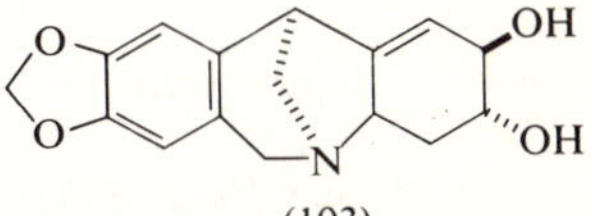

(102) R^1 = Me or H , R^2 = H

(99) and/or (100)

(103)

Two other alkaloids, manthidine and brunsvigine, have been assigned[59] to this sub-group on the basis of the similarity of their i.r. spectra to the other montanine alkaloids. Brunsvigine contains a 1,2-glycol system and it had been tentatively represented by the structure which is now attributed to pancracine. The high melting point (243–245 °C) and the insolubility of brunsvigine in solvents such as chloroform are in accord with a *trans*-glycol.

On this basis the representation of brunsvigine by the 2β,3α-dihydroxy structure (103) should be given consideration.

8.8 LACTONE AND HEMI-ACETAL ALKALOIDS RELATED TO HOMOLYCORINE

Recent progress in structural studies in this field have been rather limited and somewhat surprisingly no reports of attempts to synthesise a member of this sub-group have appeared. With the exception of the alkaloids related to clivonine which are discussed below, the lactone and hemi-acetal alkaloids of this sub-group are based upon a single stereochemical modification of the [2]benzopyrano[3,4g]indole skeleton as exemplified by the structure of

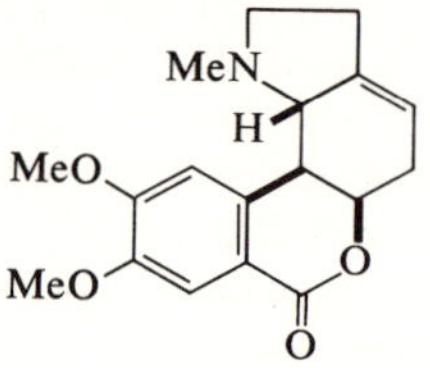

(104)

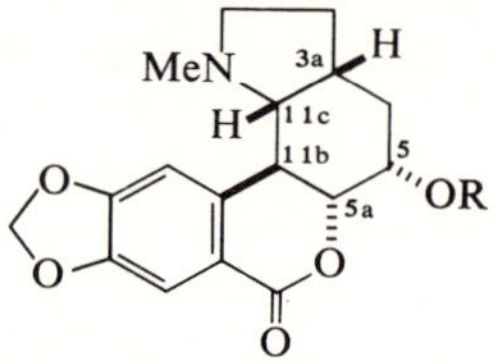

(105) R = H
(106) R = Ac

(112) R =

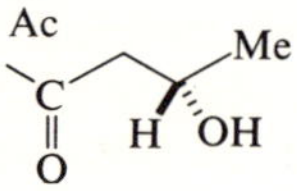

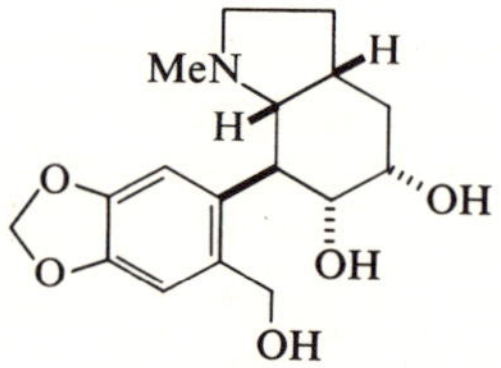

(107)

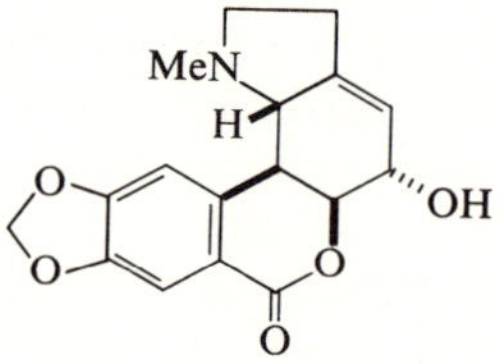

(108)

homolycorine (104). Interconversions within the sub-group and the existence of interrelationships with the lycorine-type bases has placed the absolute configuration of the lactone alkaloids of this series on a firm basis[1].

8.8.1 Clivonine

The plant *Clivia miniata* produces a variety of lactone alkaloids and it has been these alkaloids which have provided the only examples of recent structural investigations in the sub-group. Of central importance to this work has been the elucidation of the structure of clivonine (105), which is a component of several of the more complex alkaloids elaborated by this species. The early studies of clivonine which led to the proposal[63] of the tentative structure (105) (without stereochemical detail) have been extended through a detailed analysis of the n.m.r. spectra of the alkaloid and its *O*-acetyl derivative (106)[64]. The *cis*-relationship of the C-5 and C-5a oxygen functions which had been previously suggested[65] from a comparative study of the rates of periodate cleavage of the clivonine reduction product (107) and the isomeric *trans*-5,5a-glycol derived from hippeastrine (108), was supported by a comprehensive analysis of the n.m.r. spectra of the alkaloid and its *O*-acetyl derivative which has led to a full expression of the stereochemistry of clivonine as indicated. The stereochemical assignments were fully confirmed by the transformation of clivonine via the 5β-chlorolactone (109) to the known dihydrohippeastrine (110). The inversion of configuration at both C-5 and C-5a which has to occur in the conversion of the chlorolactone to dihydrohippeastrine under the reaction conditions involving successive base hydrolysis and acid treatment was rationalised by a mechanism which requires two successive intramolecular displacements; the first forms the intermediate epoxide (111) and the second produces (110) by ring opening of the protonated epoxide by the neighbouring carboxyl group.

8.8.2 Clivatine

The occurrence of clivatine, m.p. 166–169 °C, $[\alpha]_D + 54$ degrees, in *C. miniata* was first reported by Mehlis[66] who determined the molecular formula $C_{21}H_{25}NO_7$ and tentatively ascribed it to the lactone series on the basis of the close correspondence of its u.v. spectrum with that of clivonine. Further progress was made by the finding that clivatine could be converted to clivonine through base hydrolysis and re-lactonisation of the product. This suggested that clivatine should be represented as a hydroxybutanoic ester of clivonine. A detailed spectral study[67] of the alkaloid has confirmed the results of the earlier work and has led to the representation of its structure as (112). The n.m.r. spectrum of clivatine was found to bear a strong similarity to the n.m.r. spectrum of 5-*O*-acetylclivonine (106) and in addition showed a methyl doublet at δ 1.24 ($J = 7.0$ Hz). The location of a hydroxyl on the penultimate carbon of the four-carbon ester side-chain suggested by the chemical shift of the methyl doublet was supported by the occurrence of strong fragment ions at $m/e = 388$ ($M - 15$) and $m/e = 45$ ($CH_3CH\!=\!\overset{+}{O}H$) in the mass spectrum of the alkaloid. A paucity of material precluded full characterisation of the 3-hydroxybutanoic acid produced in the hydrolysis of clivatine, however, its tentative assignment as the R-isomer has been made on the basis that a solution containing the acid showed a positive rotation[68].

8.8.3 Clivimine and miniatine

The structure of clivimine as a bis-[5-O-clivonyl]diester of lutidine-3,5-dicarboxylic acid was established by Mehlis[65] in 1965 utilising the reactions depicted schematically in Figure 8.8. The subsequent revision of the stereo-

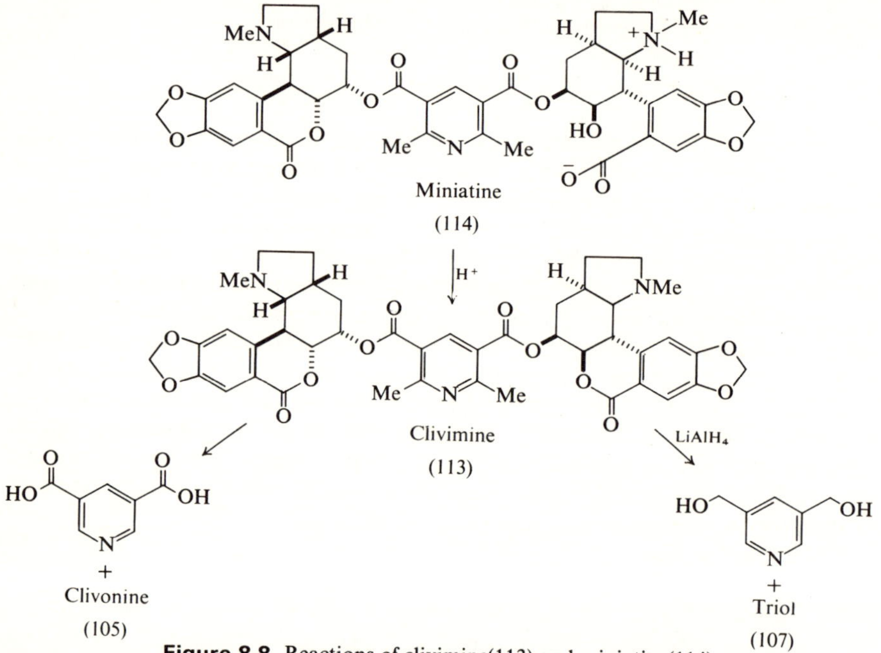

Miniatine
(114)

H^+

Clivimine
(113)

$LiAlH_4$

Clivonine
(105)

Triol
(107)

Figure 8.8 Reactions of clivimine(113) and miniatine(114)

chemistry of clivonine required that clivimine should be represented by structure (113) and n.m.r. and mass spectral results have been presented which provide confirmation of this structure[69].

Miniatine, $C_{43}H_{45}N_3O_{13}$, is closely related to clivimine as evidenced by its conversion to this alkaloid in warm dilute hydrochloric acid. In all of the other reactions which have been reported for this base, the products obtained parallel those reported for clivimine. The mass spectrum of miniatine does not show a parent ion and an ion corresponding to $M-H_2O$ at $m/e = 793$ is the highest mass peak observed in the spectrum. However, a titration against tetramethylammonium hydroxide was reported to correspond to the calculated value of 811. The above evidence has been interpreted[70] to indicate that miniatine is best represented by structure (114). The possibility that miniatine is an artifact has been raised, since chromatography of clivimine over alumina is reported to give miniatine.

8.8.4 Clividine

This alkaloid, m.p. 195–197 °C, is a minor constituent of *C. miniata* and its characterisation as a lactone alkaloid was apparent from its i.r. (CO at 1715 cm^{-1}) and u.v. spectral properties. The designation of clividine as a

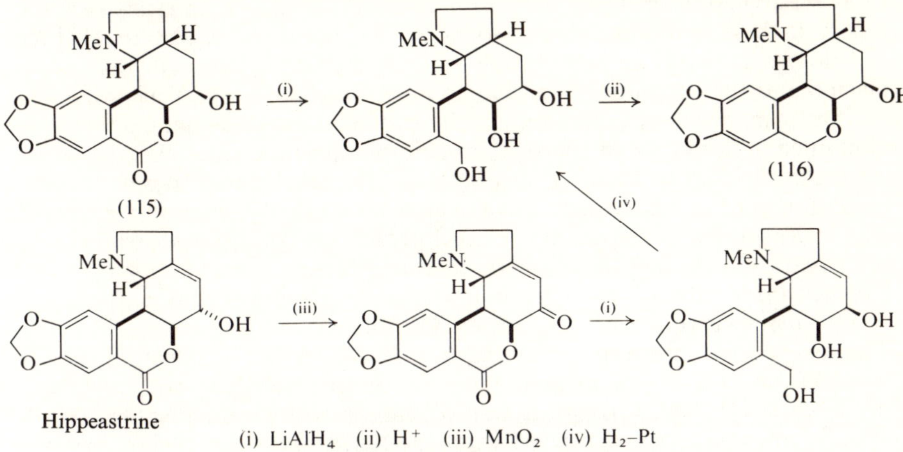

Hippeastrine

(i) LiAlH$_4$ (ii) H$^+$ (iii) MnO$_2$ (iv) H$_2$–Pt

Figure 8.9   Interrelation of hippeastrine and clividine(115)

stereoisomer of dihydrohippeastrine was suggested by the virtual identity of the mass spectra of these two compounds. A comprehensive analysis[67] of the n.m.r. spectrum of the alkaloid supported the preliminary findings and suggested that clividine is best represented by structure (115). Chemical evidence in support of this assignment has been provided by the conversion of clividine and hippeastrine to the cyclic ether (116) by the reactions shown schematically in Figure 8.9[71].

8.9 NARCICLASINE AND RELATED COMPOUNDS

8.9.1 Narciclasine, lycoricidinol and narciprimine

Independent work in Italy and Japan has led to the isolation from *Amaryllidaceae* plants of several compounds containing antimitotic properties. Two compounds, margetine and narciclasine, along with a related phenanthridone, narciprimine, which is possibly an artifact, have been isolated by the Italian group[72]. The antimitotic substances isolated by the Japanese[73], lycoricidine and lycoricidinol, show a close correspondence in their respective physical properties with margetine and narciclasine which suggests their identity with these compounds, and despite some early discrepancies in the structures assigned to narciclasine and lycoricidinol, there is now chemical evidence to support this conclusion.

Narciclasine, a non-basic phenolic compound showing a strong yellow–green fluorescence and with antimitotic properties was isolated together with a structurally related compound, narciprimine, from various *Narcissuss* species[72]. Narciclasine, $C_{14}H_{13}NO_7$, showed both strong hydroxyl absorption and an amide carbonyl band at 1675 cm^{-1} in its i.r. spectrum. The n.m.r. spectrum of narciclasine possessed absorptions characteristic of an intramolecular hydrogen bonded phenol, (δ 13.23), an amide NH (δ 7.85), a single aromatic hydrogen (δ 6.87) and a methylenedioxy group (δ 6.18). Conversion of O-methylnarciclasine to an O,O,O-triacetate and a comparison of the n.m.r. of these two compounds indicated that narciclasine contained three secondary hydroxyl groups. Catalytic hydrogenation gave dihydro-narciclasine and the accompanying change noted in the u.v. spectral properties of this compound as compared to the parent suggested that narciclasine contained a double bond conjugated with the aromatic ring. The formation of phenanthridine on zinc dust distillation of the alkaloid and permanganate oxidation of its O-methyl-derivative to cotarnic acid (117) were also reported. The arrangement of the three hydroxyl groups on contiguous carbons was indicated from the fact that O-methylnarciclasine consumed 2 mol of HIO$_4$. Further information was derived from the observation that narciclasine on treatment with dilute hydrochloric acid afforded narciprimine, $C_{14}H_9NO_5$. The u.v. spectral properties of narciprimine differed markedly from that of narciclasine and it was evident that the double dehydration which had occurred in this transformation, was accompanied by aromatisation. More specific information was derived from the n.m.r. spectrum of narciprimine which in addition to exhibiting an aromatic singlet at δ 7.50 also contained an ABX pattern in the aromatic region of a type characteristic of three adjacent hydrogens. Of the two possible structures, (118) and (119), which are consistent with the foregoing evidence, the former was selected by the Italian workers to represent narciprimine, and narciclasine was accordingly depicted by structure (120). The validity of these assignments was placed in doubt by the results of an independent study[73] in Japan on the growth inhibitory substances, lycoricidinol and lycoricidine, isolated from *Lycoris radiata*. Lycoricidinol, from its chemical and spectral properties, appears to be identical with narciclasine. Unfortunately, the Japanese authors did not report on the optical activity of their compound and in the absence of a direct

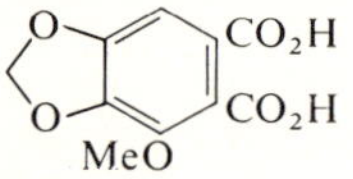

(117)

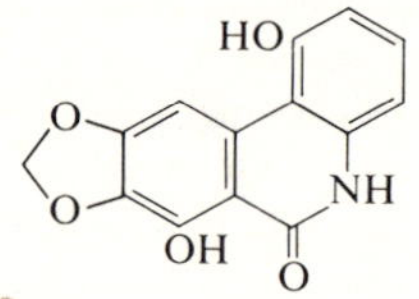

(118)

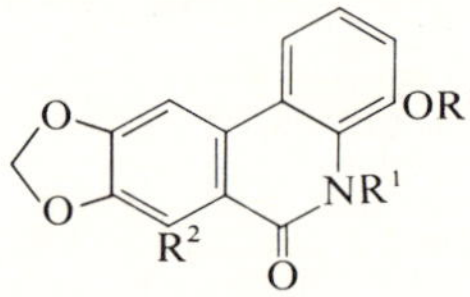

(119) R^1 = H, R^2 = OH
(122) R^1 = Me, R^2 = OMe
(123) R^1 = R^2 = H

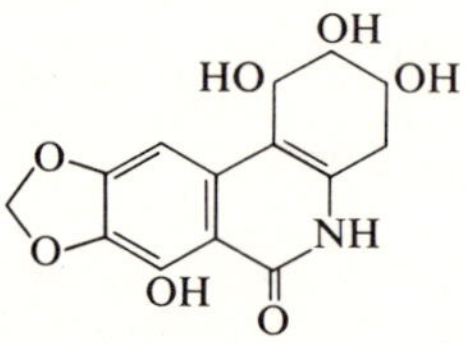

(120)

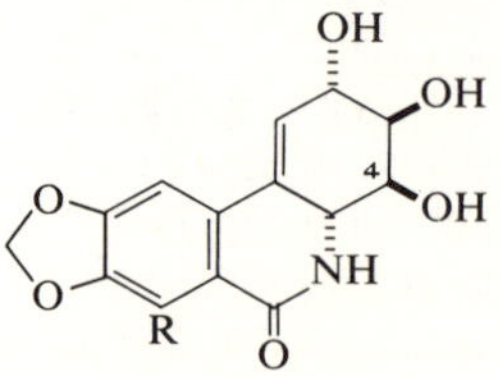

(121) R = OH
(124) R = H

comparison of samples the identity of lycoricidinol and narciclasine cannot be considered as firmly established. A key point in the elucidation of the structure of lycoricidinol proved to be the demonstration that a signal which appeared in the region of δ 6.10 in the n.m.r. spectra of the alkaloid and its derivatives was assignable to an olefinic hydrogen since its absence was noted in the spectra of compounds in the dihydro-series. The reaction of lycoricidinol with hydrochloric acid paralleled that reported for narciclasine in affording a product which corresponded in its properties to those reported for narciprimine and provided additional evidence for the identity of these compounds.

On this basis, lycoricidinol and its dehydration product were assigned the structures (121) (no stereochemical implication) and (119), respectively.

Proof of the structure of narciprimine was obtained by Mondon and Krohn[74, 75] who completed separate syntheses of both (118) and (119) and confirmed the identity of the alkaloid with the latter. The routes employed are outlined schematically in Figure 8.10. The first step in the synthesis of the

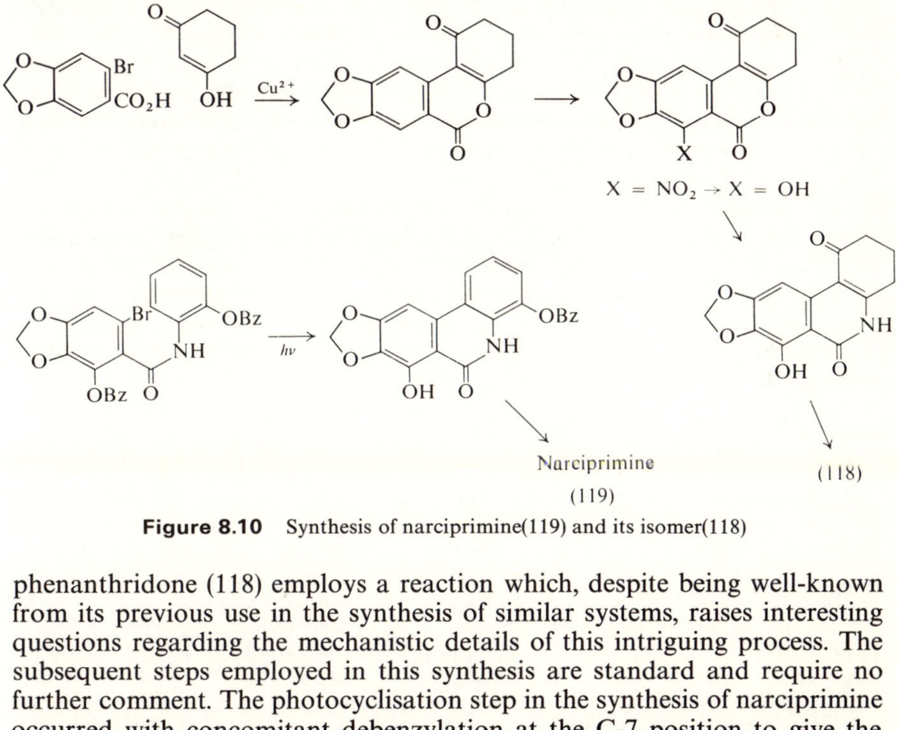

Figure 8.10 Synthesis of narciprimine(119) and its isomer(118)

phenanthridone (118) employs a reaction which, despite being well-known from its previous use in the synthesis of similar systems, raises interesting questions regarding the mechanistic details of this intriguing process. The subsequent steps employed in this synthesis are standard and require no further comment. The photocyclisation step in the synthesis of narciprimine occurred with concomitant debenzylation at the C-7 position to give the monobenzylphenanthridone, which was subsequently converted to narciprimine by hydrogenolysis. Conversion of narciprimine to its permethylated derivative (122) and synthesis of the latter via a Pschorr reaction has been recently reported[76].

Confirmation of the structure of narciclasine and elucidation of the relative stereochemistry was recently provided by a single-crystal x-ray structure analysis of narciclasine tetra-acetate[77]. These results which differ in stereochemical detail from an earlier proposal made by Mondon and Krohn[73] have been extended to include the absolute configuration as represented by structure (121) from the observation that (+)-crinine, but not its enanantiomer, serves as an intermediate in the biosynthesis of narciclasine (Section 8.11.6)[78].

8.9.2 Lycoricidine and margetine

Lycoricidine, $C_{14}H_{13}NO_4$, a non-basic substance exhibiting bands associated with hydroxyl, amide ($1660\ cm^{-1}$), and methylenedioxy groups in its i.r. spectrum, differs in one significant respect from lycoricidinol in being non-phenolic. The structural studies[73] of this alkaloid proceeded in parallel with those described for lycoricidinol. Catalytic hydrogenation of the alkaloid gave dihydrolycoricidine. A comparison of the u.v. spectra of the dihydro-derivative and the parent alkaloid confirmed the conjugated nature of the double bond in the latter. An O,O,O-triacetate was obtained on acetylation and the similarity of the n.m.r. spectra of this and the corresponding derivative of lycoricidinol indicated the presence of three vicinal hydroxyls. Lycoricidine when heated with palladium–charcoal or methanolic HCl was converted readily to a phenanthridone, which was assigned structure (123) on the basis of its u.v. spectrum and the presence of an ABX pattern in the aromatic region of its n.m.r. spectrum.

The conversion of (123) to its O-methyl-N-benzyl derivative, which was synthesised via a Pschorr reaction, provided the first chemical evidence for location of a hydroxyl group at the C-4 in the alkaloid and its dehydration product. Lycoricidine was accordingly represented as (124). The identity of lycoricidine with margetine[79], a growth inhibitory substance isolated from *Narcissus* species has been suggested from the close correspondence of the physical properties of these compounds[75].

The names narciclasine and margetine have (chronological) priority over lycoricidinol and lycoricidine and if the identities suggested are confirmed, as seems likely, the use of the names lycoricidinol and lycoricidine should be discontinued.

8.10 CHERYLLINE

Examination of the alkali-soluble alkaloid fraction of certain *Crinum* species has provided cherylline, a new phenolic base[80]. The structural assignment of cherylline as (125) was made primarily from considerations of spectral evidence. The n.m.r. spectrum of the alkaloid showed signals corresponding to an aromatic methoxyl, an N-methyl and an A_2B_2 pattern (δ 6.91 and 6.64) and a pair of 1-proton singlets (δ 6.49 and 6.23) indicative of the presence of both a 1,4-disubstituted aromatic ring and a tetra-substituted aromatic ring containing two hydrogens in a *para* relationship. Mass spectral results and elemental analysis established the molecular formula as $C_{17}H_{19}NO_3$. The phenolic character of the alkaloid was evident from the characteristic base-induced bathochromic shifts of maxima in the u.v. spectrum of cherylline and this was verified by the conversion of cherylline to an *O,O*-dimethyl derivative (126) with diazomethane. The structure of *O,O*-dimethylcherylline was confirmed by a routine synthesis of the racemate and both chiral forms. Since the alkaloid did not show the properties of a catechol, a decision which led to the assignment of the ring A phenolic

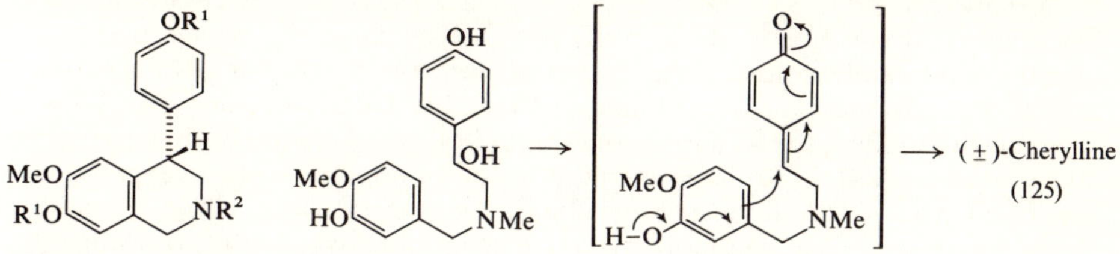

(125) R¹ = H, R² = Me
(126) R¹ = R² = Me
(127) R¹ = Me, R² = H

(128)

→ (±)-Cherylline
(125)

hydroxyl to C-7 in proposing structure (125) was made from the observation that the C-8 hydrogen signal in the n.m.r. spectrum of cherylline underwent a diamagnetic shift upon addition of base. An initial assignment of the absolute configuration as indicated in structure (125) was made from a comparison of the o.r.d. and c.d. spectra of (−)-cherylline and its *O,O*-dimethyl derivative with the data from model compounds in the 4-aryltetraline series. This structure was verified by a single-crystal x-ray analysis of a heavy atom derivative of the chiral intermediate (127) which had been employed in the synthesis of (−)-*O,O*-dimethylcherylline.

Two independent syntheses of (±)-cherylline and a synthesis of the natural, (−)-isomer have confirmed the structure of the alkaloid. Brossi and his colleagues[81] have accomplished a synthesis of the alkaloid by the route indicated schematically in Figure 8.11. The choice of hydrobromic acid to

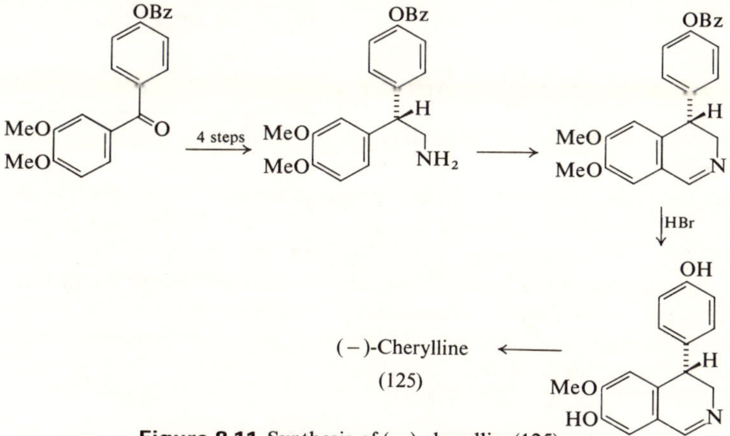

Figure 8.11 Synthesis of (−)-cherylline(125)

effect concomitant debenzylation and selective cleavage of the C-7 methoxyl in the dihydroisoquinoline intermediate was based on findings from earlier studies in which selective *O*-demethylation reactions in the isoquinoline and dihydroisoquinoline series under specific conditions had been demonstrated.

A repetition of the synthesis using a resolution step to provide both *R* and *S* forms of the phenethylamine intermediate gave (−)-cherylline from the *S*-isomer and the unnatural antipodal form of the alkaloid from the *R*-isomer.

An ingenuously simple synthesis of (±)-cherylline based upon a suggested biogenetic route has been described by Schwartz and Scott[82]. In cognisance of the general involvement of the norbelladine system in the biosynthesis of *Amaryllidaceae* alkaloids, it was suggested that the biogenesis of cherylline could be accounted for in a similar manner by the conversion of *O,N*-di-methylnorbelladine to the quinonemethide (128) which would be expected to cyclise to the alkaloid. In devising a laboratory model to test this suggestion the (±)-hydroxy-*O,N*-dimethylnorbelladine was synthesised. An extremely facile conversion of this compound to (±)-cherylline was found to occur in either acidic or basic medium and for preparative purposes refluxing ammonium hydroxide was used to give the alkaloid in 79% yield.

8.11 BIOSYNTHESIS

8.11.1 Introduction

The major details of the biosynthetic pathway by which phenylalanine and tyrosine are converted via the norbelladine system into alkaloids representing the crinine, lycorine, and galanthamine series was largely defined by tracer studies carried out in the period 1960–1965. With few exceptions more recent efforts have been concerned with stereochemical aspects of hydroxylation and protonation processes and with defining the pathways by which alkaloids of the other major sub-groups originate. The origin of these sub-groups, none of which are directly derivable by phenol-oxidative coupling of a norbelladine intermediate, are reasonably derived in the case of pretazettine and montanine types from rearrangement of the appropriate hydroxylated crinine intermediates while those of the homolycorine series can occur from rearrangement of the lycorine skeleton.

Studies reported[83] in 1964 pertaining to the biosynthesis of tazettine can now be reinterpreted in terms of the biosynthesis of pretazettine. The sequence haemanthamine ⟶ haemanthidine ⟶ pretazettine is indicated, with the last step presumably proceeding via rearrangement of the haemanthidine metho-salt (Section 8.4) rather than rearrangement of haemanthidine and subsequent-*N*-methylation. No evidence is yet available to support the derivation of the montanine-type alkaloids from the crinine system – the only report concerning this was a brief comment[1] that labelled haemanthamine was not converted to manthine in *Haemanthus coccinius*. Studies on the *in vivo* formation of the homolycorine alkaloid system have now been obtained and a discussion of these experiments is presented later in this section.

8.11.2 Galanthamine-alkaloids

8.11.2.1 *O-Methylnorbelladine a a precursor to galanthamine in L. aestivum*

Early experiments by Barton and his co-workers[84] have defined the role of norbelladine intermediates in the biosynthesis of galanthamine in the *narcissus* 'King Alfred' daffodil. The sequence norbelladine ⟶ *N*-methyl-norbelladine ⟶ *O,N*-dimethylnorbelladine was established by the appropriate labelling experiments and the importance of this order of methylation was supported by the failure to observe any significant incorporation of *O*-methylnorbelladine into galanthamine.

More recent experiments[85] with *Leucojum aestivum* have shown that *O*-methylnorbelladine, containing ^{3}H and ^{14}C-labels as indicated in structure (129) is incorporated intact into galanthamine and lycorine. The $T/^{14}C$ ratio in galanthamine was the same as in the labelled *O*-methylnorbelladine. Proof that the tritium was located at the expected C-2 and C-4 positions was obtained by oxidation of the galanthamine to narwedine which was found to lose all tritium activity after chromatography over alumina. Obviously these results imply that the biosynthetic route to galanthamine in *L. aestivum* and 'King Alfred' daffodils are different. This is a rather disconcerting finding. It is hoped that exceptions of the kind are rare, otherwise the seemingly reasonable extrapolation of results which have been made in the past for the biosynthesis of a particular alkaloid using different plants species may require re-examination.

8.11.2.2 *Conversion of narwedine, galanthamine and chlidanthine*

The structure of chlidanthine, when considered in relation to narwedine and galanthamine, suggests that one possible explanation of its biosynthesis is

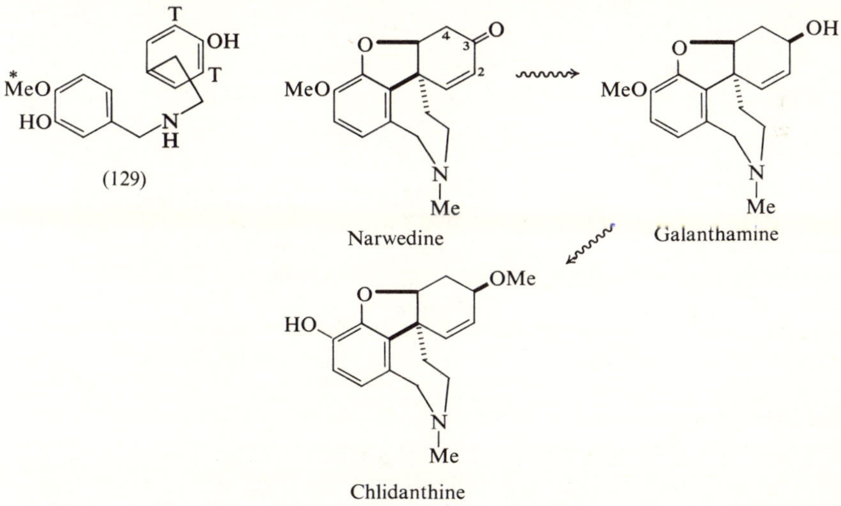

(129)

Narwedine

Galanthamine

Chlidanthine

that it is derived from galanthamine by methylation and demethylation. This proposal was substantiated by the incorporation of [^{3}H]narwedine and [^{3}H]galanthamine, each containing tritium at the C-2 and C-4 positions, into chlidanthine in *Chlidanthus fragrans*[14]. The observation that narwedine, labelled as described, was incorporated to the extent of 7.7% into galanthamine established the occurrence of the suspected sequence narwedine ⟶ galanthamine ⟶ chlidanthine. The biological demethylation of a phenol methyl ether in the formation of chlidanthine, although somewhat unusual, has precedent as a late stage process in the conversion of codeine to morphine.

8.11.3 Lycorine alkaloids

8.11.3.1 *Formation of the pyrrolo* [d, e] *phenanthridine ring system*

The results of tracer experiments prior to 1965 had led to the elucidation of the major features concerning the biosynthetic route to the lycorine system. The utilisation of *O*-methylnorbelladine as the direct progenitor of the basic ring system of this sub-group and introduction of the C-2 oxygen at a late stage in the biosynthesis of lycorine were two of the key points which have provided the basis for the further studies reported below.

One of the major areas of interest has been the elucidation of the precise details of the conversion of *O*-methylnorbelladine to norpluviine. The incorporation of *O*-methylnorbelladine, labelled as in (130), into norpluviine occurred with loss of two of the original four tritium labels[86]. Proof of the location of one of the residual tritiums in norpluviine at C-8 was provided by removal of 50% of the tritium activity by base catalysed exchange. Conversion of the labelled norpluviine (131) to the lactam (132) was shown to occur without loss of tritium and excluded any possibility of occurrence of tritium at the C-11b position. Bromination of the lactam to (133) was accompanied by a 50% loss of the label and established that the second tritium is at the C-2 position.

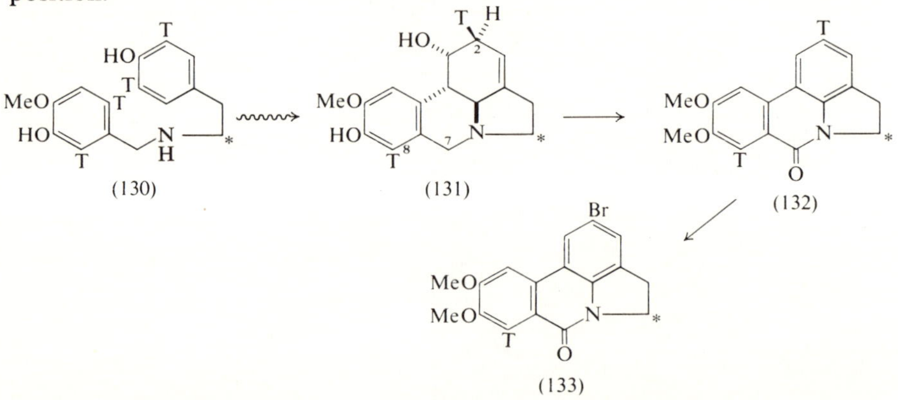

The loss of tritium from the C-11b position in the formation of norpluviine from *O*-methylnorbelladine has at least two possible explanations (Figure 8.12): the conversion may proceed by path *a* via the medium ring biphenyl and the diphenoquinone as originally suggested by Barton and Cohen[87], or if the alternative, and more expedient route *b*[88] has any validity, then it is necessary to invoke the intervention of an additional step involving enolisation towards C-11b in the intermediate enone (134) prior to reduction to norpluviine.

The required tracer experiments with the medium-ring biphenyl which would provide a means of distinguishing between the two pathways have

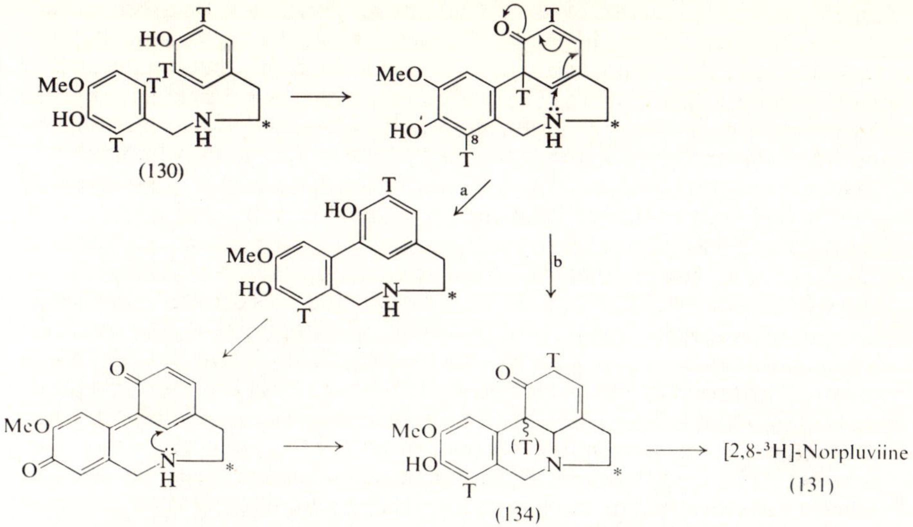

Figure 8.12 Possible biosynthetic routes from *O*-methylnorbelladine to norpluviine

still to be accomplished. In relation to this subject it is appropriate to call attention to a precedent for pathway *a* which now exists from the recent work from Barton's laboratory[89] on the biosynthesis of the erythrina alkaloids. The

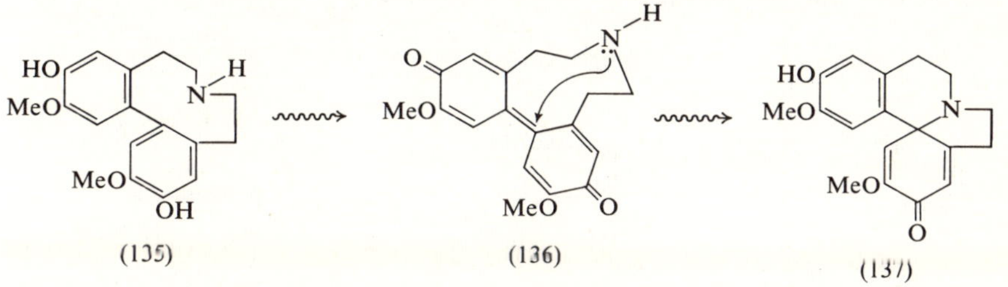

sequence involving oxidation of the biphenyl(135) to the diphenoquinone (136) and the cyclisation of the latter to the *erythrina* system (137) have been demonstrated in an elegant manner.

8.11.3.2 *Stereochemistry of protonation and hydroxylation processes*

A major point of interest concerning the incorporation of (130) into nor-pluviine concerns the stereochemical aspects of protonation at C-2 which must occur in this process. The retention of tritium at C-2 in the lactam (132) implies that a stereospecific protonation occurs at C-2 in the biosynthesis of norpluviine and that the subsequent chemical aromatisation process used in converting the alkaloid to the lactam is similarly stereospecific. Although alternative explanations are possible, further evidence in support of this suggestion has been provided[90]. Norpluviine derived from the labelled

O-methylnorbelladine (129) was converted by a previously described route[91] to the methiodide (138). The methiodide when treated with sodium ethoxide in ethanol gave the tritium free amine (139) in a reaction which was shown to follow second-order kinetics. On this evidence the reasonable assumption was made that the latter reaction involved a concerted 1,4-*cis*-elimination and the C-2β-configuration was assigned to the tritium in norpluviine.

The introduction of the C-2 hydroxyl at a late stage in the biosynthesis of lycorine had been indicated from the provisional findings that norpluviiine-[^{3}H-random] was converted to lycorine. More recent work has established this point and shown that the sequence norpluviine $\longrightarrow$ caranine $\longrightarrow$ lycorine occurs with the C-2 hydroxylation at the saturated methylene proceeding stereospecifically with inversion of configuration. The evidence on which this rests is provided by two investigations. In the first definitive study of this problem, stereospecifically labelled [2β-^{3}H]caranine, prepared from lithium aluminium [^{3}H]-hydride reduction of the epoxide (140), when administered to *Zepheranthes candida*, gave a 7.05% incorporation into lycorine[92]. The complete loss of tritium which occurred when the labelled lycorine was converted to the 2-keto-compound, jonquilline (141), established the site of the label at C-2 and suggested that hydroxylation of caranine to lycorine proceeded with at least partial, and possibly even complete, inversion of configuration. Parenthetically, the intervention of a 2-keto-intermediate in the biosynthesis of lycorine is also excluded by this result.

(129) $\leadsto$ [2-^{3}H; OMe-^{14}C]Norpluviine

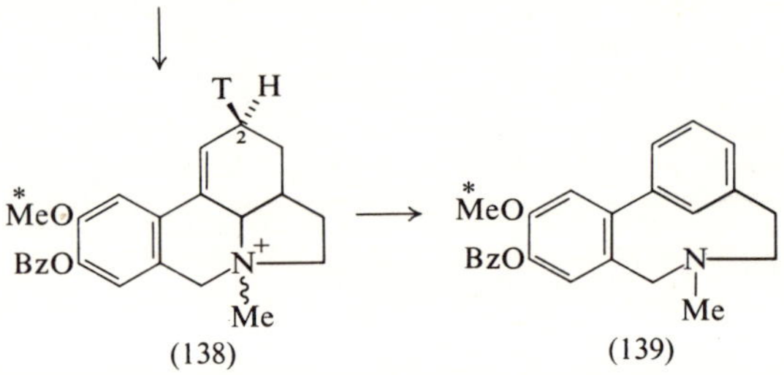

(138)   (139)

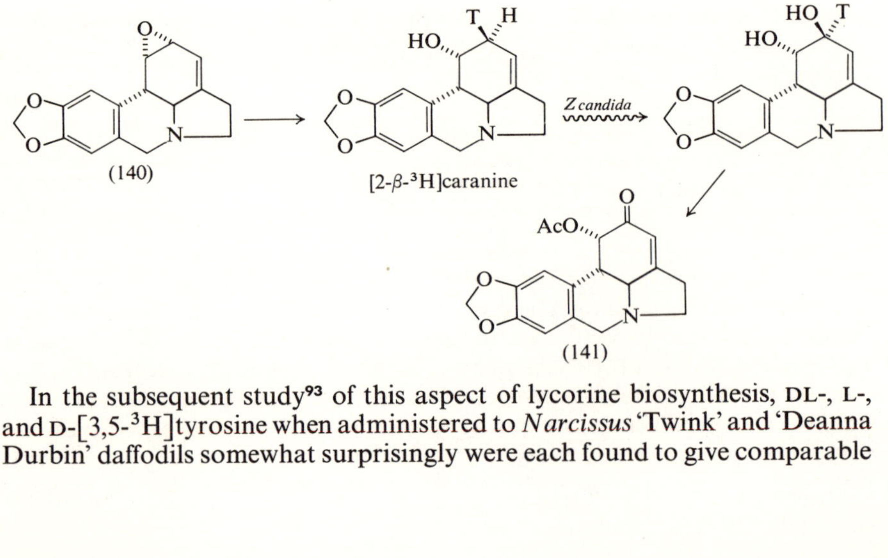

(140)

[2-β-^{3}H]caranine

(141)

In the subsequent study[93] of this aspect of lycorine biosynthesis, DL-, L-, and D-[3,5-^{3}H]tyrosine when administered to *Narcissus* 'Twink' and 'Deanna Durbin' daffodils somewhat surprisingly were each found to give comparable

incorporations into norpluviine and lycorine. Although the expected 50%
loss of the tritium occurred from each labelled precursor in the formation of
norpluviine, more importantly, the tritium content of the lycorine was shown
to be the same as that of the norpluviine for each comparable experiment.
Degradation of the labelled lycorine to (141) established that the tritium
resided exclusively at the C-2 position. Taken together these experiments
confirm the late steps in the biosynthesis as indicated above and demonstrate
the occurrence of a stereospecific hydroxylation with inversion of configuration in the conversion of caranine to lycorine.

Since all other examples of biological hydroxylations at a saturated
methylene group that have been examined are found to proceed stereospecifically with retention of configuration, this has prompted suggestions[92, 93]
that the conversion of caranine to lycorine may involve several steps and both
the 1,2α-epoxide (140) and caranine 3,3a-β-epoxide have been invoked as
possible intermediates.

8.11.3.3 *Aromatic hydroxylation*

Enzymic hydroxylation of certain aromatic substrates is known to proceed
in mammalian and bacterial systems via arene oxides by a process which
involves a 1,2-hydrogen migration (NIH shift)[94]. Illustrative of this reaction
is the conversion of [4-³H]phenylalanine to tyrosine which occurs with
migration of the tritium to the *ortho*-positions corresponding to 90%
retention of the label. A study of aromatic hydroxylation in the biosynthesis
of norpluviine has been reported[95]. Using the two labelled cinnamic acids
(142) and (143) in separate feeding experiments with 'Texas' daffodils, the loss
of tritium during their conversion to norpluviine was found in both instances
to be close to 50%. This result is consistent with a 1,2 shift of tritium during
the conversion of cinnamic acid to *p*-hydroxycinnamic acid and the complete
loss of tritium during the introduction of the second hydroxyl at one of the two
equivalent positions during its subsequent conversion to 3,4-dihyroxy-
cinnamic acid on the pathway to norpluviine. Since a H/T isotope effect is
expected to be large in comparison to the D/T isotope effect, the complete
retention of tritium in the first hydroxylation of each of these labelled
cinnamic acids requires that proton loss from the intermediate arene oxide
(144) is stereospecifically controlled rather than isotopically controlled, with
the migrating hydrogen being retained. Complementary evidence for the
stereospecificity of proton removal in the arene oxide intermediate was
provided by the observation that norpluviine derived from [4-³H, 3-¹⁴C]

(142) X = H
(143) X = D

(144) X = H or D

[7-¹⁴C; 2-³H]Norpluviine

cinnamic acid retained 28% of the original tritium. This is in good agreement with the predicted 25% retention for the occurrence of *para*-hydroxylation with hydrogen migration and 50% loss of tritium in the first hydroxylation and a subsequent 50% loss of the remaining tritium during the introduction of the second hydroxyl group.

The stereospecific, and hence enzymatically controlled, loss of hydrogen in the conversion of the arene oxide (144) to *p*-hydroxycinnamic acid is in conflict with the results obtained in mammalian[94], bacterial[94] and fungal[96] systems where in experiments with labelled substrates the loss of hydrogen is isotopically controlled. Kirby and his co-workers[95] have indicated that this discrepancy can be explained by assuming that the partially-purified enzyme systems used by other workers contained the oxygenase enzyme but may have lacked the isomerase, whereas the intact plant system contained both enzymes. This argument has been challenged but the evidence on which it is based is not yet available[96].

8.11.4   Crinine alkaloids

The major emphasis in biosynthetic studies of this group has been concerned, as in the lycorine alkaloids, with elucidation of stereochemical details of certain of the individual steps in the pathway to the alkaloids of this sub-group.

8.11.4.1   *Incorporation of protocatechuic acid and benzylic hydroxylation*

The conversion of protocatechuic aldehyde and tyramine to *O*-methyl-norbelladine occurs by reduction of the intermediate imine. Elucidation of the stereochemistry of this reduction step is possible since one of the prochiral hydrogens of the benzylic methylene group in *O*-methylnorbelladine is lost through hydroxylation in its conversion to haemanthidine. In studying[97] the stereospecificity of the reduction and hydroxylation processes, *O*-methyl-norbelladine containing a single non-stereospecific tritium label at C-1' and a ^{14}C label at C-1 was incorporated by *Sprekelia formossissima* plants into haemanthamine without loss of tritium whereas the formation of haemanthidine in the same experiment occurred with loss of close to 50% of the original tritium. This demonstrated that the benzylic hydroxylation of haemanthamine to haemanthidine is stereospecific. A biosynthetic preparation of stereospecifically labelled 6*R* or 6*S*-[6-^{3}H]haemanthamine was achieved from administering [formyl-^{3}H]protocatchuic aldehyde to 'Twink' and 'Texas' daffodils. Combination with a ^{14}C-sample of haemanthamine derived from feeding [1-^{14}C]-*O*-methylnorbelladine to the same plants gave doubly-labelled haemanthamine containing a stereospecific ^{3}H label at C-6 and a ^{14}C-label at C-12. The reintroduction of the doubly-labelled hae-manthamine to *Sprekelia formosissima*, proceeded to give labelled hae-manthidine with complete retention of the tritium label.

These results demonstrate that the hydrogen introduced in generating the C′-1 methylene group in norbelladine from the imine is the same hydrogen which is removed during hydroxylation in the formation of haemanthidine. A decision on whether it is the *pro-R* or *pro-S* hydrogen of the methylene group that is removed during hydroxylation, and the question of retention or inversion of configuration during hydroxylation, requires further study.

8.11.4.2 *Hydroxylation at C-11*

Some attention has been directed to elucidating the nature of the intermediates which lie between *O*-methylnorbelladine and haemanthamine. Since this alkaloid possesses a hydroxyl function at the C-11 position, *O*-methylnorbelladine or a subsequent intermediate has to undergo hydroxylation at a saturated carbon.

The finding[98] that the labelled crinine intermediate $(\pm)$-[2,4,4-^{3}H]oxocrinine and $(\pm)$-[3-^{3}H]crinine were each incorporated into haemanthamine by 'Texas' and 'Twink' daffodils established that hydroxylation must occur after formation of the crinine ring system. Incorporations into haemanthamine were also observed with $(\pm)$-[1,3,4b-^{3}H]normaritidine and $(\pm)$-[1,4b-^{3}H]-noroxomaritidine (Figure 8.13). This suggests that the enzyme responsible

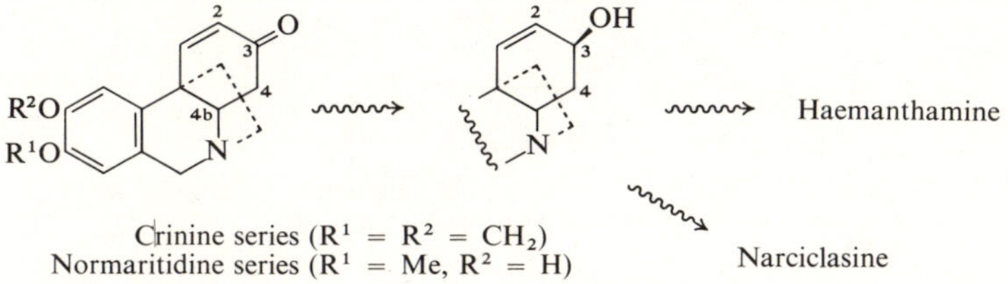

Crinine series ($R^1 = R^2 = CH_2$)
Normaritidine series ($R^1 = Me, R^2 = H$)

Haemanthamine

Narciclasine

Figure 8.13 Crinine intermediates in the biosynthesis of haemanthamine and narciclasine

for conversion of the *o*-methoxyphenol to a methylenedioxy group is not able to discriminate between substrates containing a 3-ketone or a 3β-hydroxyl in these intermediates.

Although conclusive evidence for the identification of the intermediate undergoing C-11 hydroxylation is still lacking, the importance of stereochemistry at C-3 was indicated by the observation that neither 3-*epi*crinine or the analogous 3-*epi*normaritidine were converted to haemanthamine to any appreciable extent.

A further important contribution to this problem was provided simultaneously by two independent studies which have established the stereochemistry of the hydroxylation process. Previous studies had shown that tyrosine serves as the precursor of the C_6—C_2—N hydroaromatic unit in the crinine and lycorine alkaloids and that *O*-methylnorbelladine stands between this amino acid and the alkaloids on the biosynthetic pathway. The mode of incorporation of these two precursors is such that the C-3 methylene of tyrosine and the C-2 methylene of *O*-methylnorbelladine correspond in biosynthetic terms to the C-11 methylene of the crinine system. Kirby and Michael[99] achieved an ingeniously simple synthesis of asymmetrically

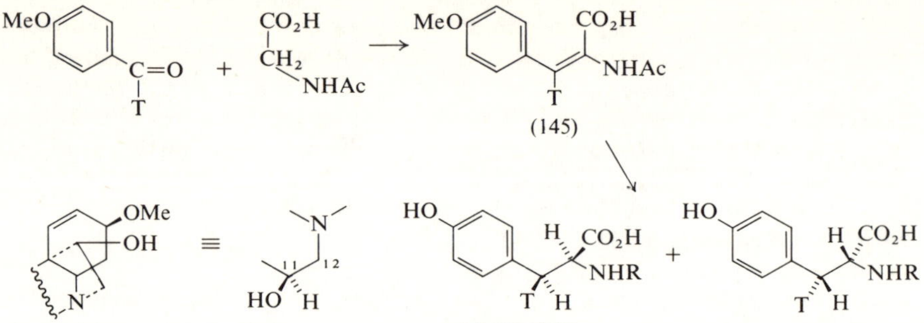

(145)

(148)

2R, [3R-³H]

(146) R = H

2S, [3S-³H]

(147) R = H

labelled DL-[3R-³H]tyrosine and DL-[3S-³H]tyrosine for the purpose of their investigation of this problem. Catalytic hydrogenation of the labelled β-acylaminocinnamic acid (145) when followed by hydrolysis of the product with HBr gave an equimolar mixture of 2R-[3R-³H]tyrosine and 2S-[3S-³H]tyrosine. Enzymatic resolution of this mixture via the N-chloroacetate and subsequent epimerisation at the C-2 centre provided, after

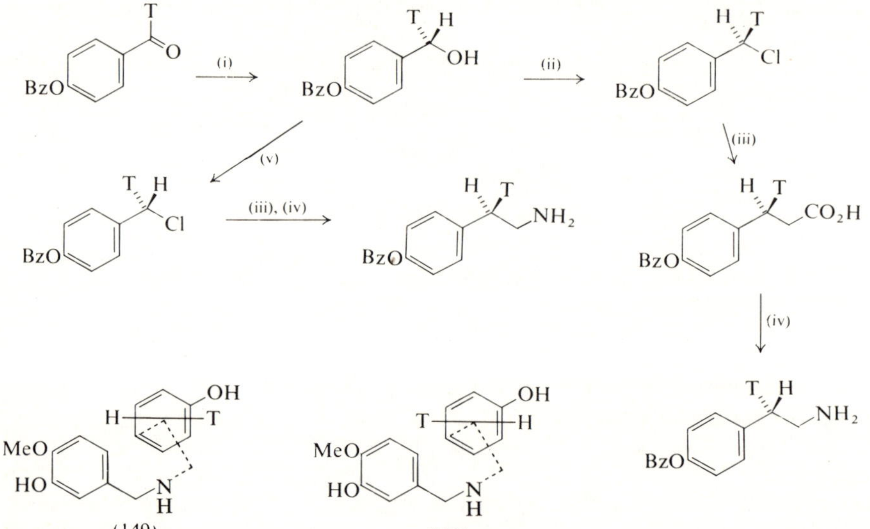

(149)

(150)

(i) Liver alcohol dehydrogenase; (ii) Ph₃P–CCl₄; (iii) Na⁺–CH(CO₂R)

(iv) (a) SOCl₂   (b) NH₃   (c) HOCl–⁻OH   (v) SOCl₂–Et₂O

Figure 8.14   Synthesis of asymetrically-labelled O-methylnorbelladines(149) and (150)

admixture with DL-[2-¹⁴C] tyrosine, the asymmetrically labelled DL-tyrosines (146) and (147) for the biosynthetic study. Feeding of (147) to 'Texas' daffodils gave haemanthanine showing 87% retention of tritium whereas haemanthamine derived from feeding the 3R-³H isomer showed a comparable loss of tritium. Oxidation of haemanthamine derived from (157) resulted in the removal of the tritium and established that its location was restricted to the expected C-11 position. The absolute stereochemistry of haemanthamine is firmly established (Section 8.6.1) and corresponds to an absolute configuration at the C-11 position as represented in the partial structure (148). Thus these results establish that hydroxylation at C-11 proceeds with retention of configuration with the *pro-R* hydrogen being removed.

Complementary results in agreement with this finding were obtained by Battersby and his co-workers[100]. *O*-Methylnorbelladine (149), asymmetrically labelled at C-2 with tritium and its enantiomer (150) were prepared by the sequence of reactions which are summarised in Figure 8.14. The enantiomeric purity in each case was determined to be *c.* 70% from a study of analogous reactions carried out with ^{2}H-labelled compounds. When (149) and (150) were mixed with ^{14}C-labelled *O*-methylnorbelladine and fed in separate experiments to 'King Alfred' daffodils, the haemanthamine derived from the 2*S*-^{3}H isomer retained 66% of the label while that obtained from the enantiomer (149) retained only 31% of the original tritium. Oxidation of each sample to the 11-ketone was used to establish the location of tritium at C-11. Thus, the conclusion derived from these experiments indicated that hydroxylation proceeds with retention of configuration in accord with the independent results of Kirby and Michael. Any possible involvement of the adjacent C-12 hydrogens in the hydroxylation process was excluded by demonstrating that non-stereospecifically labelled *O*-methyl[1-^{3}H$_2$]norbelladine was converted to haemanthanine without any significant loss of tritium.

8.11.4.3 *Stereochemistry of protonation at C-4*

A preliminary investigation[101] of the stereochemistry of the protonation at C-4 which occurs subsequent to the internal Michael addition to the intermediate dienone (151) has given inconclusive results. Doubly labelled

T
OH
T
MeO
HO
N
H
(129)
O
4
N
H
(151)
Haemanthamine

O-methylnorbelladine (129) was incorporated into haemanthamine without loss of tritium by 'Twink' and 'Deanna Durbin' daffodils. Half of the tritium was reported to be located at the expected C-2 position but attempts to establish the location of the second tritium were unsuccessful.

8.11.5 Lycorenine

The biosynthesis of alkaloids of the sub-group represented by homolycorine, and its related hemi-acetal lycorenine (152), are reasonably derived from the appropriate lycorine-type by a sequence of reactions involving benzylic hydroxylation, *N*-methylation (not necessarily in that order), ring opening and recyclisation of the C-1 hydroxyl with the C-7 aldehyde as portrayed in Figure 8.15. The correctness of this proposed route has been established recently by studies which were carried out in conjunction with the work described in Section 8.11.4.1 concerned with the transformation of protocatechuic acid to *O*-methylnorbelladine and haemanthamine, and the benzylic hydroxylation of the latter to haemanthidine[97]. The derivation of

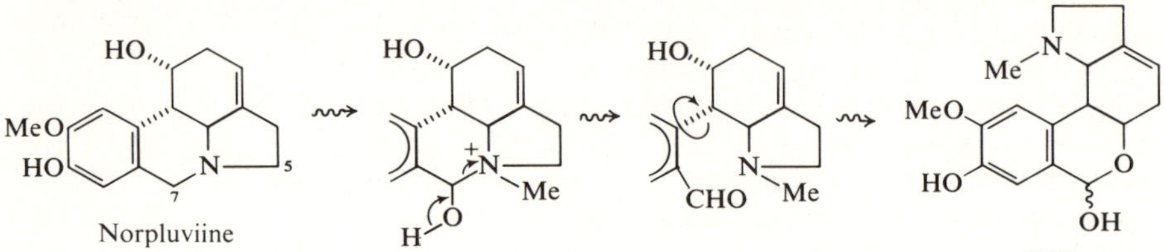

Norpluviine

(152)

Figure 8.15 Biosynthetic pathway for the conversion of norpluviine to lycorenine(152)

norpluviine from *O*-methylnorbelladine containing a single non-stereospecific tritium label at C-1′ and a ^{14}C-label at C-1 occurred without loss of tritium whereas lycorenine obtained in the same experiment was found to have lost 50 % of the original tritium. Proof of the conversion of norpluviine to lycorenine was obtained by feeding a mixture of biosynthetically prepared 7*R*- or 7*S*-[7-^{3}H]norpluviine and [5-^{14}C]norpluviine to 'Twink' daffodils. The lycorenine subsequently isolated was radio-active and showed the same T/^{14}C ratio as in the original norpluviine. This conversion demonstrates that the conversion of norpluviine involves a stereospecific hydroxylation at C-7 and that the hydrogen that is retained is the one which is derived from protocatechuic aldehyde.

8.11.6 Ismine and narciclasine

Ismine (153) and narciclasine (130) are notable in that neither contain a C_{15} skeleton characteristic of most *Amaryllidaceae* alkaloids. Of the C_6—C_1 and C_6—C_2—N fragments which together comprise the skeleton in the typical alkaloids, the latter is present as a C_6–N moiety in ismene and narciclasine. This has led to the suggestion that these two alkaloids might be derived by the natural breakdown of a representative member of one of the typical sub-groups. Experiments reported recently have substantiated this idea.

[2,3-^{3}H]Oxocrinine when administered to *Sprekelia formosissima* gave an extremely low incorporation (0.007 %) into ismine[102]. Although it is natural to question the significance of such a result, the specificity of the apparent conversion of oxocrinine to ismine was provided by the location of the tritium labels in the alkaloid at the expected 3′,5′-positions. Bromination of the labelled ismine in deuterioacetic acid proceeded to afford radio-inactive 3,5-dibromoismine (154) without any uptake of deuterium. Further evidence on the role of crinine type intermediates in the biosynthesis of ismine would seem to be necessary before this route can be considered established.

A start has been made on the investigation of the biosynthesis of narci-

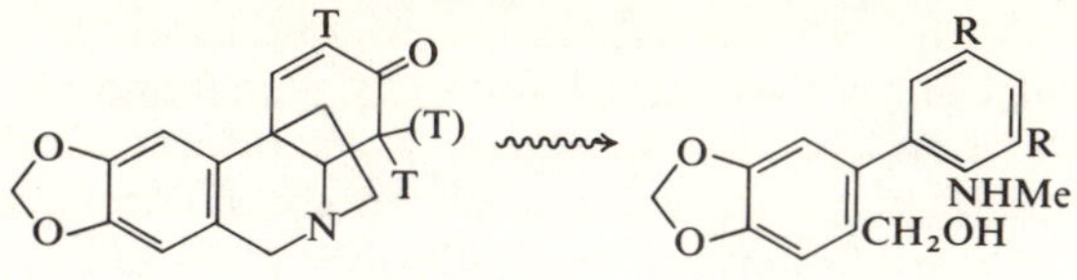

(153) R = H
(154) R = Br

clasine. *O*-Methylnorbelladines labelled as in (129) and (155) are incorporated into narciclasine by 'Texas' and 'Twink' daffodils with 75% and 100% retention of tritium, respectively[103]. The retention of 75% of the original tritium during the incorporation of (129) when compared to the observed tritium retentions observed for haemanthamine (100%) and norpluviine (49%) from the same feeding experiment, eliminated the lycorine ring system as a possible intermediate in the biosynthesis of narciclasine. The position of the tritium labels in narciclasine biosynthically derived from the doubly-labelled *O*-methylnorbelladines was established as illustrated in Figure 8.16.

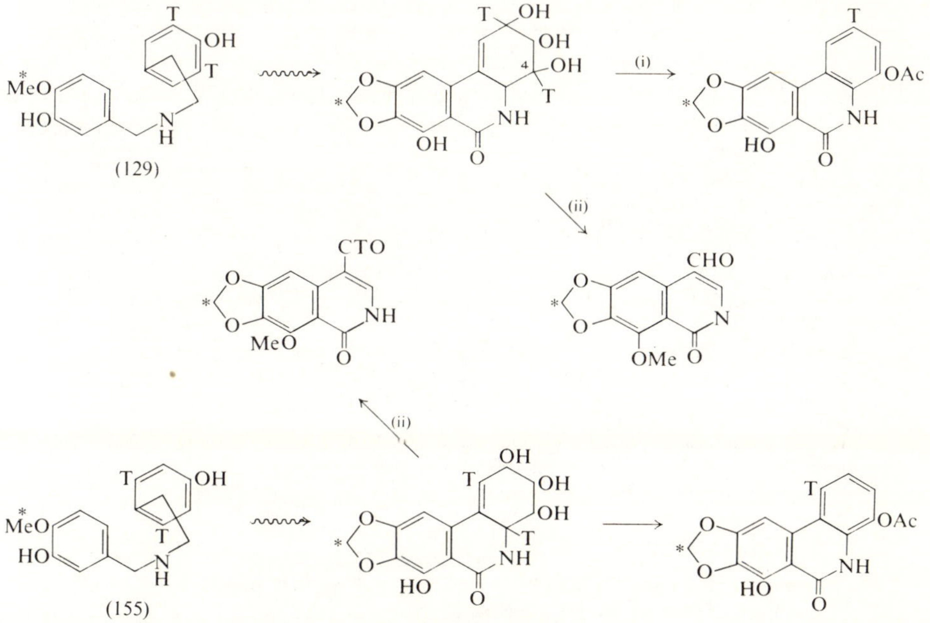

(129)

(155)

(i) H^+/D_2O; Ac_2O, (ii) CH_2N_2; HIO_4

Figure 8.16 · Incorporation of *O*-methylnorbelladines(129) and (155) into narciclasine

The loss of 25% of the tritium during the biosynthesis of narciclasine from (129) evidently occurs at the C-4 position and in the absence of further evidence this is probably best interpreted as arising by random protonation at C-4 during the formation of the initial crinine intermediate which is then followed by a subsequent stereospecific hydroxylation at this position.

Proof of the involvement of crinine-type intermediates in the biosynthesis of narciclasine came from a series of experiments which have been referred to earlier in this chapter in connection with the biosynthesis of haemanthamine[98]. Each of the crinine-type intermediates from normaritidine to crinine (Figure 8.13) which were incorporated into haemanthamine were also incorporated into narciclasine. Similarly, 3-*epi*crinine which is not incorporated into haemanthamine was not incorporated into narciclasine. The degradation of the radiolabelled narciclasine from these experiments paralleled those described in Figure 8.16 and demonstrated that tritiums at C-1, C-3 and C-4b in the labelled racemic crinine precursors were completely retained.

A recent and important extension of these results has shown that $(+)$-$[2,4-^3H]$crinine (Figure 8.13) but not its enanantiomer, is incorporated into narciclasine. The location of the tritium labels at C-2 and C-4 at the expected positions was established by parallel degradation experiments to those already described and confirmed the loss of *c.* 50% tritium activity at the C-4 position as a consequence of hydroxylation. The implication of this result in establishing the absolute stereochemistry of narciclasine has been referred to earlier.

References

1. Wildman, W. C. (1966). *The Alkaloids,* **XI**, 308. (R. M. F. Manske, editor). (New York: Academic Press)
2. Shiro, M., Sato, T. and Koyama, H. (1966). *Chem. and Ind.,* 1229
3. Fales, H. M. and Wildman, W. C. (1958). *J. Amer. Chem. Soc.,* **80,** 4395
4. Kinstle, T. H., Wildman, W. C. and Brown, C. L. (1966). *Tetrahedron Letters,* 4659
5. Clardy, J. C., Wildman, W. C. and Hauser, F. M. (1970). *J. Amer. Chem. Soc.,* **92,** 1781
6. Immirizi, A. and Fuganti, C. (1971). *J. Chem. Soc. B,* 1218
7. Hendrickson, J. B., Alder, R. W., Dalton, D. R. and Hey, D. G. (1969). *J. Org. Chem.,* **34,** 2667
8. Kametani, T., Seino, C. and Nakano, T. (1971). *Chem. Pharm. Bull.,* **19,** 1959
9. Kametani, T., Yagi, H., Kawamura, K. and Kohno. (1970). *Chem. Pharm. Bull.,* **18,** 645
10. Ueda, N., Tokuyama, T. and Sakan, T. (1966). *Bull. Chem. Soc. Japan,* **39,** 2012
11. Ganem, B. (1971). *Tetrahedron Letters,* 4105
12. Irie, H., Nishitani, Y., Sugita, M. and Uyeo, S. (1970). *Chem. Commun.,* 1313
13. Boit, H.-G. and Döpke, W. (1960). *Naturwissenschaften,* **47,** 109
14. Bhandarkar, J. G. and Kirby, G. W. (1970). *J. Chem. Soc. C,* 1224
15. Nogueiras, C., Döpke, W. and Lehmann, G. (1971). *Tetrahedron Letters,* 3249
16. Bhandarkar, J. G. and Kirby, G. W. (1970). *J. Chem. Soc. C,* 592
17. Barton, D. H. R. and Kirby, G. W. (1962). *J. Chem. Soc.,* 806
18. Abramovitch, R. A. and Takahashi, S. (1963). *Chem. and Ind.,* 1039
19. Frank, B. and Lubs, H. J. (1968). *Ann.,* **720,** 131
20. Kametani, T., Yamaki, K., Yaki, H. and Fukumoto, K. (1969). *J. Chem. Soc. C,* 2602
21. Kametani, T., Seino, C., Yamaki, K., Shibuya, S., Fukumoto, K., Kigasawa, K., Satoh, F., Huragi, M. and Hayasaka, T. (1971). *J. Chem. Soc. C,* 1043
22. Kametani, T., Shishido, K., Hayashi, E., Seino, C., Khono, T., Shibuya, S. and Fukumoto, K. (1971). *J. Org. Chem.,* **36,** 1295
23. Hazama, N., Irie, I., Mitzutani, T., Shingu, T., Takada, M., Uyeo, S. and Yoshitake, A. (1968). *J. Chem. Soc. C,* 2947
24. Misaka, Y., Mitzutani, T., Sekido, M. and Uyeo, S. (1968). *J. Chem. Soc. C,* 2954
25. Wildman, W. C. and Bailey, D. T. (1968). *J. Org. Chem.,* **33,** 3749
26. Proskurnina, N. F. (1957). *Zhur. Obshchei. Khim.,* **23,** 3365
27. Döpke, W. and Jeffs, P. W. (1968). *Tetrahedron Letters,* 1307
28. Wildman, W. C. and Bailey, D. T. (1969). *J. Amer. Chem. Soc.,* **91,** 150
29. Murphy, C. F. and Wildman, W. C. (1964). *Tetrahedron Letters,* 3857
30. Tsuda, Y. and Uyeo, S. (1961). *J. Chem. Soc.,* 2485

31. Morrison, G. A. (1967). *Progress in the Chemistry of Organic Natural Products,* **25,** 269. (L. Zechmeister, editor). (New York: Springer-Verlag)
32. Ikeda, T., Taylor, W. I., Tsuda, Y., Uyeo, S. and Yajima, H. (1956). *J. Chem. Soc.,* 4749
33. Highet, R. J. and Highet, P. F. (1968). *J. Org. Chem.,* **33,** 3105
34. Highet, R. J., Ma, J. C. N. and Highet, P. F. (1968). *J. Org. Chem.,* **33,** 3096
35. Sato, T. and Koyama, H. (1971). *J. Chem. Soc. B,* 1070
36. Clardy, J. C., Hauser, F. M., Dahm, D., Jacobson, R. A. and Wildman, W. C. (1970). *J. Amer. Chem. Soc.,* **92,** 6337
37. Jeffs, P. W., Warren, F. L. and Wright, W. G. (1960). *J. Chem. Soc.,* 1090
38. Kuriyama, K., Iwata, T., Moriyama, M., Kotera, K., Hameda, Y., Mitsui, R. and Takeda, K. (1967). *J. Chem. Soc. B,* 46
39. Snatzke, G. and Ho, P. (1971). *Tetrahedron,* **27,** 3645
40. DeAngelis, G. G. and Wildman, W. C. (1969). *Tetrahedron,* **25,** 5099
41. DeAngelis, G. G. and Wildman, W. C. (1969). *Tetrahedron Letters,* 729
42. Slabaugh, M. R. and Wildman, W. C. (1971). *J. Org. Chem.,* **36,** 3202
43. King, R. W., Murphy, C. F. and Wildman, W. C. (1965). *J. Amer. Chem. Soc.,* **87,** 4912
44. Schwartz, M. A. and Holton, R. A. (1970). *J. Amer. Chem. Soc.,* **92,** 1090
45. Muxfeldt, H., Schneider, R. S. and Mooberry, J. B. (1966). *J. Amer. Chem. Soc.,* **88,** 3760
46. Whitlock, H. W. and Smith, G. L. (1967). *J. Amer. Chem. Soc.,* **89,** 3600
47. Hendrickson, J. B., Bogard, T. L. and Fish, M. E. (1970). *J. Amer. Chem. Soc.,* **92,** 5538
48. Irie, H., Uyeo, S. and Yoshitake, A. (1968). *J. Chem. Soc. C,* 1802
49. Stevens, R. V. and DuPree, L. E. (1970). *Chem. Commun.,* 1585
50. Tsuda, Y. and Isobe, K. (1971). *Chem. Commun.,* 1555
51. Kametani, T. and Kohno, T. (1971). *Tetrahedron Letters,* 3155
52. Curphey, T. J. and Kim, H. L. (1968). *Tetrahedron Letters,* 1441
53. Stevens, R. V. and Wentland, M. P. (1968). *J. Amer. Chem. Soc.,* **90,** 5580
54. Keely, S. L. and Tahk, F. C. (1968). *J. Amer. Chem. Soc.,* **90,** 5584
55. Haugwitz, R. D., Jeffs, P. W. and Wenkert, E. (1965). *J. Chem. Soc.,* 2001
56. Kametani, T. and Kohno, T., Shibuya, S. and Fukumoto, K. (1971). *Tetrahedron,* **27,** 5441
57. Uyeo, S., Irie, H., Yoshitake, A. and Ito, A. (1965). *Chem. Pharm. Bull. (Japan),* **13,** 427
58. Hendrickson, J. B., Foote, C. and Yoshimura, N. (1965). *Chem. Commun.,* 165
59. Inubushi, Y., Fales, H. M., Warnhoff, E. W. and Wildman, W. C. (1960). *J. Org. Chem.,* **25,** 2153
60. Sandberg, F. and Michel, K.-H. (1963). *Lloydia,* **26,** 78
61. Wildman, W. C., Murphy, C. F., Michel, K.-H., Brown, C. L., Bailey, D. T., Heimer, N. and Shaffer, R. (1967). *Pharmazie,* **72,** 725
62. Wildman, W. C. and Brown, C. L. (1968). *J. Amer. Chem. Soc.,* **90,** 6439
63. Briggs, C. K., Highet, P. F., Highet, R. J. and Wildman, W. C. (1956). *J. Amer. Chem. Soc.,* **78,** 2899
64. Jeffs, P. W., Hansen, J. F., Döpke, W., Bienert, M. (1971). *Tetrahedron,* **27,** 5065
65. Melis, B. (1965). *Naturwissenschaften* **52,** 34
66. Mehlis, B. (1964). *Ph.D. Thesis,* Humboldt University
67. Jeffs, P. W. and Hansen, J. F. (1968). Unpublished results
68. Bienert, M. (1970). *Ph.D. Thesis,* Humboldt University
69. Döpke, W., Bienert, M., Burlingame, A. L., Schnoes, H. K., Jeffs, P. W. and Farrier, D. S. (1967). *Tetrahedron Letters,* 451
70. Döpke, W. and Bienert, M. (1970). *Tetrahedron Letters,* 745
71. Döpke, W. and Bienert, M. (1970). *Tetrahedron Letters,* 3245
72. Piozzi, F., Fuganti, C., Mondelli, R. and Ceriotti, G. (1968). *Tetrahedron,* **24,** 1119; Ceriotti, G. (1967). *Nature (London),* **213,** 595; Piozzi, F. Marino, M. L., Fuganti, C. and Di Matino, A. (1969). *Phytochemistry,* **8,** 1745
73. Okamoto, T., Torii, Y. and Isogai, Y. (1968). *Chem. Pharm. Bull. (Japan),* **16,** 1860
74. Mondon, A. and Krohn, K. (1970). *Tetrahedron Letters,* 2123
75. Mondon, A. and Krohn, K. (1970). *Chem. Ber.,* **103,** 2729
76. Savona, G., Piozzi, F. and Marino, M. L. (1970). *Chem. Commun.,* 1006
77. Fuganti, C. and Mazza, M. (1972). *Chem. Commun.,* 239
78. Immirzi, A. and Fuganti, C. (1972). *Chem. Commun.,* 240
79. Fuganti, C., Selva, A., Piozzi, F. (1967). *Chim. and Ind. (Milan),* **49,** 1196
80. Brossi, A., Grethe, G., Teitel, S., Wildman, W. C., Bailey, D. T. (1970). *J. Org. Chem.,* **35,** 1100

318   ALKALOIDS

81. Bross, A. and Teitel, S. (1970). *J. Org. Chem.*, **35**, 3559
82. Schwartz, M. A. and Scott, S. W. (1971). *J. Org. Chem.*, **36**, (1971)
83. Fales, H. M. and Wildman, W. C. (1964). *J. Amer. Chem. Soc.*, **86**, 294
84. Barton, D. H. R., Kirby, G. W., Taylor, J. B., Thomas, G. M. (1963). *J. Chem. Soc.*, 4545
85. Fuganti, C. (1969). *Chim. and Ind. (Milan)*, **51**, 1254
86. Kirby, G. W. and Tiwari, H. P. (1966). *J. Chem. Soc. C*, 676
87. Barton, D. H. R. and Cohen, T. (1957). *Festschrift Arthur Stoll*, 117. (Basle: Birkhäuser-Basle)
88. Barton, D. H. R. (1963). *Proc. Chem. Soc.*, 293
89. Barton, D. H. R., Boar, R. B. and Widdowson, D. A. (1970). *J. Chem. Soc. C*, 1213
90. Bruce, I. T. and Kirby, G. W. (1968). *Chimia*, **22**, 314
91. Takeda, K., Kotera, K. and Mitzukami, S. (1958). *J. Amer. Chem. Soc.*, **80**, 2562
92. Heimer, N. E. and Wildman, W. C. (1967). *J. Amer. Chem. Soc.*, **89**, 5265
93. Bruce, I. T. and Kirby, G. W. (1968). *Chem. Commun.*, 207
94. Guroff, G., Daly, J. W., Jerina, D. M., Renson, J., Udenfriend, S. and Witkop, B. (1967). *Science*, **157**, 1524
95. Bowman, W. R., Bruce, I. T. and Kirby, G. W. (1969). *Chem. Commun.*, 1075
96. Avret, R. J., Boyd, D. R., Robinson, P. M., Watson, C. G., Daly, J. W. and Jerina, D. M. (1971). *Chem. Commun.*, 1585
97. Fuganti, C. and Mazza, M. (1971). *Chem. Commun.*, 1196
98. Fuganti, C. and Mazza, M. (1971). *Chem. Commun.*, 1388
99. Kirby, G. W. and Michael, J. (1971). *Chem. Commun.*, 415
100. Battersby, A. R., Kelsey, J. E. and Staunton, J. (1971). *Chem. Commun.*, 183
101. Fuganti, C. (1969). *Chim. and Ind. (Milan)*, **51**, 1254
102. Fuganti, C. and Mazza, M. (1970). *Chem. Commun.*, 1466
103. Fuganti, C., Staunton, J. and Battersby, A. R. (1971). *Chem. Commun.*, 1154

9
The Structure and Synthesis of C₁₉-Diterpene Alkaloids

S. W. PELLETIER and S. W. PAGE
University of Georgia, Athens, U.S.A.

9.1 INTRODUCTION

The nitrogeneous bases isolated from plants of the *Aconitum* and *Delphinium* genera of Ranunculaceae, the *Garrya* genus of Garryaceae and *Inula royleana* of the Compositae have long been of interest because of their pharmacological properties and complex structures. These alkaloids are derived from tetra-cyclic or pentacyclic diterpenes which incorporate the nitrogen atom of a molecule of β-aminoethanol, methylamine or ethylamine in a heterocyclic ring between either C-17 and C-19 or between C-19 and C-20.

For convenience, the diterpene alkaloids are generally divided into two classes: (a) those based on a hexacyclic C_{19}-skeleton and (b) those based on C_{20}-skeleton. The first group are commonly called the 'Aconitines' and all possess either the lycoctonine (A) or the heteratisine (B) skeleton (see Figure 9.1). These highly toxic ester bases were the active principles of many of the

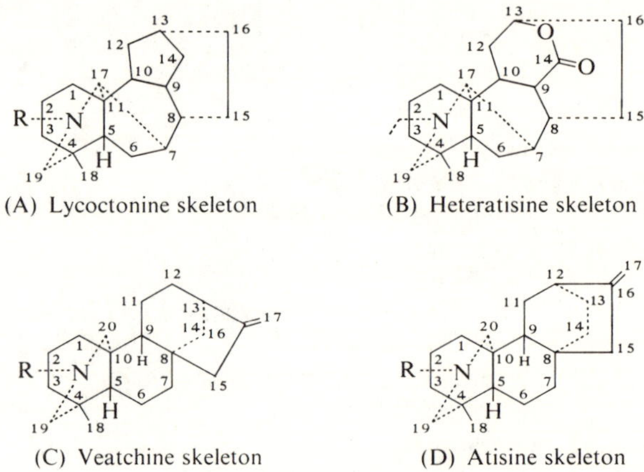

(A) Lycoctonine skeleton   (B) Heteratisine skeleton

(C) Veatchine skeleton   (D) Atisine skeleton

Figure 9.1   Skeletons of the C-19 and C-20 diterpene alkaloids

poisonous extracts used in the medieval trials by ordeal. Indeed, aconitine and pseudoaconitine are among the most toxic compounds of plant origin yet isolated. Their ancient therapeutic claims ranged from toothache to gout remedies. These ester alkaloids are generally heavily substituted with methoxyl and hydroxyl groups, with some of the hydroxyl groups existing as benzoic, veratric, or acetic esters. Hydrolysis of the Aconitines yields the relatively non-toxic parent alkamines known as the 'Aconines'.

The second class of diterpene alkaloids, modelled on a C_{20}-skeleton, are not extensively oxygenated. Some occur in the plant as monoesters of benzoic or acetic acid. These relatively nontoxic alkamines are based on either a veatchine skeleton (C) or an atisine skeleton (D) (see Figure 9.1). The *Garrya* alkaloids possess the veatchine skeleton (which incorporates the isoprenoid kaurane-type system). The atisine skeleton, based on the atisane model (which cannot be dissected in an isoprenoid fashion), appears in alkaloids isolated from the *Aconitum* and *Delphinium* species, and more recently from *Spiraea japonica*.

The foundations of diterpene alkaloid chemistry were laid by W. A.

Jacobs and his co-workers at the Rockefeller Institute in their investigations of the alkaloids of the *Aconitum* and *Delphinium* species. The New Brunswick group under the direction of K. Wiesner later initiated work on the *Garrya* alkaloids. They initially paralleled Jacobs' studies on atisine, pointing out the striking similarities in the chemistry of the two groups of alkaloids. Their subsequent discoveries greatly assisted structural progress in both alkaloid groups; the structures of the *Garrya* alkaloids were elucidated first. The Japanese groups at Hokkaido and the University of Tokyo added contributions to the field in their isolation and structural investigations of the diterpene alkaloids from the plants of the *Aconitum* species native to Japan. More recently Indian and Soviet workers have initiated isolation and structural investigations of the diterpene alkaloids from their native plants. The flourishing of synthetic activity in the diterpene alkaloid field has derived from contributions from many groups all over the world.

The structure elucidation work on the diterpene alkaloids has, as with other areas of structure determination, placed increasing reliance on x-ray crystallographic techniques. The natural-product chemist–crystallographer collaboration has proved to be a potent team as evidenced by the number of structures solved in a relatively short time frame. Moreover, the minute quantity of a compound required for crystallographic analysis allows important structural information to be derived from substances which are isolatable in very small amounts.

The chemistry of the diterpene alkaloids has previously been reviewed in detail in several publications[1]. This article surveys the work on the C$_{19}$-diterpene alkaloids reported in the literature available in our libraries and through personal communications from January 1966 to December 1971. The chemistry of the C$_{20}$-diterpene alkaloids will be reviewed in a subsequent edition in this series.

9.2   STRUCTURAL INVESTIGATIONS

The C$_{19}$-diterpene alkaloids may be subdivided into three general types: lycoctonine (1), aconitine (2) and heteratisine (3). These structural types are based on the parent alkaloid of the same name.

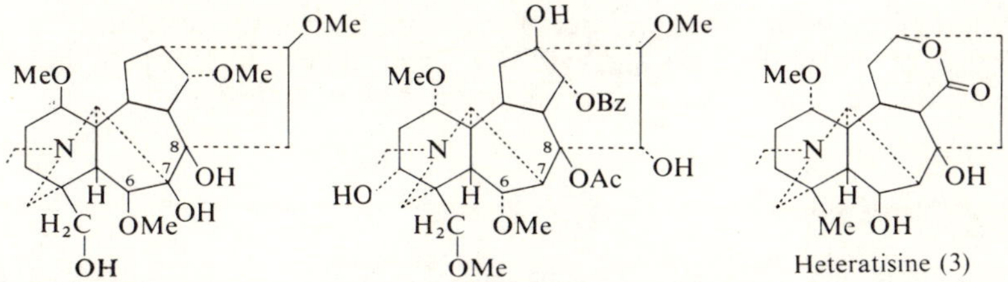

Lycoctonine (1)

Aconitine (2)

Heteratisine (3)

The most important difference between the aconitine-type and the lycoctonine-type alkaloids is that the latter contains an oxygenated group at C-7. All of the C$_{19}$-diterpene alkaloids isolated thus far possess a C-8 oxygen function. The presence of this ditertiary α-glycol system in the lycoctonine-type alkaloids is responsible for the difference in chemistry of these alkaloid types. In addition, all known lycoctonine-type alkaloids have a β-methoxyl at C-6, while in all of the aconitine-type alkaloids that contain a C-6 methoxyl, it is in the α-configuration.

The C_{19}-lactone alkaloids all occur in very small amounts. These compounds are modelled on a modified lycoctonine-type skeleton, and are related to the parent compound, heteratisine.

The earlier chemistry of the C_{19}-diterpene alkaloids has been discussed[1c, 1d, 1i, 1l]. Recent studies have clarified structural details of many of these alkaloids and several new alkaloids have been added to these classes.

9.2.1 Lycoctonine-type alkaloids

9.2.1.1 14-Dehydrobrowniine

The diterpene alkaloid 14-dehydrobrowniine, $C_{25}H_{39}NO_7$, (4) has been isolated from *Delphinium cardinale* Hook, a species native to Southern and Baja California[2]. It has been found in the aerial portions of the plant along with browniine (5), lycoctonine, hetisine and dehydrohetisine. The structure of 14-dehydrobrowniine was established by direct correlation with browniine[2]. Sodium borohydride reduction of (4) proceeded stereospecifically to give browniine as the only product. Oxidation of browniine with sodium dichromate in acetic acid gave low yields of a ketonic, basic material from which

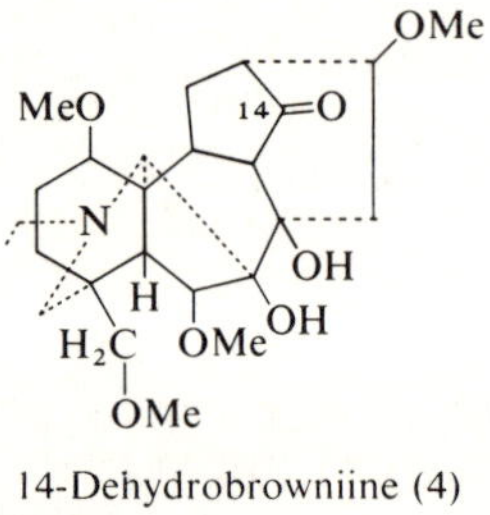

14-Dehydrobrowniine (4) 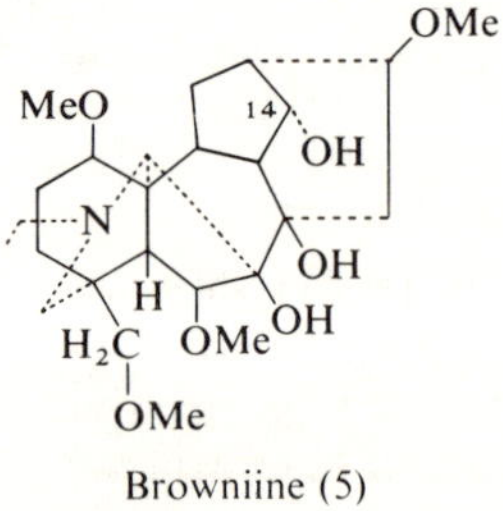   Browniine (5) 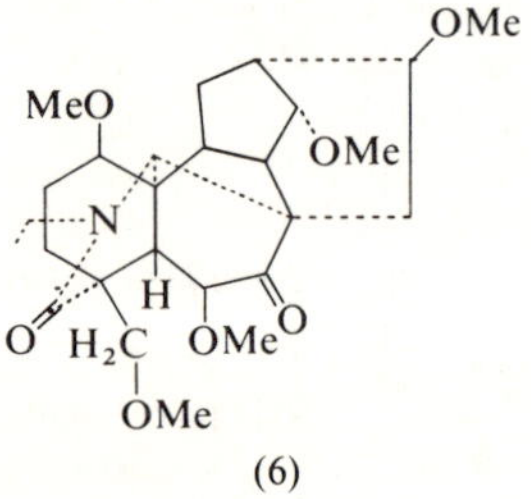   (6)

14-dehydrobrowniine was isolated by chromatography on alumina. The structure of browniine had previously been determined by a chemical correlation with lycoctonine via a common derivative (6)[3].

9.2.1.2 Delcosine

The pinacol rearrangement of delcosine (7) and a novel hydrogenation reaction with its product have been described by Amiya and Shima[4]. Treatment of delcosine with acetyl chloride effected the rearrangement giving anhydrodiacetyldelcosine (8). Similar treatment with oxodelcosine (9), followed by hydrolysis produced anhydrooxodelcosine (10), a compound also obtainable from (8) by oxidation with chromium trioxide–pyridine complex, followed by hydrolysis. Treatment of 1-dehydrooxodelcosine (11) with acetyl chloride, followed by hydrolysis and further oxidation with Sarett reagent, gave anhydro-1,14-didehydrooxodelcosine (12), a triketo-lactam also obtained by further oxidation of (10) or of the hydrolysis product of (8).

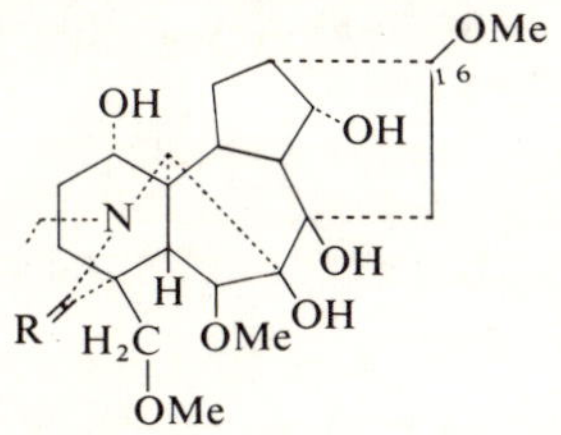

Delcosine (7) R = H$_2$
(9) R = O

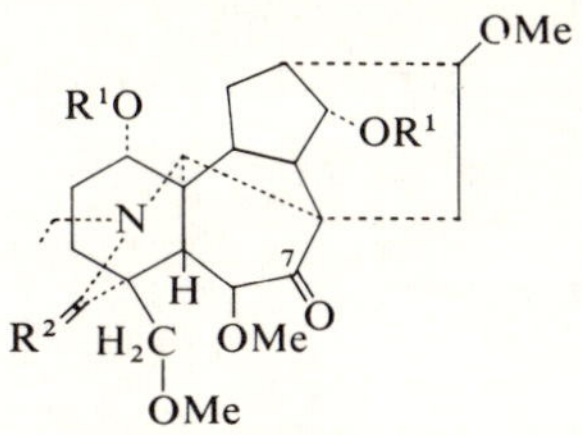

(8) R^1 = Ac, R^2 = H$_2$
(10) R^1 = H, R^2 = O

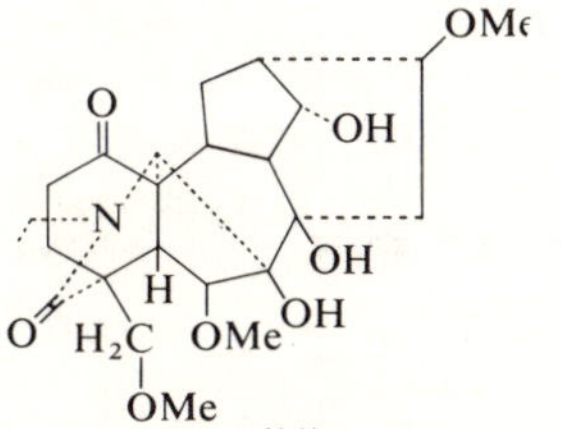

(11)

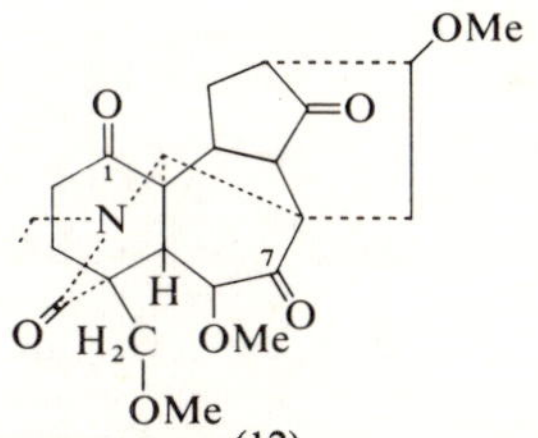

(12)

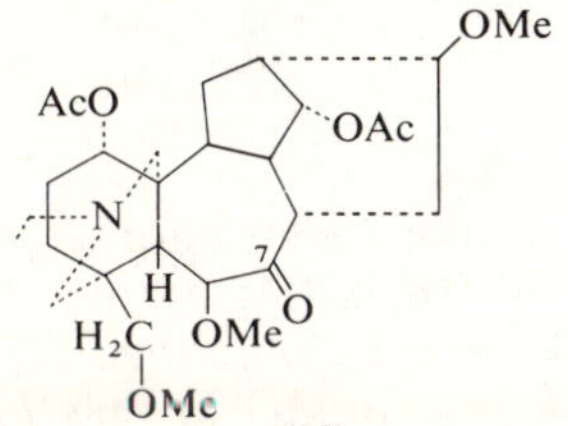

(13)

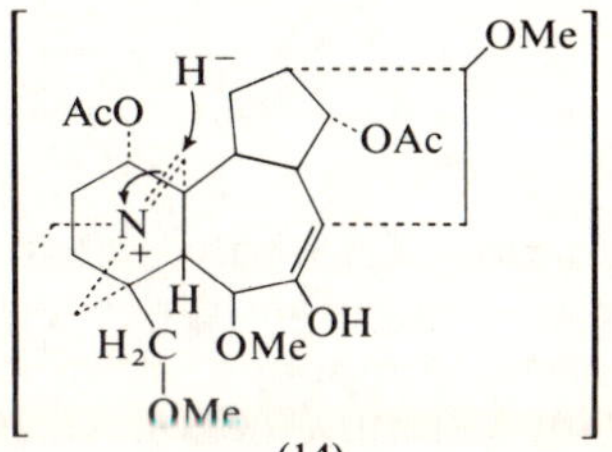

(14)

Catalytic hydrogenation of (8) over platinum yielded dihydroanhydrodiacetyldelcosine (13), presumably involving the intermediate (14). However, (10) was recovered unchanged after being subjected to the same conditions of hydrogenation. Probably the unavailability of the free electron pair of the lactam nitrogen atom prevents the formation of the intermediate (14).

Some interesting i.r. and u.v. spectral features were also observed. The C-7-carbonyl of compound (8) absorbs at 5.81 µm while in the lactam derivatives (9) and (10) this carbonyl absorption occurred at 5.78 µm. However, in the less-strained dihydro derivative (13) the C-7 ketone absorption moved to 5.93 µm. The u.v. spectrum of anhydrodiacetyldelcosine (8) was abnormal: in methanol a maximum was observed at 237 nm (log ε 3.20). However, amide formation removes the anomaly and absorption maxima appear near 300 nm and are ascribed to ketone carbonyl groups. This phenomenon has been observed in other systems and is discussed in Section 9.2.2.10.

9.2.1.3 *Delphatine*

Structure (15) for delphatine, $C_{26}H_{43}NO_7$, has been proposed by Yunusov and Yunusov on the basis of chemical and spectral data[5,6]. Delphatine has been obtained from the seeds of *Delphinium biternatum* Huth., and differs from lycoctonine only in that it has a methoxyl group at C-18, while lycoctonine has an hydroxyl group at that position.

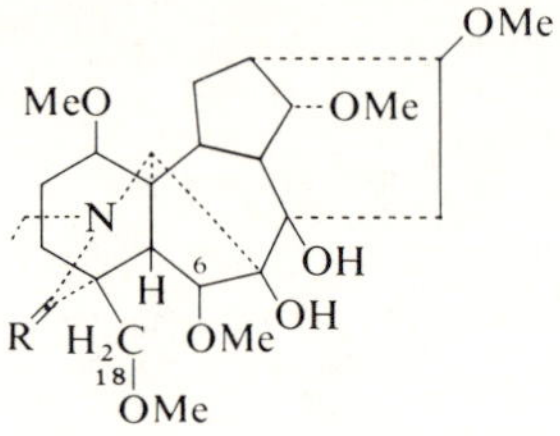

Delphatine (15) R = H_2
(16) R = O

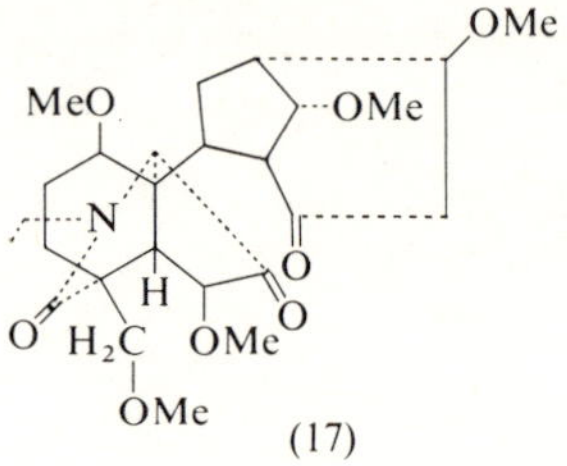

(17)

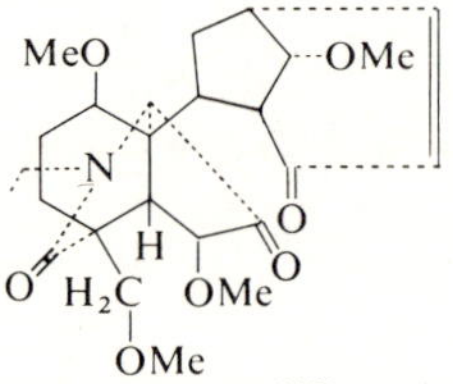

(18)

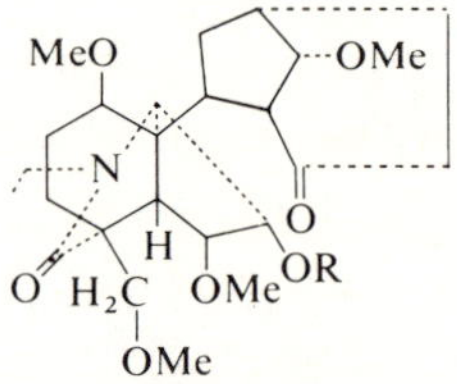

(19) R = H
(20) R = Me

Oxidation of (15) with potassium permanganate afforded the cyclic lactam (16), while oxidation of oxodelphatine (16) with periodic acid resulted in a tricarbonyl compound identified as oxosecodelphatine (17). Treatment of (17) with sulphuric acid gave an α,β-unsaturated ketone, identified as (18). Mass-spectral comparison of delphatine and its derivatives with lycoctonine (1), browniine (5), and their derivatives indicated that there were methoxyl groups at C-6 and C-18 in delphatine. Hydrogenation of (18) gave a tetra-hydro product (19) which was methylated with methyl iodide and sodium hydride. The resulting methyl ether (20) was identical with the product obtained from lycoctonine by similar procedures.

9.2.1.4 *The structures of lycoctamone and its analogues*[7]

Mild acid treatment of the lactam (21) derived from lycoctonine gave the pinacolone (22) via dehydration of the *vic*-glycol system. A more vigorous acid treatment of (21) or (22) gave lycoctamone (23), $C_{23}H_{31}NO_6$. Spectral and chemical evidence showed that this compound is still a δ-lactam, with a readily-acetylateable hydroxyl group, an allylic tertiary hydroxyl, an unsaturated aldehyde and an exocyclic methylene group which is resistant to hydrogenation.

That the primary hydroxyl group on C-18 is not involved in the transformation to lycoctamone was deduced from the fact that 18-deoxy-oxolycoctonine (24) gave a parallel product. Both 18-O-methyl-6-demethoxy-oxolycoctonine (25) and oxodelpheline (26) gave lycoctamone analogues, hence the C-6 oxygen function cannot be involved in the rearrangement. Because 1-deoxyoxodelcosine (27) and oxobrowniine (28) gave lycoctamone analogues, the hydroxyl on C-14 must be involved, while the function on C-1 is not. Thus, by elimination, the authors concluded that the other O-methyl group lost in the reaction is on C-16.

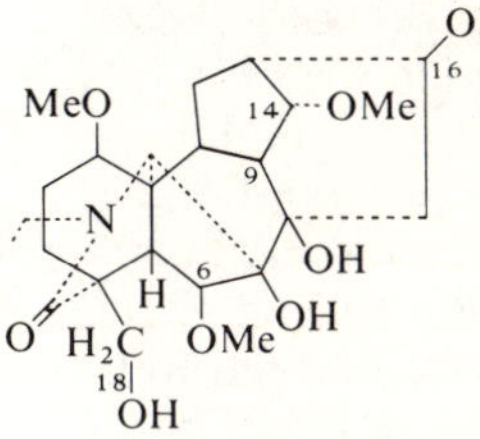

(21)

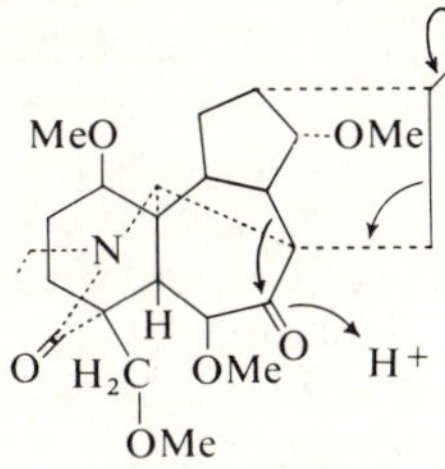

(22)

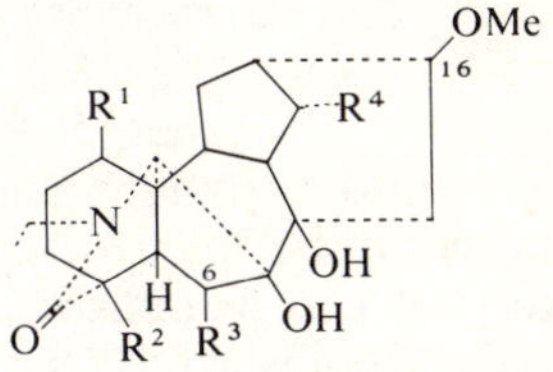

(24) $R^1 = R^3 = R^4 = OMe$, $R^2 = Me$
(25) $R^1 = R^4 = OMe$, $R^2 = CH_2OMe$, $R^3 = H$
(26) $R^1 = R^4 = OMe$, $R^2 = Me$, $R^3 = OH$
(27) $R^1 = H$, $R^2 = R^3 = OMe$, $R^4 = OH$
(28) $R^1 = R^2 = R^3 = OMe$, $R^4 = OH$

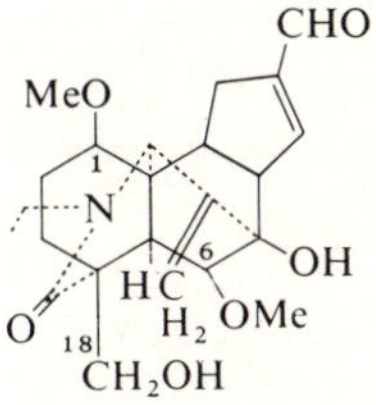

Lycoctamone (23)

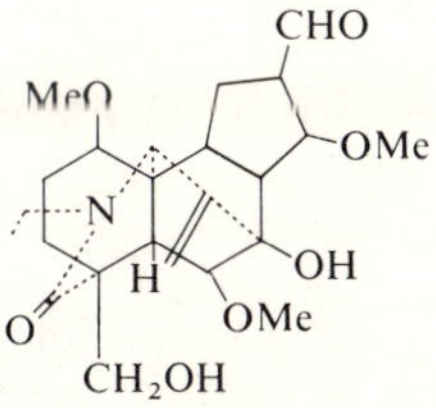

(29)

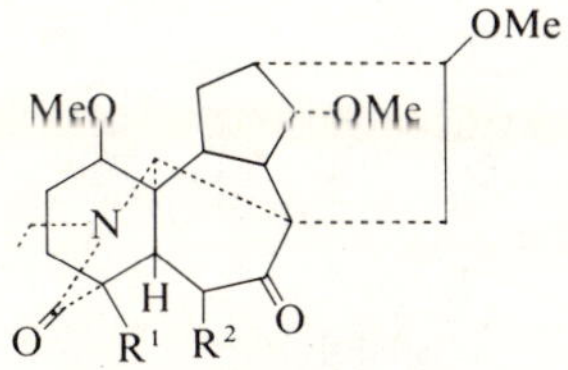

(30) $R^1 = H$, $R^2 = OMe$
(31) $R^1 = CH_2OMe$, $R^2 = H$

On the basis of extensive chemical study of lycoctamone and its analogues, it appears that the pinacolic dehydration product (22) is probably an intermediate in the transformations via (29) to lycoctamone (23).

The authors postulated that the repulsive interactions between the substituents on C-18 and the 6-hydrogen and between the 6-methoxy and the 9-hydrogen are the primary driving forces in this rearrangement. This was supported by the much more drastic conditions necessary to effect the rearrangement of (30) and (31) to their respective lycoctamone analogues and the comparatively lower yields of these products. This interesting work clearly establishes the nature and chemistry of the lycoctamone rearrangement, which were not obvious from the previous chemical and x-ray crystallographic studies of lycoctonine.

9.2.2 Aconitine-type alkaloids

9.2.2.1 *A revision of the structures of aconitine and related alkaloids*

The ring-A methoxyl at C-1 of aconitine had been assigned the configuration *trans* to the nitrogen bridge (β-axial) on the basis of conformational arguments[8]. The desmethoxy derivative (32) on oxidation gave both the β-lactone (33) and the γ-lactone (34). Under identical conditions, (35) and (36) gave only γ-lactones. It was argued that the fused β-lactone would force ring-A into a pseudo-chair conformation, and the resulting non-bonded interactions between an axial methoxyl at C-1 and the quasi-axial methoxyl-methyl at C-4 would destabilise the β-lactone. In delphinine and all of the other alkaloids which were chemically correlated with aconitine (cf. Refs. 1c and 1d), the C-1-methoxyl was accordingly assigned a β-axial configuration.

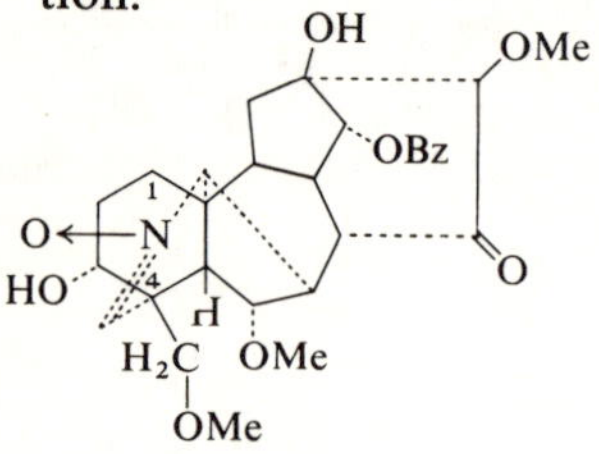

(32)

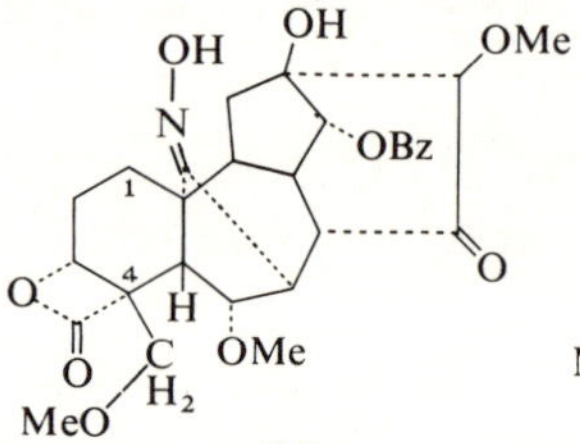

(33)

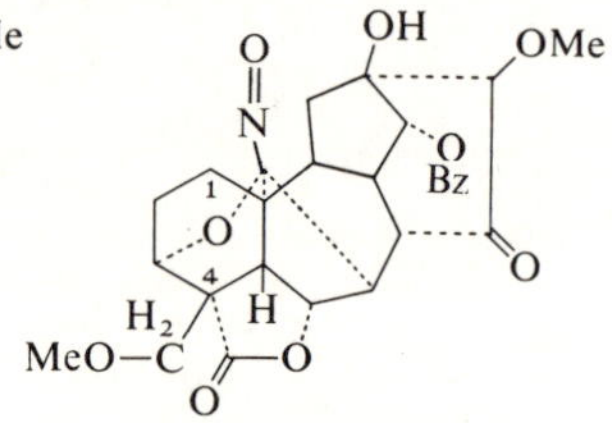

(34)

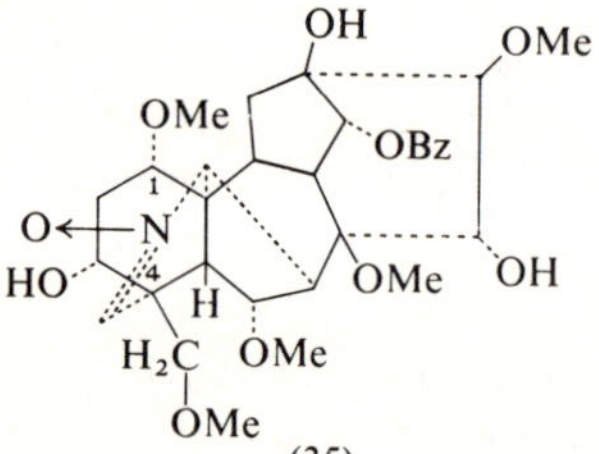

(35)

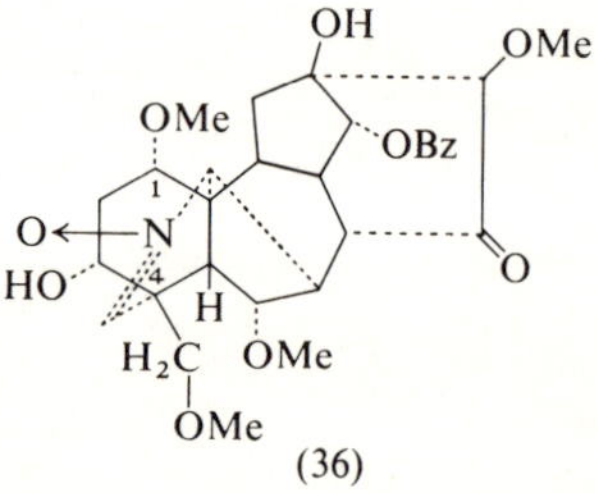

(36)

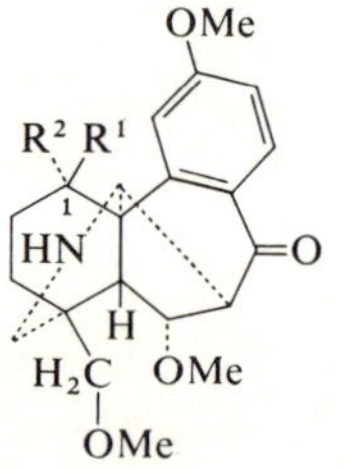

(37) R^1 = OMe, R^2 = H
(38) R^1 = H, R^2 = OMe

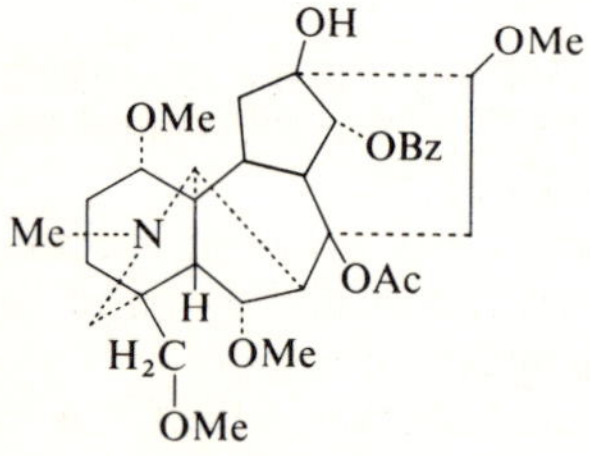

Delphinine (39)

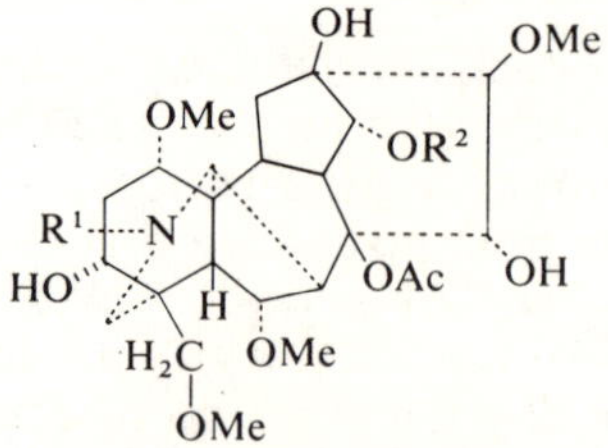

Aconitine (2) R^1 = Et, R^2 = Bz
Mesaconitine (40) R^1 = Me, R^2 = Bz
Jesaconitine (41) R^1 = Et, R^2 = CO·C$_6$H$_4$·OMe-*p*

In recent work towards the synthesis of delphinine, however, Section 9.3, Wiesner and co-workers demonstrated that the methoxyl at C-1 in delphinine has an α-equatorial configuration, i.e. *cis* to the nitrogen bridge[9]. The racemic delphinine degradation product, thought to be (37), was synthesised and proved to be identical with the 'natural' degradation product of delphinine. An x-ray crystallographic study of the acid oxalate salt of the 'identical racemeate' proved, however, that the actual structure is (38). Therefore, the structures of the following alkaloids must be revised to indicate an α-equatorial methoxyl group at C-1: delphinine (39), aconitine (2), mesaconitine (40), jesaconitine (41), indaconitine (42), pseudaconitine (43), hypaconitine (44), deoxyaconitine (45), chasmaconitine (46), bikhaconitine (47), and chasmanthinine (48).

Indaconitine   (42) R = Bz

Hypaconitine   (44) R = Me
Deoxyaconitine   (45) R = Et

Pseudaconitine (43) R = C(=O)—C$_6$H$_3$(OMe)(OMe)

Chasmaconitine   (46) R = Bz

Bikhaconitine   (47) R = C(=O)—C$_6$H$_3$(OMe)(OMe)

Chasmanthinine   (48) R = CO·CH:CHPh

9.2.2.2   Oxonitine

Oxonitine (49) is the permanganate oxidation product of aconitine (2) in which the *N*-alkyl group has been converted to an *N*-acyl group. Much of the initial chemical elucidation of aconitine was based on oxonitine (cf. Refs. 1c and 1d). The uncertainty regarding the structure of oxonitine has mainly centred on the nature of the *N*-acyl group.

Although Jacobs and Pelletier had proposed that oxonitine contained an

N-acetyl group[10], Turner later demonstrated that this compound actually contained an N-formyl group[11]. This was shown by a sequence of reactions involving removal of the N-ethyl group from aconitine, acetylation, and generation of an N-formyl group, resulting in compound (50). Comparison of this com-

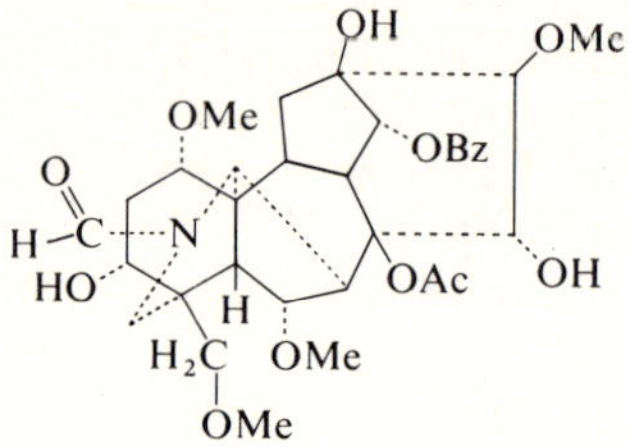

Oxonitine (49)

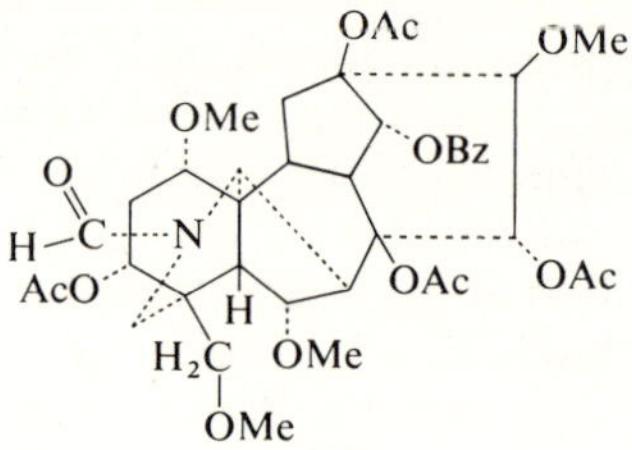

(50)

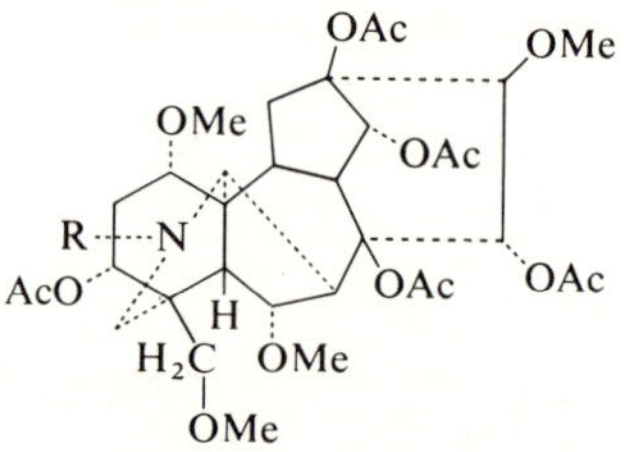

(51) R = $^{14}CH_2 \cdot CH_3$
(52) R = —CHO

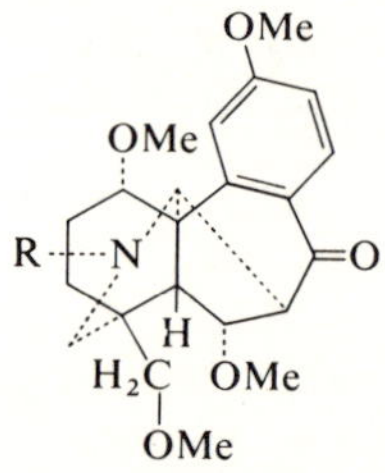

(53) R = —CHO + —COMe
(54) R = H

pound with an authentic sample of triacetyloxonitine indicated that they were identical. The most significant evidence was that the permanganate oxidation of penta-acetylaconine (51) containing an N-ethyl group labelled at the carbon adjacent to the nitrogen produced penta-acetyloxonine (52) with a residual radioactivity of only 6% of that of (51). This result clearly indicates that the N-formyl group of oxonitine does not arise from the N-ethyl group of aconitine.

A recent study by Wiesner has provided an explanation as to the origin of the problem of the nature of oxonitine[12]. The N-acyl aromatisation product (53) from oxonitine was prepared, and its mass and n.m.r. spectra showed it to be a mixture of both N-acetyl and N-formyl derivatives even though it was homogeneous on t.l.c. Hydrolysis of this mixture gave (54), which was identical with samples prepared from delphinine and by total synthesis (Section 9.3).

The authors believe that because aconitine is known to contain varying amounts of mesaconitine (40), the mixture of oxonitine may result in part from the oxidation of the N-methyl of the mesaconitine impurity to the N-formyl group.

Because it has been demonstrated that the N-formyl group does not arise from the N-ethyl group in the methanolic permanganate oxidation, some of the conflicting data may be explained on the basis of the mesaconitine impurity.

9.2.2.3 The structure of jesaconitine

This alkaloid was first isolated from a variety of the tuber *Aconitum fischeri* Reichb., a plant native to Hokkaido[13]. It has also since been found in other East Asian species, e.g. *A. subcuneatum* Nakai[14, 15], *A. sachalinense* F. Schmidt[14, 15] and *A. mitakense* Nakai[16].

Jesaconitine differs from aconitine (2) by only one ester group. Whereas, aconitine is hydrolysed to acetic acid, benzoic acid, and the parent amino alcohol, aconine (55), jesaconitine is hydrolysed to acetic acid, anisic acid (4-methoxybenzoic acid), and aconine[15]. Pyrolysis of jesaconitine was reported to give pyrojesaconitine but there was no mention of whether acetic acid or anisic acid was eliminated during the pyrolysis[15]. Because the structure of aconine was known with certainty, the problem was to locate the acetoxy and *p*-methoxybenzoyloxy moieties on the aconine skeleton. Although there are five hydroxyl groups in aconine, the number of possible combinations is considerably simplified when one recognises that one of the ester groups must be at C-8 in order to produce pyrojesaconitine.

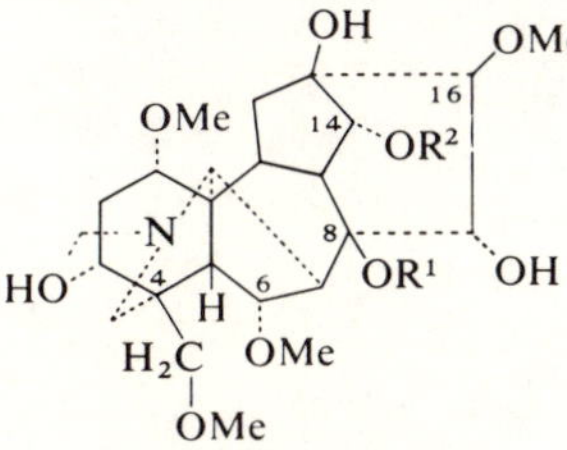

Aconitine   (2) R^1 = Ac, R^2 = Bz
Aconine   (55) R^1 = R^2 = H
Jesaconitine (41) R^1 = —Ac, R^2 = CO·C$_6$H$_4$·OMe-*p*

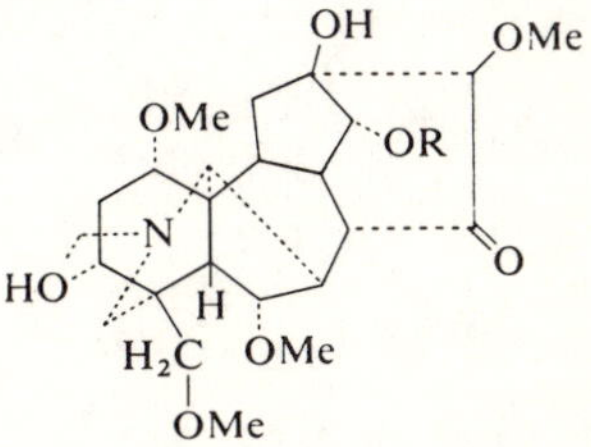

Pyrojesaconitine (56) R = CO·C$_6$H$_4$·OMe-*p*

An examination of the 100 MHz n.m.r. spectrum of jesaconitine revealed that the C-14 position was also substituted since the signal of the C-14 proton appeared at δ 4.62 (doublet, J = 4.5 Hz)[17]. When not esterified this proton appears at *c.* δ 4.2, while when esterified with a benzoyloxy or acetoxy group it usually appears at *c.* δ 4.9 and 4.8, respectively. Further proof of a C-9–C-14 substitution was the highly-shielded signal of the acetoxy protons (δ 1.36). This signal has been observed with all of the aconitine-type alkaloids examined which contain a C-14 benzoyloxy–C-8 acetoxy substitution pattern. By analogy, therefore, it was expected that the same situation exists in jesaconitine. However, to eliminate any possibility that the acetoxy and benzoyloxy groups might be interchanged, the pyrolysis was carried out on a few milligrammes of jesaconitine dissolved in glycerol and the progress of the reaction was continuously monitored by n.m.r. spectra[17]. The highly-shielded signal of the acetoxy protons was observed to slowly disappear while a new signal, due to acetic acid, appeared and grew to about the same intensity as the original shielded acetoxy signal. The signals due to the aromatic protons, however, remained completely unchanged. This pyrolysis was also repeated with aconitine and mesaconitine and similar results were observed. Thus jesaconitine is represented by structure (41) and pyrojesaconitine by (56).

9.2.2.4   *Correlation of chasmanine with browniine*

Chasmanine, $C_{25}H_{41}NO_6$, (57), has been isolated from the roots of *Aconitum chasmanthum*. After considerable structural investigations which tentatively assigned structure (57) to chasmanine[18], a chemical correlation of chasmanine with browniine (5) has been achieved[19]. The latter differs from chasmanine only in that it has a β-oriented C-6 methoxyl and a lycoctonine-type skeleton (thereby containing a C-7 hydroxyl). Its structure is known with certainty by virtue of its correlation with lycoctonine. It was reasoned that the C-6 α-methoxyl of chasmanine, being crowded by the C-4 substituents, probably could be epimerised if a 7,17-seco-7-keto derivative of chasmanine could be prepared. Accordingly diacetylchasmanine (58) was converted to the 7,17-seco-olefin (59) by treatment with lithium tri-t-butoxyaluminum hydride in diglyme. Hydration of (59) with diborane gave (60) which, in turn, gave a mixture of (61) and (62) when oxidised with chromic oxide in acetone. After separation on alumina, (62) was refluxed in basic solution to epimerise the C-6 methoxyl and a small amount of the diketo epimer (63) was obtained.

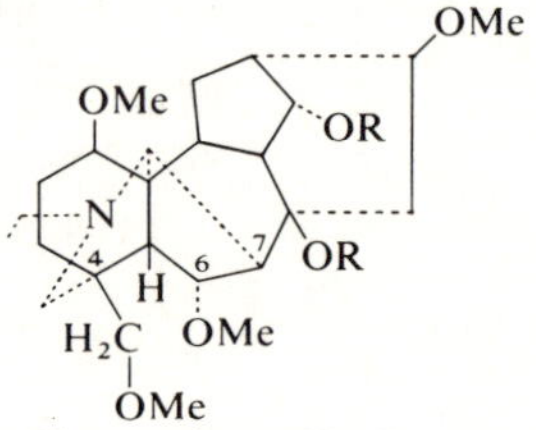

Chasmanine (57) R = H
         (58) R = Ac

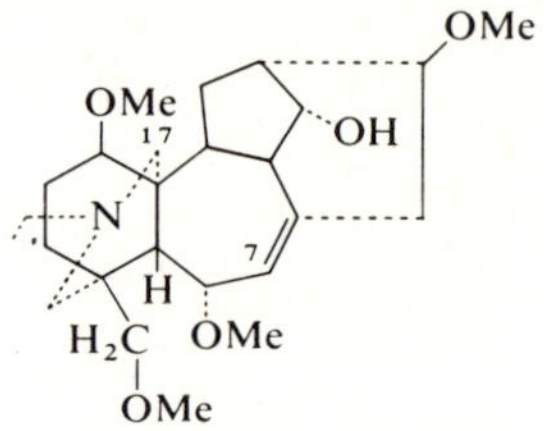

(59)

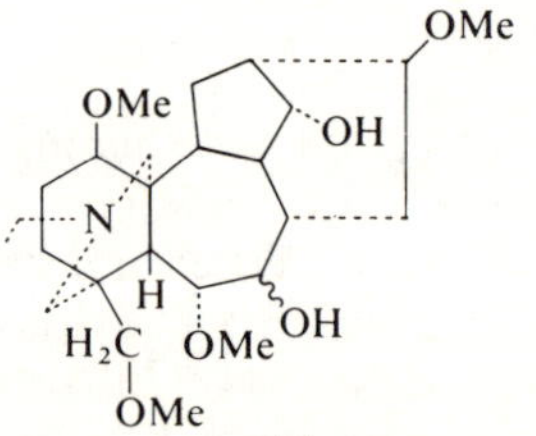

(60)

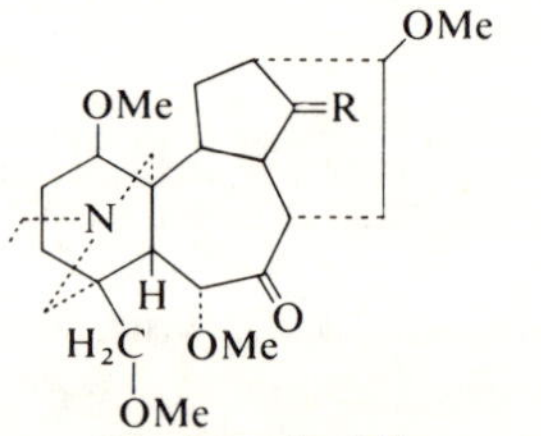

(61) R = H, OH
(62) R = O

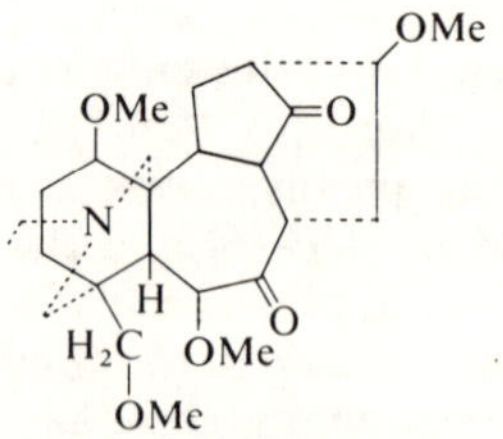

(63)

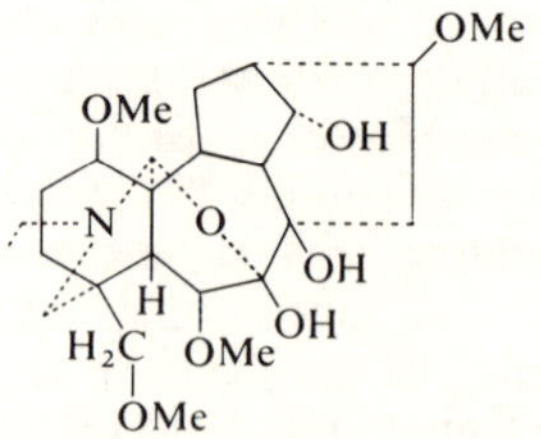

(64)
(65) Perchlorate

Oxidation of browniine (5) with lead tetra-acetate gave a nearly quantitative yield of hydroxybrowniine (64) and hydrogenation of hydroxybrowniine perchlorate (65) gave isobrowniine (66). The products of this series of reactions

$(5 \rightarrow 66)$ are exactly analogous with those of lycoctonine treated under similar conditions. A 3-min oxidation of isobrowniine with permanganate in acetone produced oxoisobrowniine (67) which, when reduced with zinc in acetic acid, gave a mixture of 8-deoxyoxoisobrowniine (68) and 8-deoxy-7-dihydro-oxoisobrowniine (69). The latter could also be prepared by borohydride reduction of the former. Reduction of either (68) or (69) with lithium aluminium hydride gave 8-deoxy-7-dihydroisobrowniine (70) which, when oxidised with chromic oxide in acetone, produced 14-dehydro-8-deoxyisobrowniine (63). This diketone was identical with the corresponding diketone obtained from chasmanine.

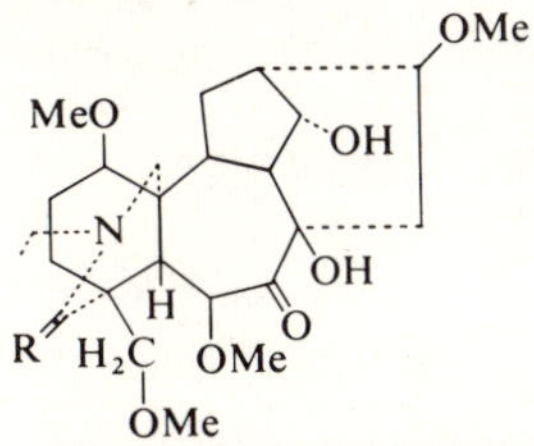

Isobrowniine (66) R = H$_2$
Oxoisobrowniine (67) R = O

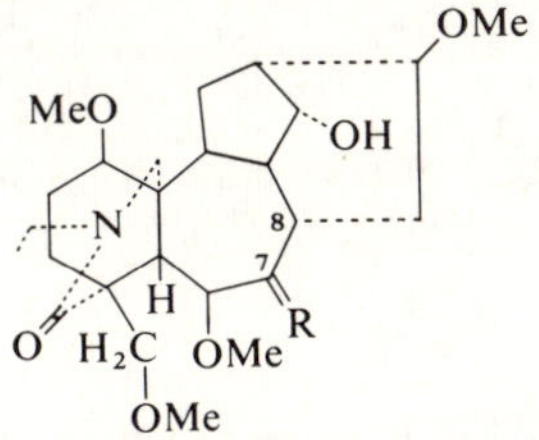

(68) R = O
(69) R = H, OH

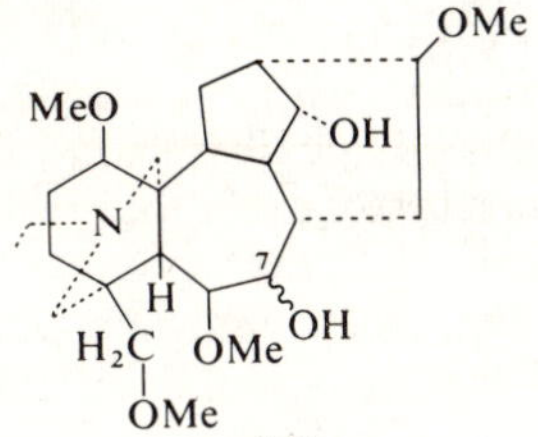

(70)

9.2.2.5 The structures of condelphine, talatizidine and isotalatizidine

The alkaloid condelphine, C$_{25}$H$_{39}$NO$_6$, (71), was first isolated in 1942 from *Delphinium confusum* Popov and was found to be an *O*-acetate derivative of isotalatizidine (72)[20]. The latter alkaloid, along with its C-1 epimer, talatizidine, (73) had been isolated two years earlier from *Aconitum talassicum* Popov[21]. Research by Soviet workers allowed the postulation of partial structures for these three closely related compounds[22] and more recent work has established the complete structures of all three[23]. The latter work was done with condelphine and isotalatizidine isolated from *Delphinium denudatum* Wall[23]. A recent x-ray analysis[24] of the hydroiodide derivative of condelphine confirmed the structure and absolute configuration as (71). From the previous correlations, the absolute configurations of isotalatizidine, talatizidine, talatizamine and cammaconine are also definitely defined.

Condelphine contains one ethyl, two methoxyl, one acetoxy and two hydroxyl groups. Isotalatizidine and talatizidine are unesterified. Acetylation of isotalatizidine or condelphine with acetic anhydride–pyridine produces the same diacetate (74)[22, 23]. Likewise, vigorous treatment of either alkaloid with acetyl chloride gives the fully acetylated derivative (75). Saponification of (74) or (75) gives back the amino alcohol (72).

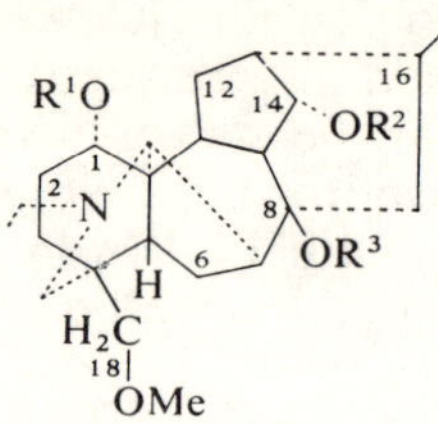

Condelphine (71) $R^1 = R^3 = H$, $R^2 = Ac$
Isotalatizidine (72) $R^1 = R^2 = R^3 = H$
(74) $R^1 = R^2 = Ac$, $R^3 = H$
(75) $R^1 = R^2 = R^3 = Ac$
(76) $R^1 = Ts$, $R^2 = Ac$, $R^3 = H$
(77) $R^1 = Ms$, $R^2 = Ac$, $R^3 = H$
(79) $R^1 = Bz$, $R^2 = Ac$, $R^3 = H$

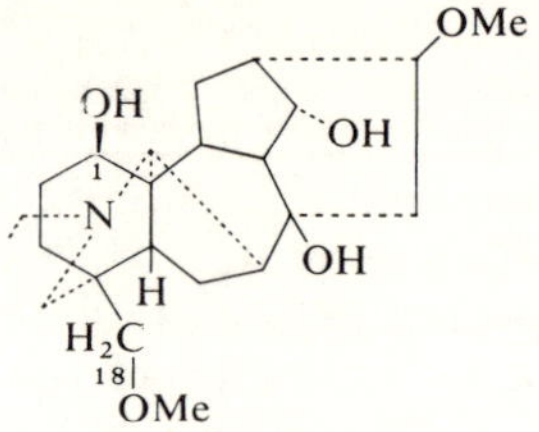

Talatizidine (73)

Chemical studies indicated that two of the hydroxyl groups of isotalatizidine must be secondary, while the third is tertiary[22, 23]. Also, the acetoxy group of condelphine is shown to be in a five-membered ring and the C-18 position must be substituted with a methoxyl rather than a primary hydroxyl group. The presence of an *N*-ethyl group in these alkaloids was indicated by n.m.r. spectral studies as well as by dealkylation followed by alkylation with ethyl iodide to regenerate the starting alkaloid[22, 23].

Proof of the C-1 position of the ring-A hydroxyl was provided by both spectroscopic and chemical evidence[22]. Since both the toluenesulphonate (76) and methanesulphonate (77) derivatives of condelphine undergo elimination to give (78) as the only product, substitution at C-3 was precluded. The n.m.r. spectrum of the hydrogen geminal to the aromatic ester in benzoyl condelphine (79) is explicable only if substitution is at C-1 or C-2. Substitution at C-3 is not possible as evidenced by the formation of inner carbinolamine ethers[22] and on the basis of the n.m.r. signals due to the C-18 methylene and methoxyl groups in (80). The α-configuration of the C-1 hydroxyl group is supported by the observed intramolecular hydrogen bonding in the infrared spectrum of condelphine and by the formation of (81) and (82) on oxidation of isotalatizidine with permanganate in acidic acetone.

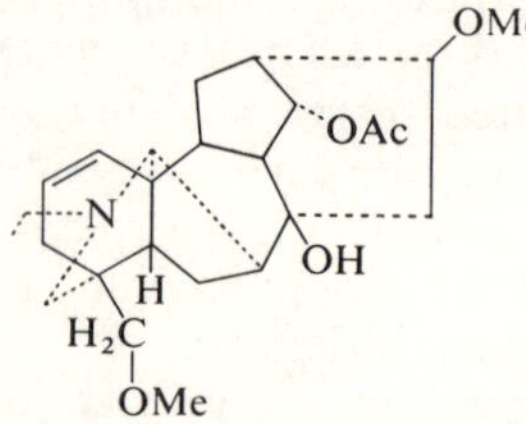

(78)

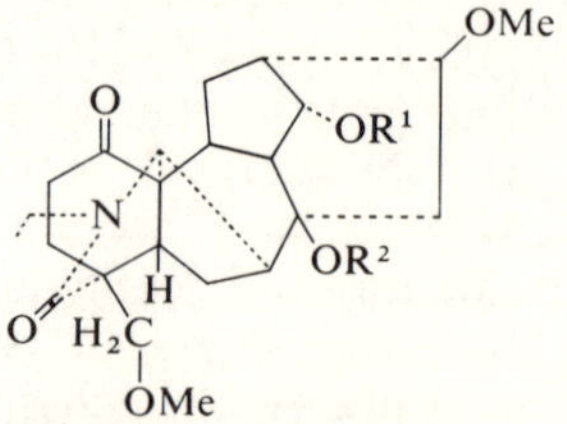

(80) $R^1 = Ac$, $R^2 = H$
(89) $R^1 = R^2 = H$
(90) $R^1 = Bz$, $R^2 = H$
(91) $R^1 = Bz$, $R^2 = Ac$

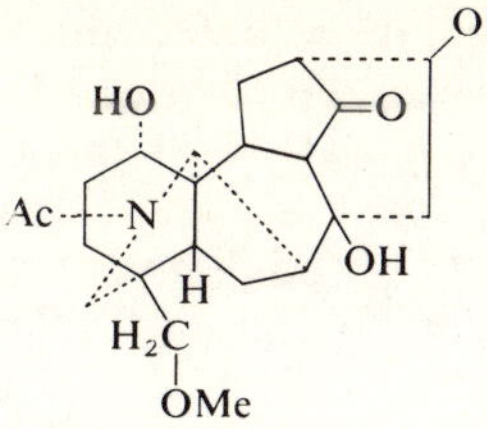

(81)

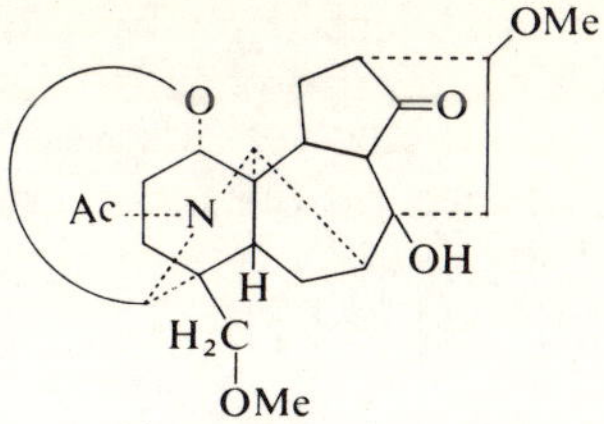

(82)

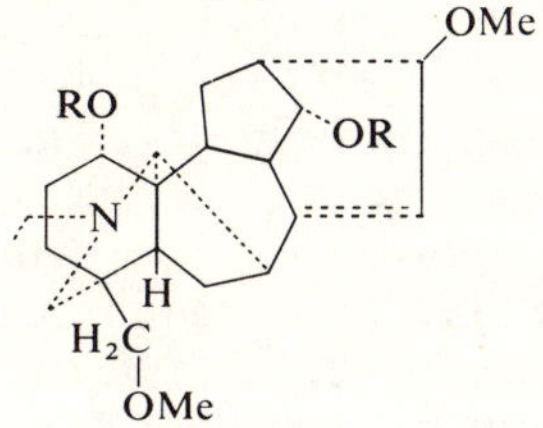

(83) R = Ac
(84) R = H

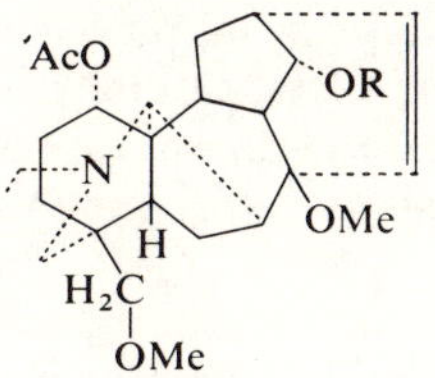

(85) R = H
(88) R = Ac

The location of the tertiary hydroxyl at C-8 follows from the pyrolysis of triacetylisotalatizidine (75) to give the corresponding pyro derivative (83). Saponification of (83) yields pyroisotalatizidine (84). Isopyroacetylisotalatizidine (85) is produced by refluxing (83) in acidic methanol. The site of the partial hydrolysis was shown by oxidising (85) to (86) followed by saponification to (87). The latter derivative contained infrared carbonyl absorption characteristic of a cyclic five-membered ketone and a lactam. Acetylation of (85) gave isopyroacetylcondelphine (88). The u.v. spectrum of pyroacetylcondelphine contains the characteristic absorption of 240 nm which disappears when it is neutralised and which is absent in the isopyro derivatives; this demonstrates the presence of an aconitine-type skeleton.

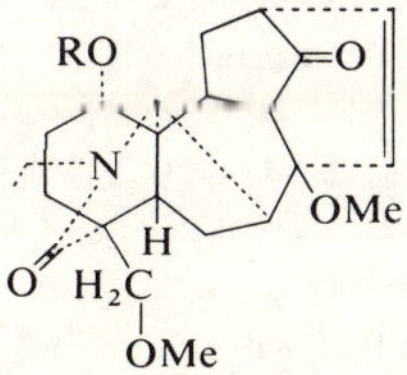

(86) R = Ac
(87) R = H

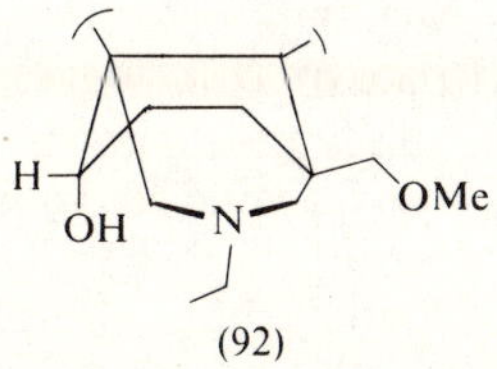

(92)

The C-15 olefinic proton of pyroacetylcondelphine is a doublet and thereby shows that C-16 is substituted[23]. Because only a methoxyl and an acetoxy group remain to be located, the remaining methoxyl was assumed to be the C-16 substituent. In addition, the n.m.r. coupling constant indicated that it was β-oriented. Both the foregoing position and configuration of this methoxyl group were shown to be correct by the isopyro rearrangement.

The remaining hydroxyl group of isotalatizidine was assigned to C-14 by the Soviet workers merely because acetylation of this hydroxyl (to give the alkaloid condelphine) does not result in a substantial lowering of basicity. Rigorous exclusion of C-6 and C-12 was achieved as shown in the reaction sequence (89) → (91)[23]. Saponification of (80) to dehydro-oxoisotalatizidine

(89) was followed successively by first benzoylation to (90) and then acetylation to (91). This result indicated that the secondary hydroxyl was at C-14 and the tertiary hydroxyl at C-8. The configuration of the C-14 oxygen function was based on the n.m.r. coupling constants and the fact that the reduction of the C-14 keto derivatives (86), (87) and (81) regenerated exclusively the hydroxyl with the same stereochemistry as in isotalatizidine and talatizidine.

Insight into the conformations of ring A in condelphine and its C-1 esters was gained from n.m.r. spectra of these compounds. The large coupling constants of the geminal proton of the C-1 esters can only be explained by a chair formation. Condelphine and isotalatizidine, however, must exist in the boat conformation (92) an appreciable amount of their time, as indicated by the intramolecular hydrogen bonding between the C-1 hydroxyl and the nitrogen. Actually, they most likely exist in a rapid equilibrium between the boat and chair conformations as evidenced by the fact that the n.m.r. spectra always show the C-1 proton as a multiplet when the C-1 hydroxyl is not esterified. Heating the sample resolves the multiplet into a triplet whose coupling constant ($J = 3\,\text{Hz}$) is in accord with that expected from this proton when ring A is in the boat conformation and the dihedral angles are nearly equal. Thus, at higher temperatures the boat form tends to predominate. It should be noted that the recent x-ray crystallographic studies have indicated that in the hydroiodide derivative, ring A of condelphine is in a boat conformation.

9.2.2.6 Talatisamine

Talatisamine, $C_{24}H_{39}NO_5$, was first isolated from the roots of *Aconitum talassicum* in 1940[21]. It has also been isolated from *A. nemorum* Popov[25] and *A. carmichaeli*[26]. More recently it has been found along with cammaconine in *A. variegatum*[27]. The structure of talatisamine has been established as (93) by correlation with isotalatizidine (72)[28].

Talatisamine had been shown to possess a lycoctonine-type skeleton with an *N*-ethyl group, one secondary and one tertiary hydroxyl group, and three methoxyl groups[29, 30]. Pyrolysis of diacetyltalatisamine (94) proceeds with the loss of acetic acid to afford pyroacetyltalatisamine (95) and isopyroacetyltalatisamine (96), products which are characteristic of aconitine-type alkaloids with a C-8-hydroxyl and a C-16-methoxyl. Hydrogenolysis of the 'pyro' derivative with lithium aluminium hydride confirmed the presence of a C-16-methoxyl, and an n.m.r. study placed this group in a β-configuration. Further n.m.r. studies located an α-hydroxyl group at C-14 and a methoxyl group at C-18. The second methoxyl was assigned to C-1 by mass spectral analysis[31].

Talatisamine yields 1,8-tri-*O*-methylisotalatizidine (97) on refluxing with methyl iodide and sodium hydride. Under the same conditions, isotalatizidine gave rise to two products, one of which was identified as 1,8,14-tri-*O*-methylisotalatizidine. This compound was found to be identical with di-*O*-methyltalatisamine. This correlation establishes the structure of talatisamine, since the structure and stereochemistry of isotalatizidine have been determined.

Talatisamine   (93) R^1 = R^4 = Me, R^2 = R^3 = H
Isotalatizidine   (72) R^1 = R^2 = R^3 = H, R^4 = Me
                   (94) R^1 = R^4 = Me, R^2 = R^3 = Ac
                   (97) R^1 = R^2 = R^3 = R^4 = Me

(95)

(96)

(98)

(99) R = ⊕
(100) R = H

The pyrolysis of diacetyltalatisamine (94) in glycerine affords a product differing from the starting material by the loss of methanol, acetic acid and an acetyl group. Spectral evidence showed the compound contains two methoxyl groups, an *N*-ethyl group and no acetate groups. The presence of a single hydroxyl was demonstrated by formation of a monoacetate, and the site was fixed at C-14 by n.m.r. spectroscopy. Structure (98) was assigned to the product. The authors suggest that the product is generated by an ionic hydrogenation process. The formation of the carbonium ion (99) is succeeded by the abstraction of hydride from the glycerine solvent to afford (100) which then proceeds to the final product[32].

9.2.2.7   *Cammaconine*

Cammaconine, C$_{23}$H$_{27}$NO$_5$, has been isolated from *Aconitum variegatum*[28]. Spectral studies indicated the presence of three hydroxyl groups, two methoxyl groups and an *N*-ethyl group. Methylation of cammaconine affords 1,8-tri-*O*-methylisotalatizidine (97). This result demonstrates that cammaconine possesses the same skeleton, oxygen location and stereochemistry as isotalatizidine but a different pattern of *O*-methylation.

Oxidation of cammaconine with Sarett reagent afforded two products, dehydro-oxocammaconine (101) and didehydro-oxocammaconine (102). Both products showed spectral characteristics for two methoxyls, an *N*-ethyl group, a hydroxyl group, a cyclopentanone moiety and a tertiary lactam in a six-membered ring. The second product also possessed an infrared band characteristic of a cyclohexanone. There was no evidence of an aldehyde proton in the n.m.r. spectra of either product.

The correlation of cammaconine with isotalatizidine establishes the tertiary hydroxyl of cammaconine at C-8 and a secondary hydroxyl at C-14. To avoid identity with isotalatizidine, cammaconine must be assigned structure (103)[28]. This is the first example of an aconitine- or lycoctonine-type alkaloid with a β-hydroxyl at C-16, instead of the usual β-methoxyl at this position.

(101) R = β-OH, H
(102) R = O

Cammaconine (103)

9.2.2.8 The structures of lappaconitine and lappaconine

Lappaconitine, $C_{32}H_{44}N_2O_8$, isolated from *Aconitum septentrionale* Koelle, is the acetylanthranilic acid ester of the alkamine lappaconine, $C_{23}H_{37}NO_6$. Chemical and spectral studies of lappaconine gave partial structural information[33] and an x-ray crystallographic study has shown its structure to be (104)[34].

Oxidation of lappaconine with potassium permanganate produced oxolappaconine (105). Oxidation of oxolappaconine with periodic acid produces seco-oxolappaconine, $C_{23}H_{33}NO_7$, (106). However, some controversy exists with regard to the cleavage of oxolappaconine by lead tetraacetate to dehydro-seco-oxolappaconine. The Canadian workers[33] have proposed that the scission produces the intermediate (107) which, in turn, undergoes an aldol condensation to afford (108), whose structure was assigned on evidence from its infrared spectrum (1645 cm^{-1}, CO—N;

1710 cm^{-1}, $>$C$=$O in a five-membered ring conjugated with a double bond).

Only a single ketonic absorption was present in the spectrum. Tel'nov and co-workers[35] have reported, however, that there is no absorption characteristic of an α,β-unsaturated ketone in the ultraviolet spectrum of this compound. On the other hand, the hydrogenation over palladium–carbon of dehydroseco-oxolappaconine resulted in the absorption of 1 mol of hydrogen and the resulting product was seco-oxolappaconine (106).

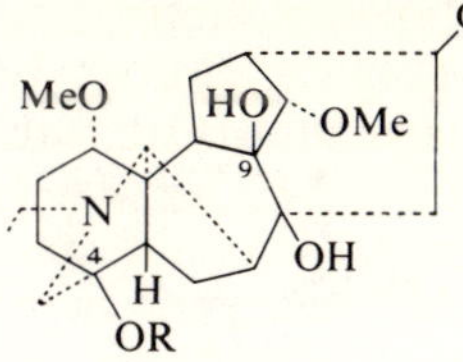

Lappaconine   (104) R = H
Lappaconitine   (110) R = OCO·C$_6$H$_4$·NH·COMe-*p*

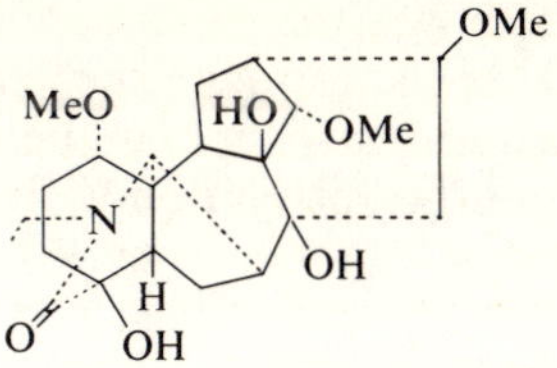

(105)

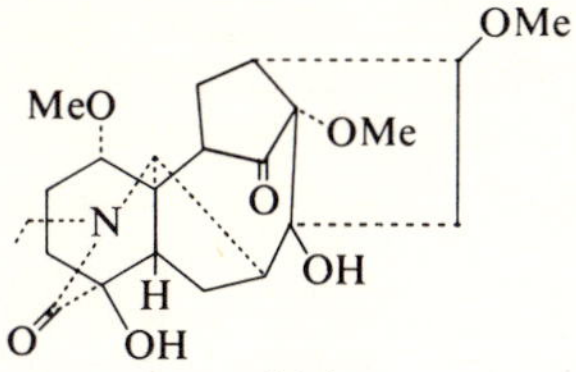

(106) 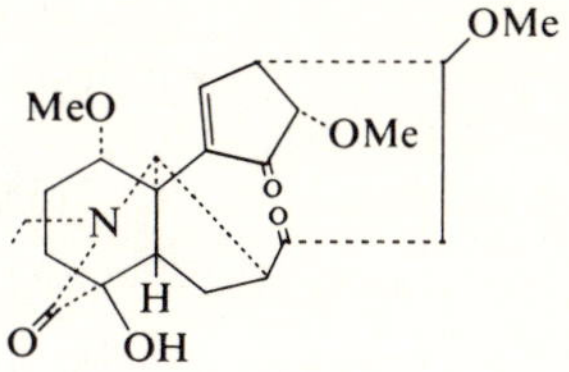   (107)

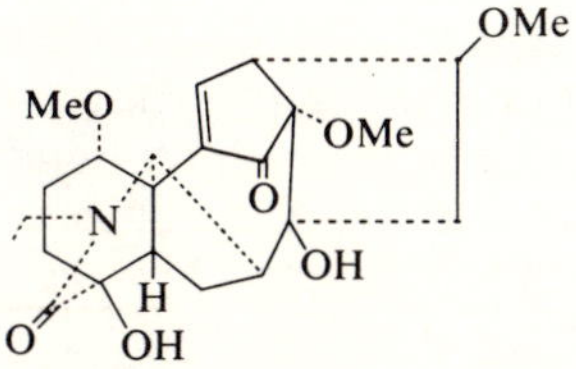

(108) 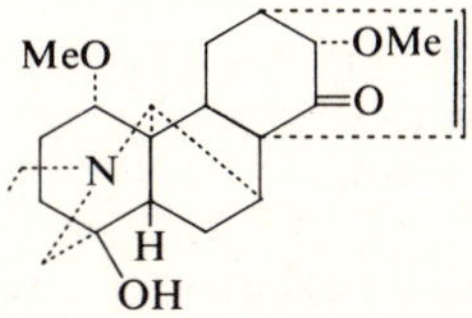   (109)

Lappaconine, on treatment with sulphuric acid was reported to give an amorphous solid whose molecular formula differs from the starting material by the loss of methanol and water. On the basis of spectral data, this product was assigned the structure (109)[35]. This compound could have resulted from a pinacolinic-type rearrangement, and possesses a possibly biogenetically-significant denudatine-type skeleton.

Lappaconitine, on oxidation with chromium trioxide in acetone followed by hydrolysis, affords a lactam containing a ketone in a five-membered ring. Reasoning that only the vicinal glycol system could give rise to such a product, the ester function of lappaconitine (110) was assigned to C-4[35].

Lappaconitine is the first example of a C$_{19}$ diterpene alkaloid with one of the carbon atoms normally attached to C-4 replaced by an oxygen atom. The oxygen function at C-9 is also unique. The attempted correlation of lappaconine with other aconitine-type alkaloids was hindered because of these different functionalities.

9.2.2.9   The structure of lapaconidine

This alkaloid, C$_{22}$H$_{35}$NO$_6$, has been isolated from *Aconitum leucostomum*[36a, 36b]. Spectral studies indicated the presence of two methoxy groups and an N-ethyl group. The presence of four hydroxyl groups was evidenced by the formation of tetra-acetate and tetramethyl derivatives. The latter was identical with the trimethyl derivative of lappaconine (111). On the basis of mass spectral comparison, one of the hydroxyl groups was assigned at C-1 and the remainder of the structure (112) follows directly from the correlation with lappaconine[36a, 36b].

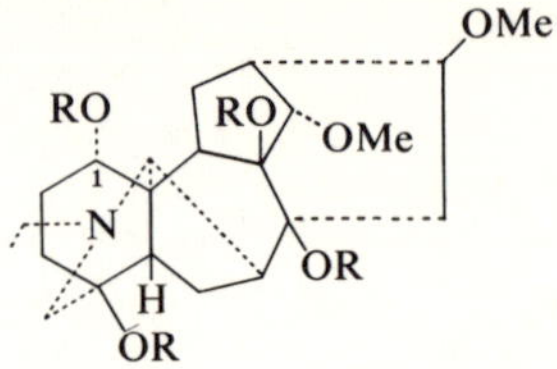

(111) R = Me
Lapaconidine (112) R = H

9.2.2.10 Further studies on the 'pyrodelphonine' chromophore

In further studies of the proposed σ-coupled π electron system[37, 38], the photoreduction of 16-desmethoxypyrodelphonine (113) has been examined[39]. In order to explain the unexpected ultraviolet absorption of pyrodelphonine (114) and pyroneoline (115) which disappear upon acidification, Wiesner had postulated the participation of the free electron pair on nitrogen, the C-7–C-19 σ-bond, and the π-system of the double bond between C-8 and C-15, with an excited state resembling (116)[37].

On the basis that the proposed excited state (116) should be reducible with sodium borohydride, (113) was irradiated in the presence of sodium borohydride to yield (117) as the major product. Since one of the new

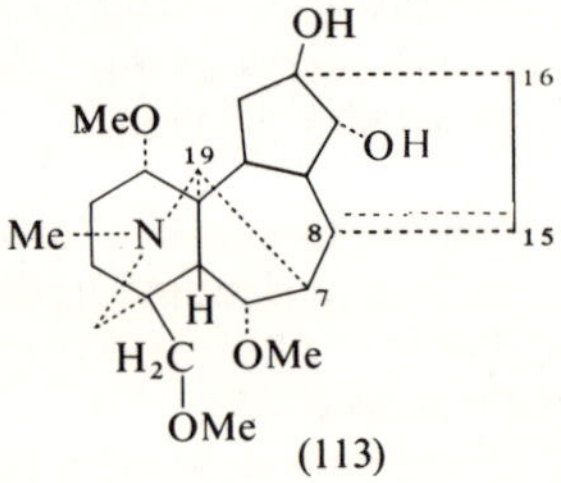

(113)

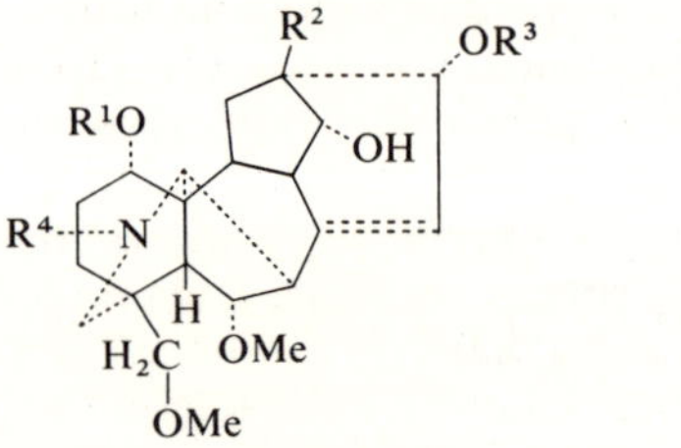

Pyrodelphonine (114) R^1 = R^3 = R^4 = Me ,
R^2 = OH
Pyroneoline (115) R^1 = R^2 = H,
R^3 = Me, R^4 = Et

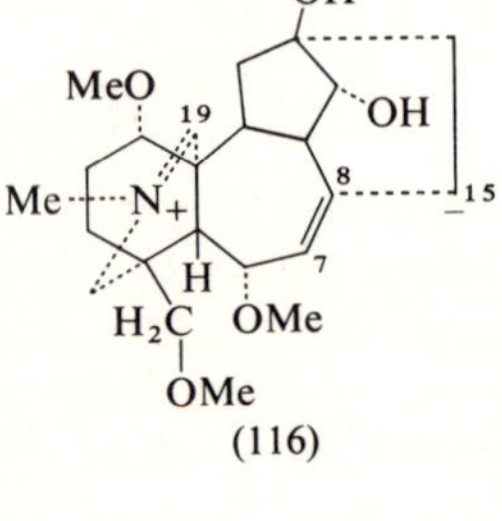

(116)

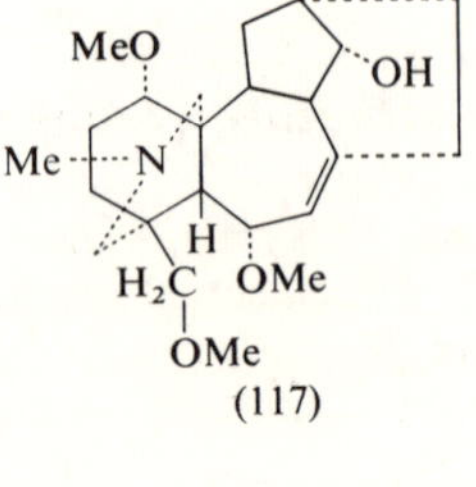

(117)

hydrogen atoms in the product must come from the borohydride and the other from the solvent (methanol) if such an excited state were involved, this experiment was repeated using sodium borodeuteride. Although the resulting product was indistinguishable from the previous product by chromatographic behaviour and i.r. and n.m.r. spectral comparison, its high-resolution mass spectrum indicated that it was the monodeuterated product. Therefore, these studies support the proposed nature of this chromophore.

9.2.2.11 Partial structure of a new alkaloid from Aconitum sachalinense

Ichinohe and Yamaguchi have reported the isolation of a new aconitine-type alkaloid from the roots of *Aconitum sachalinense* Fr. Schmidt[40]. This compound was found to be an anisic acid ester with molecular formula C$_{34}$H$_{47}$NO$_{12}$.

Chemical and spectral studies indicated the presence of four methoxyl groups, three hydroxyl groups, an acetoxy group and an *N*-methyl group. This alkaloid undergoes pyrolysis with the elimination of acetic acid, as is the case with other aconitine-type alkaloids.

Mesaconine (118)

The hydrolysis product of this new alkaloid has the same molecular formula and functional groups as does mesaconine (118). However, the pentaacetyl derivative of this alkamine and pentaacetyl mesaconine were not identical. These authors have suggested that this hydrolysis product is an epimer with regard to a secondary hydroxyl or a methoxyl group of mesaconine.

9.2.2.12 Mass spectral studies of the C$_{19}$-diterpene alkaloids

Russian workers have studied the mass spectra of condelphine, aconitine, aconine, lycoctonine, isotalatizidine, talatisamine and some of their derivatives[41a, 41b]. Their mass spectral analyses indicate that the dominant ionisation centre is the nitrogen atom, with the role of other centres being apparently negligible. In most of the spectra examined, the base peak resulted from the elimination of the substituent (hydroxyl or methoxyl) at C-1. The $M-18$ peak was one of the most abundant ions in the high-mass region of these spectra. The authors presented evidence asserting that the primary source of this fragment was the hydroxyl group at C-8. They also proposed fragmentation pathways differing from those previously postulated[47] for some of the observed ions. With the evidence from their much larger data base, convincing support was provided for most of their arguments[41a].

The observed temperature dependence of most of the mass-spectral determinations, with the previously known facile elimination of acetic acid from the C$_{19}$-diterpene alkaloids, led to further studies of the pyrolysis of these compounds[41b]. The fragmentations produced during pyrolysis of aconitine, 14-benzoxy-8-acetoxytalatisamine, 8,14-diacetoxytalatisamine and 14-dehydro-8-acetoxytalatisamine were examined. At temperatures above 100 °C pyrolytic elimination of acetic acid predominates, while at 70–80 °C pyrolytic decomposition apparently does not occur. Several fragmentation pathways were proposed to account for the observed spectra of these compounds.

This work should be valuable in the structural elucidation of other compounds with lycotonine-type skeletons.

9.2.3 The lactone lycoctonine alkaloids

Heteratisine, heterophyllisine, heterophylline, and heterophyllidine are C$_{19}$-diterpene alkaloids occurring in the mother liquors of *Aconitum heterophyllum* Wall[42]. These weak bases occur in very small amounts, with heteratisine, the most prevalent, present in a concentration of 0.03 %. The structure of heteratisine, C$_{22}$H$_{33}$NO$_5$, was determined independently by x-ray diffraction techniques[43] and by chemical and spectral analyses[44, 45], and its probable biogenetic origin has been discussed[46].

Heterophyllisine (119), heterophylline (120) and heterophyllidine (121) were isolated from the heteratisine mother liquors[47]. Their behaviour on treatment with aqueous sodium hydroxide and subsequent acidification and their strong infrared absorption at 1727–1748 cm^{-1} indicated the presence of a δ-lactone system. This, along with other data, pointed to a close structural similarity to heteratisine.

Heterophyllisine, C$_{22}$H$_{33}$NO$_4$, was shown to differ from heteratisine by a single hydroxyl group. Structure (119) was accordingly assigned heterophyllisine on the basis of mass spectral comparison.

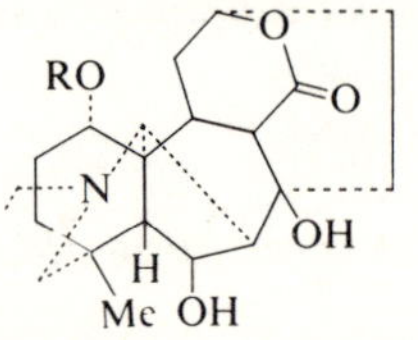

Heteratisine (3) R = Me   Heteraphyllisine (119) R = Me
Heteraphyllidine (121) R = H   Heterophylline (120) R = H

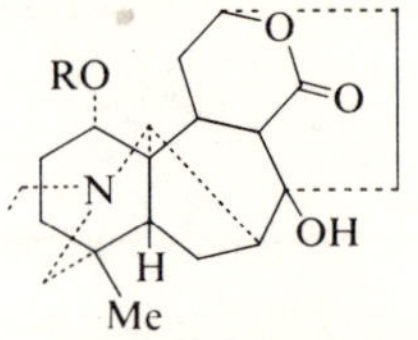

Heterophylline, C$_{21}$H$_{31}$NO$_4$, contains no methoxyl group and one less oxygen function than heteratisine. It was assigned structure (120) primarily on the basis of n.m.r. and mass spectral data.

Heterophyllidine, C$_{21}$H$_{31}$NO$_5$, was shown to contain no methoxy function, and spectral comparison clearly indicated that it possesses structure (121).

9.3 SYNTHETIC INVESTIGATIONS

In recent years Professor Wiesner's group at New Brunswick has expended considerable efforts towards the synthesis of delphinine (39) and other alkaloids of this type[48a-d]. After extensive studies with model systems, their immediate goal, the aromatisation product (38) obtained from the degradation of delphinine, has been masterfully achieved*. Since this compound can be obtained in comparatively high yield from delphinine, it is an ideal advanced relay for the synthesis of delphinine. Their studies have also served as a rigorous chemical proof of the A, B, E, and F ring systems of aconitine and delphinine[49], and have resulted in a revision of the configuration of the methoxyl group in ring A[9].

Starting with the alkylation of methoxytetralone via the Stork pyrrolidine–enamine procedure, the allyltetralone (122), was obtained. A second alkylation with benzyl chloromethyl ether and sodium hydride gave (123), which was then oxidised catalytically with osmic acid and sodium chlorate to the diastereoisomeric diols (124). Treatment of this diol system with excess

Delphinine (39)

(38)

(122) R^1 = H, R^2 = CH$_2$·CH:CH$_2$
(123) R^1 = CH$_2$·O·CH$_2$Ph,
     R^2 = CH$_2$·CH:CH$_2$
(124) R^1 = CH$_2$·O·CH$_2$Ph,
     R^2 = CH$_2$·CHOH·CH$_2$OH
(125) R^1 = CH$_2$·O·CH$_2$Ph, R^2 = CH$_2$CHO

(126) R^1 = CH$_2$Ph, R^2 = H, R^3 = O

(127) R^1 = CH$_2$Ph, R^2 = H, R^3 = (ketal)

(128) R^1 = H, R^2 = CH$_2$Ph, R^3 = (ketal)

metaperiodate afforded the aldehyde (125) in quantitative yield. A base-catalysed aldol condensation produced the hydroxy ketone (126) which was converted to the ketal (127) by standard methods. The benzyl blocking group was then transferred from the primary to the secondary alcoholic function (128) as indicated schematically in Figure 9.2. While the directness of this operation (as opposed to their suggested modification involving the use

* We wish to thank Professor Wiesner for a pre-publication draft of the full report of this work.

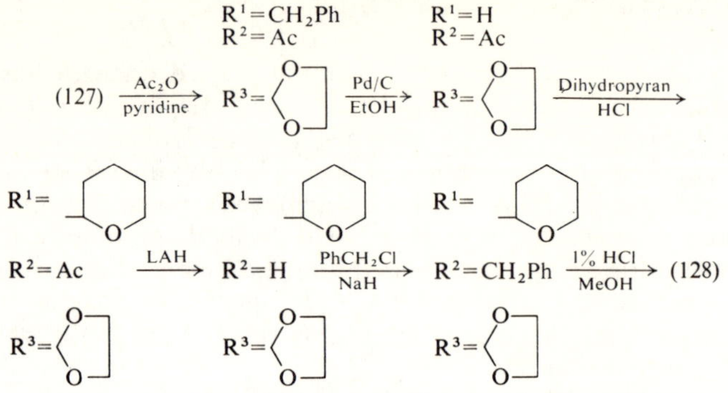

$$R^1 = CH_2Ph, \quad R^2 = Ac \qquad\qquad R^1 = H, \quad R^2 = Ac$$

$$(127) \xrightarrow[\text{pyridine}]{Ac_2O} R^3 = \text{(dioxolane)} \xrightarrow[\text{EtOH}]{Pd/C} R^3 = \text{(dioxolane)} \xrightarrow[\text{HCl}]{\text{Dihydropyran}}$$

$$R^1 = \text{(tetrahydropyranyl)} \qquad R^1 = \text{(tetrahydropyranyl)} \qquad R^1 = \text{(tetrahydropyranyl)}$$

$$R^2 = Ac \xrightarrow{LAH} R^2 = H \xrightarrow[\text{NaH}]{PhCH_2Cl} R^2 = CH_2Ph \xrightarrow[\text{MeOH}]{1\% \text{ HCl}} (128)$$

$$R^3 = \text{(dioxolane)} \qquad R^3 = \text{(dioxolane)} \qquad R^3 = \text{(dioxolane)}$$

Figure 9.2

of an acid-sensitive *p*-methoxybenzyl group in compound (127)) might leave something to be desired, the overall yield from (127) to (128) of 65% is very satisfactory.

The primary alcohol (128) was then oxidised to the aldehyde (129) with chromium trioxide–pyridine in dichloromethane. The alcohol (130) was prepared from (129) by treatment with Grignard reagent (131) (prepared as outlined schematically in Figure 9.3). This compound was discovered to have

$$MeO \cdot CH_2CH:CH \cdot CN \xrightarrow[\text{Na}]{PhCH_2OH} (132) \xrightarrow[CH_3OH]{10\% \; H_2SO_4} R = -CO_2Me$$

$$\downarrow LAH$$

$$R = -CH_2Br \xleftarrow{LiBr} R = -CH_2OTs \xleftarrow[\text{pyridine}]{TsCl} R = -CH_2OH$$

Figure 9.3

the epimeric configuration of the natural delphinine. Therefore, (130) was oxidised to the ketone (133) which was then reduced with lithium aluminium hydride. The desired epimer (134) was formed in a ratio 7:3 over (130) and was preferentially acetylated and separated by chromatography. The recovered epimer (130) was recycled in the oxidation–reduction sequence. The pure epimer was obtained by saponification of the acetate with methanolic potassium hydroxide, then methylated with sodium hydride and excess methyl iodide. The ketal was hydrolysed with acetic acid. The resulting compound (135) was reated with Raney nickel in methanolic ammonia, and the resulting crude amine (136) was acetylated. The benzyl protective groups were removed by hydrogenolysis and the resulting hydroxyl groups were oxidised with chromium trioxide in pyridine. The diketone (137) was treated with potassium cyanide in aqueous ethanol to obtain the desired lactamol (138). This reaction was suggested to proceed via a base-catalysed aldol condensation of (137) to an α,β-unsaturated ketone, which then adds cyanide. The nitrile group is subsequently hydrolysed to a primary amide, which tautomerises to the lactamol (138).

This lactamol was converted to the ketolactam (139) by heating with methanol and concentrated hydrochloric acid. The keto lactam was reduced with lithium aluminium hydride to yield the desired exo-alcohol (140) and

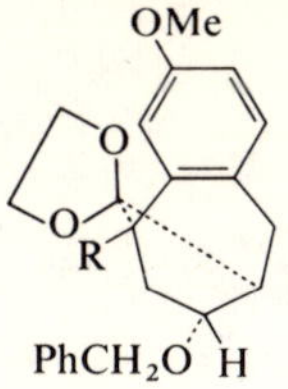

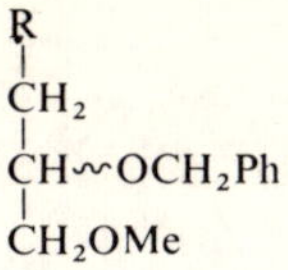

(129) R = CHO

R = CH$_2$ / CH$\sim$OCH$_2$Ph / CH$_2$OMe

(131) R = ·CH$_2$·MgBr
(132) R = C:N

(130) R =

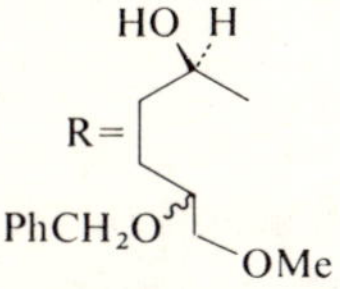

(133) R =

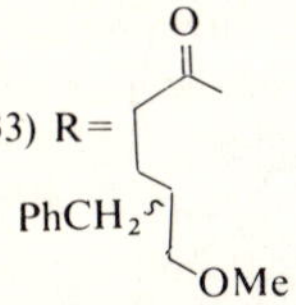

(134) R =

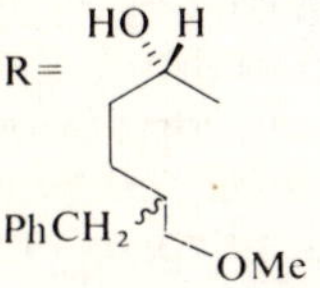

its epimer in a 1:1 ratio. These epimers were separated by chromatography. The endo epimer was oxidised to the ketone (142) with Jones reagent, then this ketone was reduced with sodium in absolute ethanol to yield the alcohols (140) and (141) in a more favourable 7:3 ratio.

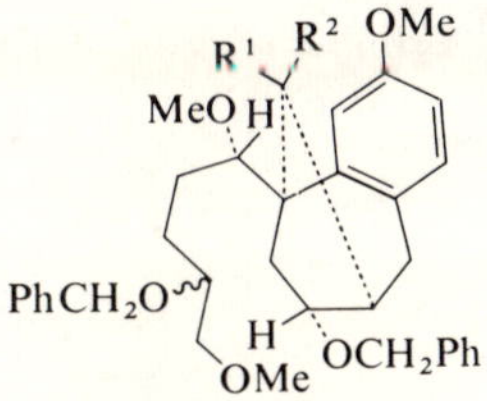

(135) R^{1}R^2 = O
(136) R^1 = NH$_2$, R^2 = H

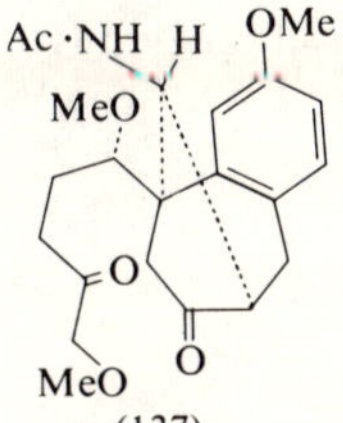

(137)

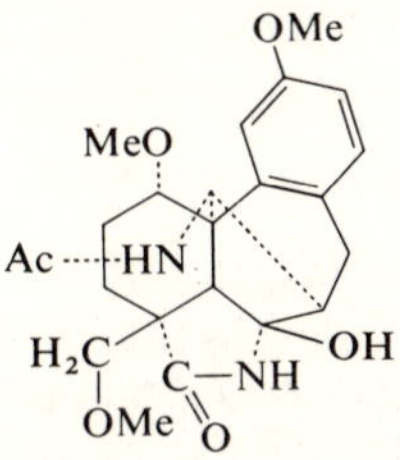

(138)

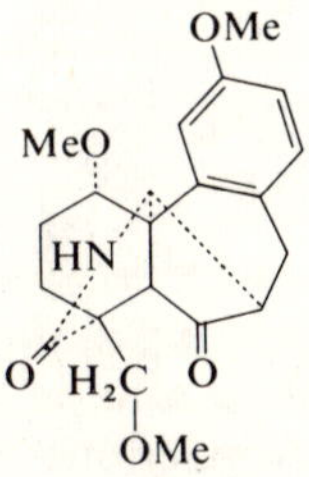

(139)

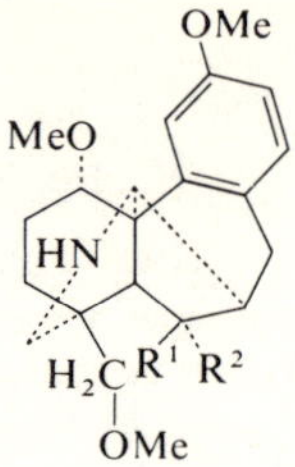

(140) R¹ = H, R² = OH
(141) R¹ = OH, R² = H
(142) R¹R² = =O

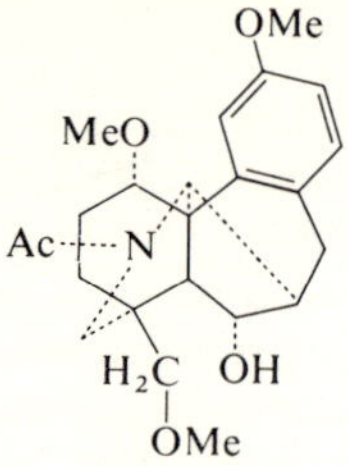

(143)

The exo-alcohol was acetylated with acetic anhydride in pyridine and the resulting product was hydrolysed with dilute methanolic potassium hydroxide to yield the *N*-acetate alcohol (143). This compound was then methylated with sodium hydride and methyl iodide and converted to the desired intermediate (144) by oxidation with Jones reagent. The synthetic racemate was found to be indistinguishable from the 'natural' degradation product of delphinine by t.l.c., i.r., n.m.r and mass spectral analysis.

At this point an x-ray crystallographic determination of the oxalate of the synthetic racemate (38) was undertaken to definitely establish the configuration of the methoxyl in ring A[9]. This study corroborated all of the features of the molecule and revealed the configuration of the ring-A methoxyl to be not β, as previously assigned, but α (*cis* to the nitrogen bridge).

In order to complete the preparation of totally synthetic optically-active (38), the racemate *dl* (38) was resolved into its optical antipodes by the selective reaction of the undesired antipode with L-camphor-sulphonyl chloride in pyridine[48d]. The unreacted free base was collected, converted to the acid oxalate derivative, and recrystallised. This material was found to be identical with the 'naturally' derived base oxalate derivative. The totally synthetic free base was liberated from the oxalate and after recrystallisation, was found to be identical with the 'natural' free base in all respects.

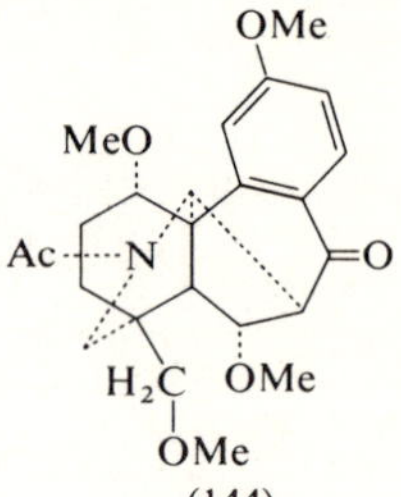

(144)

The synthesis of (38) provides an intermediate which is ideal for the completion of the synthesis of delphinine and other similar alkaloids. The *p*-methoxyacetophenone groups provide the necessary active positions for constructing the CD ring systems.

Acknowledgement

We thank The Chemical Society, London, for permission to reproduce in this chapter certain sections of my review on the diterpene alkaloids which were published in *The Alkaloids*, ed. J. E. Saxton (*Specialist Periodical Reports*), The Chemical Society, London, 1972, vol. 2, chapter 13. We also gratefully acknowledge permission from the Academic Press to utilise certain sections of material appearing in my Introduction and Chapters 1 and 2 of volume 12 of R. H. Manske's *The Alkaloids* (1970).

References

1. The following reviews of diterpene alkaloid chemistry have been published:
   (a) Pelletier, S. W. and Wright, L. H. (1972). *Specialist Periodical Reports: The Alkaloids*, Vol. 2, *Recent Developments in Diterpene Alkaloid Chemistry*, 247 (London: The Chemical Society)
   (b) Edwards, O. E. (1971). *Specialist Periodical Reports: The Alkaloids*, Vol. 1, *Diterpenoid Alkaloids*, 343 (London: The Chemical Society)
   (c) Pelletier, S. W. and Keith, L. H. (1970). *Chemistry of the Alkaloids*, 503, (New York: Van Nostrand Reinhold)
   (d) Pelletier, S. W. and Keith, L. H. (1969). *Alkaloids*, **12**, 1
   (e) Pelletier, S. W. (1967). *Quart. Rev. Chem. Soc.*, **21**, 525
   (f) Pelletier, S. W. (1964). *Experientia*, **20**, 1
   (g) Pelletier, S. W. (1961). *Tetrahedron*, **14**, 76
   (h) Boit, H. G. (1961). *Ergebnisse der Alkaloid-Chemie bis 1960*, 851, 1009, (Berlin: Academie-Verlag)
   (i) Pinder, A. R. (1960). *Chemistry of Carbon Compounds, Vol. 4-C*, 2019, (Amsterdam: Elsevier)
   (j) Stern, E. S. (1960). *Alkaloids*, **7**, 473
   (k) Wiesner, K. and Valenta, Z. (1958). *Progr. Chem. Org. Nat. Prod.*, **16**, 26
   (l) Stern, E. S. (1954). *Alkaloids*, **4**, 275
2. Benn, M. H. (1966). *Can. J. Chem.*, **44**, 1
3. Benn, M. H., Cameron, M. A. M. and Edwards, O. E. (1963). *Can. J. Chem.*, **41**, 477
4. Amiya, T. and Shima, T. (1967). *Bull. Chem. Soc. Jap.*, **40**, 1957
5. Yunusov, M. S. and Yunusov, S. Yu. (1969). *Dokl. Akad. Nauk SSSR* **188(5)**, 1077
6. Yunusov, M. S. and Yunusov, S. Yu. (1970). *Khim. Prir. Soedin.*, **6(3)**, 334
7. Benn, M. H., Connolly, J. D., Edwards, O. E., Marion, L. and Stojanac, Z. (1971). *Can. J. Chem.*, **49**, 425
8. Bachelor, F. W., Brown, R. F. C. and Büchi, G. (1960). *Tetrahedron Letters*, **10**, 1
9. Birnbaum, K. B., Wiesner, K., Jay, E. W. K. and Jay, L. (1971). *Tetrahedron Letters*, **13**, 867
10. Jacobs, W. A. and Pelletier, S. W. (1954). *J. Amer. Chem. Soc.*, **76**, 4048
11. Turner, R. B., Yeschke, J. P. and Gibson, M. S. (1960). *J. Amer. Chem. Soc.*, **82**, 5182
12. Wiesner, K. and Jay, L. (1971). *Experientia*, **27**, 758
13. Makoshi, K. (1909). *Arch. Pharm.* (Weinheim), **247**, 243
14. Suginome, H. and Imato, S. (1950). *J. Fac. Sci. Hokkaido Univ., Ser. III*, **4**, 33
15. Majima, R., Suginome, H. and Morio, S. (1924). *Ber.*, **57B**, 1486
16. Ochiai, E., Okamoto, T. and Sakai, S. (1955). *J. Pharm. Soc. Jap.*, **75**, 545
17. Keith, L. H. and Pelletier, S. W. (1968). *J. Org. Chem.*, **33**, 2497
18. Achmatowicz, O., Jr., Tsuda, Y., Marion, L., Okamoto, T., Natsume, M., Chang, H. H. and Kajima, K. (1965). *Can. J. Chem.*, **43**, 825
19. Edwards, O. E., Fonzes, L. and Marion, L. (1966). *Can. J. Chem.*, **44**, 583
20. Rabinovich, M. S. and Konovalova, R. A. (1942). *Zh. Obshch. Khim.*, **29**, 329
21. Konovalova, R. A. and Orekhov, A. P. (1940). *Bull. Soc. Chim.*, **7**, 95
22. Kuzovkov, A. D. and Platonova, T. F. (1961). *J. Gen. Chem. USSR*, **31**, 1286
23. Pelletier, S. W., Keith, L. H. and Parthasarathy, P. C. (1967). *J. Amer. Chem. Soc.*, **89**, 4146

24. Pelletier, S. W., Herald, D. L. and Newton, M. G. (1971). Unpublished results
25. Platonova, T. F., Kuzovhov, A. D. and Massagetov, P. S. (1958). *J. Gen. Chem. USSR,* **28,** 3157
26. Ch'en, Y. Chu, Y. L. and Chu, J. H. (1965). *Yao Hsueh Hsueh Pao,* **12,** 435
27. Khaimova, M. A., Palamareva, M. D., Grozdanova, L. G., Mollov, N. M. and Panov, P. P. (1967). *Compt. Rend. Acad. Bulgare Sci.,* **20,** 193
28. Khaimova, M. A., Palamareva, M. D., Mollov, N. M. and Krestev, V. P. (1971). *Tetrahedron,* **27,** 819
29. Konovalova, R. A. and Orekhov, A. P. (1940). *Zhur. Obshch. Khim.,* **10,** 745
30. Yunusov, S. Yu., Sichkova, E. V. and Potiemkin, G. F. (1954). *Zhur. Obshch. Khim.,* **24,** 2237
31. Yunusov, M. S. and Yunusov, S. Yu. (1970). *Khim. Prir. Soedin.,* **6,** 90
32. Yunusov, M. S., Telnov, V. A. and Yunusov, S. Yu. (1970). *Khim. Prir. Soedin.,* **6,** 774
33. Mollov, M., Tada, M. and Marion, L. (1969). *Tetrahedron Letters.,* **26,** 2189
34. Birnbaum, G. (1969). *Tetrahedron Letters.,* **26,** 2193
35. Tel'nov, V. A., Yunusov, M. S. and Yunusov, S. Yu. (1970). *Khim. Prir. Soedin.,* **6,** 583
36a. Tel'nov, V. A., Yunusov, M. S. and Yunusov, S. Yu. (1970). *Khim. Prir. Soedin.,* **6,** 639
36b. Tel'nov, V. A., Yunusov, M. S. Rashkes, Ya. V. and Yunusov, S. Yu. (1971). *Khim. Prir. Soedin.,* **7,** 622
37. Wiesner, K., Brewer, H. W., Simmons, D. L., Babin, D. R., Bickelhaupt, F., Kallos, J. and Bogri, T. (1960). *Tetrahedron Lett.,* **3,** 17
38. Cookson, R. C., Henstock, J. and Hudec, J. (1966). *J. Amer. Chem. Soc.,* **88,** 1060
39. Wiesner, K. and Inaba, T. (1969). *J. Amer. Chem. Soc.,* **91,** 1036
40. Ichinohe, Y. and Yamaguchi, M. (1969). *Bull. Chem. Soc. Jap.,* **42,** 3038
41a. Yunusov, M. S., Rashkes, Ya. V., Tel'nov, V. A. and Yunusov, S. Yu. (1969). *Khim. Prir. Soedin.,* **5,** 515
41b. Yunusov, M. S., Rashkes, Ya. V. and Yunusov, S. Yu. (1971). *Khim. Prir. Soedin.,* **7,** 626
42. Jacobs, W. A. and Craig, L. C. (1942). *J. Biol. Chem.,* **143,** 605
43. Przybylska, M. (1963). *Can. J. Chem.,* **41,** 2911
44. Aneja, R. and Pelletier, S. W. (1964). *Tetrahedron Letters.,* **12,** 669
45. Aneja, R. and Pelletier, S. W. (1965). *Tetrahedron Letters.,* **3,** 215
46. Edwards, O. E. and Ferrari, C. (1964). *Can. J. Chem.,* **42,** 172
47. Pelletier, S. W. and Aneja, R. (1967). *Tetrahedron Letters,* **6,** 557
48a. Wiesner, K. and Santroch, J. (1966). *Tetrahedron Letters,* **47,** 5939
48b. Wiesner, K., Kao, W. and Santroch, J. (1969). *Can. J. Chem.,* **47,** 2431
48c. Wiesner, K., Jay, E. W. K., Demerson, C., Kanno, T., Krepinsky, J., Poon, L., Tsai, T. Y. R., Vilim, A. and Wu, C. S. (1970). *Experientia,* **26,** 1030
48d. Wiesner, K., Jay, E. W. K. and Jay, L. (1971). *Experientia,* **27,** 363
50. Wiesner, K., Kao, W. and Jay, E. W. K. (1969). *Can. J. Chem.,* **47,** 2734